Geophysical Monograph Series

Geophysical Monograph Series

235 **Electric Currents in Geospace and Beyond** *Andreas Keiling, Octav Marghitu, and Michael Wheatland (Eds.)*
236 **Quantifying Uncertainty in Subsurface Systems** *Celine Scheidt, Lewis Li, and Jef Caers (Eds.)*
237 **Petroleum Engineering** *Moshood Sanni (Ed.)*
238 **Geological Carbon Storage: Subsurface Seals and Caprock Integrity** *Stephanie Vialle, Jonathan Ajo-Franklin, and J. William Carey (Eds.)*
239 **Lithospheric Discontinuities** *Huaiyu Yuan and Barbara Romanowicz (Eds.)*
240 **Chemostratigraphy Across Major Chronological Eras** *Alcides N. Sial, Claudio Gaucher, Muthuvairavasamy Ramkumar, and Valderez Pinto Ferreira (Eds.)*
241 **Mathematical Geoenergy: Discovery, Depletion, and Renewal** *Paul Pukite, Dennis Coyne, and Daniel Challou (Eds.)*
242 **Ore Deposits: Origin, Exploration, and Exploitation** *Sophie Decree and Laurence Robb (Eds.)*
243 **Kuroshio Current: Physical, Biogeochemical and Ecosystem Dynamics** *Takeyoshi Nagai, Hiroaki Saito, Koji Suzuki, and Motomitsu Takahashi (Eds.)*
244 **Geomagnetically Induced Currents from the Sun to the Power Grid** *Jennifer L. Gannon, Andrei Swidinsky, and Zhonghua Xu (Eds.)*
245 **Shale: Subsurface Science and Engineering** *Thomas Dewers, Jason Heath, and Marcelo Sánchez (Eds.)*
246 **Submarine Landslides: Subaqueous Mass Transport Deposits From Outcrops to Seismic Profiles** *Kei Ogata, Andrea Festa, and Gian Andrea Pini (Eds.)*
247 **Iceland: Tectonics, Volcanics, and Glacial Features** *Tamie J. Jovanelly*
248 **Dayside Magnetosphere Interactions** *Qiugang Zong, Philippe Escoubet, David Sibeck, Guan Le, and Hui Zhang (Eds.)*
249 **Carbon in Earth's Interior** *Craig E. Manning, Jung-Fu Lin, and Wendy L. Mao (Eds.)*
250 **Nitrogen Overload: Environmental Degradation, Ramifications, and Economic Costs** *Brian G. Katz*
251 **Biogeochemical Cycles: Ecological Drivers and Environmental Impact** *Katerina Dontsova, Zsuzsanna Balogh-Brunstad, and Gaël Le Roux (Eds.)*
252 **Seismoelectric Exploration: Theory, Experiments, and Applications** *Niels Grobbe, André Revil, Zhenya Zhu, and Evert Slob (Eds.)*
253 **El Niño Southern Oscillation in a Changing Climate** *Michael J. McPhaden, Agus Santoso, and Wenju Cai (Eds.)*
254 **Dynamic Magma Evolution** *Francesco Vetere (Ed.)*
255 **Large Igneous Provinces: A Driver of Global Environmental and Biotic Changes** *Richard. E. Ernst, Alexander J. Dickson, and Andrey Bekker (Eds.)*
256 **Coastal Ecosystems in Transition: A Comparative Analysis of the Northern Adriatic and Chesapeake Bay** *Thomas C. Malone, Alenka Malej, and Jadran Faganeli (Eds.)*
257 **Hydrogeology, Chemical Weathering, and Soil Formation** *Allen Hunt, Markus Egli, and Boris Faybishenko (Eds.)*
258 **Solar Physics and Solar Wind** *Nour E. Raouafi and Angelos Vourlidas (Eds.)*
259 **Magnetospheres in the Solar System** *Romain Maggiolo, Nicolas André, Hiroshi Hasegawa, and Daniel T. Welling (Eds.)*
260 **Ionosphere Dynamics and Applications** *Chaosong Huang and Gang Lu (Eds.)*
261 **Upper Atmosphere Dynamics and Energetics** *Wenbin Wang and Yongliang Zhang (Eds.)*
262 **Space Weather Effects and Applications** *Anthea J. Coster, Philip J. Erickson, and Louis J. Lanzerotti (Eds.)*
263 **Mantle Convection and Surface Expressions** *Hauke Marquardt, Maxim Ballmer, Sanne Cottaar, and Jasper Konter (Eds.)*
264 **Crustal Magmatic System Evolution: Anatomy, Architecture, and Physico-Chemical Processes** *Matteo Masotta, Christoph Beier, and Silvio Mollo (Eds.)*
265 **Global Drought and Flood: Observation, Modeling, and Prediction** *Huan Wu, Dennis P. Lettenmaier, Qiuhong Tang, and Philip J. Ward (Eds.)*
266 **Magma Redox Geochemistry** *Roberto Moretti and Daniel R. Neuville (Eds.)*
267 **Wetland Carbon and Environmental Management** *Ken W. Krauss, Zhiliang Zhu, and Camille L. Stagg (Eds.)*
268 **Distributed Acoustic Sensing in Geophysics: Methods and Applications** *Yingping Li, Martin Karrenbach, and Jonathan B. Ajo-Franklin (Eds.)*
269 **Congo Basin Hydrology, Climate, and Biogeochemistry: A Foundation for the Future (English version)** *Raphael M. Tshimanga, Guy D. Moukandi N'kaya, and Douglas Alsdorf (Eds.)*
269 **Hydrologie, climat et biogéochimie du bassin du Congo: une base pour l'avenir (version française)** *Raphael M. Tshimanga, Guy D. Moukandi N'kaya, et Douglas Alsdorf (Éditeurs)*
270 **Muography: Exploring Earth's Subsurface with Elementary Particles** *László Oláh, Hiroyuki K. M. Tanaka, and Dezsö Varga (Eds.)*
271 **Remote Sensing of Water-Related Hazards** *Ke Zhang, Yang Hong, and Amir AghaKouchak (Eds.)*
272 **Geophysical Monitoring for Geologic Carbon Storage** *Lianjie Huang (Ed.)*
273 **Isotopic Constraints on Earth System Processes** *Kenneth W. W. Sims, Kate Maher, and Daniel P. Schrag (Eds.)*
274 **Earth Observation Applications and Global Policy Frameworks** *Argyro Kavvada, Douglas Cripe, and Lawrence Friedl (Eds.)*
275 **Threats to Springs in a Changing World: Science and Policies for Protection** *Matthew J. Currell and Brian G. Katz (Eds.)*
276 **Core-Mantle Co-Evolution: An Interdisciplinary Approach** *Takashi Nakagawa, Madhusoodhan Satish-Kumar, Taku Tsuchiya, and George Helffrich (Eds.)*
277 **Compressional Tectonics: Plate Convergence to Mountain Building (Tectonic Processes: A Global View, Volume 1)** *Elizabeth J. Catlos and Ibrahim Çemen (Eds.)*
278 **Extensional Tectonics: Rifting and Continental Extension (Tectonic Processes: A Global View, Volume 2)** *Ibrahim Çemen and Elizabeth J. Catlos (Eds.)*
279 **Strike-Slip Tectonics: Oceanic Transform Faults to Continental Plate Boundaries (Tectonic Processes: A Global View, Volume 3)** *Ibrahim Çemen and Elizabeth J. Catlos (Eds.)*
280 **Landscape Fire, Smoke, and Health: Linking Biomass Burning Emissions to Human Well-Being** *Tatiana V. Loboda, Nancy H. F. French, and Robin C. Puett (Eds.)*
281 **Clouds and Their Climatic Impacts: Radiation, Circulation, and Precipitation** *Sylvia Sullivan and Corinna Hoose (Eds.)*
282 **Fast Processes in Large-Scale Atmospheric Models: Progress, Challenges, and Opportunities** *Yangang Liu and Pavlos Kollias (Eds.)*
283 **Helicities in Geophysics, Astrophysics, and Beyond** *Kirill Kuzanyan, Nobumitsu Yokoi, Manolis K. Georgoulis, and Rodion Stepanov (Eds.)*
284 **Noisy Oceans: Monitoring Seismic and Acoustic Signals in the Marine Environment** *Gaye Bayrakci and Frauke Klingelhoefer (Eds.)*
285 **Alfvén Waves Across Heliophysics: Progress, Challenges, and Opportunities** *Andreas Keiling (Ed.)*
286 **Salt in the Earth Sciences: Evaporite Rocks and Salt Deposition** *Webster Mohriak*
287 **Salt in the Earth Sciences: Basin Analysis and Salt Tectonics** *Webster Mohriak*
288 **Microanalysis of Atmospheric Particles: Techniques and Applications** *Joseph M. Conny and Peter R. Buseck (Eds.)*

Geophysical Monograph 288

Microanalysis of Atmospheric Particles

Techniques and Applications

Joseph M. Conny
Peter R. Buseck
Editors

This Work is a co-publication of the American Geophysical Union and John Wiley and Sons, Inc.

AGU
ADVANCING EARTH
AND SPACE SCIENCE

WILEY

This edition first published 2025

Published under the aegis of the AGU Publications Committee

Matthew Giampoala, Vice President, Publications
Steven A. Hauck, II, Chair, Publications Committee
For details about the American Geophysical Union visit us at www.agu.org.

The right of Joseph M. Conny and Peter R. Buseck to be identified as the editors of this work has been asserted in accordance with law.

Registered Office
John Wiley & Sons, Inc., 111 River Street, Hoboken, NJ 07030, USA

Editorial Office
111 River Street, Hoboken, NJ 07030, USA

For details of our global editorial offices, customer services, and more information about Wiley products visit us at www.wiley.com.

Wiley also publishes its books in a variety of electronic formats and by print-on-demand. Some content that appears in standard print versions of this book may not be available in other formats.

Library of Congress Cataloging-in-Publication Data Applied for:
Hardback 9781119554349

Cover Design: Wiley
Cover Image: Courtesy of Kouji Adachi; Spherical “tar balls” produced during aging of forest-fire smoke. (Transmission electron microscope image using a lace-like carbon substrate.)

Set in 10/12pt Times New Roman by Straive, Pondicherry, India

CONTENTS

LIST OF CONTRIBUTORS

Kouji Adachi
Department of Atmosphere, Ocean, and Earth System Modeling Research
Meteorological Research Institute
Tsukuba, Japan

Peter A. Alpert
PSI Center for Energy and Environmental Sciences
Paul Scherrer Institute
Villigen, Switzerland

Markus Ammann
PSI Center for Energy and Environmental Sciences
Paul Scherrer Institute
Villigen, Switzerland

Veronica Berta
Scripps Institution of Oceanography
University of California at San Diego
La Jolla, CA, USA

Sébastien Bonhommeau
Institute of Molecular Sciences
University of Bordeaux
CNRS, UMR 5255
Talence, France

Gary S. Casuccio
RJ Lee Group
Pittsburgh, PA, USA

Joseph M. Conny
Materials Measurement Science Division
National Institute of Standards and Technology (retd.)
Gaithersburg, MD, USA

Pablo Corral Arroyo
Laboratory for Physical Chemistry
Swiss Federal Institute of Technology Zurich
Zurich, Switzerland

Karine Deboudt
Laboratory of Physical Chemistry of the Atmosphere
University of the Littoral Opal Coast
Dunkirk, France

Arnaud Desmedt
Institute of Molecular Sciences
University of Bordeaux
CNRS, UMR 5255
Talence, France

Evelyn Freney
Laboratory of Physical Meteorology
UMR 6016 CNRS/UBP
University of Clermont Auvergne
Aubière, France

Daniel A. Knopf
School of Marine and Atmospheric Sciences
Stony Brook University
Stony Brook, NY, USA

Hansol D. Lee
Department of Chemistry
University of Iowa
Iowa City, IA, USA

Weijun Li
Department of Atmospheric Sciences
Zhejiang University
Hangzhou, China

Lei Liu
Department of Atmospheric Sciences
Zhejiang University
Hangzhou, China

Lynn M. Russell
Scripps Institution of Oceanography
University of California at San Diego
La Jolla, CA, USA

Sophie Sobanska
Institute of Molecular Sciences
University of Bordeaux
CNRS, UMR 5255
Talence, France

David Talaga
Institute of Molecular Sciences
University of Bordeaux
CNRS, UMR 5255
Talence, France

Alexei V. Tivanski
Department of Chemistry
University of Iowa
Iowa City, IA, USA

Cynthia H. Twohy
NorthWest Research Associates
Redmond, WA, USA
and
Scripps Institution of Oceanography
University of California at San Diego
San Diego, CA, USA

Liang Xu
Department of Atmospheric Sciences
Zhejiang University
Hangzhou, China

PREFACE

Most of what is visible in the atmosphere—be it pollution, windblown dust, haze, fog, or clouds—arises from aerosol particles with sizes in the micrometer and nanometer range. These small atmospheric particles, in their solid or liquid states, play a fundamental role in air quality and climate change. Assessing air quality and how climate is changing requires understanding the behavior of both primary particles from terrestrial and anthropogenic sources and secondary particles formed within the atmosphere. This book presents the study of these atmospheric particles from a microscopic perspective.

It may seem a stretch of scientific faith to extrapolate from single particles at the microanalytical scale to a geophysical scale, such as the global atmosphere or a regional airmass. However, the detailed information derived from microscopic techniques can reveal how particles affect various physicochemical phenomena in the atmosphere. For example, together with particle shape and size, microscopy may be used to identify thin hydrophilic coatings on otherwise hydrophobic particles, which allow these particles to take up water and, therefore, act as cloud condensation nuclei. As this book shows, individual-particle analyses offer unique insights into the behavior of the particulate atmosphere, whether the scientific goal is determining urban air quality, global climate change over time, climate change in specific regions, or cloud behavior with respect to climate change.

This book describes the most relevant microscopic techniques for studying atmospheric particles and presents specific applications of those techniques for understanding particle behavior. Static single-particle (microscopic) techniques provide uniquely important information about particle origins and atmospheric processes. The methods are categorized as electron microscopies (scanning and transmission), Raman microspectroscopy, atomic force microscopy (AFM), and scanning transmission X-ray microscopy (STXM).

The book's 12 chapters are divided into three parts. In Part I, "Overview of Techniques and Applications," Chapters 1–3, respectively, provide general introductions to the techniques ("Analysis of Individual Atmospheric Particles: An Overview of Techniques"), their application to air quality studies ("Importance of Microanalysis in Air Quality Studies"), and their application to climate change studies ("Importance of Microanalysis in Climate Studies of Atmospheric Aerosol Particles").

In Part II, "Microanalytical Techniques," four chapters provide an in-depth treatment of selected techniques. Chapter 4 ("Particle Population Analysis by Automated Scanning Electron Microscopy") discusses the use of scanning electron microscopy (SEM) for determining the distribution of particle sizes, morphologies, and elements in particle populations. The technique, known as computer-controlled SEM, automates the analysis of thousands of particles in a sample. Chapter 5 ("Characterization of Atmospheric Particles by Electron Energy Loss Spectroscopy") shows how inelastic scattering of beam electrons in transmission electron microscopy (TEM) produces a spectrum of absorption edges, the result of electron energy loss spectroscopy (EELS), which elucidates chemical bonding within individual particles. Chapter 6 ("Raman Microspectroscopy of Particles") demonstrates how laser excitation in Raman spectroscopy, whereby spectra exhibit vibrational states due to light-scattering rather than light absorption, reveals chemical information about individual particles. Chapter 7 ("Atomic Force Microscopy of Individual Particles") illustrates how the cantilever probe in AFM is used to determine particle morphology, mixing state, hygroscopicity, phase state, and surface tension. Sea spray aerosol particles serve as examples.

In Part III, "Particle Types and Transformations," five chapters discuss specific types of particles and atmospheric processes involving particles. Chapter 8 ("Transmission Electron Microscopy for Studying Atmospheric Soot Particles") shows how imaging and electron diffraction with TEM are used to examine the size, morphology, mixing state, fractal dimensionality, and crystallography of soot particles. In addition, techniques such as energy dispersive X-ray spectroscopy (EDX) and EELS associated with TEM provide information on soot composition and chemical bonding. Chapter 9 ("Microscopy Methods for Organic Composition of Atmospheric Aerosol Particles") describes how the organic chemistry of particles is mapped with STXM and near-edge X-ray absorption fine structure (NEXAFS), electron microscopy with EDX, and Raman microspectroscopy. Limitations of these techniques are discussed, as are emerging techniques for organic particles, such as tip-enhanced and surface-enhanced Raman spectroscopies, AFM with infrared spectroscopy, and optical photothermal infrared spectroscopy.

The remaining chapters in Part III discuss particle transformations. Chapter 10 ("Microanalytical Studies of Multiphase Reactions and Particle Aging") focuses on the use of microscopic techniques to understand the chemical transformations, including aging, that particles undergo in ambient air with gases and photochemically. The chapter examines in-situ studies of heterogeneous reactions with various techniques, particularly STXM-NEXAFS. Chapter 11 ("Microanalysis Techniques to Study Atmospheric Ice Nucleation and Ice Crystal Growth") illustrates how ice nucleation proceeds, primarily heterogeneously, with a nucleating particle as substrate. The chapter covers ice nucleation pathways and theory as well as the complexities of ice crystals during nucleation, further growth, and sublimation using various microscopic methods. Chapter 12 ("Studying Aerosol Hygroscopicity Using Environmental Electron Microscopy") shows at the microscopic level how crystalline particles take up water to become droplets (deliquescence) during cloud formation. The chapter focuses primarily on electron microscopies with humidity-controlled instrumentation that allow for observing particle hygroscopicity using techniques known as environmental SEM (ESEM) and environmental TEM (ETEM).

We thank everyone who contributed chapters and the reviewers who provided much helpful criticism. We are grateful to everyone who ensured the relevance and thoroughness of the scientific content. We thank Ritu Bose, Lesley Fenske, and Noel McGlinchey at Wiley for, respectively, proposing the topic, managing the project, and assisting with chapter submission. We also thank AGU's Publications Director, Jenny Lunn, for advice and encouragement, Geeta Persad of the University of Texas at Austin for her work as subject editor, and copyeditor Tiffany Taylor. In addition, we thank the National Institute of Standards and Technology (NIST) for support.

Joseph M. Conny
National Institute of Standards and Technology USA (retd.)
Peter R. Buseck
Arizona State University, USA

Part I
Overview of Techniques and Applications

1

Analysis of Individual Atmospheric Particles: An Overview of Techniques

Joseph M. Conny

ABSTRACT

Investigations into atmospheric aerosols are key to understanding how chemical and physical processes within the atmosphere contribute to climate change, affect public health, and impact visibility. Determinations of atmospheric particulate matter at the microscopic, or single-particle, level offer distinct advantages over determinations of particulate matter in bulk form. These advantages include chemical composition with respect to particle size, shape, and mixing state along with the physical properties of particles, particularly surface properties, as they affect the formation and lifetime of clouds and haze. This chapter is an overview of the various static techniques that allow the detailed physicochemical interrogation of single atmospheric particles and the benefit of these techniques in atmospheric studies. A general description of techniques is followed by their use in studying atmospheric particles. Included are scanning and transmission electron microscopies and associated techniques such as X-ray spectroscopy, electron diffraction, and electron energy loss spectroscopy; micro-Raman spectroscopy and associated techniques such as optical tweezers, surface-enhanced Raman, and tip-enhanced Raman; atomic force microscopy; and scanning transmission X-ray microscopy. Dynamic single-particle techniques such as aerosol time-of-flight mass spectrometry, which can be appreciated only from the analysis of particle ensembles, are briefly mentioned.

1.1. INTRODUCTION

We all enjoy a clear, cloudless sky. Yet, more often than not, there are features in the daylight sky that, for better or worse, make it more interesting. These features include clouds, rainbows, colorless and colored haze, contrails, smoke from fires, steam and smoke from stacks, and soot from tailpipes. Virtually all these features are due to microscopic particles, either as solids or liquids, suspended in the gaseous ambient air, which we know as aerosols.

Particulate matter in the atmosphere is of far greater consequence, however, than keeping our interest in sky-watching. To assess air quality and regulate harmful emissions, breathable particulate matter is commonly dichotomized as particles <2.5 μm (PM2.5) and <10 μm (PM10) in aerodynamical size. Numerous epidemiological studies have revealed the health hazards of PM2.5 and PM10 for vulnerable populations (e.g. Kim et al., 2015; Miller et al., 2007; Pope et al., 2002, 2009). Especially revealing is the rate of mortality from lung cancer and cardiopulmonary disease in cities due to exposure to PM2.5 (Dockery et al., 1993; Hoek et al., 2013). Reduced visibility from particulate matter is also a concern (Hyslop, 2009). More than an annoyance, the impact of

Materials Measurement Science Division, National Institute of Standards and Technology, Gaithersburg, MD, USA

Microanalysis of Atmospheric Particles: Techniques and Applications, Geophysical Monograph 288, First Edition.
Edited by Joseph M. Conny and Peter R. Buseck.

DOI: 10.1002/9781119554318.ch1

particulate matter on visibility can be critically important in road traffic safety (Pauley et al., 1996). Air traffic is also affected by reduced visibility or, worse, jet engine malfunction. Massive injections of particulate matter from volcanic eruptions in 1982 in Indonesia and 1989 in Alaska caused jet engines to fail before being restarted in flight. In 2010 and 2011, emissions from Iceland's Eyjafjalla and Grimsvötn volcanos, respectively, were responsible for closing airspace to busy air traffic (Brooker, 2010). However, dust storms, albeit intense, appear more likely to affect airline on-time performance than endanger passengers (Baddock et al., 2013).

Aerosols have also been intensely studied for their effect on climate. Particulate matter shifts Earth's radiative balance, measured as radiative forcing in watts per meter squared ($J\,s^{-1}\,m^{-2}$), either by absorbing or scattering solar and longwave radiation directly or affecting physical processes within clouds and serving by cloud-condensation or ice nuclei (Hansen et al., 1997; Haywood & Boucher, 2000; Lohmann & Feichter, 2005; Twomey, 1974). Uncertainties in the behavior of particulate matter in the atmosphere, particularly regarding clouds, cause by far the largest gap in our understanding of how atmospheric aerosols affect climate (Boucher et al., 2013).

At the microscopic level, individual atmospheric particles are fascinating in the breadth of variation in their size, morphology, and chemical composition. The tremendous variation in particle properties provides for the rich geographic and temporal diversity of atmospheric aerosols. This diversity is manifested by volcanic forces, wind-driven processes such as weathering of Earth's crust and release of sea spray, and the chemical processing of natural and anthropogenic emissions within the atmosphere. Variation in cloud formation and precipitation are also closely related to the prevalence and character of aerosols in the atmosphere.

1.1.1. Bulk Analysis Versus Microanalysis

Atmospheric particulate matter is normally analyzed either as a bulk material or at the scale of individual particles. The approaches are often complementary: for example, in studying heterogeneous aerosol chemistry (e.g., Laskin et al., 2005). Inductively coupled plasma mass spectrometry (ICP-MS) and gas and liquid chromatography with mass spectrometry (GC/MS and LC/MS, respectively), optical techniques such as Fourier transform IR spectroscopy, fluorescence spectrometry, and magnetic resonance techniques such as ^{13}C-NMR are powerful bulk-analysis techniques for chemically speciating organic and metallic constituents in particles from emission sources (Noziere et al., 2015). X-ray diffraction techniques can help identify crystalline constituents in samples (Jeong, 2008). X-ray fluorescence and inductively coupled plasma mass spectrometry can quickly reveal elemental composition, the latter isotopic composition as well (Majestic et al., 2009; Mazzei et al., 2008; Pekney & Davidson, 2005). However, these techniques cannot reveal chemical composition with respect to the shapes of particles, their size distribution, surface versus subsurface layers, or whether chemical components are mixed within individual particles (internal mixing) or associated with separate particles (external mixing). These inquiries are better addressed with microanalysis.

Because of the importance of understanding the properties of individual particles, microanalytical methods for atmospheric particles have become as numerous as conventional bulk methods, perhaps more so. Over the last decade or so, a number of comprehensive reviews involving individual particle techniques have been published (Ault & Axson, 2017; Bzdek et al., 2012; Elmes & Gasparon, 2017; Fletcher et al., 2011; Laskin et al., 2012, 2019; Li et al., 2016; Ott et al., 2021; Pósfai & Buseck, 2010; Pratt & Prather, 2012b; Tang et al., 2019). Interestingly, advances in the technology of sample introduction can allow a bulk analysis technique to evolve into a microanalysis technique. This transition is often driven by a need for real-time or near-real-time data to indicate particle mixing state – that is, whether particles are internally or externally mixed – and particle surface properties such as the effect of coatings from oxidized gas-phase species such as sulfate, nitrate, and semivolatile organic carbon on hygroscopicity.

An example of the transition from bulk analysis to microanalysis is the mass spectrometry of aerosols. For decades, mass spectrometry has been an important aerosol methodology, initially with bulk extracts injected into a gas chromatograph followed by electron ionization of the eluate and ions detected by a quadrupole or magnetic sector spectrometer. As aerosol mass spectrometry (AMS) came to be known, aerosols were introduced into a mass spectrometer as a focused particle stream to facilitate real-time time analysis. However, AMS is still considered a bulk technique (Pratt & Prather, 2012a) because portions of the particle stream are collected on a surface for subsequent vaporization, typically by resistive heating, and ionization, typically by electrons. Conventional AMS with a quadrupole detector has time-of-flight (TOF) functionality, but in this case, it is the time of flight of particles to the collection surface (PToF) rather than the time of flight of ions to the detector (Canagaratna et al., 2007). Numerous variations of real-time bulk mass spectrometric methods for aerosols exist, with the common connection that they employ the TOF spectrometer. However, as instruments were designed to ionize individual particles by laser in the focused aerosol stream, which required an ion TOF detector, the mass spectrometry of aerosols evolved into a single-particle methodology known now as aerosol TOF mass spectrometry (AToFMS) (McKeown et al., 1991; Pratt & Prather, 2012b). Along with the TOF spectrometer (either linear or reflectron), the key to real-time single-particle mass spectrometry is a laser system to detect, size, and pulse ionize individual particles (laser desorption/ionization) in the particle stream.

AMS and aerosol time-of-flight techniques constitute a large analytical field that has been reviewed in a number of comprehensive publications (e.g., Canagaratna et al., 2007; Johnston, 2000; Noble & Prather, 2000; Pratt & Prather, 2012a, 2012b; Zelenyuk & Imre, 2009) and therefore is not covered separately in this book.

1.1.2. Dynamic Versus Static Microanalysis

Microanalysis techniques can be categorized as dynamic or static. Dynamic techniques provide momentary snapshots of particles in a stream. Static techniques operate on lengthier time scales and therefore allow for the physicochemical interrogation of individual particles. Data from dynamic techniques such as AToFMS can only be appreciated from an ensemble of particles, much like a movie is appreciated only as a collection of individual frames. A specific analytical example concerns the particle mixing state. AToFMS can inform if a particle ensemble can be categorized as internally or externally mixed. As a static technique, electron microscopy allows us to examine how chemical phases in a single internally mixed particle are arranged from element mapping with energy-dispersive X-ray spectroscopy or phase identification with selected-area electron diffraction (SAED). The mixing state of actual individual particles helps to identify the appropriate particle mixing-state model used in climate studies, such as the concentric core-shell configuration for absorbing and nonabsorbing phases (Fuller et al., 1999; Jacobson, 2001). Other configurations might be nonconcentric inclusion-within-matrix and adduct-upon-matrix configurations (Jeong & Nousiainen, 2014; Kahnert et al., 2014; Muinonen et al., 2009).

1.1.3. Selection of Microanalytical Techniques

In this book, the authors report a host of modern microanalytical techniques and their applications that allow for the detailed physicochemical interrogation of single atmospheric particles. This chapter provides a description of techniques followed by examples of their specific use in studying particles. The various techniques can be categorized into six static "microscopies" involving a variety of analytical probes:

- Electrons: scanning electron microscopy (SEM) and transmission electron microscopy (TEM)
- Photons: micro-Raman spectroscopy and scanning transmission X-ray microscopy (STXM)
- Ions: time-of-flight secondary ion mass spectrometry (TOF-SIMS)
- Deflection of a mechanical arm: atomic force microscopy (AFM)

In electron microscopy, the specimen is probed by a beam of electrons generated by an electron gun. Electrons are propelled toward the specimen and focused by a series of electromagnetic lenses. Information about particle shape, topography, and composition is provided by the detection of beam electrons that have been elastically or inelastically scattered and by diffraction patterns of the beam electrons. The detection of secondary electrons and X-rays from the specimen also provides essential information in electron microscopy. In micro-Raman spectroscopy, photons in the visible region probe the specimen to reveal different chemical species as the polarization of chemical bonds by visible light shifts the energies of the photons as they interact with atoms. In STXM, intense synchrotron-generated X-rays probe the specimen to reveal organic functional groups that absorb highly monochromatic photons. The edge of absorption peaks in STXM spectra reveal bonding-specific details. In TOF-SIMS, pulsed primary ions, typically oxygen, argon, cesium, or gallium, are used to sputter surface material from a specimen. The ions probe the specimen surface or at depth by dislodging characteristic secondary ions, which are then rapidly mass-separated and time-detected in a flight tube by mass spectrometry. In AFM, specimen surfaces are probed by detecting the movement of a high-resolution micro-cantilever above the substrate surface. Cantilever movement is due to mechanical contact with the substrate or atomic interactions between cantilever and substrate, such as van der Waals attractions and electrostatic forces.

The choice to focus on the six methodologies listed is based in part on their prevalence in the analytical literature for atmospheric aerosols. Figure 1.1 shows the number of citations for each methodology retrieved from Web of Science over five decades. Before 1971, microscopy of aerosols was mainly limited to light microscopes. In the decade from 1971 to 1980, a few reported studies involved SEM and micro-Raman spectroscopy. Although TEM had been developed prior to SEM, TEM studies of atmospheric aerosols began to appear between 1981 and 1990, along with TOF-SIMS. It wasn't until the 1990s that microanalysis of single particles began to take off. In that decade, AFM appeared for the first time and the use of TEM, TOF-SIMS, and micro-Raman for atmospheric aerosols increased severalfold, but particularly with SEM. In the early 2000s, the initial application of STXM to atmospheric aerosols appeared in the literature.

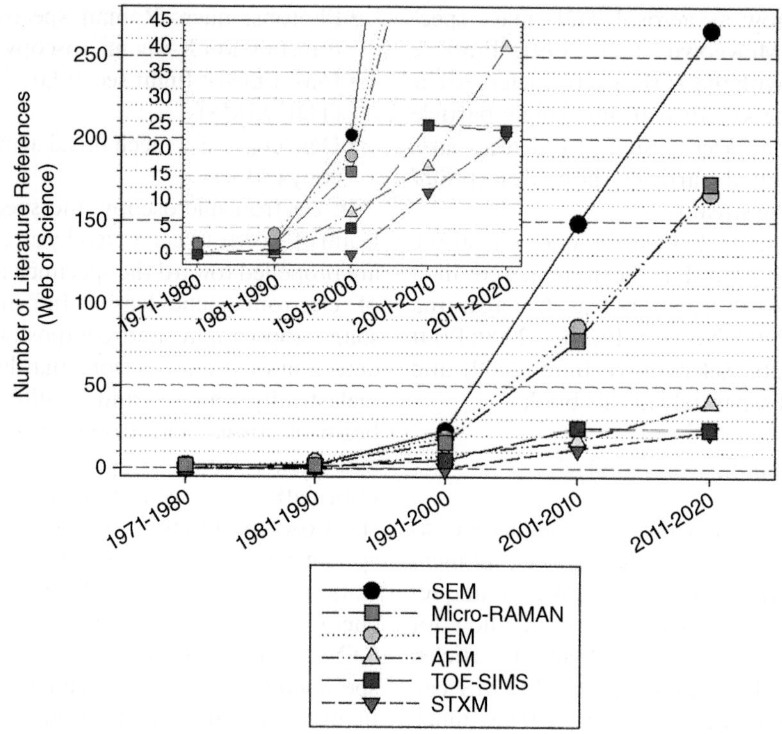

Figure 1.1 Number of references in the atmospheric aerosol literature over five decades from Web of Science for the most prominent microanalysis techniques used to investigate individual atmospheric aerosol particles. The inset magnifies the lower end of the number range to better show the prevalence of techniques in the literature over time.

During the decade 2011–2020, the literature shows that all microscopies except one saw increased use for analyzing atmospheric aerosols. The exception is TOF-SIMS. Nevertheless, it is a useful analytical technique for studying surface chemistry in particular. As the need to understand particle surface properties and heterogeneous reactions increases, the use of TOF-SIMS for atmospheric aerosols will likely continue.

SEM appears to be the most-referenced microscopy (possibly excluding light microscopy) in the atmospheric aerosol literature for the past three decades. Since 2001, micro-Raman spectroscopy and TEM have been the next-most-referenced microscopies. Notwithstanding the need to travel to a specialized facility to perform it, STXM will likely see increased use in future years, particularly for studying individual organic aerosol species in particles. AFM use will also likely increase due to the need to understand how particle topography affects hygroscopicity and reactions on particle surfaces.

Associated with each of the six microscopies (and spectroscopies) are various techniques that can be performed. Table 1.1 shows the techniques, the types of signals acquired, and detection properties. What follows is an overview of the microscopic techniques and examples of applications. (TOF-SIMS is included when combined with other microscopic techniques.) It is not intended that the applications and associated references reported here are all-inclusive; rather, they are some salient examples of how the techniques are applied in aerosol studies.

1.2. ELECTRON MICROSCOPY

In electron microscopy, beam electrons that probe the specimen are set to a specific acceleration energy ranging from 500 eV to 30 keV in SEM and from 80 to 400 keV in TEM. The beam current can be adjusted to a specific current from 1 pA to 100 nA or more.

In older SEM and TEM instruments, the electron source is a hairpin tungsten filament that serves as the electron gun's cathode. To provide a brighter beam, later instruments utilized a single, sharp-tipped crystal of lanthanum boride (LaB_6) as the electron source. The tungsten filament and LaB_6 electron guns required high heating to overcome the work function to release electrons from the gun tip (Goldstein et al., 2018). State-of-the-art instruments commonly have a thermally assisted field-emission electron gun (tFEG). The cathode emitter, in this case, is typically a sharp-tipped tungsten wire heated to around 1600 K and subjected to an intense electric

Table 1.1 Characteristics and Properties of the Various Techniques Associated with the Six Microscopies for Particle Microanalysis

Technique	Probe	Detected signal	Particle property determined	Relative sensitivity (%)[1]	Minimum detectable amount[1]	Probe lateral resolution[1]
SEM EPMA[2] ESEM[3] VPSEM[4] FIB-SEM[5]	Electrons, Ions (FIB-SEM)	Secondary electrons	Size, morphology from electron image	0.1	10^{-16} g	1–3 nm
		Backscattered electrons	Size, topography, composition (qualitative) from TOPO/COMP image			
		Backscatter electron diffraction	Crystallinity			
		Secondary ions (FIB-SEM)	Size, morphology from ion image			
		Characteristic X-rays with EDX, WDS	Element composition >Beryllium Element spatial distribution (map)			
		Luminescence	Cathodoluminescence			
CTEM[6] STEM[7]	Electrons	Elastically scattered electrons	Size, shape from bright-field image (low-angle scattering) Size, shape from dark-field image (high-angle scattering) Crystallinity, grain morphology, specimen thickness from diffraction	0.1	Atom	0.15–0.3 nm
		Inelastically scattered electrons	Oxidation state, chemical bonding from electron energy loss spectroscopy			
		Secondary electrons (STEM)	Size, morphology from electron image			
		Selected-area electron diffraction	Crystallinity, grain structure			
		Convergent-beam electron diffraction	Specimen thickness, small area crystallinity, grain structure			
		Characteristic X-rays with EDX	Element composition >Beryllium Element spatial distribution (map)			
		Luminescence	Cathodoluminescence			
Micro-Raman SERS[8] TERS[9]	Photons (VIS)	Photons red-shifted by molecular vibration and bond polarizability	Molecular composition Molecular functional groups	1	10^{-16} g	0.8 μm (micro-Raman) 0.5 μm (SERS)[10] 0.1 μm (TERS)[10]
AFM AFM-IR	Mechanical cantilever	Force on oscillating cantilever	Size, volume, morphology Surface area, surface tension Hygroscopicity IR absorption (AFM-IR)	NA	NA	10 nm

(Continued)

Table 1.1 (Continued)

Technique	Probe	Detected signal	Particle property determined	Relative sensitivity (%)[1]	Minimum detectable amount[1]	Probe lateral resolution[1]
TOF-SIMS NanoSIMS	Pulsed ions (TOF-SIMS) Ion stream (NanoSIMS)	Secondary ions (m/z)	Surface composition (elemental/isotopic/molecular) Depth profile	3.1 to 0.1% (1–1000 ppm)	10^{-16} g	40 nm to 1 μm (TOF-SIMS) >40 nm (NanoSIMS)
STXM	Monochromatic (synchrotron) X-rays	Transmitted and secondary X-rays	Elemental/molecular composition and bonding (NEXAFS)			0.06 μm

[1] Fletcher et al., "Microscopy and Microanalysis of Individual Collected Particles," in *Aerosol Measurement: Principles, Techniques, and Applications*, P. Kulkarni, P.A. Baron, & K. Willeke, eds., New York: John Wiley & Sons (2011). Relative sensitivity is the fraction of measured mass for a component (element or molecule) measured analytically by a microscopic technique: for example, X-ray spectroscopy in SEM. Relative sensitivity is based on a flat polished specimen.
[2] Electron probe microanalysis.
[3] Environmental SEM.
[4] Variable-pressure SEM.
[5] Focused ion beam SEM.
[6] Conventional TEM.
[7] Scanning TEM.
[8] Surface-enhanced Raman spectroscopy.
[9] Tip-enhanced Raman spectroscopy.
[10] Ault & Axson (2017), *Anal. Chem.*, *89*, 430–452.

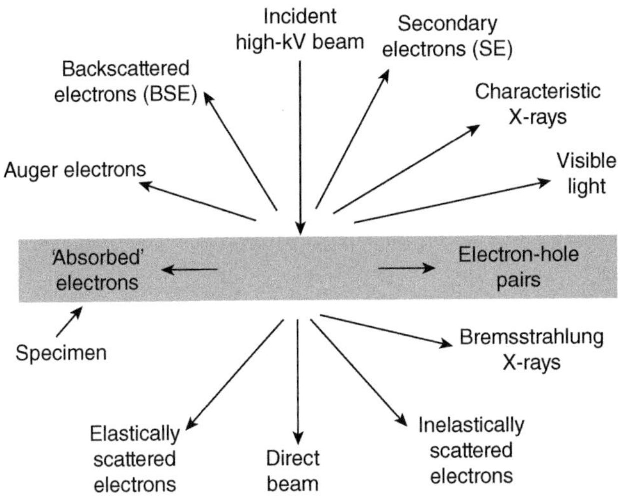

Figure 1.2 Signals generated when a high-energy beam of electrons interacts with a thin specimen. Williams and Carter, 1996a/with permission of Springer Nature.

field whereby electrons "tunnel" from the emitter tip (Goldstein et al., 2018). A variation on the tFEG is the Schottky emitter, whereby the tungsten tip is thinly coated with zirconium oxide. The tip is heated to around 1800 K.

Figure 1.2 shows the signals that emanate from a specimen exposed to an electron beam (primary electrons). The figure depicts signals from TEM. However, signals detected from above the plane of the specimen with SEM (as well as technically with TEM) include backscattered primary electrons, secondary electrons, characteristic X-rays, and visible light (cathodoluminescence). Signals detected below the plane of the specimen with TEM include elastically and inelastically scattered primary electrons.

The principal drawbacks to using electron microscopy to study atmospheric particles relate to the relatively high energies and currents of the incident electron beam and having the specimen under vacuum. When the specimen and its substrate are insufficiently grounded electrically, the incident beam may statically charge a particle and cause it to move from its position on the substrate. The incident beam may also damage a fragile particle before sufficient time exists to acquire analytical signals: for example, characteristic X-rays. Although particles with mineral phases are typically undamaged, fragile particles such as organics often undergo damage, which may be observed in the microscope as it occurs. With the high vacuum that is required for conventional SEM and TEM, water, volatile organics, and ammonium nitrate are removed under vacuum, and thus SEM and TEM are often unsuitable for particles with these phases.

1.2.1. Scanning Electron Microscopy

In SEM with a field-emission gun, the electron probe is focused to approximately a 1 nm spot on the specimen (Table 1.1) and then rastered across the specimen. However, the volume within a thick specimen that is sampled by the probe at each point is much larger than the probe itself. Figure 1.3 shows the extent that primary electrons travel and deposit energy within a specimen of the NIST K-412 glass (former NIST Standard Reference Material No. 470) using a beam energy of 15 keV (Ritchie, 2020). Also shown is the lateral extent for 90% of characteristic X-rays that are emitted for the various elements in the glass.

Along with higher magnification (as much as 300,000×) and the concomitant increase in resolution, a significant advantage of SEM over light microscopy is a large depth of field. Magnetic lenses are used to first demagnify the electron beam (condenser lens) and, further down the column, focus the beam to converge on a spot (objective lens). Imaging is performed by detecting secondary electrons released from the specimen or primary (beam)

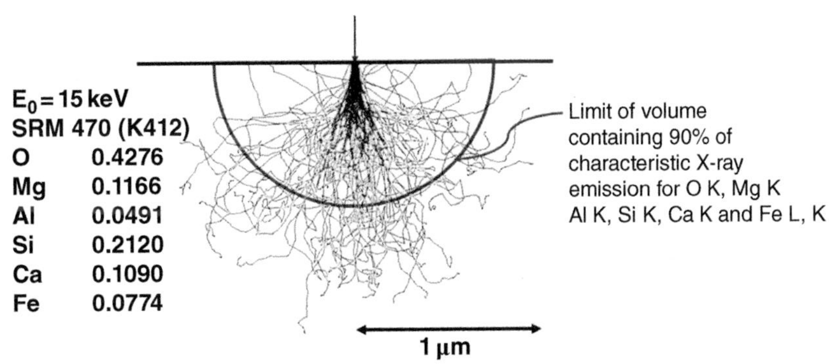

Figure 1.3 Monte Carlo simulation of the interaction volume for glass K412 (SRM 470) containing several elements probed by a 15 keV electron beam in SEM. Element numbers indicate fractional weight. From NIST DTSA-II (Ritchie, 2020), public-domain software for electron-excited X-ray energy-dispersive spectrometry. https://www.nist.gov/services-resources/software/nist-dtsa-ii.

electrons that are backscattered. Backscattered electrons typically lose some energy upon entering the specimen, but the process primarily involves elastic scattering at large angles (Goldstein et al., 2018). Secondary electrons are typically counted by a scintillation device, whereby electrons interact with a scintillating material and light is emitted and then detected. Backscattered electrons may also be counted by a scintillation detector, but a small solid-state detector placed close to the specimen provides much higher collection efficiency. Configuration of the SEM lenses can also affect how secondary and backscattered electrons are detected.

SEM provides fine detail regarding the topography and composition of a specimen from the strength of secondary electron and backscattered electron signals above the specimen, as shown in Figure 1.2. Electron imaging with SEM (as well as TEM) can provide quantitative information regarding the size and morphology (e.g., aspect ratio) of atmospheric particles. However, the value of imaging is largely qualitative.

Secondary electron emission occurs when energy from the inelastic scattering of beam electrons ejects valence electrons from specimen atoms. Because valence atoms are weakly bound, secondary electrons have low kinetic energy, typically less than 50 eV. Both specimen structure and composition affect the emission of secondary electrons. Secondary electrons from a particle near its surface form an image that mainly reveals the particle's two-dimensional (2D) shape and topographic features. Maximum resolution is on the order of the beam spot size, which is typically around 1 nm.

Backscattered electrons can have a wide range of energies depending on the specimen's elements. As a result, backscatter electron imaging can reveal qualitative element differences in a particle's composition as well as its shape and topography. For heavier elements in the specimen, the energy of backscattered electrons is a sizable fraction of the beam energy. Thus, energies of backscattered electrons are typically much higher than secondary electron energies.

With both secondary and backscattered electrons, detailed imaging depends on the energies of the emitted electrons, the angle of the electron beam entering the specimen, and the angle of the electrons exiting above the plane of the specimen. Other factors include the type, configuration, and placement of the electron detector within the instrument. Backscatter electrons reveal contrast in atomic number because higher-Z atoms are more likely to scatter back electrons at a larger angle relative to the direction of the beam electrons than will lower-Z atoms.

Topographic information is more dependent on the angles at which the beam enters the specimen and secondary or backscattered electrons exit. In addition, secondary electron emission is enhanced at the edges of the specimen, where the valence electrons energized by the primary beam can escape the specimen surface. Secondary electrons in direct line of sight to the detector tend to provide greater topographic detail than backscattered electrons because with their low energies, secondary electrons can be drawn by a positive electric potential to the detector with greater efficiency.

X-ray Spectroscopy with Electron Microscopy

Additional analytical devices associated with SEM instrumentation provide essential compositional information on atmospheric particles. A technique that is typically integral to SEM and commonly associated as well with TEM instrumentation is X-ray spectroscopy. When energetic electrons enter a material and liberate core-shell electrons, the ionized states of lower-Z atoms relax primarily by emitting Auger electrons, whereas ionized states of higher-Z atoms relax primarily by emitting X-rays. In addition to emission by target atoms, X-rays can be absorbed by other atoms within the material without further X-ray emission, or they can result in X-ray emission by other atoms: that is, fluorescence.

Both X-ray absorption and X-ray fluorescence can be corrected algorithmically for specimens with surfaces that are not perfectly flat. In seminal work reported by Armstrong and Buseck, the theory behind X-ray absorption and fluorescence in individual particles with surface topography was developed (Armstrong & Buseck, 1975). That and subsequent early work (Bradley et al., 1981) led the way for the quantitative elemental analysis of atmospheric particles by energy-dispersive X-ray spectroscopy (EDX or EDS)-associated SEM and TEM. However, the quantitative analysis of particles in the size range to produce Mie scattering (particle diameters on the order of incident light wavelengths in the actinic region) remains problematic, particularly when measurement reference particles are not used in the calibration process. Studies of pollution, as well as solar radiative forcing by aerosols, focus on particles in the Mie-scattering size range.

Another historic development in the X-ray microanalysis of atmospheric aerosols was automated SEM imaging and X-ray analysis for determining the sizes and compositions of thousands of particles collected on a filter substrate (Byers et al., 1971; Casuccio et al., 1983; Germani & Buseck, 1991; Li et al., 2023). In generating distributions for the sizes and compositions of large numbers of particles, automated particle analysis, or computer-controlled SEM (CCSEM), links the analysis of an individual particle by SEM-EDX with the population of aerosols in an airshed.

In newer instruments, EDX with silicon-drift detectors provides rapid, high-throughput counting of characteristic X-rays from elements with atomic numbers greater than that of boron. In analyzing atmospheric particles, electron probe microanalysis (EPMA) is often synonymous with SEM-EDX (e.g., Ro et al., 1999). However, materials scientists and geochemists typically associate EPMA with wavelength-dispersive X-ray spectroscopy (WDS). The technique is important for studying mineral samples and is therefore relevant with respect to atmospheric mineral dust. With WDS, a series of crystals is used to collect the X-ray photons rather than the silicon-drift detector. The crystals are mechanically aligned to diffract the emitted photons toward a proportional counter. Due to the mechanical alignment of crystals, WDS is more time consuming than EDX but provides far better resolution of X-ray energies and quantification (Goldstein et al., 2018).

When the electron-probe beam enters the specimen, X-rays produced within the specimen fill a volume (X-ray interaction volume) that is larger than the volume of the probe itself (probe interaction volume). The size of the X-ray interaction volume depends in large part on the density and thickness of the material. However, even with the disparity in the X-ray and probe interaction volumes, rastering of the probe allows for element mapping of a particle to reveal the qualitative compositional heterogeneity of atmospheric particles with rather high resolution. This is because X-ray emission is greatest from the part of the interaction volume nearest where the beam enters the specimen. A typical minimum resolution for X-ray mapping is around 20 nm (Fletcher et al., 2011). As an example, the resolution for a 1 μm particle filling a 60-pixel by 60-pixel scan frame is 17 nm.

Studies have looked at ways to improve EDX analysis of lighter elements (low-Z elements) in aerosols and how to improve substrates to avoid inaccuracies. The analysis of carbon, nitrogen, and oxygen by EDX without a conventional beryllium spectrometer window (i.e., an ultrathin window or windowless spectrometer) was demonstrated along with the importance of assessing the absorption of soft X-rays of light elements by the matrix to interpret spectral intensities (Ro et al., 1999, 2000). Boron substrates in lieu of commonly used polycarbonate membranes, in combination with a reverse Monte Carlo procedure for correcting the X-ray volume, were used to improve CCSEM analysis involving low-Z elements (Choel et al., 2005, 2007).

Applications in Air Quality and Climate Studies

SEM-EDX has been used in a wide variety of atmospheric aerosol studies. Single-particle studies include physicochemical processes and reactions in the atmosphere, characterization of local and regional emission sources, source apportionment, particle shape, and composition effects in climate-related radiative transfer. Particle types include aerosols from urban and rural ambient airsheds, primary and secondary organics, biomass burning, sea spray, mineral dust, indoor aerosols, glacial snowpack, and ice cores. Although much work has been reported on SEM and atmospheric particles, references cited in the following paragraphs exemplify work on many of these sample types.

Many studies of atmospheric particles combine SEM-EDX with other individual particle techniques such as TEM, micro-Raman, and STXM. A few examples are the combination of CCSEM-EDX with TEM in analyzing tar balls from biomass burning (J. L. Hand et al., 2005), SEM-EDX with micro-Raman and TOF-SIMS for determining emissions from a lead battery recycling facility (Dappe et al., 2018), and CCSEM-EDX with STXM for characterizing the outflow of carbonaceous particles from Mexico City (Moffet et al., 2010). CCSEM-EDX has also been used with micro-Raman spectroscopy and near-edge X-ray absorption fine structure associated with STXM (STXM-NEXAFS) in observing secondary sulfate in sea spray aerosol in the Arctic (Kirpes et al., 2018). The optimization of experimental parameters has been demonstrated for parallel analyses of individual particles with SEM-EDX and micro-Raman spectroscopy (Worobiec et al., 2010). Ice particles from Asian dust outflow have been studied with AFM and micro-Raman in combination with SEM-EDX (Iwata & Matsuki, 2018). SEM-EDX and total reflection Fourier transform IR spectroscopy were used to study spectroscopic properties of deliquescing ammonium sulfate aerosols with oleic acid (Najera & Horn, 2009).

The greatest use of SEM is with EDX for the element composition analysis of particle populations on a substrate, with particle counts often in the several to many thousands. For example, populations of 29,331–53,105 particles from a large U.S. city were analyzed for iron content in coarse particle mode (Ault et al., 2012). Compositions of more than 60,000 particles in six samples were determined to assess North African dust reaching Madrid, Spain (Coz et al., 2009). Compositions of 27,000 particles from a rural site in Germany influenced by the urban Rein-Main area were used to calculate size-resolved complex refractive indices for the air masses (Ebert et al., 2004). The combination of CCSEM with environmental scanning electron microscopy (ESEM) and STXM provided a comprehensive determination of particle phase-state variability in the free troposphere during summertime (Cheng et al., 2022).

CCSEM is ideally suited for studying potential errors in the manual microanalysis of atmospheric particles. A comparison of the compositions of particle populations by CCSEM versus operator-controlled (i.e., manually controlled) SEM, shown in Figure 1.4, suggested large differences between the two approaches in the abundance of various particle groups (Hammer et al., 2019).

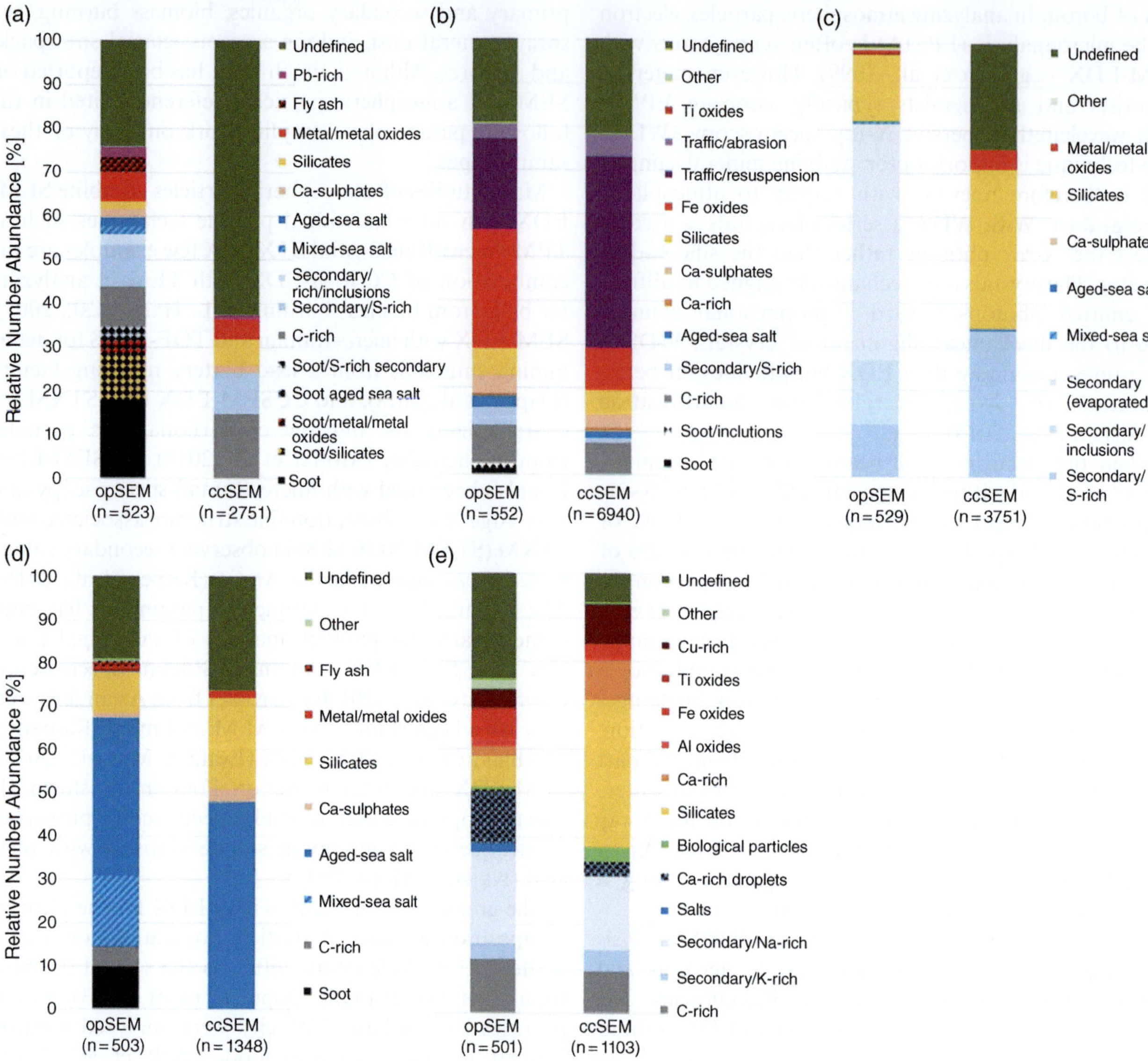

Figure 1.4 Percent of particles assigned to particle types determined by operator-controlled SEM (opSEM) and automated SEM (ccSEM). (a) Background urban air (Mainz, Germany). (b) Traffic site (Essen, Germany). (c) Free troposphere (Swiss Alps). (d) Marine/rural background (Cyprus). (e) Rural background (Germany). Hammer et al., 2019/with permission of Springer Nature.

Heterogeneous reactions have been extensively studied to demonstrate, for example, how the uptake of water might affect particle optical properties and formation of cloud-condensation nuclei. Differences have been found in how mineral dusts from different regions react with nitric acid. Calcium-containing dusts such as calcite and dolomite were found to react the most with HNO_3, whereas sulfate-containing gypsum did not react (Krueger et al., 2004; Laskin et al., 2005). Solid $CaCO_3$ particles reacting with HNO_3 in a humid environment were shown to form spherical droplets (Krueger et al., 2003). Calcium nitrate particles were shown to deliquesce at relative humidities as low as 9–11% (Laskin et al., 2005; Liu et al., 2008). Also studied was deliquescence in the uptake of HNO_3 by NaCl and a mixture of NaCl with MgCl to resemble sea salt (Liu et al., 2007).

Various compounds were found to be adsorbed on mineral dust particles. Adsorption of chlorine-containing gas-phase compounds was believed to be the source of chlorine in dust particles collected in southwestern Japan (Zhang & Iwasaka, 2001). Adsorption of semivolatile organic compounds such as polycyclic aromatic hydrocarbons and pesticides has also been observed on mineral dust particles. Differences in the distribution of PAHs and pesticides were found between submicron and larger particles (Falkovich et al., 2004).

The largest application of SEM-EDX is the characterization of local and regional ambient aerosol compositions and source emissions. Metal-containing particles have been the focus of studies of ambient air in the vicinity of steelmaking factories. Time-resolved injections of fine particles containing oxides of iron, manganese, and zinc were observed near a steelworks (Choel et al., 2010). Iron oxide particle compositions, including isotopic compositions from ICP-MS, that resemble those of iron ores enriched from the manufacturing process were observed in ambient air near a steelworks (Flament et al., 2008). Particle element profiles were determined from the component processes in iron and steelmaking (i.e., sintering, blast furnace operation, steelmaking, slag desulfurization) (Hleis et al., 2013). A study of emissions from sintering in the steelmaking process involving SEM-EDX and inductively coupled plasma emissions mass spectrometry revealed emissions factors for particles with a complement of toxic elements, including arsenic, cadmium, and lead (Zhao et al., 2017). SEM-EDX was used along with X-ray photoelectron spectroscopy to assess ambient air around a large opencast coal mining region of Columbia (Huertas et al., 2012). Coarse spherical aluminosilicate particles and carbon and sulfur-containing particles with vesicular morphologies were observed in emissions from an oil-fueled power plant (Hays et al., 2009).

Single-particle investigations with SEM-EDX have looked at urban emission sources that historically have attracted little attention. Emissions factors were determined for particles in the PM10, PM2.5, and PM1 ranges from building demolition by monitoring particulate concentrations at various distances from a building site during demolition (Azarmi & Kumar, 2016). PM2.5 particles from the combustion of various wastes (sawdust, plastics, textile, and food) were characterized by SEM-EDX for elemental composition, and temperatures of maximum PM2.5 release were simulated (Zhu et al., 2017). Dust from electronic waste recycling workshops in Chennai, India, was characterized by SEM-EDX and further analyzed for polychlorinated biphenyls by gas chromatography/mass spectrometry (Chakraborty et al., 2016).

SEM-EDX studies have also focused on ambient air in various locations to assess emission sources and the fate of emitted particles over time. Analyses of samples were used as input to the UNMIX multivariate receptor model to determine source contributions in Phoenix, Arizona, from gasoline engines, diesel engines, secondary sulfate, crustal minerals, and vegetative burning (Lewis et al., 2003). Due to the extent of pollution, Mexico City has been the focus of field campaigns to understand the fate of soot emissions, primarily from traffic and biofuels burning. Freshly emitted soot was found to rapidly form internally mixed particles, mainly with ammonium sulfate (Johnson et al., 2005). Iron-containing particles from a polluted air mass originating from the Asian mainland and collected in Okinawa, Japan, were shown to be internally mixed with sulfate, soot, and organic carbon (Moffet et al., 2012). A number of studies have assessed the composition of ambient air particles with SEM-EDX by looking at the surface of tree leaves (e.g., Tomasevic et al., 2005). Individual particles from indoor air in schools has also been investigated with SEM-EDX (Fromme et al., 2008).

SEM-EDX has also been useful in determining the compositions of heterogeneous samples containing biomass-burning particles. In a study of internal mixing of African dust and biomass-burning aerosols (Figure 1.5), particles containing sulfur, sodium, and potassium dominated the biomass burning fraction and were distinguished from mineral dust by having little aluminosilicate and no chloride (Hand et al., 2010). In a laboratory study, emissions from the burning of coniferous tree boughs, agricultural waste, and perennial plants such as grasses were found to contain water-soluble fluoride compounds (Jayarathne et al., 2014).

As input to optical property calculations of aerosols and radiative forcing, SEM has been instrumental in providing particle size, element composition, and morphology, primarily from the particle aspect ratio. Size, morphology, and element composition from SEM-EDX were used to determine complex refractive indices for ambient air in a rural region of Germany that was influenced by air from an urban area and Eastern Europe (Ebert et al., 2004). The imaginary part of the complex refractive index was 0.034–0.08 for urban-influenced air and 0.001–0.021 for rural ambient air. In contrast, the real part of the complex refractive index was lower for urban-influenced air. In a study of Saharan dust collected at Tinfuo, Morocco, during the Saharan Mineral Dust Experiment, CCSEM on 74,000 particles was used to identify three size regimes: $<500\,nm$ for sulfates and mineral dust with a median aspect ratio of 1.3, 500 nm to $50\,\mu m$ for mainly mineral silicates with a median aspect ratio of 1.6, and $>50\,\mu m$ for particles with substantial quartz (Kandler et al., 2009). Aspect ratios were used to calculate size-resolved complex refractive indices of 1.51–1.55 for the real part and 0.0008–0.006 for the imaginary part at 550 nm (Otto et al., 2009). SEM imaging was used to design realistic dust shapes consisting of spheres, spheroids, Chebyshev forms, and cylinders for calculating the single scattering albedo of dust from the Indian subcontinent (Mishra & Tripathi, 2008). Particle sizes and shapes from SEM of various mineral dust types were used with polarization instrumentation to determine scattering properties, of which backscattering was found to vary by as much as a factor of seven (Glen & Brooks, 2013). Particle size and shape distributions from

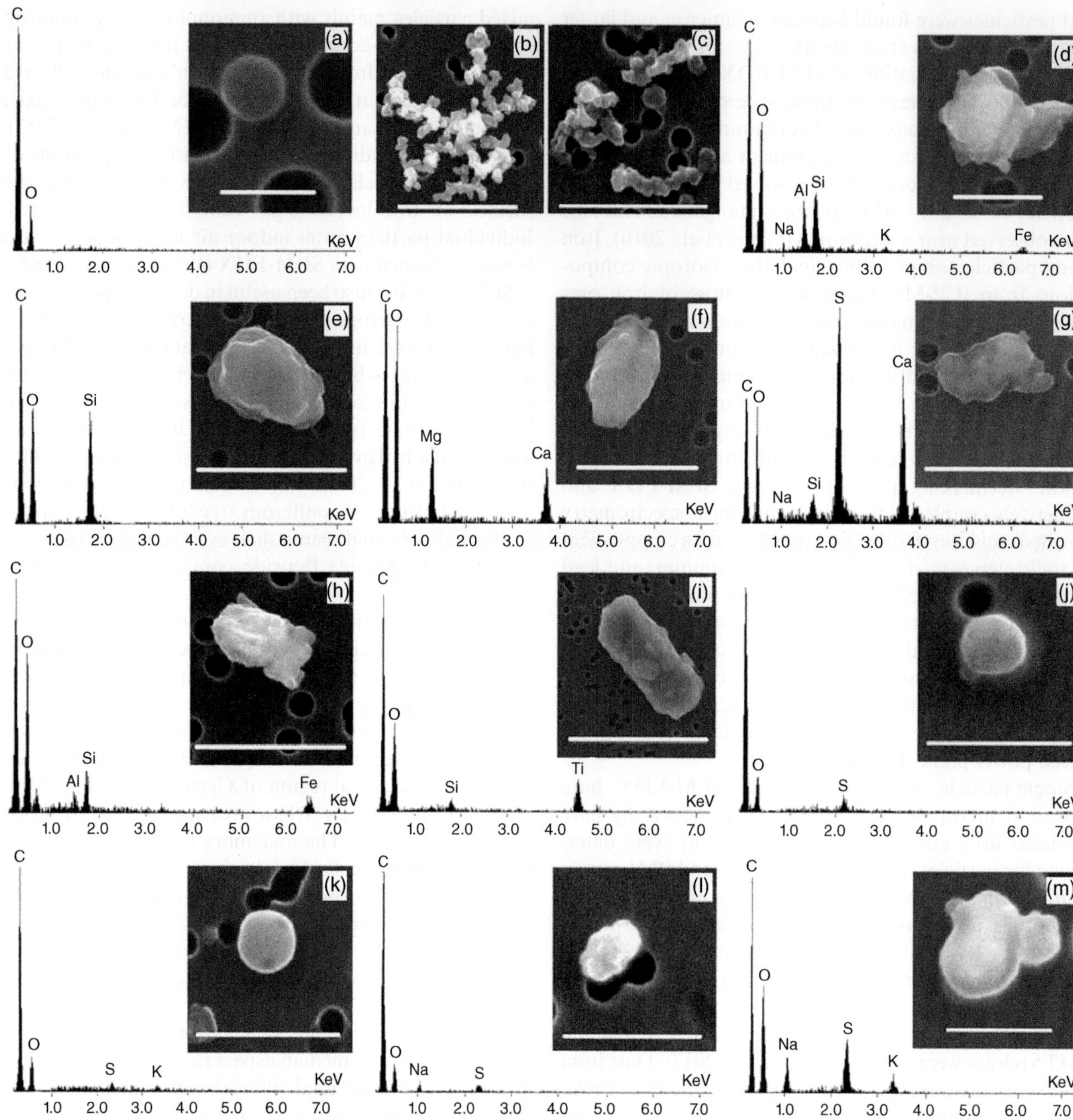

Figure 1.5 EDX spectra and SEM (ESEM) images of selected particles from biomass burning and African dust. (a) Tar ball, biomass burning; (b) soot, mixed; (c) soot, biomass burning; (d) aluminosilicate, dust; (e) Si-rich, dust; (f) Ca-Mg carbonate, dust; (g) Ca sulfate, dust; (h) Fe-rich, dust; (i) Ti-rich, dust; (j) S-rich, biomass burning; (k) S-K particle, biomass burning; (l) Na-S particle, biomass burning; (m) Na-S-K particle, biomass burning. Scale bars: (a, j, l, m) 500 nm; (d, k) 1 μm; (b, c, e–h) 2 μm; (i) 5 μm. Hand et al. (2010)/with permission of John Wiley & Sons.

SEM were also used to determine mass absorption coefficients and complex refractive indices for spectral wavelengths from 300 to 2500 nm for the 2010 eruption of the Eyjafjallajökull volcano in Iceland (Rocha-Lima et al., 2014).

Also important in climate research are characterizations of particles in ice cores and snow. Seven classes of particles from a snow pit on the northern slope of Mt. Everest (Mt. Qomolangma) were characterized by monsoon and nonmonsoon periods with SEM-EDX including soot, aluminosilicates, fly ash, calcium sulfates, calcium/magnesium carbonates, metal oxides, and nonelemental carbonaceous particles (Cong et al., 2009). SEM-EDX analysis of dust in snow on a glacier in the Qilian Mountains of northern China suggested that the snowpack was influenced by dust from the Badain Jaran Desert and Qaidam Basin (Dong et al., 2014). Analyses with SEM-EDX of ice cores from three locations on the Tibetan Plateau indicated long-range transport of quartz, clay (mainly illite and chlorite), and feldspar minerals (Wu et al., 2016). Aspect ratios were 1.48–1.53.

Among other climate change studies, SEM-EDX has also been used to understand the formation of cloud ice nuclei. CCSEM-EDX with STXM-NEXAFS analysis was used to show that particles with organic material collected in Mexico City potentially served as ice nuclei in the formation of cirrus clouds (Knopf et al., 2010). In a study of Asian dust particles, irregular clay particles formed ice nuclei at relatively high temperatures (>−30 °C) more than other particle types (Iwata & Matsuki, 2018).

Ancillary SEM Techniques and Applications to Atmospheric Aerosols

Cathodoluminescence and electron backscatter diffraction. Two SEM techniques that offer significant analytical reward but have yet to find much application to atmospheric aerosols are cathodoluminescence spectrometry and electron backscatter diffraction (EBSD). Cathodoluminescence is the emission of light by a material that has been exposed to an electron beam. In releasing the beam energy within the material, valence electrons can be promoted to the conduction band, producing an electron-hole pair. Recombination occurs when the electron in the conduction band relaxes back to the valence band, at which event a photon may be emitted (Parish & Russell, 2007). Many materials emit a wide spectrum of light, from UV to near IR, which allows for the spectroscopic analysis of the material. The main drawback to cathodoluminescence is the difficulty in interpreting spectra. Cathodoluminescence spectra exhibit not only compositional differences within a specimen but grain structure, boundaries, and defects as well (Holt, 1992; Parish & Russell, 2007). Nevertheless, cathodoluminescence and X-ray spectroscopy can be complementary techniques for atmospheric particles, as they are for mineralogical study (Muller et al., 2003). The analysis of urban haze particles with cathodoluminescence and EDX showed the benefit of the former for detecting nitrates in heterogeneous particles (Wang et al., 2018a). In this case, cathodoluminescence appears advantageous over EDX. Along with carbon, nitrogen in nitrate is often difficult to measure because its X-rays are strongly absorbed by the polymer window of the detector.

If a detector is positioned at a highly oblique angle from the plane of the specimen, it is possible to detect a pattern of diffracted backscattered electrons that form two cones emanating from the specimen toward the detector. Each pair of diffraction cones corresponds to the leading and trailing edges of an atomic plane in a crystal. The cones are registered by the detector as a pair of parallel lines, often known as Kikuchi lines. The pattern of line pairs for multiple atomic planes in the crystal represents the D-spacings and planes in the crystal lattice for a specific material.

EBSD can be used to identify crystalline phases in particles of a few micrometers or less (Small & Michael, 2001). For atmospheric aerosols, the technique has been used in combination with EDX (Figure 1.6) to identify phases in workplace aerosol from ferromanganese production (Ervik et al., 2018) and to identify phases in urban haze in combination with cathodoluminescence and EDX (Wang et al., 2018a).

Variable-pressure SEM and environmental SEM. Often, a drawback of conventional SEM (and TEM) is that specimens must be kept at relatively high vacuum. In conventional SEM, the sample chamber is evacuated to around 10^{-3} Pa (7.5×10^{-6} Torr) or less to circumvent the scattering of the probe electrons by residual gas. Variable-pressure SEM (VPSEM) instruments and the proprietary ESEM instrument were developed to operate with a sample chamber gas pressure of around 27–2700 Pa (0.2–20 Torr). The purpose is twofold. First, increasing the pressure in the sample chamber allows for the imaging and analysis of vacuum-sensitive specimens, such as aqueous aerosols, or those with semivolatile species, such as organic carbon and ammonium nitrate. Second, for specimens that are vulnerable to charging by the electron beam, gas in the sample chamber (typically water vapor) dissipates the buildup of charge in nonconducting specimens and substrates.

VPSEM instruments typically operate with a maximum chamber pressure of 267 Pa (2 Torr), whereas the ESEM instrument operates at maximum chamber pressure of 2700 Pa (20 Torr). Due to their relatively low energies, secondary electrons from the specimen readily interact with the chamber gas. As a result, secondary electron imaging is problematic, and a specialized detector is necessary. The detector, in fact, relies on the secondary electrons to ionize the gas molecules. Ionization of the gas generates a current at the detector as the specimen is scanned. The residual gas in the sample chamber commonly affects X-ray analysis to a greater extent than imaging. However, in this case, spatial resolution in the element mapping of particles is degraded due to broadening of the electron beam by the interaction of electrons with the residual gas (Goldstein et al., 2018). Beam broadening in the VPSEM and ESEM instruments

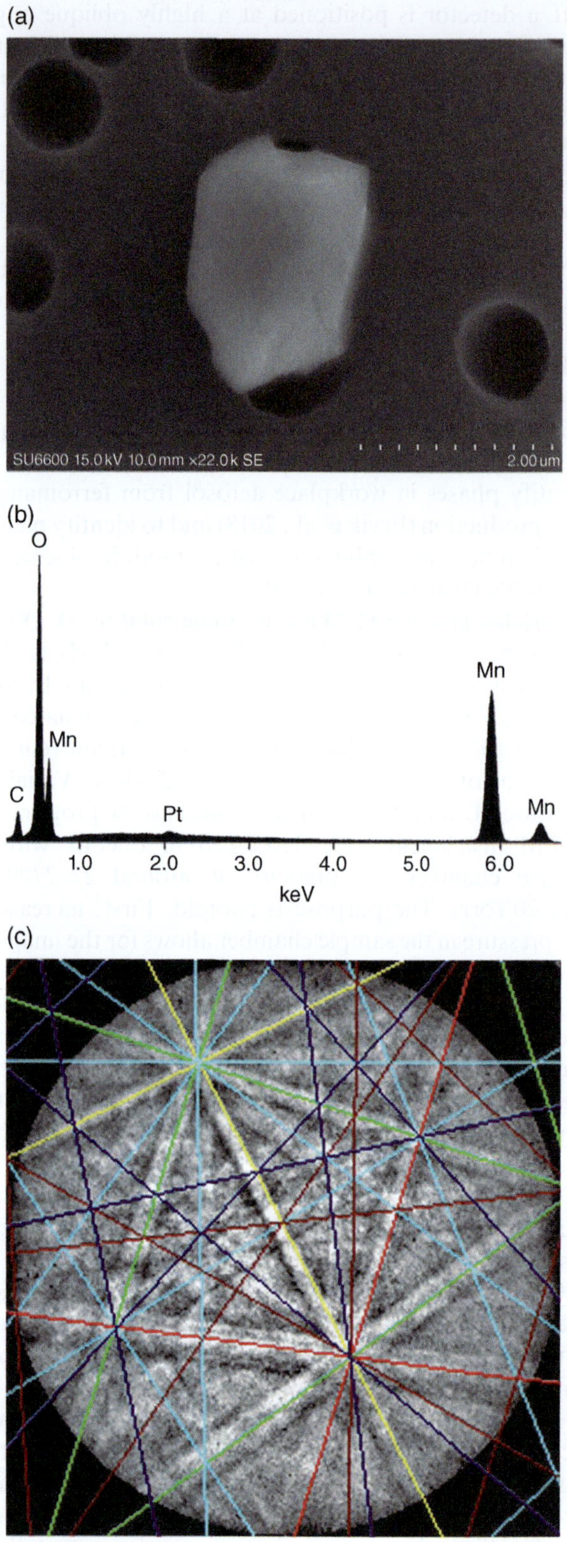

Figure 1.6 Secondary electron image (a), EDX spectrum (b), and the indexed EBSD pattern (c) following background subtraction for a MnO particle. EBSD indexing is performed with EDAX Orientation Imaging Microscopy software (EDAX Inc., Mahwah, NJ, USA). Ervik et al. (2018)/with permission of Springer Nature.

exacerbates the existing compromise in X-ray mapping resolution because the X-ray interaction volume is larger than the probe interaction volume.

The introduction of water vapor in the SEM sample chamber to investigate how individual aerosol salt particles take up water, or effloresce, has led to important insights concerning cloud-condensation nuclei and ice crystals in the formation and climatology of clouds. Martin previously provided a lengthy review of aqueous aerosol physics and chemistry and its importance in meteorology and climatology (Martin, 2000). Due to its higher maximum chamber pressure, ESEM has been used exclusively in SEM studies rather than VPSEM. Instrumentation typically includes a Peltier or cryogenic stage to cool the specimen: for example, to have the particles at 100% relative humidity or supersaturation. Early studies with ESEM observed the deliquescence of 0.1–20 μm sized NaCl, ammonium sulfate, sodium sulfate, and ammonium nitrate particles at 5 °C (Ebert et al., 2002; Hoffman et al., 2004). Oxidation was found to make soot much more hydrophilic than freshly emitted soot, allowing aged soot to act as cloud-condensation nuclei (Zuberi et al., 2005). Deliquescence observed with ESEM (and environmental TEM) of potassium salts emitted with biomass burning was observed to deliquesce and effloresce (Figure 1.7) at relative humidities of 82–>90% and 52–58%, respectively (Freney et al., 2009). ESEM was also used to observe the hygroscopicity of calcium nitrate particles following chemical conversion from less hygroscopic calcium carbonate with implications for the production of aqueous aerosols from China loess (Laskin et al., 2005). Physical changes in particles with water uptake – for example, rupturing of fungal spores – have also been observed (China et al., 2016). Ice nucleation has been observed in the ESEM at temperatures as low as −23 °C with individual silver iodide and clay particles (kaolinite and montmorillonite) (Zimmermann et al., 2007). Use of a custom-built sample cell with precise control of temperatures down to −66 °C allowed the observation of how ice nucleates around kaolinite particles (Wang et al., 2016).

Focused ion beam SEM. If the SEM instrument includes an ion beam column, the instrument becomes a dual-beam, and the technique is focused ion beam SEM (FIB-SEM). FIB instruments, typically utilizing Ga^+, provide the versatility of multifunctionality that is not often found in microanalytical instrumentation. In addition to analytical tools in SEM, such as EDX and EBSD, the ion beam is used for structure building, etching, and milling at the nanoscale as well as imaging. Structure fabrication with FIB-SEM requires the addition of a gas-phase material in the sample chamber, a gas such as a platinum organometallic that is injected close to the specimen and then decomposed by the ion beam to a solid material,

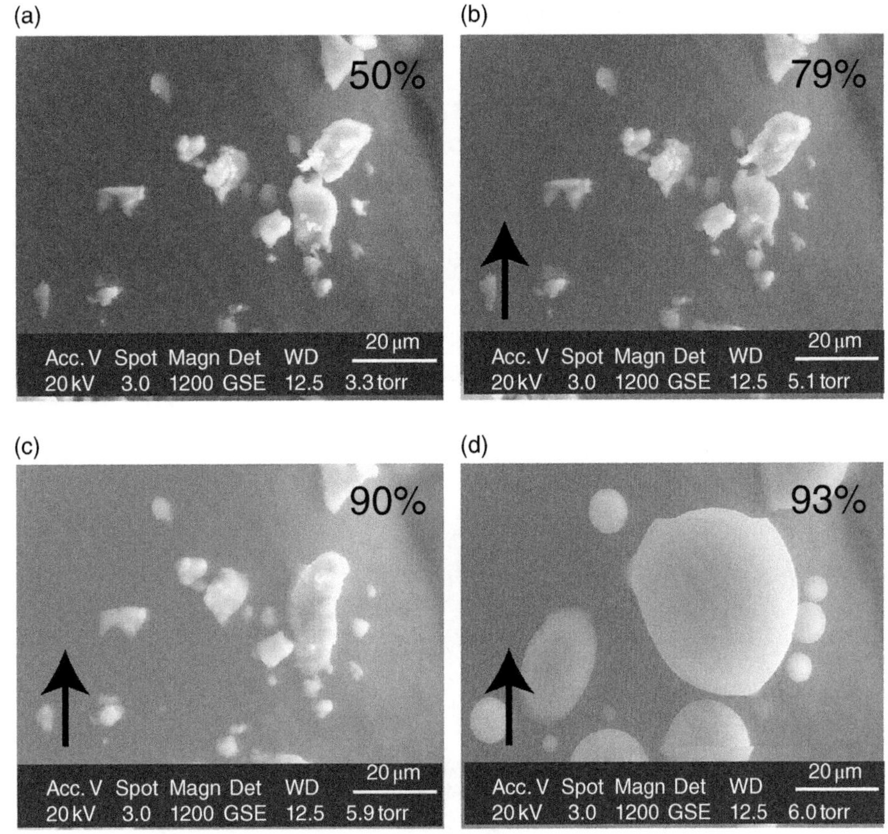

Figure 1.7 ESEM images of KNO3 particles at the onset of deliquescence at 93% relative humidity. Percent relative humidity is indicated in (a–d). Freney et al. (2009)/with permission of Taylor and Francis.

such as metallic Pt. Milling of the particle in uniform slices exposes the particle's internal structure (Figure 1.8). Imaging can be accomplished with the electron beam (with the emission of secondary and backscatter electrons) or with the ion beam. The impact of Ga^+ ions on the specimen releases secondary electrons as well as ions from the specimen, both of which are detected by FIB-SEM instrumentation.

For decades, FIB instruments have been used in the microelectronics industry for circuit fabrication, modification, testing, device failure analysis, and device repair (*Introduction to Focused Ion Beams*, 2005; Stevie et al., 1995). A major application of FIB-SEM is the production of specimen lamellae, thin sections of materials, including minerals (Heaney et al., 2001) for analysis by TEM and other techniques such as SIMS (Stevie et al., 2001). Recently, however, FIB-SEM and its versatility have been used in a few studies of individual atmospheric particles. Following methods for producing specimens for TEM, ion milling has been employed to produce thin sections of atmospheric dust particles for TEM analysis (Conny et al., 2014; Kaegi & Gasser, 2006).

The principal drawback to using FIB-SEM on individual atmospheric particles is that the particle must be "rugged" enough to withstand sequential milling by the Ga^+ beam. Particles that can be imaged with SEM may not be suitable for FIB-SEM. Solid mineral dust particles are typically suitable for milling, whereas liquid particles and fragile organic particles are not. Soot particles, however, can typically withstand milling with FIB-SEM.

The most useful feature of FIB-SEM for studying atmospheric particles is the ability to mill through a single particle in 15–20 nm increments to examine its interior (Adler et al., 2013; Conny, 2013). The technique was used to distinguish from background aerosol dust particles from the eruption of the Grimsvötn volcano in Iceland in 2011 collected downwind in Copenhagen, Denmark (Lieke et al., 2013). Milled slices of the volcanic ash particles showed K, Mg, Fe, and Ti distributed heterogeneously within the particles. The interior of coal fly ash exposed through FIB milling showed how iron phases within particles might shift due to acidic atmospheric processing (Chen et al., 2013). FIB milling can also reveal the porosity of aerosol particles due to in-cloud freeze-drying processes (Adler et al., 2013).

Three-dimensional (3D) spatial modeling of the interior of a mineral-like specimen following automated milling with FIB-SEM was first demonstrated on a CaMgTiOx

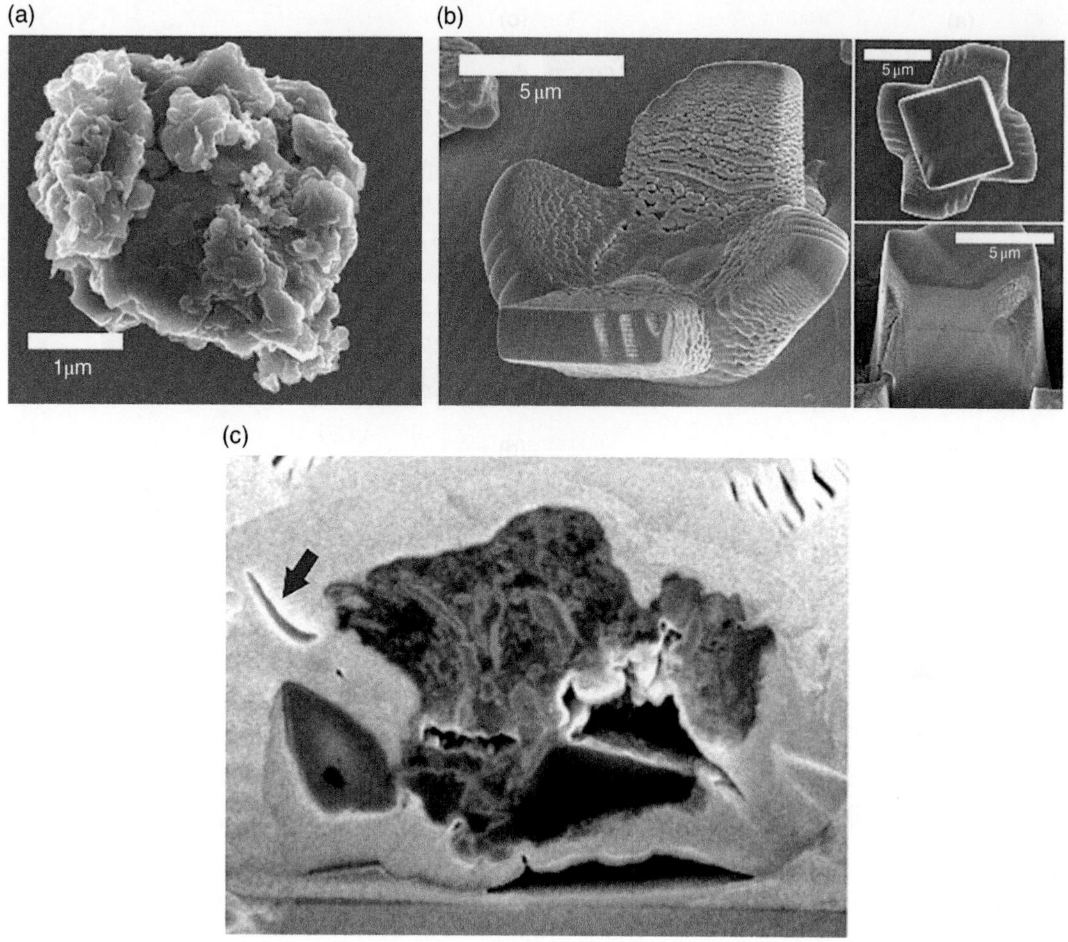

Figure 1.8 Process of milling and imaging a mineral-like dust particle (a) with FIB-SEM. (b) Particle is embedded in platinum by decomposing an organometallic gas to form wedges and a cap above the wedges with Pt. The structure is then trimmed to form a Pt block. (c) Milled slice of Pt-embedded particle showing a grayscale distinction between the particle and the surrounding Pt block. Arrow in (c) is a void in the Pt block. Conny et al. (2014)/with permission of American Chemical Society.

ceramic (Schaffer et al., 2007). The collection of slice images from FIB-SEM combined with X-ray element mapping is known as FIB tomography (Uchic et al., 2007). FIB tomography of individual atmospheric particles has been performed to show the exact shape of the particle and locations of individual element phases within the particle (Conny & Ortiz-Montalvo, 2017). The exact shapes and interior components of particles from FIB tomography have been used to calculate particle optical properties (Conny et al., 2019) and show how dust optical properties compare with those of geometric particle shapes (Conny et al., 2020).

1.2.2. Transmission Electron Microscopy

Higher resolution associated with TEM compared to SEM can be related theoretically to the use of higher electron beam acceleration voltages in TEM. In his wave equation, de Broglie demonstrated the wave nature of electrons and showed that an electron's wavelength is inversely related to its speed. However, as with light microscopy, resolution in electron microscopy is limited by wave diffraction. In TEM, using higher accelerating voltages up to 400 kV increases the speed of electrons and decreases the amount of electron scattering by the specimen, which results in less diffraction. Electron acceleration with 100 kV can theoretically result in a lower resolution limit of 4 p.m.: that is, much smaller than the diameter of an atom (Williams & Carter, 1996a).

In TEM, as in SEM, the actual resolution of the instrument is limited by the electron lenses. However, resolution with TEM is approximately a factor of 10 better than with SEM (Table 1.1). In conventional transmission electron microscopy (CTEM), the electron beam is defocused and made coherent (wave behavior parallel) when it reaches the specimen, and the transmitted electrons are then

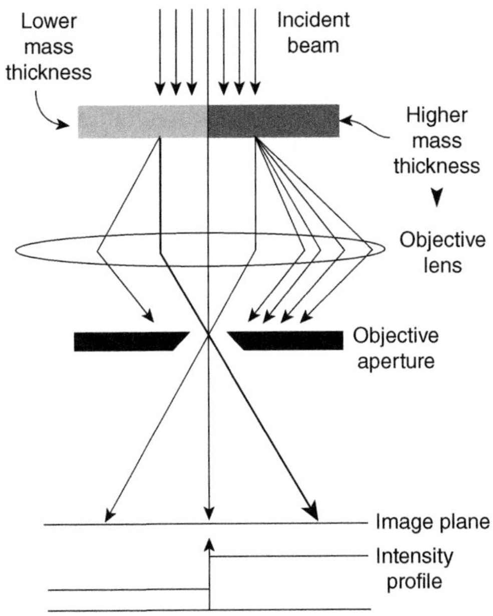

Figure 1.9 Schematic showing the flooding of a specimen with different mass thicknesses by a coherent electron beam CTEM. Differences in the specimen mass thickness are indicated by the intensity of transmitted electrons at the image plane. Mass thickness is the material density times the specimen thickness. Williams and Carter, 1996b/with permission of Springer Nature.

collected and detected below the specimen plane by either a disk-like detector in line with the primary beam for bright-field imaging or an annular detector for dark-field imaging. Figure 1.9 shows how the electron beam floods the specimen, in this case with varying material density or thickness, and the transmitted electrons are focused on the image plane by the objective lens and objective aperture. Scanning TEM (STEM) is similar to SEM in that the electron beam is commonly demagnified first and then made convergent to a spot at the specimen. With a field-emission gun, the STEM spot size is 0.3 nm or smaller (Williams & Carter, 1996b). Transmitted electrons are then detected below the specimen plane while the primary beam is rastered across the specimen.

Diffraction of the transmitted electrons and the analytical information it provides is as fundamental to TEM as imaging. Diffraction in TEM provides information on crystalline structure and grain boundaries within particles. Electron optics in CTEM and STEM are such that diffraction patterns and specimen images (bright-field and dark-field) are observed by adjusting the instrument's electron lenses and apertures rather than necessarily having separate detectors.

In CTEM, the primary imaging mode is with transmitted electrons that are scattered elastically. If the electron beam is well focused, the specimen's thickness approaches the mean free path for transmitted electrons, and transmitted electrons are not absorbed by the specimen, then the whole specimen thickness is typically in focus. Bright-field imaging occurs when the instrument's objective lens aperture allows the beam to directly pass and there is minimal interaction between the transmitted electrons and specimen atoms, whereby scattering angles are small (Figure 1.9). With dark-field imaging, the objective aperture blocks the direct beam, and the dark-field image is due to beam diffraction. In this case, the image is detected off-axis with an annular detector.

STEM instruments are often equipped to image secondary electrons. Backscatter electron imaging is also possible in STEM but offers little advantage over the other imaging modes and, therefore, is not commonly used in STEM.

Transmitted electrons may also scatter inelastically to allow for the detection of emitted X-rays (EDX) and electrons that have lost energy through the specimen. EDX and spectroscopies of inelastically scattered electrons (energy-filtered TEM [EFTEM] and electron energy loss spectroscopy [EELS]) can be performed in CTEM and STEM modes.

TEM best provides information on small particles with thicknesses of less than two times the mean free path of beam electrons, whereby an inelastic scattering event with the primary beam is less likely. The mean free path is inversely proportional to the density of the particle's material as well as its scattering cross section. The mean free path is on the order of tens of nanometers, so ideally, atmospheric particles studied with TEM would have thicknesses in that general range. Nevertheless, TEM is commonly performed on particles 1 μm and larger. For these particles, the most useful information may be acquired at the particle's edge, where it is thinnest. In addition, as described in "Diffraction with TEM," diffraction with TEM is commonly performed on particles having thicknesses much larger than the mean free path. Although SEM may be more versatile in that imaging can be performed on atmospheric particles of various sizes and thicknesses, the main advantage of SEM imaging is morphological information. The main advantage of TEM imaging is structural information, which allows for the quantitative study of polyphase grains.

Applications of TEM to Atmospheric Particles

TEM has been used in a variety of research areas concerning atmospheric aerosols. TEM-centered reviews include the effects of aerosols on climate (Pósfai & Buseck, 2010) and airborne mineral particles (Buseck et al., 2000). From early on, TEM has been used to study heterogeneous aerosol chemistry, water uptake, and particle growth. Examples include the relative changes in chloride and sodium in sea-salt particles (McInnes et al., 1994), sulfate and nitrate chemistry of coastal aerosols (Zhang et al., 2000), the reaction of nitric acid with calcium carbonate in Arctic aerosols (Chi et al., 2015), and nitrous acid formation with urban (Shanghai) aerosols (Cui et al., 2018). The shape and composition of sea-salt particles in urban air was studied with TEM in the context of particle aging (Adachi & Buseck, 2015). Particle hygroscopicity has been studied from the effect of nitrate on mineral dust as cloud-condensation nuclei (Li & Shao, 2009a) and the mixing of particulate organic compounds with inorganic particles (Li & Shao, 2010). A tandem differential mobility analyzer was used to isolate hygroscopic particles (McMurry et al., 1996) including ultrafine particles (Park et al., 2009) by electrostatic mobility followed by TEM analysis. TEM was used to investigate the growth of organic aerosols from gas-phase isoprene (Gunsch et al., 2018). Mineralogical data of size, aspect ratio, and chemical composition from both TEM and SEM were used to determine an average complex refractive index for Saharan dust sampled at Izana, Tenerife, Spain (Kandler et al., 2007).

TEM-based analyses have also focused on the characterization of various types of ambient aerosols. For example, studies have focused on aerosols from urban environments (Barone & Zhu, 2008; Hou et al., 2018; Li et al., 2017; Li & Shao, 2009b; Vester et al., 2007; Wang et al., 2017), the Saudi Arabian desert (Pósfai et al., 2013), sub-Saharan Africa during the monsoon season (Chou et al., 2008), southern Africa during the biomass burning season (Li et al., 2003), and Indonesia during the massive 1997 forest fire (Ikegami et al., 2001). Aerosol particles from the upper troposphere and lower stratosphere have also been investigated with TEM (Xu et al., 2001). Li et al. (2016) provided a review of studies on individual ambient air particles from East Asia.

TEM has also been useful in the analysis of particles in glacial snowpack and ice cores. Comparisons of individual particles in snowpack and the atmosphere of the Tibetan Plateau suggest that black carbon particles in the snowpack are larger and less fractal than particles in the atmosphere (Zhang et al., 2017). The number of particles with sulfate and nitrate coatings was lower in the snowpack than in the atmosphere, suggesting a 17–34% decrease in the snow albedo compared to the surface albedo based on particles in the atmospheric (Dong et al., 2018). Individual particles in Antarctic ice cores dating from the mid-18th century revealed individual particles with a complexity similar to those in the modern atmosphere, such as particles internally mixed with mineral-dust metals and coated with nitrogen-containing organic compounds (Ellis et al., 2016).

Diffraction with TEM. Because diffraction is an integral technique with TEM, analyses of atmospheric particles often include diffraction data, with some studies relying heavily on diffraction information. There are two modes of acquiring data from electron diffraction in TEM: SAED and convergent-beam electron diffraction (CBED). In SAED, the electron beam is defocused to flood a selected area of the specimen. Thus, it is performed with conventional TEM rather than STEM. By contrast, with CBED, the electron beam is focused on the specimen and thus is commonly associated with STEM.

For the most part, atmospheric particles have been analyzed with SAED rather than with CBED. An example is shown in Figure 1.10. Here, SAED and EDX were used to show that aggregates of spherical black carbon and dust in Daejeon, Korea, contained significant amounts of lead chromate (Lee et al., 2016). Each spot in the diffraction patterns in Figure 1.10 indicates a diffraction plane, denoted by its Miller index – for example, (011) (111) – which corresponds to atomic planes within a structure. Each diffraction pattern shows the spots corresponding to reciprocal space and associated with a zone axis: for example, $[0\bar{1}1]$ [120], which is the direction of diffraction at each spot.

Electron diffraction with TEM may be used in combination with EDX to provide confirmation of the identity of chemical species in atmospheric particles,

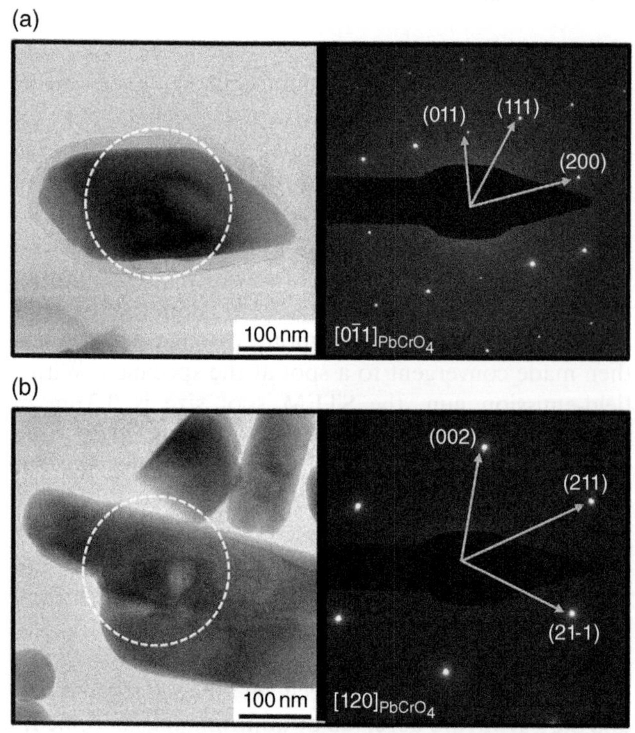

Figure 1.10 Bright-field images (left) and SAED patterns (right) for nanoparticles of monoclinic $PbCrO_4$. (a) Diffraction pattern from the $[0\bar{1}1]$ zone axis direction; (b) diffraction pattern from the [120] zone axis direction. Lee et al. (2016)/Springer Nature/CC BY 4.0.

particularly if grain outlines are less known. This was shown, for example, a study of urban air in Beijing after air masses had picked up emissions from agricultural biomass burning sites and heavy industrial areas (Li & Shao, 2009b). SAED with EDX has been important in understanding the hygroscopic properties of aerosols: for example, the reactivity of urban haze particles with acidic gases resulting in more hygroscopic nitrate-coated particles as cloud-condensation nuclei (Li & Shao, 2009a). SAED and EDX were also used to help elucidate the composition, extent of aggregation, and hygroscopicity of desert dust particles sampled by aircraft in Saudi Arabia (Pósfai et al., 2013).

Figure 1.11 is instructive for a few reasons. First, the image in (a) shows a fairly large clay particle. The particle is much thicker than the mean free path for the electron beam, which, in turn, suggests that imaging of transmitted electrons is less useful. On one hand, TEM imaging does not provide topographic information. However, particle thickness helps to provide a strong EDX signal in (c). More importantly, the diffraction pattern in (b) indicates that particle thickness is not problematic. Even though beam electrons likely undergo many inelastic collisions within the particle, enough transmitted electrons are diffracted to produce an interpretable pattern. In this case, the rings in the diffraction pattern indicate the spacing (labeled in angstroms) of the crystallographic planes from multiple crystallites in the mineral smectite.

Energy-Filtered TEM and Applications to Atmospheric Particles

Beam electrons that penetrate the specimen but have lost specific energy due to inelastic scattering are also useful signals in TEM. Imaging to reveal composition differences within a particle is accomplished by filtering the energies of inelastically scattered electrons, a technique known as energy-filtered TEM (EFTEM). Defocused beam electrons flood the particle as in CTEM. In the image plane below the specimen, an energy slit allows inelastically scattered electrons with specific energies to be detected at each pixel. Qualitative EFTEM imaging can be more advantageous than X-ray mapping for light elements such as carbon because the X-ray fluorescence yield of light elements is low. In addition, X-rays emit isotropically, whereas transmitted electrons scatter forward toward the detector. EFTEM images have been used to show the presence of sulfur and carbon in individual particles collected by aircraft in the midlatitudes of the troposphere and lower stratosphere and the intertropical convergence zone (Nguyen et al., 2008). Nevertheless, the filtering of energies in EFTEM restricts the amount of elemental information available. To maximize the amount of spatial information in EFTEM imaging, principal components analysis can be performed to identify statistically correlated features in particles (Conny et al., 2014).

Particle thickness in EFTEM, as in all imaging, is more of a concern than in SAED. If the particle's thickness largely exceeds the mean free path, then multiple inelastic scattering events will shift electron energies, and changes in specimen thickness may result in image aberration. Figure 1.12 shows a slice of a dust particle produced from focused ion beam milling (a) and the EFTEM image of a small area of the slice (b). The sharp contrast in colors of the matrix between the left and right sides of the EFTEM image is due to an abrupt shift in thickness from the milling process (Conny et al., 2014).

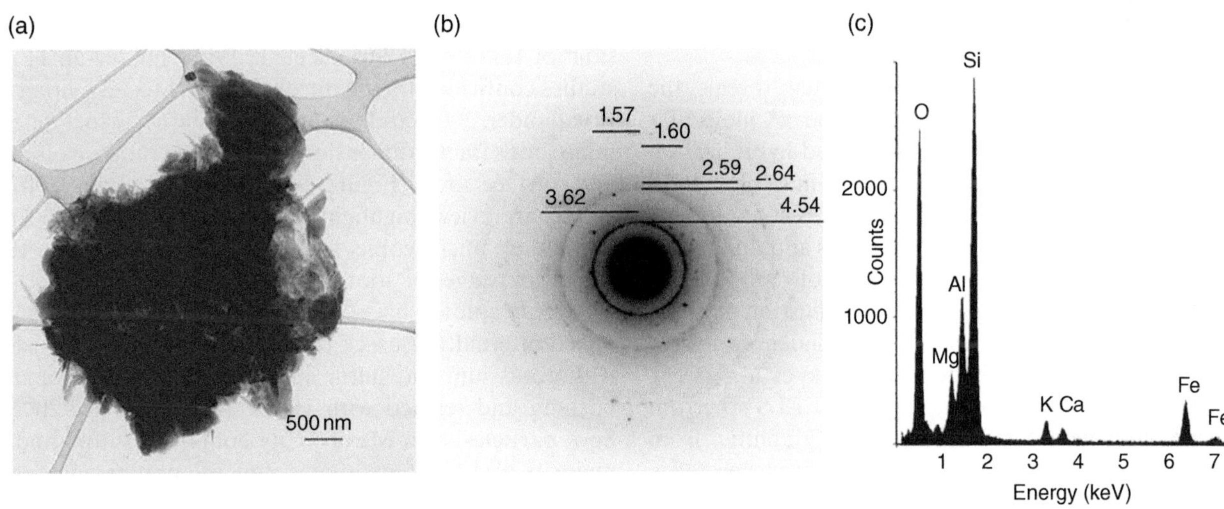

Figure 1.11 (a) Bright-field TEM image, (b) SAED pattern, and (c) EDX spectrum of a typical clay particle sampled on a lacey carbon TEM grid. The sample was collected by aircraft in Riyadh, Saudi Arabia, during April 2007. Pósfai et al. (2013)/with permission of Elsevier.

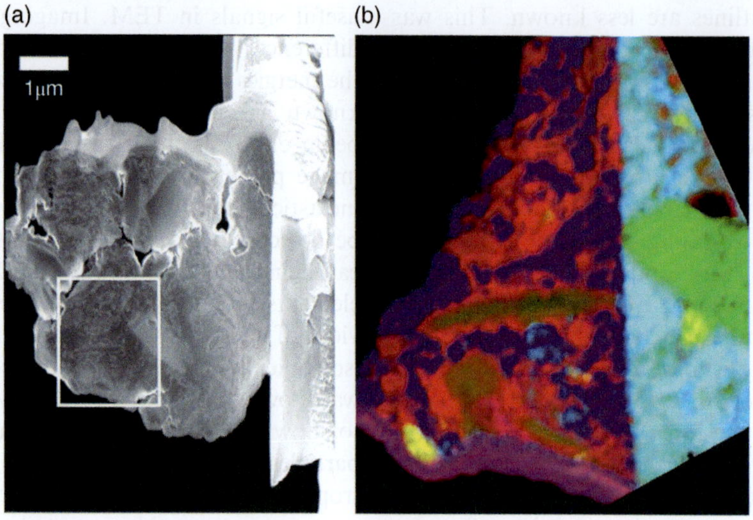

Figure 1.12 (a) Secondary electron and (b) EFTEM false-color images of a slice from the focused ion beam milling of a mineral-like particle. The contrast in color between the left and right sides of (b) is an aberration due to a straight-line shift in the slice thickness. Conny et al. (2014)/with permission of American Chemical Society.

Electron Energy Loss Spectroscopy and Applications to Atmospheric Particles

The collection of an energy spectrum of inelastically scattered electrons, known as EELS, reveals chemical bonding as well as element composition of the particle. If operated in STEM mode, a spectrum is generated at each pixel. As with EFTEM, an image can be constructed from specific energies in the EELS spectrum. Because EELS provides the information content of an energy spectrum, with particular detail near ionization edges where core-electron excitations occur, EELS imaging can provide more elemental and bonding information than EFTEM. A major advantage of EELS is the information it provides for light elements. Particle thickness should also be on the order of the mean free path to minimize multiple inelastic scattering events.

For atmospheric particles, EELS often reveals the presence of electron excitation to σ* and π* molecular orbitals, suggesting less or more sp2 bond hybridization and, therefore, the presence of single or multiple bonds. In what was likely the first reported EELS study of carbonaceous aerosols, π*/σ* intensity ratios and differences in intensities were used to show a range of carbon bond lengths, which suggested aggregated combustion particles with elemental carbon spherules and condensed hydrocarbons around the spherules (Katrinak et al., 1992). The shape of the carbon σ* peak in the EELS spectrum is often sufficient for distinguishing graphitic from amorphous carbon in soot and tar balls, respectively (Adachi & Buseck, 2011). For the latter, nitrogen was also detected in the EELS spectrum. In addition to carbon information, sulfur, potassium, and nitrogen were identified by EELS in individual particles collected in Mexico City (Adachi & Buseck, 2008).

In addition to carbon, EELS has been useful for speciating the oxidation states of iron oxides using the Fe L2-3 ionization edge. For internally mixed aluminosilicate dust originating in the north-west Sahara and sampled in Senegal, as shown in Figure 1.13, Fe^{+3} was the predominant oxidation state substituted throughout each particle (77%), but mainly in the distinct iron oxide inclusions (90%) (Deboudt et al., 2012).

Analysis of Carbonaceous Particles from Combustion

A substantial body of TEM literature exists on soot from the incomplete combustion of carbonaceous fuels either as lone particles or in mixtures with other particle types. In one of the earliest TEM studies of urban particles, soot was described as fractal-like with fractal dimensions of 1.35–1.89 (Katrinak et al., 1993). Follow-up TEM studies confirmed fractal dimensions of 1 to >2 (Xiong & Friedlander, 2001) and used tomography to further determine soot's fractal dimension, radius of gyration, volume, and surface area (Figure 1.14) (Adachi et al., 2007). A recent review highlighted the importance of TEM and other microscopic techniques in determining the formation stages of soot (Baldelli et al., 2020).

Several studies have shown that soot is commonly mixed internally (Buseck & Pósfai, 1999). Soot particles with potassium and sulfur appeared to be from biomass burning and reacted with H_2SO_4 (Okada et al., 2005). Soot particles from Mexico City contained sulfur, and a majority also contained potassium and organic material, suggesting the prevalence there of biomass and biofuel burning (Figure 1.15). Interestingly, the internal mixing of soot from Mexico City did not compact the particles as did atmospheric aging (Adachi & Buseck, 2008). From particles collected near Antarctica, the presence of

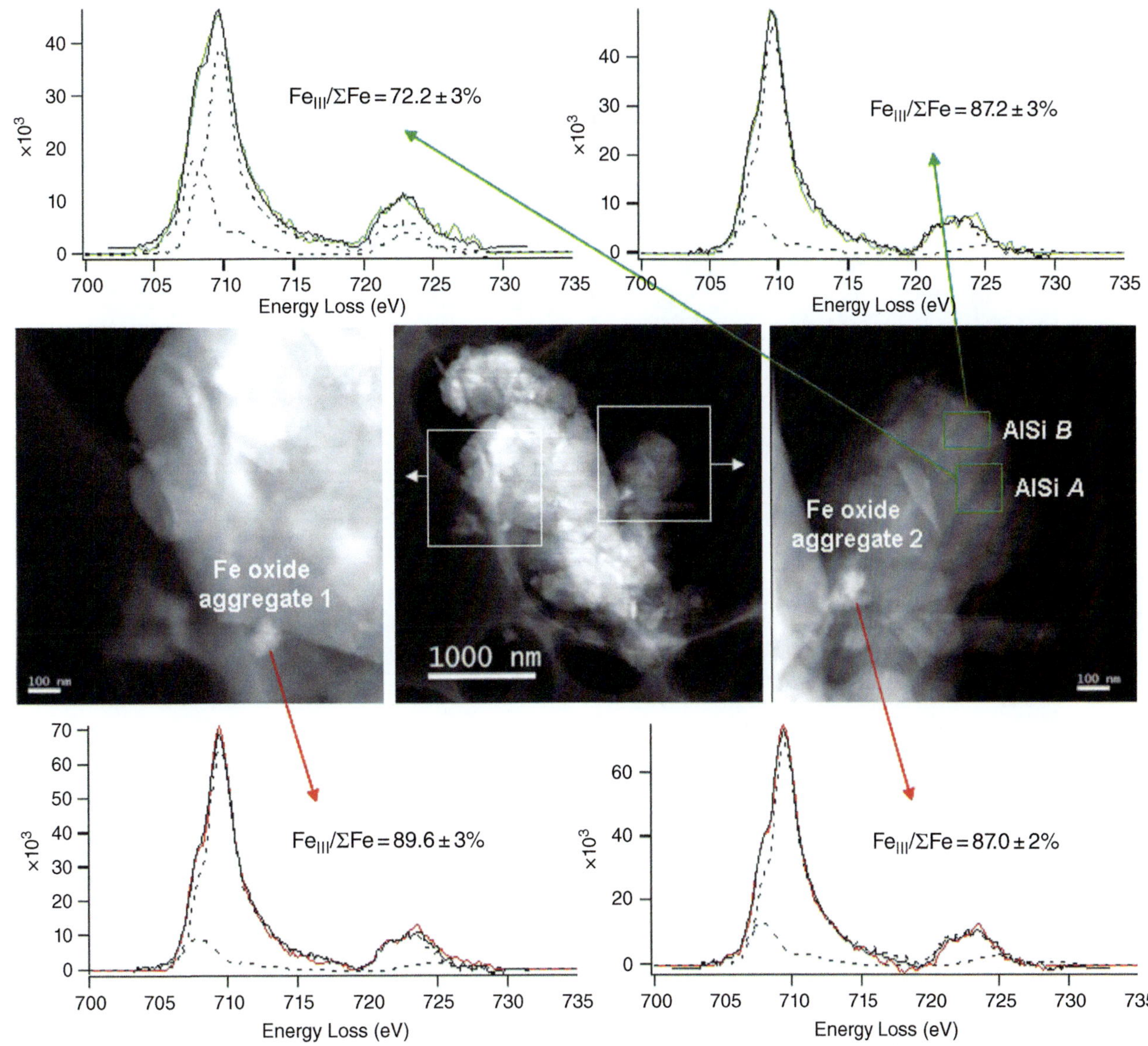

Figure 1.13 Electron energy-loss spectra and mathematical curvefits of an iron-containing aluminosilicate particle. Uncertainty associated with the Fe^{+3} content is the error in the corresponding curvefits. Dashed lines correspond to the contributions of Fe^{+2} and Fe^{+3}. Deboudt et al. (2012)/with permission of John Wiley & Sons.

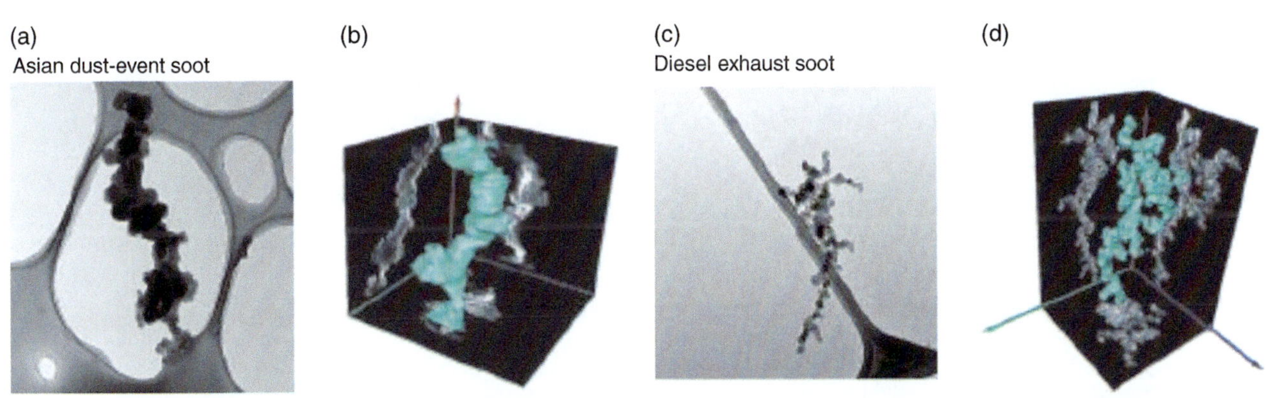

Figure 1.14 TEM images (a, c) of soot from Asian dust and diesel exhaust. The corresponding three-dimensional projections (b, d) were used to determine for particles (a) and (c) the fractal dimension (2.4 and 2.2, respectively), radius of gyration (539 and 186 nm, respectively), surface area/volume (2.7×10^{-7} and 16.4×10^{-7} m^{-1}, respectively), and structural coefficient (0.66 and 0.78, respectively). Adachi et al. (2007)/with permission of John Wiley & Sons.

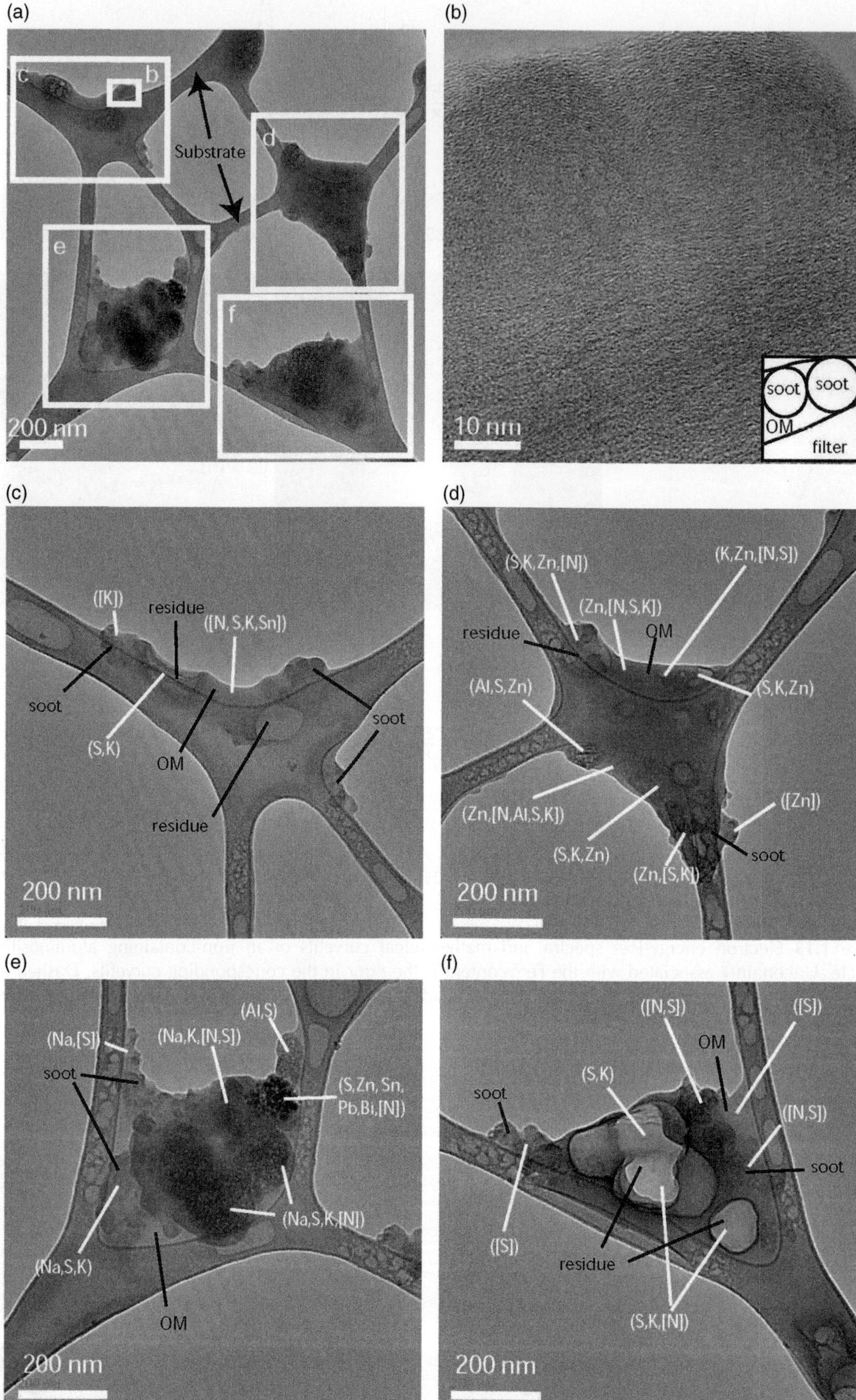

Figure 1.15 Bright-field TEM images of internally mixed urban particles (Mexico City) on lacey carbon grids. Also shown (in white) are minor to trace amounts of elements from EDS analyses. (a) Low-magnification image of particles (c–f). (b) High-resolution image of soot and organic matter (OM). (c–f) Typical internally mixed particles. Elements in square brackets indicate <1 wt. %. Adachi & Buseck (2008)/Copernicus Publications/CC BY 3.0.

non-sea-salt sulfate and sulfite suggested the mixing of far-transported soot with products of marine dimethyl sulfoxide oxidation (Ueda et al., 2018).

Atmospheric tar balls are nonaggregated spherical carbonaceous particles from biomass and biofuel burning (Li et al., 2003; Pósfai et al., 2004; Pósfai et al., 2003). The spheres are amorphous and formed from the condensation of combustion gases shortly after release from burning or possibly from the release of tar-like liquid droplets (Tóth et al., 2014), unlike soot, which is formed during pyrolysis. Thus, tar balls are considered organic (brown) carbon aerosols rather than soot-like elemental (black) carbon aerosols. Compositionally, tar balls contain only trace amounts of silicon, sulfur, and potassium and therefore are unlikely to mix internally with other materials. Minimal detected amounts of oxygen and nitrogen are from condensed organic compounds (Adachi et al., 2019). Absorption in the visible (e.g., 550 nm) by tar balls is stronger than for humic substances but less strong than for soot (Alexander et al., 2008). Tar balls sampled from wildfire sites in South Africa, East Asia, California, and Mexico exhibited average particle diameters ranging from around 100 to 400 nm (Adachi & Buseck, 2011; Alexander et al., 2008; Hand et al., 2005; Pósfai et al., 2003). Centuries-old tar balls have been observed in Antarctic ice cores (Ellis et al., 2016). Environmental TEM (analogous to ESEM) has shown that freshly formed tar balls take up virtually no water (Semeniuk et al., 2007). However, in another study using ESEM, tar balls were shown to be weakly hygroscopic at high relative humidities (e.g., 83%) (J. L. Hand et al., 2005).

1.3. RAMAN SPECTROSCOPY

A significant advantage in using Raman spectroscopy over X-ray spectroscopy is the ability to acquire molecular information on individual atmospheric particles. When a molecule scatters polarized visible or near IR radiation from a laser, a portion of the scattered light may be shifted to a different energy. If the difference in the energies of the incident light and scattered light resembles an energy transition in the infrared, then the molecule is Raman active. This inelastic scattering phenomenon was first reported by its eponym, C.V. Raman, and co-worker K.S. Krishnan in 1928 (Raman & Krishnan, 1928).

In conventional Raman scattering, the energy of the incident beam does not coincide with an absorption transition for the material. Molecular excitation occurs but without a quantized transition. Rather, excitation occurs from the ground electronic state to a virtual excited state below the first electronic state. This virtual excited state corresponds to deformation of the molecular bond by the incident radiation, resulting in a transient

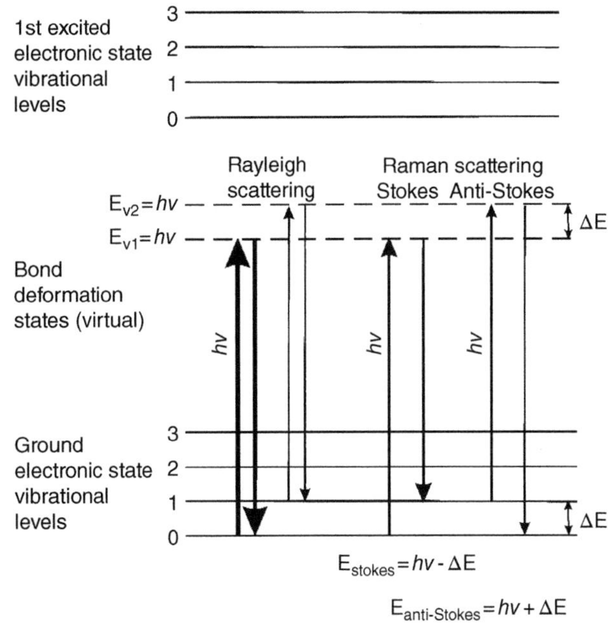

Figure 1.16 Schematic of the three energy-relaxation processes for a polarizable molecule: Rayleigh scattering, Stokes Raman scattering, and anti-Stokes Raman scattering. Adapted from Skoog et al. (1998).

dipole moment. The extent of deformation is the polarizability of the bond.

As shown in Figure 1.16, when incident radiation deforms the molecular bond to a bond-deformation virtual state above ground level, three relaxation processes occur. First, Rayleigh scattering occurs if the molecule relaxes to the same vibrational state the molecule was in prior to excitation to the virtual state. The relaxation energy is equal to the excitation energy ($h\nu$). Second, Stokes Raman scattering occurs if the molecule relaxes to the ground electronic state but at an elevated vibrational level. In this case, the energy of the Raman transition is less than the excitation energy (also the Rayleigh transition energy). Third, Raman scattering may also occur when the molecule is at an elevated vibrational level in the ground electronic state prior to excitation to the virtual state. After excitation to the virtual state, if relaxation is to a lower vibrational level in the ground state, then scattering is anti-Stokes, and the Raman transition energy is greater than the excitation energy.

1.3.1. Particle Levitation and Atmospheric Particles

A 1977 paper by Rosen and Novakov (1977) is commonly cited as the first report of Raman spectra from atmospheric aerosols, in this case, emissions from diesel and gasoline engines. However, earlier work by Rosasco et al. (1975) along with that of Delhaye and

Dhamelincourt (1975) demonstrated a laser-equipped Raman spectrometer as a microscope (i.e., Raman microprobe) for acquiring molecular information from individual particles such as Na_2SO_4, NH_4NO_3, and $(NH_4)_2SO_4$. Although these earlier efforts looked at particles on surfaces, much published work on individual atmospherically relevant particles with Raman spectroscopy from early decades to the present centers on levitated particles. A significant advantage to levitating the particle while collecting Raman spectra is the lack of an interfering effect from a substrate.

Numerous methodologies have been employed for trapping individual particles in place, such as electrostatics, laser optics, and acoustics. In some cases, particle isolation was facilitated with a vibrating orifice aerosol generator (VOAG) to produce a chain of particles whereby a Raman spectrum was taken fleetingly on each particle (Zhang et al., 1994).

Electrodynamic Balance

Many early studies focused on the fundamental physical processes of levitated microdroplets using electrodynamic balance (Davis, 1997), a technique based largely on subjecting the particle to a high-voltage AC field within a quadrupole ion trap (Paul & Raether, 1955). Numerous variations of electrodynamic balance evolved early on, such as the standard hyperboloidal electrodynamic levitator, which subjected particles to high-voltage AC and DC fields (Davis, 1997). The electrodynamic balance cell was used, for example, in measuring Raman scattering cross sections on nitrate and sulfate particles 10–15 μm in diameter (Fung & Tang, 1991). The minimum particle size and sulfate detection limits for Raman spectroscopy were determined on electrodynamically levitated particles (Vehring et al., 1998). The VOAG was also used with the electrodynamic balance to determine nitrate and sulfate detection limits (Fung et al., 1994).

Early studies involving the electrodynamic balance also took into account morphology-dependent optical resonances (MDRs) associated with elastic scattering within particles to understand interactions between Raman and Mie scattering: for example, by Chan et al. (1991). MDRs enhance elastically scattered light because the particle acts as an optical cavity where resonance modes within the particles are coherent with incident light, like coherence of stimulated emission in a laser. By accounting for elastic scattering in this way, the composition of multiple-component aerosols involving NaCl was teased out of Raman spectra (Buehler et al., 1991).

An early Raman study of single-particle chemical reactions looked at the heterogeneous reaction of SO_2 with electrodynamically levitated aqueous NaOH droplets from solid NaOH particles (Aardahl & Davis, 1996). Particle hygroscopicity and the presence of new metastable solid states were investigated from Raman spectra of levitated particles (Tang et al., 1995). Also determined by Raman spectroscopy were evaporation rates for levitated supersaturated magnesium sulfate droplets (Zhang & Chan, 2000).

A recent review and tutorial by Kohli et al. (2023) discusses the various versions of the electrodynamic balance and recent advances in the use of the technique with spectroscopies for studying particle hygroscopic growth, volatility, viscosity, and surface tension along with particle morphology and phase transformations.

Optical Levitation and Optical Tweezers

Later developments in particle levitation centered on the optical trapping of single particles by radiation pressure resulting from the momentum of photons on a particle, a technology that was pioneered by Ashkin and coworkers (1986, 1971). Thurn and Kiefer (1984) were the first to report Raman spectra from individual optically levitated particles using radiation pressure balanced by gravitational pull on the particle. Results from Thurn and Kiefer (1985) are shown in Figure 1.17. Figure 1.17a is the Raman spectrum of a cell filled with a water–glycerol mixture. Figure 1.17b, with data adapted by Davis (1997), compares a portion of the Raman spectrum for the bulk solution with the measured Raman spectrum of a water–glycerol droplet (diameter ≈30 μm). Included is the calculated Raman spectrum derived from inelastic scattering and Mie elastic scattering. The series of narrow peaks in the measured droplet spectrum in Figure 1.17b are morphology-dependent resonances in elastic scattering that are verified in the calculated spectrum.

The two well-known single-beam optical trapping techniques, optical levitation and optical tweezers, rely on the laser beam's radiation pressure forces. Single-particle trapping based on radiative pressure can involve one or more laser beams but is applicable to particles that only scatter rather than absorb light. For light-absorbing particles, trapping relies on harnessing the theoretically complex photophoretic force, which accounts for inhomogeneous heating of the particle as well as the transfer of thermal energy between surrounding gas and particle surface. A third type of optical trap, the universal optical trap, typically employs two lasers with non-Gaussian beam profiles. The arrangement provides versatility in capturing variously shaped absorbing as well as nonabsorbing particles. Optical traps often utilize two lasers for better instrument design to allow for spectroscopic analysis while maintaining optimal particle stability. Variations on the laser optics from a conventional Gaussian profile include a hollow-core beam profile, a Bessel profile with a nondiffracting core, confocal optics, and holographic beam properties (Gong et al., 2016, 2018; Manek et al., 1998; McGloin et al., 2008). Figure 1.18

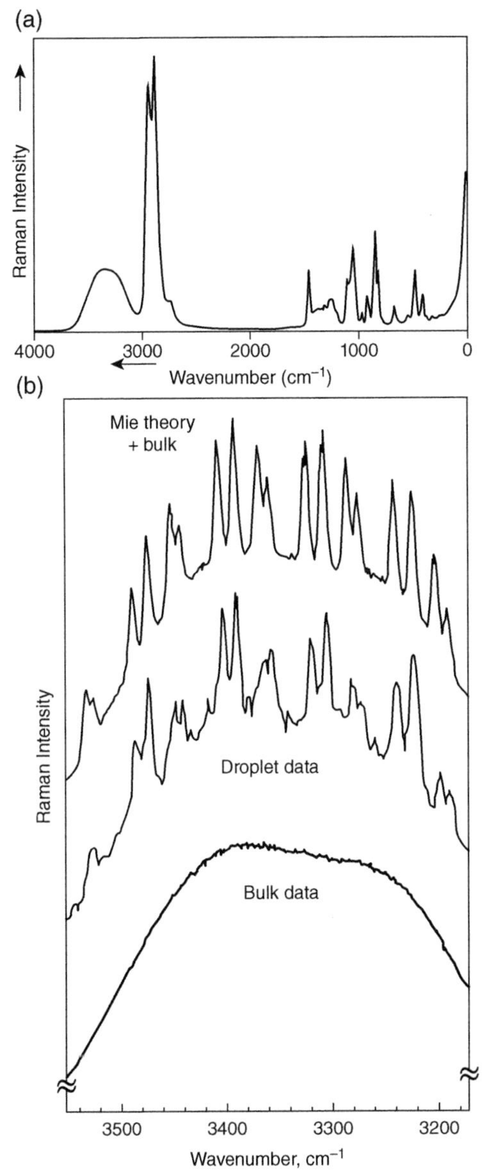

Figure 1.17 (a) Measured Raman spectrum of a 1 : 6 water–glycerol solution. Thurn and Keifer, 1985/with permission of Optical Society of America. (b) Comparison of measured and calculated spectra for a 30 µm diameter water–glycerol droplet for the 3180–3540 cm^{-1} segment of the Raman spectrum. Davis (1997).

shows schematics of single optical beam trapping devices for suspending nonabsorbing particles in air.

For decades, Raman spectra have been routinely collected on particles trapped by optical levitation and optical tweezers. Radiation pressure has two component forces: a scattering force and a gradient force. The scattering force is in the direction of the incident light beam and is proportional to the scattering cross section of the particle. The gradient force follows the beam's lateral intensity, causing the particle to move laterally toward the focal point of the laser beam. This intuitive conception of radiative pressure holds for large particles. For particles with a radius much smaller than the laser wavelength or that near the laser wavelength (i.e., Mie-sized particles), the optics of radiative pressure become more complicated, involving, for example, the particle's polarizability, the refractive indices of the particle and its surrounding medium, and Lorentz-Mie scattering theory (Davis, 1997; Gong et al., 2018; McGloin et al., 2008; Mitchem & Reid, 2008). Optical tweezers have been instrumental in understanding particle viscosity from the determination of the diffusion coefficient of water with the technique (Chu et al., 2018; Davies & Wilson, 2016).

The principal difference between optical levitation and single-beam optical tweezers is that the former involves a loosely focused beam with a low numerical aperture objective lens, whereas the latter involves a tightly focused beam with typically a high-numerical-aperture objective lens (Gong et al., 2018). Because optical levitation relies on a less-defined focal point, the gradient force is less important than the scattering force, which is balanced by the gravitational force on the particle. The gradient force is much more important with optical tweezers because the laser beam has a sharply defined focal point. In this case, gravitational pull on the particle becomes less important. Variations on the laser beam for optical trapping may be modified from a conventional Gaussian profile to have a hollow-core profile, Bessel profile with a nondiffracting core, or a beam with holographic behavior (Gong et al., 2018; Manek et al., 1998; McGloin et al., 2008).

The balance between scattering and gravitational forces with optical levitation provides for only limited control over the positional and temporal stability of the particle (Gong et al., 2018). Electro-optical feedback control is typically needed for stability (Davis, 1997). In addition, the laser beam is vertical with optical levitation, which limits instrument design. Because optical tweezers rely much more on the gradient force, there is more positional and temporal stability for collecting Raman spectra than with optical levitation. A disadvantage of single-beam optical tweezers is that the objective lens has a short working distance due to the high numerical aperture of the objective lens, which often hinders the optimal positioning of spectroscopic instrumentation. A common workaround is the use of dual laser beams. The optimal configuration of the two beams, propagating exactly opposite one another (counterpropagating), can take better advantage of the scattering force rather than rely primarily on the gradient force. Two vertically configured beams in optical tweezers have allowed for the Raman spectroscopy of spherical mineral-like particles (silica, alumina, and titania) (Rkiouak et al., 2014).

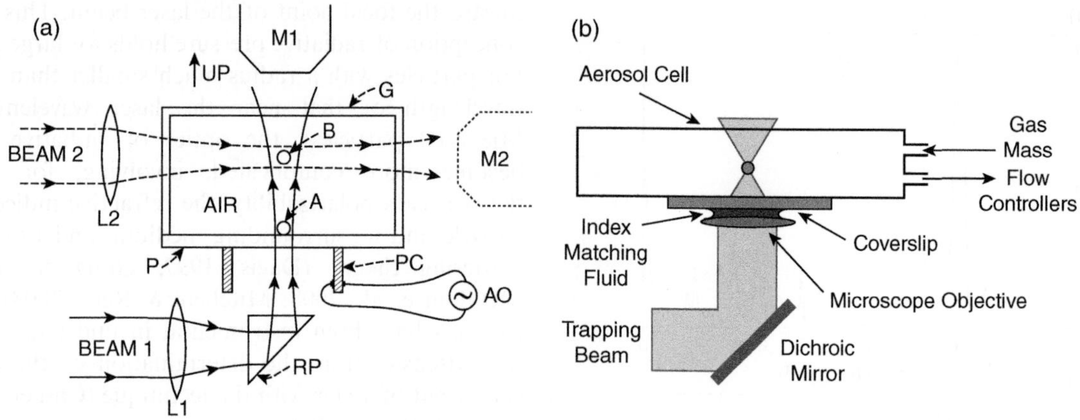

Figure 1.18 (a) Optical levitation diagram. Ashkin et al. (1971) with permission of American Institute of Physics. The device has a low numerical aperture. (b) Optical tweezers diagram reported by Hopkins et al. (2004). Hopkins et al. (2004); Gong et al. (2018).

Spectroscopy of Levitated Particles

Raman spectra have been collected from single levitated atmospheric particles in numerous studies. Early work showed how particle size could be determined from the changes in Raman intensity with the slow evaporation of a microdroplet trapped by a laser beam (Schweiger, 1991). Phase transitions within aerosol droplets levitated optically were observed from Raman spectra (Buehler et al., 1991; Trunk et al., 1997). In a study of isolated particles from a VOAG, stimulated Raman scattering involving morphology-dependent resonances was used to study hydrogen bonding at the water-particle interface (Zhang et al., 1994) and to determine trace nitrate and sulfate detection limits (Fung et al., 1994). In the latter study, Raman spectra were also collected with the particle suspended electrodynamically. Pollen particles levitated electrodynamically have also been studied with Raman (Laucks et al., 2000). Optical tweezers and Raman spectroscopy were used in determining the composition and mixing state of two coagulating droplets with immiscible phases (decane and water) (Mitchem et al., 2006) as well as the growth of aerosol droplets from gas-phase components (Mitchem et al., 2006).

Among studies of particle optical properties, a dual-beam optical trap was used to determine the size and complex refractive index of weakly absorbing particles, with particle size and the real part of the refractive index derived from cavity-enhanced Raman spectra. Electromagnetic heating in the trap was used to determine the imaginary part of the refractive index (Rafferty & Preston, 2018). Mie resonances have been combined with cavity-enhanced Raman spectroscopy of optically trapped particles to also determine particle size and refractive index (Bain et al., 2018; Lew et al., 2018).

In a demonstration of how 3D manipulation with optical tweezers may be used to study particle coagulation and heterogeneous processes, cavity-enhanced Raman spectroscopy was performed on aqueous NaCl droplets 4–14 μm in size (Hopkins et al., 2004). The effects of water uptake on gel formation and acetic acid volatility in magnesium acetate particles levitated with optical tweezers were observed from cavity-enhanced Raman spectra and changes in particle morphology and refractive index (Lv et al., 2018). Spectral whispering gallery modes from cavity-enhanced Raman spectroscopy with optical tweezers were used to distinguish coagulated core-shell organic droplets from single-phase droplets (Gorkowski et al., 2016). An efficient algorithm was also developed for analyzing spectral whispering gallery modes of core-shell droplets (Gorkowski et al., 2018).

A number of studies involving Raman spectroscopy have centered on the effect of humidity and hygroscopic properties of particles, including a pioneering study by Tang et al. involving the electrodynamic balance (1995). Raman spectroscopy was used to determine the volatility of ammonium nitrate at different relative humidities for individual particles captured with optical tweezers (Lu et al., 2019). The uptake and removal of water were studied from single electrodynamically levitated aqueous particles of H_2SO_4 and NH_3, including solid $(NH_4)_3H(SO_4)_2$ (Colberg et al., 2004). Hygroscopic changes were determined in electrodynamically levitated droplets of metallic ammonium sulfate due to various ion contact pairs (Zhang & Chan, 2002). Also measured with Raman spectroscopy was the hygroscopicity of electrodynamically levitated ammonium sulfate droplets coated with glutaric acid as a model aerosol organic phase (Chan et al., 2006). Optical levitation with Raman and Mie scattering was used to determine the hygroscopic growth of ammonium sulfate-glutaric acid droplets with various composition proportions (Jordanov & Zellner, 2006). Raman spectra from both the electrodynamic balance and optical tweezers were compared in observing the change in size with 1 nm accuracy of a NaCl droplet in equilibrium with varying relative humidity (Hargreaves et al., 2010).

Particle levitation has been used in a few studies involving ice nuclei. Relevant to stratospheric particle chemistry, levitated aqueous H_2SO_4 droplets were supercooled and analyzed by Raman spectroscopy to determine how particle nucleation is affected by acid concentrations at low temperatures (Mund & Zellner, 2003). The freezing of ammonium sulfate particles was also demonstrated from Raman spectra and optical trapping (Ishizaka et al., 2011). Relevant to cirrus cloud formation, nucleation rates were determined from Raman spectra for aqueous NaCl droplets and sea-salt particles levitated with the thermodynamic balance at temperatures from 241 to 250 K (Peckhaus et al., 2016).

Raman spectroscopy has been used to study pathways and products of atmospheric reactions with single levitated particles, in particular the reaction with ozone. Electrodynamically levitated oleic acid particles reacted with ozone, resulting in the formation of products with peroxidic, carbonyl, and hydroxyl functionality (Lee & Chan, 2007). In particles with maleic acid levitated with optical tweezers, measurements were made of the particles' refractive index during ozonolysis at varying relative humidities (Dennis-Smither et al., 2014). Figure 1.19 shows the change in the cavity-enhanced Raman spectra of droplets with maleic acid and NaCl after oxidative aging with O_3, whereas the relative humidity dropped sharply from around 80% to <10%.

In studies on the dynamics of mixed-phase particles with optical tweezers and Raman spectroscopy, phase partitioning and the rate of water evaporation were observed from aqueous NaCl particles with a decane shell layer (Buajarern et al., 2007a). Changes in the size, composition, and formation of a microgel were observed from droplets of NaCl and the surfactant sodium dodecyl sulfate (Buajarern et al., 2007b). Insoluble palmitic acid was observed to coalesce on the surface of an aqueous NaCl droplet or as an inclusion (Laurain & Reid, 2009). Aqueous particles containing ammonium nitrate and ammonium sulfate as atmospheric aerosols were shown with the electrodynamic balance and Raman to form metastable salts (Ling & Chan, 2007). Internally mixed particles of ammonium sulfate and various organic acids (malonic acid, glutaric acid, or succinic acid) were observed to deliquesce and crystallize heterogeneously with changes in water uptake and evaporation (Ling & Chan, 2008).

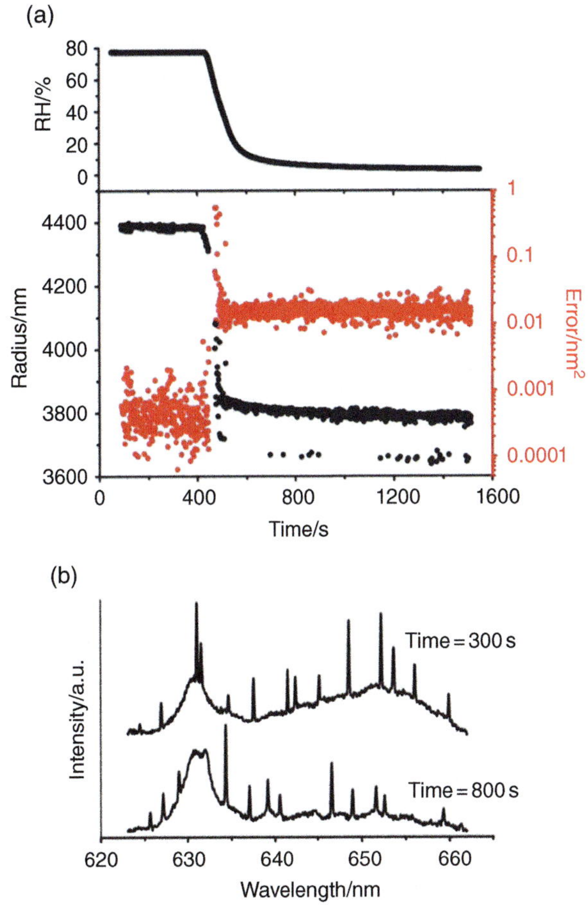

Figure 1.19 (a) Optical tweezers experiment showing O_3 oxidation of droplets containing maleic acid and NaCl as relative humidity is sharply reduced to <10%. (b) Cavity-enhanced Raman spectra at 300 and 800 s into experiment. Also shown in (a) are the change in droplet radius and error in the best fit of the measured and predicted whispering gallery modes in the Raman spectra. Dennis-Smither et al. (2014)/with permission of American Chemical Society.

1.3.2. Spectroscopy of Atmospheric Particle Surfaces

The capture of single particles on surfaces has allowed many studies to take advantage of confocal micro-Raman spectroscopy. Typically, as a Raman spectrometer is paired with a laser confocal microscope, wavelengths of Raman scattered light above or below the microscope's focal plane are removed from detection, and therefore, resolution in the image and Raman spectrum is enhanced. The technique is commonly known as micro-Raman spectrometry or Raman microspectrometry.

Among early Raman studies of atmospheric particles with confocal microscopy were those relevant to polar stratospheric cloud formation. Experimental homogeneous ice nucleation rates for hydrated forms of nitric acid captured on hydrophobically coated quartz plates (Knopf et al., 2002), and the dissociation constant for bisulfate in sulfuric acid aerosols were compared with theoretical nucleation rates (Knopf et al., 2003).

Studies have continued to use micro-Raman spectrometry to study ice nucleation relevant to upper cloud formation. Heterogeneous ice nucleation by ammonium sulfate was shown to be a more likely process in cirrus cloud formation compared with nucleation involving glutaric acid particles (Baustian et al., 2010). In a comparison of Raman spectra and spectral maps of two coarse-mode particles, one of which was an organic-coated ammonium sulfate particle and the other a phase mixture of organic material, the organic particles appeared to have sufficient ice nuclei only as coarse particles containing oxygen (Baustian et al., 2012). However, ice may nucleate on an organic shell around an ammonium sulfate particle if the organic is semi-solid or glassy (Schill & Tolbert, 2013). Ice nucleation by internally mixed ammonium sulfate and palmitic acid suggests that particles with an organic phase have little effect on ice nucleation in the tropics (Wise et al., 2010). Investigated by the same research group was the potential of NaCl as ice nuclei from sea-salt aerosol (Wise et al., 2012). Heterogenous ice nucleation was observed by mineral dusts (feldspars, Arizona test dust, and source Asian dust) to occur at higher temperatures compared to homogeneous ice nucleation by the dusts. Particularly active ice nuclei were irregularly shaped particles with clay minerals (Iwata & Matsuki, 2018). A study of volcanic ash suggests that fine ash is a significant global source of ice nuclei (Schill et al., 2015).

Liquid particle phases have been studied with micro-Raman spectrometry and micrometer-sized polar droplets on hydrophobic surfaces, such as Teflon or silanized quartz, and less polar droplets on hydrophilic surfaces, typically quartz. Raman spectra exhibited how and by what mechanism phases of ammonium sulfate and polyethylene glycol in aqueous droplets separated with changes in humidity and the phase ratio (Ciobanu et al., 2009). For droplets of mixed magnesium sulfate and glutaric acid, phase separation occurred at high relatively humidities with droplets on both hydrophobic and hydrophilic surfaces (Wu et al., 2018). Droplets of mixed ammonium sulfate and each of various dicarboxylic acids exhibited phase separation at controlled humidities and various organic/inorganic dry mass ratios (Song et al., 2012). Experimental results were compared with phase separation models.

Particle hygroscopicity relevant to cloud processing was demonstrated from micro-Raman spectrometry in simulations of secondary organic aerosols (SOAs) containing glyoxal, methylglyoxal, glycolaldehyde, and hydroxyacetone, as well as small amine particles on hydrophobic quartz substrates (Hawkins et al., 2014). Hygroscopicity was also studied from the interaction between glycerol and magnesium nitrate in particles on hydrophobic quartz substrates (Wang et al., 2015).

Among the studies of heterogeneous reactions of atmospheric particles, the oxidation of SO_2 by NO_2 on individual $CaCO_3$ particles was demonstrated from micro-Raman spectrometry (Zhao et al., 2018). From micro-Raman and other techniques, model particles of an acidic sulfate core and a SOA shell reacted with a gas-phase oxidized isoprene product (isoprene epoxydiol) showed the dynamic physicochemical state of SOA-containing particles (Olson et al., 2019). Raman spectra and density functional theory calculation were used to study various organosulfate forms that contribute to SOA from isoprene oxidation (Bondy et al., 2018). Vibrational modes from Raman spectra were used to develop a method for determining the pH of individual particles (Rindelaub et al., 2016).

Primary atmospheric particles have also been studied by micro-Raman spectrometry. Various types of elemental carbon have been characterized, including graphite, industrial carbon blacks, soot from diesel and spark-discharge engines, and a soot reference material (NIST SRM 1650) (Sadezky et al., 2005). Also studied from micro-Raman spectra have been the structure and reactivity with nitrogen oxides and oxygen of soot from a spark-discharge generator and a light-duty diesel engine (Figure 1.20) (Ivleva et al., 2007).

The distinction between urban and rural types of soot in association with mineral phases has been demonstrated with urban soot particles appearing mostly crystalline (Catelani et al., 2014). In one study with micro-Raman and diffuse reflectance FTIR, ambient urban continental aerosols were shown to consist of soot coated with organic species (Gaffney et al., 2015). In a study of the structure and protein composition of pollen species in urban and rural areas, soot adsorbed on pollen was observed (Guedes et al., 2009). Ambient marine aerosols were found to contain internally mixed organics and inorganics with long-chain organics prevalent during Fall and Spring along the Eastern Pacific coast and throughout the Pacific Ocean (Deng et al., 2014).

The capture of particles on substrates allows for additional analysis on the same individual particles such as SEM, TEM, and AFM. The combination of micro-Raman spectrometry and SEM-EDX helped elucidate the composition of particles from urban air and industrial emissions, in particular lead and zinc near a smelter in northern France (Choel et al., 2006) and iron species in particles from subway stations (Eom et al., 2013). The internal structure and physicochemical properties of nearly 100 Asian dust particles were elucidated by EPMA and micro-Raman (Sobanska et al., 2012). The mixing state of African dust particles influenced by biomass

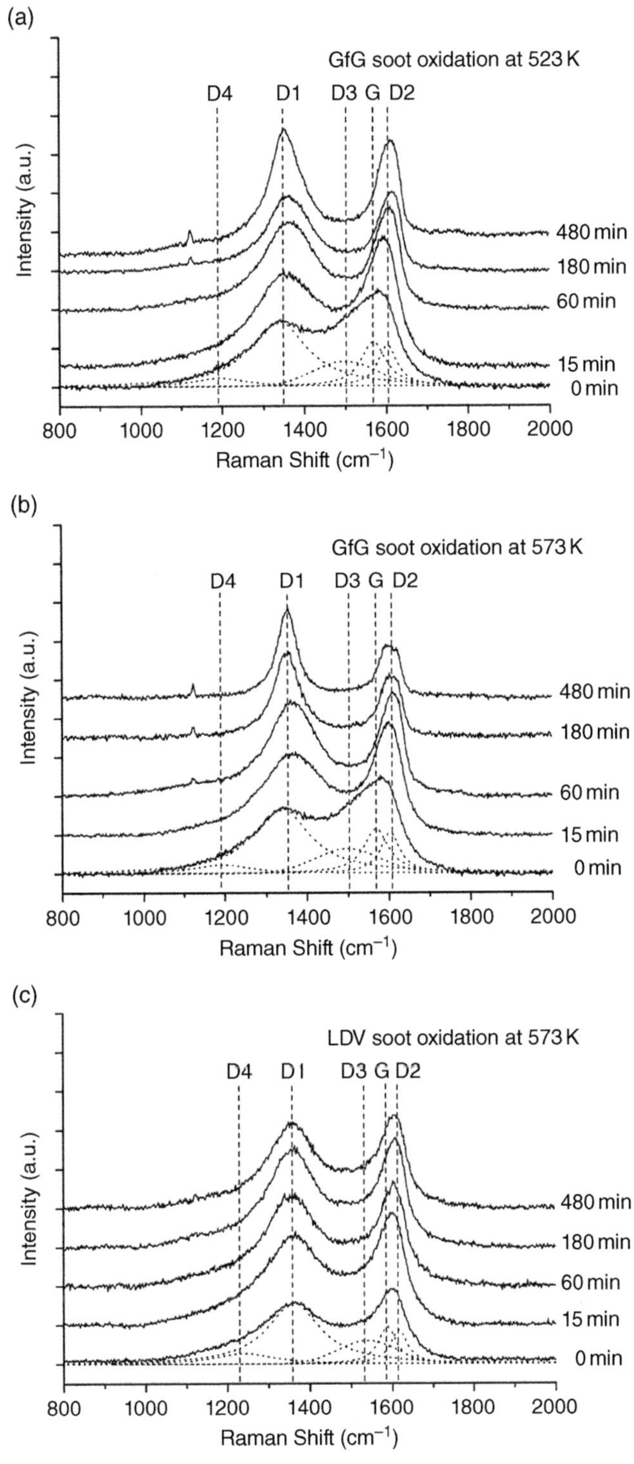

Figure 1.20 Raman spectra of soots from a graphite spark-discharge generator (GfG) (a, b) after partial oxidation at 523 and 573 K and (c) from a light-duty diesel vehicle (LDL) after partial oxidation at 573 K. Ivleva et al. (2007)/with permission of American Chemical Society.

burning and urban pollution was determined from SEM-EDX and TEM-EELS along with micro-Raman (Deboudt et al., 2010). Aerosols from Antarctica during both the Austral summer and winter seasons were analyzed by EPMA and micro-Raman (Eom et al., 2016). SEM was used to screen for metal-rich urban particles, which was followed by micro-Raman mapping of particles and processing of maps by multivariate curve resolution (MCR) to determine the chemical heterogeneity of particles (Batonneau et al., 2006). Results from micro-Raman mapping agreed with WDS-X-ray mapping.

In studies of particle dynamics, structural and hygroscopic changes of model diesel soot (carbon black) in reaction with ozone were observed with TEM and micro-Raman spectrometry (Liu et al., 2010). SEM-EDX and micro-Raman were used to investigate the deliquescence of $MgCl_2$ and NaCl-$MgCl_2$ particles as proxies for sea-salt aerosols, with the binary-phase particles exhibiting deliquescence in two stages and two stages for efflorescence as well (Gupta et al., 2015). In a field study of aerosol formation from volatile emissions within a mobile chamber above a salt lake, SEM-EDX and mass spectrometry along with micro-Raman were used to detect SOAs of oxidized organics and sulfur-containing organics (Kamilli et al., 2016).

Several studies have combined AFM and micro-Raman spectrometry on atmospheric particles. The characterization of internally mixed particles of ammonium sulfate and organics by AFM and micro-Raman showed how the solubility of the organic component and changes to the internal structure of the particles are related (Freedman et al., 2010). Heterogeneous and homogeneous ice nucleation by mineral dust was studied from a combination of AFM, SEM-EDX, and micro-Raman (Iwata & Matsuki, 2018). A study that included AFM, TOF-SIMS, ESEM-EDX, and micro-Raman of ambient particles near an industrial site showed the influence of marine air in internally mixed particles of iron oxides and marine components, such as sea salts and nitrates coated with an organic layer (Sobanska et al., 2014).

In addition to AFM and TOF-SIMS, micro-Raman spectrometry of particles has been paired with synchrotron X-ray analysis. X-ray computed tomography and micro-Raman spectrometry were used to measure the surface area and volume of individual calcite particles $<6.5\,\mu m$ (Wang et al., 2018b). In a revealing study based on STXM-NEXAFS, automated SEM, and micro-Raman results, Arctic sea spray aerosol collected near Barrow, Alaska, was found to be internally mixed with anthropogenic secondary sulfate from Alaskan oil fields (Kirpes et al., 2018).

1.3.3. Instrument Automation and Processing of Spectral Data

Automation of micro-Raman spectrometry has allowed the compositional mapping of far more particles in a sample than can be achieved by selecting particles manually. The large data sets from automation have required multivariate data analysis techniques to determine the relative mass contributions of various species within particles. In an early study, a self-modeling chemometric technique known as SIMPLe-to-use Interactive Self-modeling Mixture Analysis (SIMPLISMA) was used on Raman spectra to determine the contributions of individual components in lead and zinc ore dust particles (Batonneau et al., 2001). In a later automated study on airborne particles collected in an urban area of a former coal mining region, SIMPLISMA was used to resolve luminescence spectra and composition components in Raman spectra. This was followed by MCR with alternating least squares to further refine the contributions of compounds detected in the particles (Sobanska et al., 2006). In a related study, MCR of Raman spectra was used to elucidate products resulting from the water-mediated reaction of micrometer-sized $PbSO_4$ emitted from the smelting of lead with calcite (Falgayrac et al., 2012). Expanding upon the application of MCR-alternating least squares to Raman spectra of metal-rich particles sampled from mines, hierarchical cluster analysis and principal components analysis were used to determine the number of particle types within a sample based on the particles' molecular composition (Siepka et al., 2018).

Further development of computational techniques for the hyperspectral micro-Raman imaging of atmospheric particles has focused on improving the spatial resolution of micro-Raman image maps, determinations of particle composition, and classification of particle types. A super-resolution algorithm was employed that improved spatial resolution of the confocal micro-Raman spectrometer by 65%. Raman spectra from improved spatial resolution were then processed by MCR-alternating least squares to identify particle types (Offroy et al., 2015). The analysis was performed on ambient metal-containing particles in the vicinity of a battery recycling facility. Micro-Raman and SEM-EDX data sets were combined in a hyperspectral data cube for processing the cube with principal component analysis, hierarchical cluster analysis, k-means clustering, and vertex component analysis (Ofner et al., 2015).

The Raman microspectrometer has also been modified to increase analysis throughput and automate both sampling and analysis. Software designed to detect particles automatically and process the data with multivariate analysis algorithms was improved to allow for the analysis of up to 100 particles per hour (Craig et al., 2017). A Raman spectrometer designed for the detection of airborne bioagents was modified to automatically alternate the collection of ambient aerosols with collection of Raman spectra (Doughty & Hill, 2017).

1.3.4. Ancillary Raman Techniques and Applications to Atmospheric Particles

Surface-Enhanced Raman Scattering

Surface-enhanced Raman scattering (SERS) occurs when a Raman active analyte molecule is located on or near a textured metal surface such that the metal's free electrons exhibit localized plasmon resonance from excitation by visible laser light. Roughness of the textured surface is on the order of 10–100 nm (Campion & Kambhampati, 1998). The surface may be roughened from islands in a thin film, colloidal nanoparticles, or nanospheres deposited on a substrate. The localized plasmon resonance is strongest for textured surfaces of silver, gold, and copper when laser light is in the visible, as is used for Raman spectroscopy.

The occurrence of localized plasmon resonance amplifies the incident laser light at the textured metal's surface. If the analyte molecule is near enough to the surface when laser light exits, the induced dipole in the analyte molecule increases in magnitude. The physics is known as electromagnetic enhancement. By this mechanism, the Raman signal enhancement for an analyte molecule near a surface containing gold or silver spheroidal particles may range from 10^6 to 10^{11} (Kneipp et al., 2002). Signal enhancement by SERS necessarily also allows Raman spectroscopy to overcome the Abbe diffraction limit, the minimum length in a microscopic field whereby features can be resolved (typically around one-half the wavelength of light used) in conventional microscopy in the visible. However, electromagnetic enhancement also predicts that intensities of the Raman signals fall off with increasing vibrational frequency. For example, the signal from C-H stretches tends to be relatively weak (Campion & Kambhampati, 1998). In addition, the size of the metallic nanospheres affects the enhancement strength. Localized plasmon resonance extends further out for larger-radii nanospheres than for smaller-radii nanospheres.

If the analyte molecule is adsorbed on the surface, an additional enhancement in the Raman signal is explained by a mechanism that depends on the chemistry of the analyte molecule. In this case, there appears to be a transfer of charge between the substrate's Fermi level and the analyte molecule's valence orbitals, either the lower unoccupied level or the higher occupied level (Campion & Kambhampati, 1998; Kneipp et al., 2002). Raman signal enhancement by this "chemical" effect appears to be a factor of 10^1 to 10^2 (Kneipp et al., 2002), far smaller than the enhancement from localized plasmon resonance.

SERS was used to examine the complexity of individual micrometer-sized ambient heterogeneous particles containing secondary organic and inorganic compounds (Craig et al., 2015). Particles were collected on quartz substrates coated with silver nanoparticles. The enhancement of peak signals over conventional micro-Raman was as much as a factor of 14 for ambient particles around 1 µm or smaller in size. SERS was also used to determine the mixture state of ammonium aerosols with sulfate and nitrate that are responsible for haze formation in urban areas of China (Sun et al., 2019). In this study, PM2.5 particles were collected in Beijing during a winter haze weather event. Haze particles were observed mainly as internal mixtures of double nitrate/sulfate salts: for example, $2(NH_4NO_3)\cdot(NH_4)_2SO_4$ and $3(NH_4NO_3)\cdot(NH_4)_2SO_4$.

Tip-Enhanced Raman Scattering

Tip-enhanced Raman scattering (TERS) can be considered an extension of SERS whereby both electrostatic and chemical, or charge transfer, enhancement effects are involved, but the surface is confined to a metallic tip. In most cases, a cantilever tip in the AFM is used, which may also have attached a metallic particle at the end. A limitation of SERS is that roughness of the surface results in variation in the electrostatic field enhancement in the vicinity of the analyte particle. TERS theoretically overcomes this limitation by having a high curvature at the tip with lateral dimensions of 10–100 nm, which confines enhancement of the field to the tip. The high curvature allows for maximum field enhancement compared to the commonly lower curvature features of a SERS surface (Bailo & Deckert, 2008). Due to the increased Raman signal enhancement of TERS, lateral resolution of analyte particles has been reported at <5 nm with tip sizes of 15–30 nm (Deckert et al., 2015). However, compared with SERS, TERS requires additional considerations for the laser optics and the orientation of the laser beam relative to the enhancement tip. For example, polarization of the laser beam is important if the laser has a low numerical aperture. In this case, electrostatic enhancement occurs only if the electric field is parallel rather than perpendicular to the axis of the tip (Bailo & Deckert, 2008).

Nanoparticles emitted from a simulated salt lake inside a smog chamber were analyzed with TERS to show the technique's potential for studying organic nucleation mode particles and their chemical aging (Ofner et al., 2016). The study was the first of its kind involving TERS on atmospheric particles <100 nm. Figure 1.21 shows AFM images and TERS spectra for three particles assigned to the Aitken and accumulation

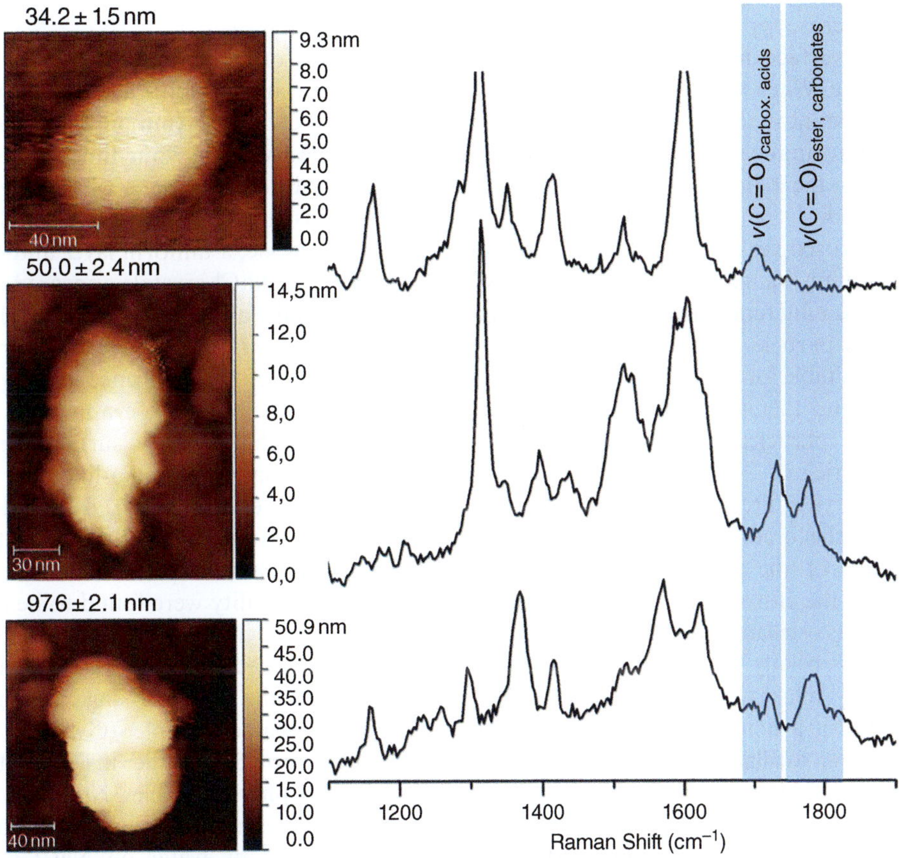

Figure 1.21 TERS analyses of atmospheric particles <100 nm emitted from a simulated salt lake inside a smog chamber. (Left) Images from AFM with the average lateral dimension indicated above each image. The Raman spectra point to carboxylic acid, ester, and carbonate species as indicated in the C=O vibrational features. Ofner et al. (2016)/with permission of American Chemical Society.

particle-size modes. The AFM tip in this experiment was coated with silver nanoparticle "islands." Vibrational features in the spectra were attributed to the C=O stretch of carboxylic acids, esters, and carbonates.

1.4. ATOMIC FORCE MICROSCOPY

AFM, also known as scanning force microscopy, characterizes a microscopic particle by measuring the movement of a cantilevered stylus as it is rastered over the surface of the particle. Due to the flexibility of the cantilever arm, reflected laser light off the arm allows for the detection of minute changes in the force between the stylus and sample surface. In typical practice, the force between stylus and sample surface is maintained by adjustment to the height of the sample and hence the height of the particle. The force exerted by the stylus in close contact with the surface includes a repulsive force component that depends on the compositions of the stylus and particle (Al-Abadleh & Grassian, 2003; Barkay et al., 2005; Kollensperger et al., 1997).

AFM is a type of scanning probe microscopy, along with scanning tunneling microscopy, which also relies upon the spatial movement of a fine-tipped stylus. Rather than close contact with the sample surface as with AFM, scanning tunneling microscopy maintains a gap between the stylus and sample surface due to a voltage difference between tip and surface. When the gap is maintained at a few nanometers, a constant tunneling current exists from tip to surface. By monitoring change in the tunnel current, movement of the tip in the z direction, and thus particle thickness, is determined.

Scanning probe microscopies provide lateral and depth resolution in the 0.1 nm range. The main difference between AFM and scanning tunneling microscopy is that the latter requires the particle surface to be electrically conductive. AFM is suitable for nonconducting as well as conducting particles and therefore applicable to a wide range of atmospheric particles such as mineral dusts, salts, and soot-like particles.

Because the AFM stylus directly touches the particle, the potential exists for the force from the stylus to damage or distort the surface of the particle. The problem is avoided by operating the instrument in tapping mode, whereby the cantilever is oscillated at a few hundred cycles per second. The stylus briefly taps the particle surface and then momentarily relaxes from the surface. AFM may also be operated in a noncontact and nontapping mode with the cantilever oscillating. An AFM technique for determining the adhesion force of a particle was developed as a potential way to elucidate differences among particles in chemical composition and aging (Ono et al., 2020).

1.4.1. Applications to Atmospheric Particles

A significant benefit of AFM is its use to acquire information on particle volume and morphology. Early on, the calculation of particle volume was shown from 3D AFM data, which allowed particle change with absorbed water to be quantified (Kollensperger et al., 1997). Also studied was the effect on particle shape as humidity-saturated particles land on substrate surfaces (Wittmaack & Strigl, 2005). AFM volume determinations have also been used to derive particle mass (Barkay et al., 2005). AFM was used to determine at the surface the size distribution of particles in a study of aerosol deposition in snow toward reconstructing the composition of past atmospheres in ice cores (Udisti et al., 2004). The fractal character of soot particles from 3D AFM determinations is believed to be more accurate than 2D determinations from AFM or electron microscopy (Pósfai & Buseck, 2010). In addition to 3D morphology, AFM has been used to determine phase state and viscoelastic properties of component systems found in sea spray aerosols (Ray et al., 2019). The work involved multiple organic acids, saccharides, and lipopolysaccharides as single-component systems and as binary systems with NaCl or $MgSO_4$.

Hygroscopicity

From early on, AFM has also been used to study particle volume and structural changes with exposure to humidity and ozone. Ammonium sulfate particles were found to have volumes four times larger when hydrated than when particles lost water while imaged under vacuum in a TEM instrument (Pósfai et al., 1998). Using AFM in noncontact mode, a uniform layer and steps of water on the surface of NaCl crystals appeared after exposure to relative humidity above 35% (Dai et al., 1997). The abrupt disappearance of the water step structure occurred with the onset of deliquescence at 73% RH. Water condensation and subsequent evaporation from submicron particles with characteristic surface chemical properties caused the formation of chain-like structures on particles (Kollensperger et al., 1999). Here, the formation of clump-like structures from chain-like structures accompanied a decrease in particle volume. Humidity and component volatility were found to be factors in the discrepancy in particle size distributions based on AFM and SEM measurements. The work was based on response functions for the multiple-stage micro-orifice uniform deposition impactor sampler (Gwaze et al., 2007).

AFM has provided further insights into the hygroscopicity of atmospheric particles. Noncontact AFM exposed to humidity (described as environmental AFM) showed that the uptake of water on NaCl nanoparticles was reversible well below the relative humidity at which deliquescence occurred (75%) (Bruzewicz et al., 2011). A 2–5 nm layer of water formed on the nanoparticles beginning at 70% RH. Hygroscopic growth of heterogeneous particles consisting of ammonium sulfate, succinic acid, and chlorides was shown to depend on particle size from observations of 100 nm and 6 μm particles (Laskina et al., 2015). By considering the 3D change in particle shape, hygroscopic growth factors were determined by AFM for single component particles of NaCl and malonic acid (Morris et al., 2016). Included in the study were binary mixtures of NaCl with malonic acid and NaCl with nonanoic acid. Results were compared with hygroscopic growth factors determined from 2D parameters and with a hygroscopic tandem differential mobility analyzer (Figure 1.22). Hygroscopic growth was determined by AFM and compared with growth models for heterogeneous particle systems containing simple and complex carbohydrates (Estillore et al., 2017). Particles of NaCl mixed with organics exhibited a cubic NaCl core and organic shell from AFM.

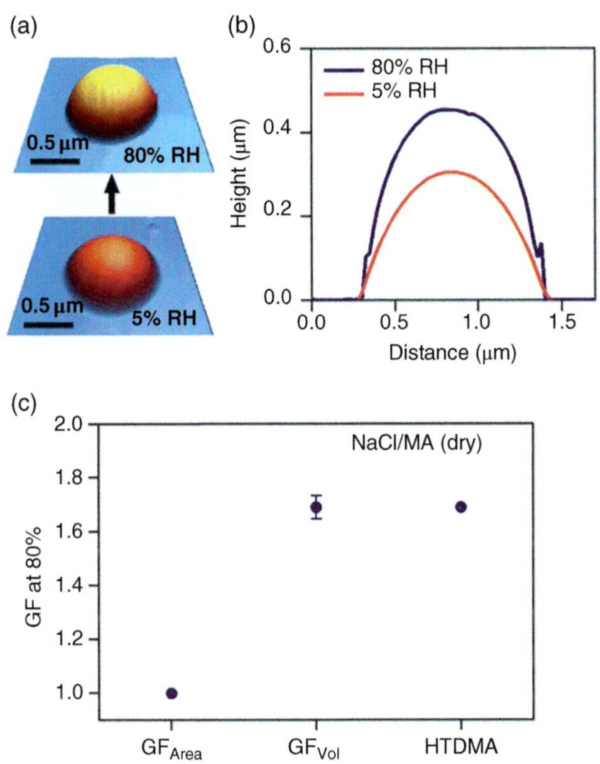

Figure 1.22 Water uptake by an initially dry particle containing NaCl and malonic acid. (a) AFM images at 5% and 80% relative humidity; (b) change in height of particle at two humidities; (c) comparison of growth factors based on area and volume with factor derived from hygroscopic tandem differential mobility analysis. Morris et al. (2016)/with permission of American Chemical Society.

Related to AFM studies on particle hygroscopicity is the use of the technique to determine surface tension. At the deliquescence humidity and higher, surface tension versus relative humidity was measured on droplets of NaCl, malonic acid, and glutaric acid 200–500 nm in diameter (Morris et al., 2015). Surface tension for particles as mixtures and single components with dicarboxylic acids and saccharides as well as electrolytes was found to vary with the proportion of solute and viscosity along with humidity (Lee et al., 2017).

Oxidation and Reactivity

The utility of AFM has also been shown for studying atmospheric particle oxidation and reactivity. For example, ozone was shown to cause the chain-like structures of soot to collapse. Organic particles as well as soot shrank in volume upon repeated exposure to ozone or reactive oxygen, whereas no change was observed for inorganic particles (Lehmpuhl et al., 1999). AFM has also been used to study the process of nucleation in soot formation (Schulz et al., 2019).

AFM was used to show the formation of homogeneously nucleated particles smaller than 300 μm upon the reaction of isoprene with ozone (Ramirez-Aguilar et al., 1999). The work demonstrated how chemical reaction information can be acquired with AFM by monitoring morphology change in a particle formed from reactive gases. More recent work involving AFM with SEM focused on the epoxydiol oxidation product of isoprene and sulfate particles coated (Y. Zhang et al., 2018). Here, the uptake of the trans-b version of isoprene epoxydiol on an acidic sulfate coated with a-pinene was observed.

Tapping-mode AFM was used to detect morphological changes to the surface of freshly cleaved MgO and $CaCO_3$ crystals as they were exposed to water and nitric acid (Krueger et al., 2005). Also shown was deliquescence of nitrate salt features on particle surfaces. In a study of fine silicate ash from volcanic eruptions, AFM revealed <10 nm coatings of soluble sulfate and halides on particle surfaces (DelMelle et al., 2007).

Recent work with AFM was conducted to determine the pH of individual aerosol particles (Ault, 2020). In one approach employed in this study, the change in individual particles was observed as particle acidity was used to catalyze the degradation of a thin organic film (poly(ε-caprolactone)) on a silicon substrate.

1.4.2. Techniques in Combination and Applications to Atmospheric Particles

AFM has often been combined with other techniques, in most cases to provide information on mixing state, composition, and hygroscopicity. AFM was combined with TEM and AMS to investigate the mixing state of complex urban aerosols consisting of ammonium and sodium sulfates, nitrates, organics, and soot in the Rhein-Main area of Germany (Vester et al., 2007). The shape, structure, and composition of mixtures of ammonium sulfate and organic components as model particles were investigated with tapping-mode AFM and micro-Raman (Figure 1.23) (Freedman et al., 2010). Here, investigators found that solubility of the organic component and perhaps internal structure can be elucidated with AFM. Changes in the physicochemistry of particles from an industrial area and influenced by marine air were investigated by AFM, ESEM with EDX, and TOF-SIMS (Sobanska et al., 2014). Particles exhibited as sea salt with nitrates and a thin organic layer. Industrial particles included internal mixtures of iron oxides, marine components, and an organic shell. Nanoscale SIMS has been used with TEM and tapping-mode AFM to show the sources, size, morphology, composition, and atmospheric aging of aerosols that produce urban haze in northeastern China (Zhang et al., 2020).

From AFM and FTIR, water was shown to dissociate and molecularly absorb on a metal oxide surface at specific sites under hygroscopic conditions relevant to the ambient atmosphere (Rubasinghege & Grassian, 2013). From AFM, micro-FTIR spectroscopy, and X-ray spectroscopy, hygroscopic properties of NaCl, malonic acid, and glutaric acid as pure components were shown to be different from particles as a mixture and dependent largely on the molar proportions of the organic components (Ghorai et al., 2014). Micro-Raman spectroscopy with AFM was used to show that the hydrophobicity or hydrophilicity of a substrate used for sampling sea spray aerosol is a factor in determining the hygroscopicity of particulate inorganic salts (Laskina et al., 2015).

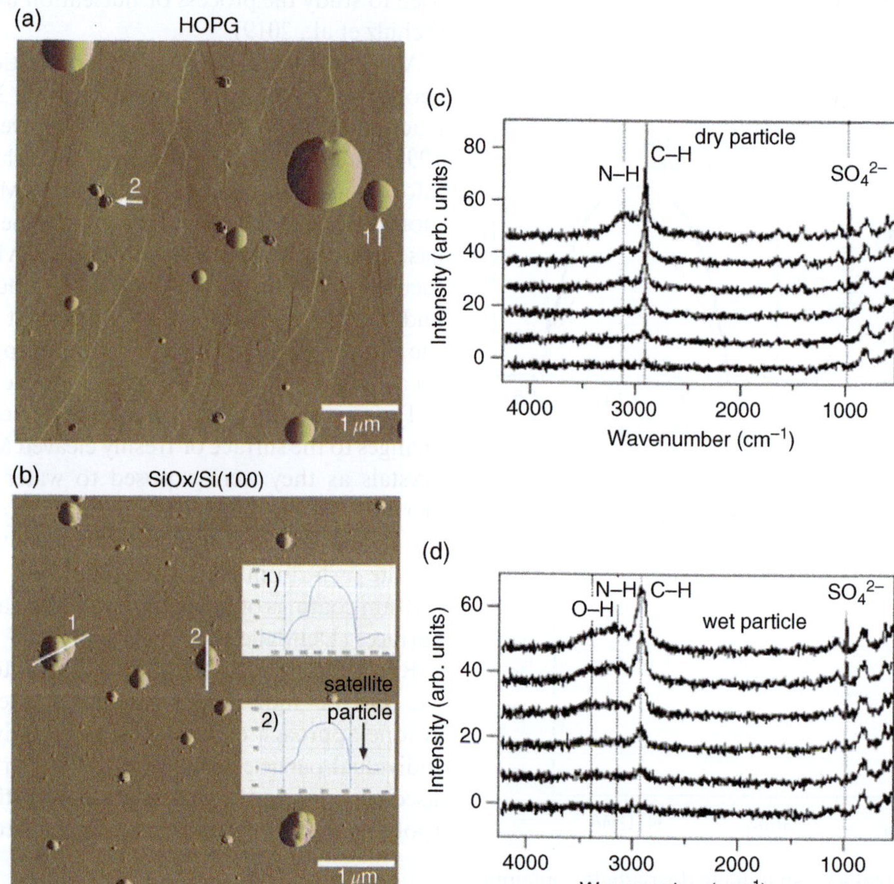

Figure 1.23 AFM images and Raman spectra of particles with ammonium sulfate and glutaric acid. (a) Particles with glutaric acid at 0.75 weight fraction on highly ordered pyrolytic graphite (HOPG) substrate and (b) at 0.50 weight fraction on a silica/silicon substrate. Raman spectra for 0.75 weight fraction (c) dry particle and (d) wet particle. Spectra 1–6 correspond to the position of the laser beam through the particle in each case. The laser beam is parallel to the substrate surface. Freedman et al. (2010)/with permission of American Chemical Society.

AFM combined with X-ray photoelectron spectroscopy (XPS) has been used to investigate volcanic emissions. Rapid dissolution of fine silicate particles from volcanic eruptions was found to occur in the presence of fluoride within the volcanic plume (DelMelle et al., 2007). Ash particles sampled near Iceland's Eyjafjallajökull volcano during two related eruptions were analyzed with AFM, SEM, EPMA, and XPS to investigate the inhalation risk from fallout (Gislason et al., 2011). Particles from the more explosive first eruption maintained the sharpness in their shapes after treatment in the lab with water to simulate a cloud environment, suggesting the ash maintained its sharpness after reaching Europe.

Recently, AFM and IR spectroscopy have been integrated into a single instrument. A pulsed laser in the mid-IR is used to locally heat and expand the sample, which then causes the AFM cantilever to oscillate with thermal expansion. Typically, a second laser reflected off the surface of the cantilever tip allows for tracking the vertical movement of the tip. The amplitude of oscillation of the cantilever tip is proportional to the IR absorbed locally by the material and therefore the material's absorption coefficient (Figure 1.24) (Gruszecki et al., 2015). AFM-IR has a spatial resolution typical of AFM, on the order of tens of nanometers, rather than resolution associated with the Abbe diffraction limit for wavelengths in the mid-IR used to heat the sample (Dazzi & Prater, 2017). Polarization of the IR excitation wavelength in a particular direction has allowed for the investigation of molecular orientation in nanofibers (Dazzi et al., 2012). In internally mixed particles with AFM-IR, trace organic and inorganic components were detected spatially with resolution on the order of 150 nm (Bondy et al., 2017). In this study, absorption features were identifiable for ammonium sulfate, $NaNO_3$, succinic acid, and sucrose in model particles.

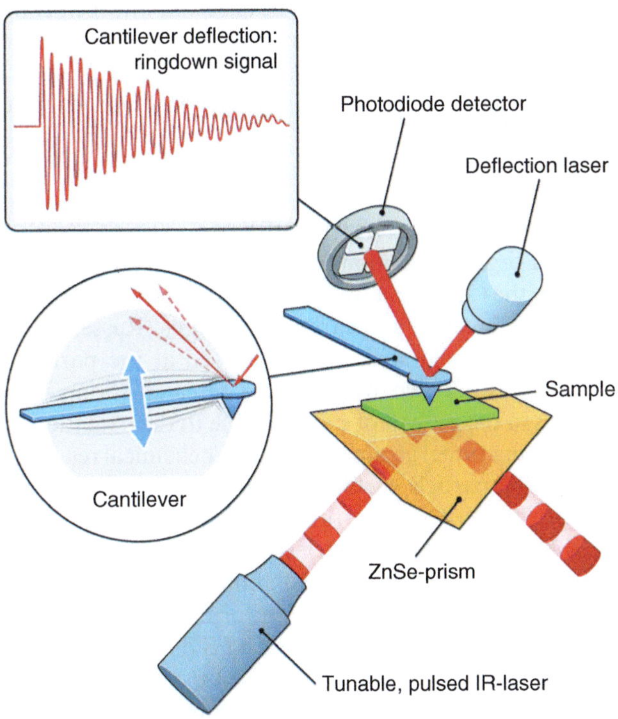

Figure 1.24 Schematic of conventional AFM-IR instrumentation. IR laser for heating sample is pulsed. A photodiode detects visible light reflected off the cantilever tip from a continuous wave laser. Gruszecki et al. (2015)/with permission of Royal Society of Chemistry.

Further modification of AFM-IR has resulted in the technique known as optical photothermal infrared spectroscopy (O-PTIR). Here, a continuous wave laser, typically in the visible, is used as a means to detect thermal expansion of the sample due to absorption of IR from the pulsed IR laser. Thermal expansion caused the visible light from the continuous wave laser to scatter elastically, thus producing a characteristic response proportional to the sample's IR absorption coefficient (Olson et al., 2020).

1.5. SCANNING TRANSMISSION X-RAY MICROSCOPY

STXM exposes a sample with a highly intense and focused beam of polarized X-rays that are generated from the acceleration of electrons at nearly the speed of light in a synchrotron. The X-ray beam line that serves STXM experiments consists of softer X-rays typically of up to 1 keV or, for some instruments, 2 keV (Moffet et al., 2011). After passing through a grating monochromator that provides energy resolution of <0.1 eV, X-rays are focused on the specimen within the STXM instrument by adjusting the distance of a Fresnel zone plate (circular diffraction grating) to the sample (Kirz, 1974; Warwick et al., 1998). Because X-rays have wavelengths on the order of less than 10 nm, the Abbe diffraction limit for X-rays is far lower than that for visible light, which allows the beam to be focused on an extremely small spot, around 25 nm or less in diameter. The small spot size therefore permits the mapping of individual particles with fairly high spatial resolution.

The spectroscopy employed with STXM, near-edge X-ray absorption fine structure (NEXAFS), largely relies on the excitation of the specimen's core-shell electrons, in the L and M shells as well as the K-shell, to the lowest unoccupied pi or sigma molecular orbitals (π^* and σ^*). The result is that the absorption spectrum exhibits an absorption edge. The great utility of NEXAFS with STXM is that the absorption edge and spectral features near the edge are highly affected by the local molecular environment. Therefore, spectra provide molecular information on the specimen, such as the presence of organic functional groups and oxidation states of inorganic constituents. Molecular information from NEXAFS is similar to spectral information from TEM-EELS due to inelastic scattering of primary beam electrons by the specimen. A significant advantage of NEXAFS over TEM-EELS is that the former imparts less energy to the specimen, which results in less beam damage to particles. In addition, the energy from soft X-rays in STXM goes toward a spectral transition, which results in sharp spectra detail of the molecular environment. In contrast, much of the energy from the TEM electron beam ionizes atoms rather than promoting a core-shell electron to an unoccupied molecular orbital (Hitchcock et al., 2008). A result is less spectral detail from TEM of the molecular environment (Braun et al., 2005). However, EELS provides better spatial resolution than NEXAFS.

1.5.1. Applications to Atmospheric Particulate Carbon

Soot and Black Carbon

NEXAFS is especially useful in the analysis of particulate carbon species due to the relatively stable promotion of a carbon *K*-shell electron at energies between 280 and 320 eV. Promotion to the π^* orbital in a C=C bond at around 285 eV typically results in a sharp absorption edge. Promotion to the σ^* orbital at a higher energy typically results in a broader absorption peak, which, with a peak around 291 eV, may indicate the sigma bond in C=C (Braun, 2005).

In purportedly the first application of STXM-NEXAFS on single diesel soot particles, graphitic C=C bonding at 285 eV and hydrocarbon C—H bonding around 288 eV were observed within the same particle (Braun et al., 2004). Early work using STXM-NEXAFS also focused on distinguishing woodsmoke carbon from diesel soot. More sp^2 bonding appears in spectra from fossil fuel combustion than from biomass burning (Hopkins et al., 2007b). Relative to the graphitic C=C peak at 285 eV, woodsmoke produced a strong peak at 287 eV associated with phenolic carbon (Vernooij et al., 2009). Peaks in spectra from woodburning emissions were associated with organic C=O, C—OH, and COOH bonding in addition to the C=C bond (Braun, 2005). An increase in the COOH/C-OH peak ratio suggested a correlation with increased water content of the fuel. Reference materials (e.g., NIST Diesel soot SRM 1650, NIST Urban Dust SRM 1659a, and the Suwanee River humic acid 2S101H) have been useful in understanding the absorption fine structure of black carbons and biomass burning particles from STXM (Hopkins et al., 2007b). Results here revealed proportionately more oxygenated functional groups as well as less aromatic carbon from woodburning emissions compared with black carbon materials. In addition to spectral transitions of the carbon 1s electron, NEXAFS spectra for the oxygen *K*-edge transition in the 525–550 eV energy range allowed for a determination of the contributions of C and O atoms in a particulate material (Moffet et al., 2011).

Biomass Burning Aerosols and Tar Balls

In a study that included the optical properties of biomass burning aerosols, emissions from the burning of different types of plant fuels were characterized with respect to the carbon and oxygen ratios and bonding sp^2 hybridization from NEXAFS spectra (Hopkins et al., 2007a). Particles were categorized by the extent of sp^2 hybridization and associated with single scattering albedo and Angstrom absorption values.

As with bonding information from TEM-EELS, NEXAFS from STXM provides important insights into the chemical character and formation of tar balls. Tar balls were shown to be highly spherical with uniform density and a composition of around 55% carbon and 45% oxygen. Also found were carbonyl groups of carboxylic acids and ketones and oxygen-substituted alkyl groups. The carbonyl groups were associated with oxygenated interfaces within tar balls (Tivanski et al., 2007). Formation of tar balls was shown to proceed from primary organic particles within 3 h of burn emission. Carboxylic acid and nitrogen compounds appear to be important intermediates in spherical ball formation (Adachi et al., 2019). Figure 1.25 shows tar ball NEXAFS 1s spectra for carbon, nitrogen, and oxygen (Tivanski et al., 2007).

Organic Carbon

In the first reported STXM-NEXAFS study of spatially resolved organic functional groups, marine particles collected in the Caribbean were shown to contain compounds with alkyl, ketonic, and carboxylic groups (Russell et al., 2002). Carboxylic acids were observed at the particles' surface. Also shown by the same research group were changes in the growth of particles due to the oxidation of organic functional groups and whether chemical reactions were at the particle surface or within its volume (Maria et al., 2004). Organic functional group analysis by NEXAFS was later used to classify ambient particles from various field campaigns (Takahama et al., 2007). Work that included particle shape information along with chemical composition from NEXAFS on mainly heterogeneous ambient particles addressed optical and cloud-nucleating properties of particles (Takahama et al., 2010). Using the multivariate analysis technique Positive Matrix Factorization on NEXAFS spectra, submicron particles collected in Mexico City during the MILAGRO campaign were shown to contain contributions from soot, secondary organics, and biomass burning (Liu et al., 2009). NEXAFS

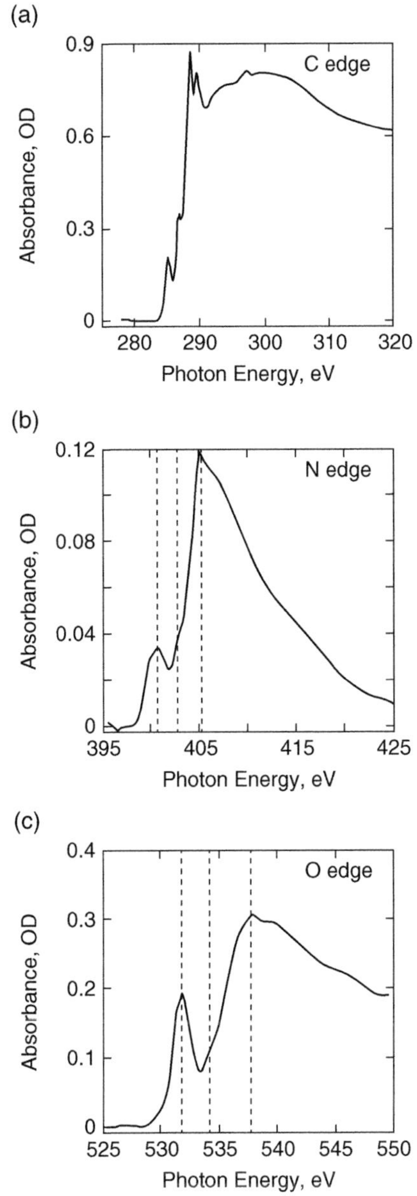

Figure 1.25 NEXAFS spectra of a tar ball collected during the Yosemite Aerosol Characterization Study in summer 2002. K-edge spectra for (a) carbon, (b) nitrogen, and (c) oxygen. Dotted vertical lines are at 400.7, 402.7, and 405.7 eV for nitrogen; 531.8, 534.5, and 538 eV for oxygen. Adapted from Tivanski et al. (2007).

was also used to distinguish airborne soil organic particles from biomass burning tar balls in samples collected in the U.S. Southern Great Plains (Fraund et al., 2020). Further studies involving STXM-NEXAFS on organics in particles have helped elucidate ambient particle sources after long-distance transport (Day et al., 2009; Leaitch et al., 2009).

STXM-NEXAFS has also found utility in analyzing biologically derived species in atmospheric particles. The technique was used to determine the composition of sea-spray particles containing components from phytoplankton blooms (Pham et al., 2017). Also determined were types of organic particles containing proteins, calcareous phytoplankton, and types of polysaccharides collected in the marine boundary layer of the Arctic and southeastern Pacific (Hawkins & Russell, 2010). Figure 1.26 shows that NEXAFS spectra from polysaccharide particles from the Arctic, containing carboxylic acids and associated with carbonate salts (Ps I), are substantially different from polysaccharide particles without carboxylic acids collected elsewhere in the Pacific (Ps II).

The application of STXM-NEXAFS for determining organics in heterogeneous particles has led to its combination with complementary techniques, often to provide inorganic information. AToFMS was used with STXM to examine particle morphologies and chemistries in particles collected at urban and coastal sites (Bahadur et al., 2010). CCSEM and STXM were used to analyze particles collected during the MILAGRO campaign (Moffet et al., 2010) and from the southeastern U.S. (Bondy et al., 2018). SEM-EDX and STXM were used to analyze particles from sampling sites in the Amazon basin (Fraund et al., 2017). Fourier transform IR and X-ray fluorescence spectroscopies were used with STXM to study biogenic emissions from a mountain forest site (Schwartz et al., 2010).

1.5.2. Applications to Particulate Inorganic Species

In addition to organics, STXM-NEXAFS provides spatial resolution of inorganic species in particles. Particularly useful is the determination of metallic oxidation states. For example, different types of Cr(III) compounds were distinguished from Cr(VI) (Werner et al., 2007). In fly ash from oil and coal combustion, hazardous nickel sulfides, Cr(VI), and As(III) were identified with STXM-NEXAFS (Huggins et al., 2004). In samples from various field campaigns, iron was found to exist as Fe(II) or Fe(II) and Fe(III) in ratios of up to 0.73 (Takahama et al., 2008).

1.5.3. Studies of Hygroscopicity and Ice Nucleation

STXM-NEXAFS has been used to study hygroscopic properties of particles and phase transitions within particles. Submicrometer particles containing sodium salts of chlorine, bromine, iodine, and nitrate were monitored with STXM-NEXAFS while exposed to a range of relative humidities (Ghorai & Tivanski, 2010). Phase transitions at deliquescence and efflorescence agreed well with previous approaches in the literature. The separation of liquid phases in particles containing ammonium sulfate and one of various organic phases was observed with

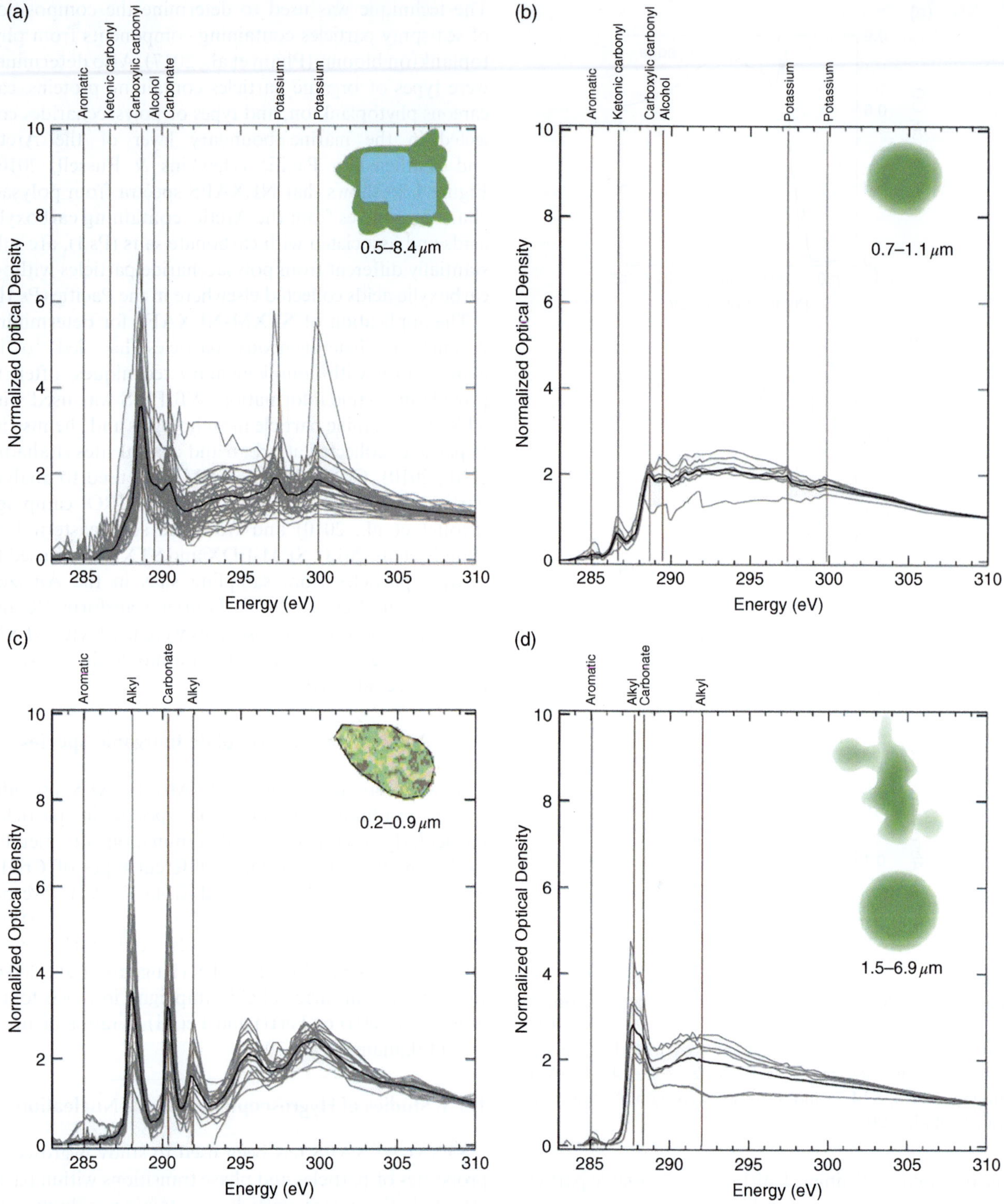

Figure 1.26 NEXAFS spectra from marine organic particle types: (a) polysaccharide with carboxylic acids (Ps_I), (b) polysaccharide without carboxylic acid (Ps_II), (c) phytoplankton fragment containing calcium carbonate, (d) protein-containing particle. Hawkins and Russell (2010)/Hindawi Publishing Corporation/CC BY 3.0.

STXM-NEXAFS and ESEM (O'Brien et al., 2015). For particles containing, for example limonene, homogeneous mixing did not occur at all relative humidities above deliquescence. However, particles containing polyethylene glycol did become homogeneous and phase separated at around 90% relative humidity.

Based on spectra from NEXAFS and CCSEM-EDX, heterogeneous ice nucleation was found to occur with particles containing organic species (Knopf et al., 2010). From particles collected in and around Mexico City, ice nucleation was found to depend upon particle temperature, relative humidity, nucleation mode, and aging-influenced particle chemistry. A recent study showed how ice nucleating particles can vary seasonally and with altitude (Knopf et al., 2023). Soot was shown to nucleate ice as it became hydrophilic in the atmospheric aging process (Mahrt et al., 2020).

1.6. CONCLUSION

An overview of microscopy-based imaging and analytical techniques was presented, along with prominent studies on atmospheric particles that depended on these techniques. The techniques center on electron microscopy (SEM, TEM, and associated techniques), Raman scattering (micro-Raman spectroscopy and associated techniques), scanning probe microscopy (primarily AFM), and synchrotron X-ray analysis (STXM and NEXAFS). The techniques and their relevance in atmospheric studies are presented in more detail in subsequent chapters.

AVAILABILITY STATEMENT

All data presented here were published previously.

REFERENCES

Aardahl, C. L., & Davis, E. J. (1996). Gas aerosol chemical reactions in the NaOH-SO2-H2O system. *Applied Spectroscopy*, *50*(1), 71–77. WOS:A1996TQ78100010

Adachi, K., & Buseck, P. R. (2008). Internally mixed soot, sulfates, and organic matter in aerosol particles from Mexico City. *Atmospheric Chemistry and Physics*, *8*(21), 6469–6481. WOS:000260927800011. doi:10.1029/2010jd015102

Adachi, K., & Buseck, P. R. (2011). Atmospheric tar balls from biomass burning in Mexico. *Journal of Geophysical Research-Atmospheres*, *116*, 7. WOS:000288324700003

Adachi, K., & Buseck, P. R. (2015). Changes in shape and composition of sea-salt particles upon aging in an urban atmosphere. *Atmospheric Environment*, *100*, 1–9. WOS:000347266700001

Adachi, K., Chung, S. H., Friedrich, H., & Buseck, P. R. (2007). Fractal parameters of individual soot particles determined using electron tomography: Implications for optical properties. *Journal of Geophysical Research-Atmospheres*, *112*(D14), 10. WOS:000248285700007. doi:10.1029/2006JD008296

Adachi, K., Sedlacek, A. J., Kleinman, L., Springston, S. R., Wang, J., Chand, D., et al. (2019). Spherical tarball particles form through rapid chemical and physical changes of organic matter in biomass-burning smoke. *Proceedings of the National Academy of Sciences of the United States of America*, *116*(39), 19336–19341. WOS:000487532900026

Adler, G., Koop, T., Haspel, C., Taraniuk, I., Moise, T., Koren, I., et al. (2013). Formation of highly porous aerosol particles by atmospheric freeze-drying in ice clouds. *Proceedings of the National Academy of Sciences of the United States of America*, *110*(51), 20414–20419. WOS:000328548600029

Al-Abadleh, H. A., & Grassian, V. H. (2003). Oxide surfaces as environmental interfaces. *Surface Science Reports*, *52*(3–4), 63–161. WOS:000187039700001

Alexander, D. T. L., Crozier, P. A., & Anderson, J. R. (2008). Brown carbon spheres in east Asian outflow and their optical properties. *Science*, *321*(5890), 833–836. WOS:000258261000045

Armstrong, J. T., & Buseck, P. R. (1975). Quantitative chemical-analysis of individual microparticles using electron-microprobe—theoretical. *Analytical Chemistry*, *47*(13), 2178–2192. WOS:A1975AU25800024

Ashkin, A., & Dziedzic, J. M. (1971). Optical levitation by radiation pressure. *Applied Physics Letters*, *19*(8), 283–285. WOS:A1971K529100014

Ashkin, A., Dziedzic, J. M., Bjorkholm, J. E., & Chu, S. (1986). Observation of a single-beam gradient force optical trap for dielectric particles. *Optics Letters*, *11*(5), 288–290. WOS:A1986C119800008

Ault, A. P. (2020). Aerosol acidity: Novel measurements and implications for atmospheric chemistry. *Accounts of Chemical Research*, *53*(9), 1703–1714. WOS:000572833000001

Ault, A. P., & Axson, J. L. (2017). Atmospheric aerosol chemistry: Spectroscopic and microscopic advances. *Analytical Chemistry*, *89*(1), 430–452. WOS:000391346600024

Ault, A. P., Peters, T. M., Sawvel, E. J., Casuccio, G. S., Willis, R. D., Norris, G. A., & Grassian, V. H. (2012). Single-particle SEM-EDX analysis of iron-containing coarse particulate matter in an urban environment: Sources and distribution of iron within Cleveland, Ohio. *Environmental Science & Technology*, *46*(8), 4331–4339. WOS:000302850300013

Azarmi, F., & Kumar, P. (2016). Ambient exposure to coarse and fine particle emissions from building demolition. *Atmospheric Environment*, *137*, 62–79. WOS:000377320200007

Baddock, M. C., Strong, C. L., Murray, P. S., & McTainsh, G. H. (2013). Aeolian dust as a transport hazard. *Atmospheric Environment*, *71*, 7–14. WOS:000318384900002

Bahadur, R., Russell, L. M., & Prather, K. (2010). Composition and morphology of individual combustion, biomass burning, and secondary organic particle types obtained using urban and coastal ATOFMS and STXM-NEXAFS measurements. *Aerosol Science and Technology*, *44*(7), 551–562. WOS:000278589900007

Bailo, E., & Deckert, V. (2008). Tip-enhanced Raman scattering. *Chemical Society Reviews*, *37*(5), 921–930. WOS:000255444700005

Bain, A., Rafferty, A., & Preston, T. C. (2018). Determining the size and refractive index of single aerosol particles using angular light scattering and Mie resonances. *Journal of Quantitative Spectroscopy & Radiative Transfer*, *221*, 61–70. WOS:000452576900008

Baldelli, A., Trivanovic, U., Sipkens, T. A., & Rogak, S. N. (2020). On determining soot maturity: A review of the role of microscopy-and spectroscopy-based techniques. *Chemosphere*, *252*, 15. WOS:000534377000064

Barkay, Z., Teller, A., Ganor, E., Levin, Z., & Shapira, Y. (2005). Atomic force and scanning electron microscopy of atmospheric particles. *Microscopy Research and Technique*, *68*(2), 107–114. WOS:000233026000008

Barone, T. L., & Zhu, Y. F. (2008). The morphology of ultrafine particles on and near major freeways. *Atmospheric Environment*, *42*(28), 6749–6758. WOS:000260265000005

Batonneau, Y., Laureyns, J., Merlin, J. C., & Bremard, C. (2001). Self-modeling mixture analysis of Raman microspectrometric investigations of dust emitted by lead and zinc smelters. *Analytica Chimica Acta*, *446*(1-2), 23–37. WOS:000172309400004

Batonneau, Y., Sobanska, S., Laureyns, J., & Bremard, C. (2006). Confocal microprobe Raman imaging of urban tropospheric aerosol particles. *Environmental Science & Technology*, *40*(4), 1300–1306. WOS:000235478700034

Baustian, K. J., Cziczo, D. J., Wise, M. E., Pratt, K. A., Kulkarni, G., Hallar, A. G., & Tolbert, M. A. (2012). Importance of aerosol composition, mixing state, and morphology for heterogeneous ice nucleation: A combined field and laboratory approach. *Journal of Geophysical Research-Atmospheres*, *117*, 13. WOS:000302237100003. doi:10.1029/2011jd016784

Baustian, K. J., Wise, M. E., & Tolbert, M. A. (2010). Depositional ice nucleation on solid ammonium sulfate and glutaric acid particles. *Atmospheric Chemistry and Physics*, *10*(5), 2307–2317. WOS:000275505500015

Bondy, A. L., Bonanno, D., Moffet, R. C., Wang, B. B., Laskin, A., & Ault, A. P. (2018). The diverse chemical mixing state of aerosol particles in the southeastern United States. *Atmospheric Chemistry and Physics*, *18*(16), 12595–12612. WOS:000443195200003

Bondy, A. L., Craig, R. L., Zhang, Z. F., Gold, A., Surratt, J. D., & Ault, A. P. (2018). Isoprene-derived organosulfates: Vibrational mode analysis by Raman spectroscopy, acidity-dependent spectral modes, and observation in individual atmospheric particles. *Journal of Physical Chemistry A*, *122*(1), 303–315. WOS:000422813700033

Bondy, A. L., Kirpes, R. M., Merzel, R. L., Pratt, K. A., Holl, M. M. B., & Ault, A. P. (2017). Atomic force microscopy-infrared spectroscopy of individual atmospheric aerosol particles: Subdiffraction limit vibrational spectroscopy and morphological analysis. *Analytical Chemistry*, *89*(17), 8594–8598. WOS:000410014900004

Boucher, O., D. Randall, P. Artaxo, C. Bretherton, G. Feingold, P. Forster, V.-M. Kerminen, Y. Kondo, H. Liao, U. Lohmann, P. Rasch, S.K. Satheesh, S. Sherwood, B. Stevens and X.Y. Zhang, 2013: Clouds and Aerosols. In: Climate Change 2013: The Physical Science Basis. Contribution of Working Group I to the Fifth Assessment Report of the Intergovernmental Panel on Climate Change [Stocker, T.F., D. Qin, G.-K. Plattner, M. Tignor, S.K. Allen, J. Boschung, A. Nauels, Y. Xia, V. Bex and P.M. Midgley (eds.)]. Cambridge University Press, Cambridge, United Kingdom and New York, NY, USA.

Bradley, J. P., Goodman, P., Chan, I. Y. T., & Buseck, P. R. (1981). Structure and evolution of fugitive particles from a copper smelter. *Environmental Science & Technology*, *15*(10), 1208–1212. WOS:A1981MJ20900017

Braun, A. (2005). Carbon speciation in airborne particulate matter with C (1s) NEXAFS spectroscopy. *Journal of Environmental Monitoring*, *7*(11), 1059–1065. WOS:000232846300021

Braun, A., Huggins, F. E., Shah, N., Chen, Y., Wirick, S., Mun, S. B., et al. (2005). Advantages of soft X-ray absorption over TEM-EELS for solid carbon studies—A comparative study on diesel soot with EELS and NEXAFS. *Carbon*, *43*(1), 117–124. WOS:000225591500014

Braun, A., Shah, N., Huggins, F. E., Huffman, G. P., Wirick, S., Jacobsen, C., et al. (2004). A study of diesel PM with X-ray microspectroscopy. *Fuel*, *83*(7-8), 997–1000. WOS:000220387300026

Brooker, P. (2010). Fear in a handful of dust: Aviation and the Icelandic volcano. *Significance*, *7*(3), 112–115. https://rss.onlinelibrary.wiley.com/doi/abs/10.1111/j.1740-9713.2010.00436.x

Bruzewicz, D. A., Checco, A., Ocko, B. M., Lewis, E. R., McGraw, R. L., & Schwartz, S. E. (2011). Reversible uptake of water on NaCl nanoparticles at relative humidity below deliquescence point observed by noncontact environmental atomic force microscopy. *Journal of Chemical Physics*, *134*(4), 10. WOS:000286897600105

Buajarern, J., Mitchem, L., & Reid, J. P. (2007a). Characterizing the formation of organic layers on the surface of inorganic/aqueous aerosols by Raman spectroscopy. *Journal of Physical Chemistry A*, *111*(46), 11852–11859. WOS:000250967800016

Buajarern, J., Mitchem, L., & Reid, J. P. (2007b). Manipulation and characterization of aqueous sodium dodecyl sulfate/sodium chloride aerosol particles. *Journal of Physical Chemistry A*, *111*(50), 13038–13045. WOS:000251615700022

Buehler, M. F., Allen, T. M., & Davis, E. J. (1991). Microparticle Raman-spectroscopy of multicomponent aerosols. *Journal of Colloid and Interface Science*, *146*(1), 79–89. WOS:A1991GE46800007

Buseck, P. R., Jacob, D. J., Posfai, M., Li, J., & Anderson, J. R. (2000). Minerals in the air: An environmental perspective. *International Geology Review*, *42*(7), 577–593. WOS:000088666700001

Buseck, P. R., & Posfai, M. (1999). Airborne minerals and related aerosol particles: Effects on climate and the environment. *Proceedings of the National Academy of Sciences of the United States of America*, *96*, 3372–3379.

Byers, R. L., Davis, J. W., White, E. W., & McMillan, R. E. (1971). Computerized method for size characterization of atmospheric aerosols by scanning electron microscope. *Environmental Science & Technology*, *5*(6), 517. WOS:A1971J475000013

Bzdek, B. R., Pennington, M. R., & Johnston, M. V. (2012). Single particle chemical analysis of ambient ultrafine aerosol: A review. *Journal of Aerosol Science*, *52*, 109–120. WOS:000307796800010

Campion, A., & Kambhampati, P. (1998). Surface-enhanced Raman scattering. *Chemical Society Reviews*, *27*(4), 241–250. WOS:000075103200002

Canagaratna, M. R., Jayne, J. T., Jimenez, J. L., Allan, J. D., Alfarra, M. R., Zhang, Q., et al. (2007). Chemical and microphysical characterization of ambient aerosols with the aerodyne aerosol mass spectrometer. *Mass Spectrometry Reviews*, *26*(2), 185–222. WOS:000244352000005

Casuccio, G. S., Janocko, P. B., Lee, R. J., Kelly, J. F., Dattner, S. L., & Mgebroff, J. S. (1983). The use of computer-controlled scanning electron-microscopy in environmental studies. *Journal of the Air Pollution Control Association*, *33*(10), 937–943.

Catelani, T., Pratesi, G., & Zoppi, M. (2014). Raman characterization of ambient airborne soot and associated mineral phases. *Aerosol Science and Technology*, *48*(1), 13–21. WOS:000328245400002

Chakraborty, P., Prithiviraj, B., Selvaraj, S., & Kumar, B. (2016). Polychlorinated biphenyls in settled dust from informal electronic waste recycling workshops and nearby highways in urban centers and suburban industrial roadsides of Chennai city, India: Levels, congener profiles and exposure assessment. *Science of the Total Environment*, *573*, 1413–1421. WOS:000390071000136

Chan, C. K., Flagan, R. C., & Seinfeld, J. H. (1991). Resonance structures in elastic and raman-scattering from microspheres. *Applied Optics*, *30*(4), 459–467. WOS:A1991EW64400024

Chan, M. N., Lee, A. K. Y., & Chan, C. K. (2006). Responses of ammonium sulfate particles coated with glutaric acid to cyclic changes in relative humidity: Hygroscopicity and Raman characterization. *Environmental Science & Technology*, *40*(22), 6983–6989. WOS:000241941700021

Chen, H. H., Grassian, V. H., Saraf, L. V., & Laskin, A. (2013). Chemical imaging analysis of environmental particles using the focused ion beam/scanning electron microscopy technique: Microanalysis insights into atmospheric chemistry of fly ash. *Analyst*, *138*(2), 451–460. WOS:000312074300010

Cheng, Z. Z., Morgenstern, M., Zhang, B., Fraund, M., Lata, N. N., Brimberry, R., et al. (2022). Particle phase-state variability in the North Atlantic free troposphere during summertime is determined by atmospheric transport patterns and sources. *Atmospheric Chemistry and Physics*, *22*(13), 9033–9057. WOS:000823511000001

Chi, J. W., Li, W. J., Zhang, D. Z., Zhang, J. C., Lin, Y. T., Shen, X. J., et al. (2015). Sea salt aerosols as a reactive surface for inorganic and organic acidic gases in the Arctic troposphere. *Atmospheric Chemistry and Physics*, *15*(19), 11341–11353. WOS:000362971000026

China, S., Wang, B. B., Weis, J., Rizzo, L., Brito, J., Cirino, G. G., et al. (2016). Rupturing of biological spores as a source of secondary particles in Amazonia. *Environmental Science & Technology*, *50*(22), 12179–12186. WOS:000388155000014

Choel, M., Deboudt, K., & Flament, P. (2007). Evaluation of quantitative procedures for X-ray microanalysis of environmental particles. *Microscopy Research and Technique*, *70*(11), 996–1002. WOS:000251074800010

Choel, M., Deboudt, K., & Flament, P. (2010). Development of time-resolved description of aerosol properties at the particle scale during an episode of industrial pollution plume. *Water Air and Soil Pollution*, *209*(1-4), 93–107. WOS:000277450300008

Choel, M., Deboudt, K., Flament, P., Lecornet, G., Perdrix, E., & Sobanska, S. (2006). Fast evolution of tropospheric Pb- and Zn-rich particles in the vicinity of a lead smelter. *Atmospheric Environment*, *40*(24), 4439–4449. WOS: 000239257200001

Choel, M., Deboudt, K., Osan, J., Flament, P., & Van Grieken, R. (2005). Quantitative determination of low-Z elements in single atmospheric particles on boron substrates by automated scanning electron microscopy—Energy-dispersive X-ray spectrometry. *Analytical Chemistry*, *77*(17), 5686–5692. WOS:000231652300046

Chou, C., Formenti, P., Maille, M., Ausset, P., Helas, G., Harrison, M., & Osborne, S. (2008). Size distribution, shape, and composition of mineral dust aerosols collected during the African monsoon multidisciplinary analysis special observation period 0: Dust and biomass-burning experiment field campaign in Niger, January 2006. *Journal of Geophysical Research-Atmospheres*, *113*(D17), 17. WOS:000259000700004. doi:10.1029/2008jd009897

Chu, Y. X., Evoy, E., Kamal, S., Song, Y. C., Reid, J. P., Chan, C. K., & Bertram, A. K. (2018). Viscosity of erythritol and erythritol-water particles as a function of water activity: New results and an intercomparison of techniques for measuring the viscosity of particles. *Atmospheric Measurement Techniques*, *11*(8), 4809–4822. WOS:000441888400001

Ciobanu, V. G., Marcolli, C., Krieger, U. K., Weers, U., & Peter, T. (2009). Liquid-liquid phase separation in mixed organic/inorganic aerosol particles. *Journal of Physical Chemistry A*, *113*(41), 10966–10978. WOS:000270538800013

Colberg, C. A., Krieger, U. K., & Peter, T. (2004). Morphological investigations of single levitated H2SO4/NH3/H2O aerosol particles during deliquescence/efflorescence experiments. *Journal of Physical Chemistry A*, *108*(14), 2700–2709. WOS: 000220640000017

Cong, Z. Y., Kang, S. C., & Qin, D. H. (2009). Seasonal features of aerosol particles recorded in snow from Mt. Qomolangma (Everest) and their environmental implications. *Journal of Environmental Sciences*, *21*(7), 914–919. WOS:000267908700008

Conny, J. M. (2013). Internal composition of atmospheric dust particles from focused ion-beam scanning electron microscopy. *Environmental Science & Technology*, *47*(15), 8575–8581. WOS:000323013400061

Conny, J. M., Collins, S. M., & Herzing, A. A. (2014). Qualitative multiplatform microanalysis of individual heterogeneous atmospheric particles from high-volume air samples. *Analytical Chemistry*, *86*(19), 9709–9716. WOS:000343017100054

Conny, J. M., & Ortiz-Montalvo, D. L. (2017). Effect of heterogeneity and shape on optical properties of urban dust based on three-dimensional modeling of individual particles. *Journal of Geophysical Research-Atmospheres*, *122*(18), 9816–9842. WOS:000416388000017. doi:10.1002/2017jd026488

Conny, J. M., Willis, R. D., & Ortiz-Montalvo, D. L. (2019). Analysis and optical modeling of individual heterogeneous Asian dust particles collected at Mauna Loa observatory. *Journal of Geophysical Research-Atmospheres*, *124*(5), 2702–2723. WOS:000462139800019. doi:10.1029/2018jd029387

Conny, J. M., Willis, R. D., & Ortiz-Montalvo, D. L. (2020). Optical modeling of single Asian dust and marine air particles: A comparison with geometric particle shapes for remote sensing. *Journal of Quantitative Spectroscopy & Radiative Transfer*, *254*, 107197.

Coz, E., Gomez-Moreno, F. J., Pujadas, M., Casuccio, G. S., Lersch, T. L., & Artinano, B. (2009). Individual particle characteristics of North African dust under different long-range transport scenarios. *Atmospheric Environment*, *43*(11), 1850–1863. WOS:000264349300003

Craig, R. L., Bondy, A. L., & Ault, A. P. (2015). Surface enhanced Raman spectroscopy enables observations of previously undetectable secondary organic aerosol components at the individual particle level. *Analytical Chemistry*, *87*(15), 7510–7514. WOS:000359277900003

Craig, R. L., Bondy, A. L., & Ault, A. P. (2017). Computer-controlled Raman microspectroscopy (CC-Raman): A method for the rapid characterization of individual atmospheric aerosol particles. *Aerosol Science and Technology*, *51*(9), 1099–1112. WOS:000416683600010

Cui, L. L., Li, R., Zhang, Y. C., Meng, Y., Fu, H. B., & Chen, J. M. (2018). An observational study of nitrous acid (HONO) in Shanghai, China: The aerosol impact on HONO formation during the haze episodes. *Science of the Total Environment*, *630*, 1057–1070. WOS:000432467700104

Dai, Q., Hu, J., & Salmeron, M. (1997). Adsorption of water on NaCl (100) surfaces: Role of atomic steps. *Journal of Physical Chemistry B*, *101*(11), 1994–1998. WOS:A1997WM88600010

Dappe, V., Uzu, G., Schreck, E., Wu, L., Li, X., Dumat, C., et al. (2018). Single-particle analysis of industrial emissions brings new insights for health risk assessment of PM. *Atmospheric Pollution Research*, *9*(4), 697–704. WOS:000437044600011

Davies, J. F., & Wilson, K. R. (2016). Raman spectroscopy of isotopic water diffusion in ultraviscous, glassy, and gel states in aerosol by use of optical tweezers. *Analytical Chemistry*, *88*(4), 2361–2366. WOS:000370454000054

Davis, E. J. (1997). A history of single aerosol particle levitation. *Aerosol Science and Technology*, *26*(3), 212–254. WOS: A1997WK86800003

Day, D. A., Takahama, S., Gilardoni, S., & Russell, L. M. (2009). Organic composition of single and submicron particles in different regions of western North America and the eastern Pacific during INTEX-B 2006. *Atmospheric Chemistry and Physics*, *9*(15), 5433–5446. WOS:000268876600004

Dazzi, A., & Prater, C. B. (2017). AFM-IR: Technology and applications in nanoscale infrared spectroscopy and chemical imaging. *Chemical Reviews*, *117*(7), 5146–5173. WOS:000399354300009

Dazzi, A., Prater, C. B., Hu, Q. C., Chase, D. B., Rabolt, J. F., & Marcott, C. (2012). AFM-IR: Combining atomic force microscopy and infrared spectroscopy for nanoscale chemical characterization. *Applied Spectroscopy*, *66*(12), 1365–1384. WOS:000312236500003

Deboudt, K., Flament, P., Choel, M., Gloter, A., Sobanska, S., & Colliex, C. (2010). Mixing state of aerosols and direct observation of carbonaceous and marine coatings on African dust by individual particle analysis. *Journal of Geophysical Research-Atmospheres*, *115*, 14. WOS:000285463300002. doi:10.1029/2010jd013921

Deboudt, K., Gloter, A., Mussi, A., & Flament, P. (2012). Red-ox speciation and mixing state of iron in individual African dust particles. *Journal of Geophysical Research-Atmospheres*, *117*, 13. WOS:000305628800001. doi:10.1029/2011jd017298

Deckert, V., Deckert-Gaudig, T., Diegel, M., Gotz, I., Langeluddecke, L., Schneidewind, H., et al. (2015). Spatial resolution in Raman spectroscopy. *Faraday Discussions*, *177*, 9–20. WOS:000353034300001

Delhaye, M., & Dhamelincourt, P. (1975). Raman microprobe and microscope with laser excitation. *Journal of Raman Spectroscopy*, *3*(1), 33–43. WOS:A1975AA63800004

DelMelle, P., Lambert, M., Dufrene, Y., Gerin, P., & Oskarsson, N. (2007). Gas/aerosol-ash interaction in volcanic plumes: New insights from surface analyses of fine ash particles. *Earth and Planetary Science Letters*, *259*(1–2), 159–170. WOS:000248637700014

Deng, C. H., Brooks, S. D., Vidaurre, G., & Thornton, D. C. O. (2014). Using Raman microspectroscopy to determine chemical composition and mixing state of airborne marine aerosols over the Pacific Ocean. *Aerosol Science and Technology*, *48*(2), 193–206. WOS:000328877600009

Dennis-Smither, B. J., Marshall, F. H., Miles, R. E. H., Preston, T. C., & Reid, J. P. (2014). Volatility and oxidative aging of aqueous maleic acid aerosol droplets and the dependence on relative humidity. *Journal of Physical Chemistry A*, *118*(30), 5680–5691. WOS:000339930000011

Dockery, D. W., Pope, C. A., Xu, X. P., Spengler, J. D., Ware, J. H., Fay, M. E., et al. (1993). An association between air-pollution and mortality in 6 United-States cities. *New England Journal of Medicine*, *329*(24), 1753–1759. WOS: A1993MK09600001

Dong, Z. W., Kang, S. C., Qin, D. H., Shao, Y. P., Ulbrich, S., & Qin, X. (2018). Variability in individual particle structure and mixing states between the glacier-snowpack and atmosphere in the northeastern Tibetan Plateau. *The Cryosphere*, *12*(12), 3877–3890. WOS:000453078800001

Dong, Z. W., Qin, D. H., Kang, S. C., Ren, J. W., Chen, J. Z., Cui, X. Q., et al. (2014). Physicochemical characteristics and sources of atmospheric dust deposition in snow packs on the glaciers of western Qilian Mountains, China. *Tellus B: Chemical and Physical Meteorology*, *66*, 15. WOS:000330117600001

Doughty, D. C., & Hill, S. C. (2017). Automated aerosol Raman spectrometer for semi-continuous sampling of atmospheric aerosol. *Journal of Quantitative Spectroscopy & Radiative Transfer*, *188*, 103–117. WOS:000392566300013

Ebert, M., Inerle-Hof, M., & Weinbruch, S. (2002). Environmental scanning electron microscopy as a new technique to determine the hygroscopic behaviour of individual aerosol particles. *Atmospheric Environment*, *36*(39–40), 5909–5916. WOS:000179920200004

Ebert, M., Weinbruch, S., Hoffmann, P., & Ortner, H. M. (2004). The chemical composition and complex refractive index of rural and urban influenced aerosols determined by individual particle analysis. *Atmospheric Environment*, *38*(38), 6531–6545. WOS:000225269900009

Ellis, A., Edwards, R., Saunders, M., Chakrabarty, R. K., Subramanian, R., Timms, N. E., et al. (2016). Individual particle morphology, coatings, and impurities of black carbon aerosols in Antarctic ice and tropical rainfall. *Geophysical Research Letters*, *43*(22), 11875–11883. WOS: 000393208100046. doi:10.1002/2016gl071042

Elmes, M., & Gasparon, M. (2017). Sampling and single particle analysis for the chemical characterisation of fine atmospheric particulates: A review. *Journal of Environmental Management*, *202*, 137–150. WOS:000408784000014

Eom, H. J., Gupta, D., Cho, H. R., Hwang, H. J., Do Hur, S., Gim, Y., & Ro, C. U. (2016). Single-particle investigation of summertime and wintertime Antarctic Sea spray aerosols using low-Z particle EPMA, Raman microspectrometry, and ATR-FTIR imaging techniques. *Atmospheric Chemistry and Physics*, *16*(21), 13823–13836. WOS:000387453600001

Eom, H. J., Jung, H. J., Sobanska, S., Chung, S. G., Son, Y. S., Kim, J. C., et al. (2013). Iron speciation of airborne subway particles by the combined use of energy dispersive electron probe X-ray microanalysis and Raman microspectrometry. *Analytical Chemistry*, *85*(21), 10424–10431. WOS:000326711400064

Ervik, T. K., Benker, N., Weinbruch, S., Skogstad, A., Thomassen, Y., Ellingsen, D. G., & Berlinger, B. (2018). Phase identification of individual crystalline particles by combining EDX and EBSD: Application to workplace aerosols. *Analytical and Bioanalytical Chemistry*, *410*(11), 2711–2721. WOS:000429327900006

Estillore, A. D., Morris, H. S., Or, V. W., Lee, H. D., Alves, M. R., Marciano, M. A., et al. (2017). Linking hygroscopicity and the surface microstructure of model inorganic salts, simple and complex carbohydrates, and authentic sea spray aerosol particles. *Physical Chemistry Chemical Physics*, *19*(31), 21101–21111. WOS:000407763700080

Falgayrac, G., Sobanska, S., & Bremard, C. (2012). Particle-particle chemistry between micrometer-sized PbSO4 and CaCO3 particles in turbulent flow initiated by liquid water. *Journal of Physical Chemistry A*, *116*(27), 7386–7396. WOS:000306264100016

Falkovich, A. H., Schkolnik, G., Ganor, E., & Rudich, Y. (2004). Adsorption of organic compounds pertinent to urban environments onto mineral dust particles. *Journal of Geophysical Research-Atmospheres*, *109*(D2), 19. WOS:000188866900006. doi:10.1029/2003jd003919

Flament, P., Mattielli, N., Aimoz, L., Choel, M., Deboudt, K., de Jong, J., et al. (2008). Iron isotopic fractionation in industrial emissions and urban aerosols. *Chemosphere*, *73*(11), 1793–1798. WOS:000261600000014

Fletcher, R. A., Ritchie, N. W. M., Anderson, I. M., & Small, J. A. (2011). Analysis of individual collected particles. In P. A. Baron & K. Willeke (Eds.), *Aerosol measurement: Principles, techniques, and applications* (3rd ed.). New York: John Wiley & Sons, Inc.

Fraund, M., Bonanno, D. J., China, S., Pham, D. Q., Veghte, D., Weis, J., et al. (2020). Optical properties and composition of viscous organic particles found in the Southern Great Plains. *Atmospheric Chemistry and Physics*, *20*(19), 11593–11606. WOS:000578990000001

Fraund, M., Pham, D. Q., Bonanno, D., Harder, T. H., Wang, B. B., Brito, J., et al. (2017). Elemental mixing state of aerosol particles collected in Central Amazonia during GoAmazon2014/15. *Atmosphere*, *8*(9), 28. WOS: 000411527500018

Freedman, M. A., Baustian, K. J., Wise, M. E., & Tolbert, M. A. (2010). Characterizing the morphology of organic aerosols at ambient temperature and pressure. *Analytical Chemistry*, *82*(19), 7965–7972. WOS:000282257400016

Freney, E. J., Martin, S. T., & Buseck, P. R. (2009). Deliquescence and efflorescence of potassium salts relevant to biomass-burning aerosol particles. *Aerosol Science and Technology*, *43*(8), 799–807. WOS:000267073700006

Fromme, H., Diemer, J., Dietrich, S., Cyrys, J., Heinrich, J., Lang, W., et al. (2008). Chemical and morphological properties of particulate matter (PM10, PM2.5) in school classrooms and outdoor air. *Atmospheric Environment*, *42*(27), 6597–6605. WOS:000259931400004

Fuller, K. A., Malm, W. C., & Kreidenweis, S. M. (1999). Effects of mixing on extinction by carbonaceous particles. *Journal of Geophysical Research-Atmospheres*, *104*(D13), 15941–15954. 000081535000004. doi:10.1029/1998jd100069

Fung, K. H., Imre, D. G., & Tang, I. N. (1994). Detection limits for sulfates and nitrates in aerosol-particles by Raman-spectroscopy. *Journal of Aerosol Science*, *25*(3), 479–485. WOS:A1994NQ27900004

Fung, K. H., & Tang, I. N. (1991). Relative Raman-scattering cross-section measurements with suspended particles. *Applied Spectroscopy*, *45*(5), 734–737. WOS:A1991FV08300004

Gaffney, J. S., Marley, N. A., & Smith, K. J. (2015). Characterization of fine mode atmospheric aerosols by Raman microscopy and diffuse reflectance FTIR. *Journal of Physical Chemistry A*, *119*(19), 4524–4532. WOS:000354911000026

Germani, M. S., & Buseck, P. R. (1991). Automated scanning electron-microscopy for atmospheric particle analysis. *Analytical Chemistry*, *63*(20), 2232–2237. WOS: A1991GJ96200008

Ghorai, S., & Tivanski, A. V. (2010). Hygroscopic behavior of individual submicrometer particles studied by X-ray spectromicroscopy. *Analytical Chemistry*, *82*(22), 9289–9298. WOS: 000284080500025

Ghorai, S., Wang, B. B., Tivanski, A., & Laskin, A. (2014). Hygroscopic properties of internally mixed particles composed of NaCl and water-soluble organic acids. *Environmental Science & Technology*, *48*(4), 2234–2241. WOS:000331774100018

Gislason, S. R., Hassenkam, T., Nedel, S., Bovet, N., Eiriksdottir, E. S., Alfredsson, H. A., et al. (2011). Characterization of Eyjafjallajokull volcanic ash particles and a protocol for rapid risk assessment. *Proceedings of the National Academy of Sciences of the United States of America*, *108*(18), 7307–7312. WOS:000290203100013

Glen, A., & Brooks, S. D. (2013). A new method for measuring optical scattering properties of atmospherically relevant dusts using the cloud and aerosol spectrometer with polarization (CASPOL). *Atmospheric Chemistry and Physics*, *13*(3), 1345–1356. WOS:000315406100016

Goldstein, J., Newbury, D. E., Michael, J. R., Ritchie, N. W. M., Scott, J. H. J., & Joy, D. C. (2018). *Scanning electron microscopy and X-Ray microanalysis* (4th ed.). New York: Springer.

Gong, Z. Y., Pan, Y. L., Videen, G., & Wang, C. J. (2018). Optical trapping and manipulation of single particles in air: Principles, technical details, and applications. *Journal of Quantitative Spectroscopy & Radiative Transfer*, *214*, 94–119. WOS:000436883400009

Gong, Z. Y., Pan, Y. L., & Wang, C. J. (2016). Optical configurations for photophoretic trap of single particles in air. *Review of Scientific Instruments*, *87*(10), 10. WOS:000387661900011

Gorkowski, K., Beydoun, H., Aboff, M., Walker, J. S., Reid, J. P., & Sullivan, R. C. (2016). Advanced aerosol optical tweezers chamber design to facilitate phase-separation and equilibration timescale experiments on complex droplets. *Aerosol Science and Technology*, *50*(12), 1327–1341. WOS: 000393463100006

Gorkowski, K., Donahue, N. M., & Sullivan, R. C. (2018). Emerging investigator series: Determination of biphasic core-shell droplet properties using aerosol optical tweezers. *Environmental Science: Processes & Impacts*, *20*(11), 1512–1523. WOS:000456720000003

Gruszecki, W. I., Kulik, A. J., Janik, E., Bednarska, J., Luchowski, R., Grudzinski, W., & Dietler, G. (2015). Nanoscale resolution in infrared imaging of protein-containing lipid membranes. *Nanoscale*, *7*(35), 14659–14662. WOS:000360400700011

Guedes, A., Ribeiro, N., Ribeiro, H., Oliveira, M., Noronha, F., & Abreu, I. (2009). Comparison between urban and rural pollen of *Chenopodium alba* and characterization of adhered pollutant aerosol particles. *Journal of Aerosol Science*, *40*(1), 81–86. WOS:000263007400008

Gunsch, M. J., Schmidt, S. A., Gardner, D. J., Bondy, A. L., May, N. W., Bertrnan, S. B., et al. (2018). Particle growth in an isoprene-rich forest: Influences of urban, wildfire, and biogenic air masses. *Atmospheric Environment*, *178*, 255–264. WOS:000428104200028

Gupta, D., Eom, H. J., Cho, H. R., & Ro, C. U. (2015). Hygroscopic behavior of NaCl-MgCl2 mixture particles as nascent sea-spray aerosol surrogates and observation of efflorescence during humidification. *Atmospheric Chemistry and Physics*, *15*(19), 11273–11290. WOS:000362971000022

Gwaze, P., Annegarn, H. J., Huth, J., & Helas, G. (2007). Comparison of particle sizes determined with impactor, AFM and SEM. *Atmospheric Research*, *86*(2), 93–104. WOS:000250887100001

Hammer, S. E., Ebert, M., & Weinbruch, S. (2019). Comparison of operator- and computer-controlled scanning electron microscopy of particles from different atmospheric aerosol types. *Analytical and Bioanalytical Chemistry*, *411*(8), 1633–1645. WOS:000463594600015

Hand, J. L., Malm, W. C., Laskin, A., Day, D., Lee, T., Wang, C., et al. (2005). Optical, physical, and chemical properties of tar balls observed during the Yosemite Aerosol Characterization Study. *Journal of Geophysical Research-Atmospheres*, *110*(D21), 14. WOS:000233355400003. doi:10.1029/2004jd005728

Hand, V. L., Capes, G., Vaughan, D. J., Formenti, P., Haywood, J. M., & Coe, H. (2010). Evidence of internal mixing of African dust and biomass burning particles by individual particle analysis using electron beam techniques. *Journal of Geophysical Research-Atmospheres*, *115*, 11. WOS:000279824400001. doi:10.1029/2009jd012938

Hansen, J., Sato, M., & Ruedy, R. (1997). Radiative forcing and climate response. *Journal of Geophysical Research-Atmospheres*, *102*(D6), 6831–6864. A1997WR43300015. doi:10.1029/96jd03436

Hargreaves, G., Kwamena, N. O. A., Zhang, Y. H., Butler, J. R., Rushworth, S., Clegg, S. L., & Reid, J. P. (2010). Measurements of the equilibrium size of supersaturated aqueous sodium chloride droplets at low relative humidity using aerosol optical tweezers and an electrodynamic balance. *Journal of Physical Chemistry A*, *114*(4), 1806–1815. WOS:000273948100026

Hawkins, L. N., Baril, M. J., Sedehi, N., Galloway, M. M., De Haan, D. O., Schill, G. P., & Tolbert, M. A. (2014). Formation of semisolid, oligomerized aqueous SOA: Lab simulations of cloud processing. *Environmental Science & Technology*, *48*(4), 2273–2280. WOS:000331774100022

Hawkins, L. N., & Russell, L. (2010). Polysaccharides, proteins, and phytoplankton fragments: Four chemically distinct types of marine primary organic aerosol classified by single particle spectromicroscopy. *Advances in Meteorology*, *2010*, 14. WOS:000208403000027

Hays, M. D., Beck, L., Barfield, P., Willis, R. D., Landis, M. S., & Stevens, R. K. (2009). Physical and chemical characterization of residual oil-fired power plant emissions. *Energy & Fuels*, *23*(5–6), 2544–2551. WOS:000267416600028

Haywood, J., & Boucher, O. (2000). Estimates of the direct and indirect radiative forcing due to tropospheric aerosols: A review. *Reviews of Geophysics*, *38*(4), 513–543. 000165171200004. doi:10.1029/1999rg000078

Heaney, P. J., Vicenzi, E. P., Giannuzzi, L. A., & Livi, K. J. T. (2001). Focused ion beam milling: A method of site-specific sample extraction for microanalysis of Earth and planetary materials. *American Mineralogist*, *86*(9), 1094–1099. WOS: 000170969600017

Hitchcock, A. P., Dynes, J. J., Johansson, G., Wang, J., & Botton, G. (2008). Comparison of NEXAFS microscopy and TEM-EELS for studies of soft matter. *Micron*, *39*(3), 311–319. WOS:000254601500011

Hleis, D., Fernandez-Olmo, I., Ledoux, F., Kfoury, A., Courcot, L., Desmonts, T., & Courcot, D. (2013). Chemical profile identification of fugitive and confined particle emissions from an integrated iron and steelmaking plant. *Journal of Hazardous Materials*, *250*, 246–255. WOS:000317878400031

Hoek, G., Krishnan, R. M., Beelen, R., Peters, A., Ostro, B., Brunekreef, B., & Kaufman, J. D. (2013). Long-term air pollution exposure and cardio- respiratory mortality: A review. *Environmental Health*, *12*, 16. WOS:000320065300001

Hoffman, R. C., Laskin, A., & Finlayson-Pitts, B. J. (2004). Sodium nitrate particles: Physical and chemical properties during hydration and dehydration, and implications for aged sea salt aerosols. *Journal of Aerosol Science*, *35*(7), 869–887. WOS:000222832800006

Holt, D. B. (1992). New directions in scanning electron microscopy: cathodoluminescence microcharacterization. *Scanning Microscopy*, *6*(1), 1–21.

Hopkins, R. J., Lewis, K., Desyaterik, Y., Wang, Z., Tivanski, A. V., Arnott, W. P., et al. (2007a). Correlations between optical, chemical and physical properties of biomass burn aerosols. *Geophysical Research Letters*, *34*(18), 5. WOS:000249683700003. doi: 10.1029/2007gl030502

Hopkins, R. J., Mitchem, L., Ward, A. D., & Reid, J. P. (2004). Control and characterisation of a single aerosol droplet in a single-beam gradient-force optical trap. *Physical Chemistry Chemical Physics*, *6*(21), 4924–4927. WOS:000224749900018

Hopkins, R. J., Tivanski, A. V., Marten, B. D., & Gilles, M. K. (2007b). Chemical bonding and structure of black carbon reference materials and individual carbonaceous atmospheric aerosols. *Journal of Aerosol Science*, *38*(6), 573–591. WOS:000248775500001

Hou, C., Shao, L. Y., Hu, W., Zhang, D. Z., Zhao, C. M., Xing, J. P., et al. (2018). Characteristics and aging of traffic-derived particles in a highway tunnel at a coastal city in southern China. *Science of the Total Environment*, *619*, 1385–1393. WOS:000424144200139

Huertas, J. I., Huertas, M. E., & Solis, D. A. (2012). Characterization of airborne particles in an open pit mining region. *Science of the Total Environment*, *423*, 39–46. WOS:000303231200006

Huggins, F. E., Huffman, G. P., Linak, W. P., & Miller, C. A. (2004). Quantifying hazardous species in particulate matter derived from fossil-fuel combustion. *Environmental Science & Technology*, *38*(6), 1836–1842. WOS:000220193900036

Hyslop, N. P. (2009). Impaired visibility: The air pollution people see. *Atmospheric Environment*, *43*(1), 182–195. WOS: 000262147400018

Ikegami, M., Okada, K., Zaizen, Y., Makino, Y., Jensen, J. B., Gras, J. L., & Harjanto, H. (2001). Very high weight ratios of S/K in individual haze particles over Kalimantan during the 1997 Indonesian forest fires. *Atmospheric Environment*, *35*(25), 4237–4243. WOS:000170749200002

Giannuzzi, L., & Stevie, F. A. (Eds.) (2005). *Introduction to focused ion beams*. New York: Springer.

Ishizaka, S., Wada, T., & Kitamura, N. (2011). In situ observations of freezing processes of single micrometer-sized aqueous ammonium sulfate droplets in air. *Chemical Physics Letters*, *506*(1–3), 117–121. WOS:000288987600021

Ivleva, N. P., Messerer, A., Yang, X., Niessner, R., & Poschl, U. (2007). Raman microspectroscopic analysis of changes in the chemical structure and reactivity of soot in a diesel exhaust aftertreatment model system. *Environmental Science & Technology*, *41*(10), 3702–3707. WOS:000246371800054

Iwata, A., & Matsuki, A. (2018). Characterization of individual ice residual particles by the single droplet freezing method: A case study in the Asian dust outflow region. *Atmospheric Chemistry and Physics*, *18*(3), 1785–1804. WOS:000424327800002

Jacobson, M. Z. (2001). Strong radiative heating due to the mixing state of black carbon in atmospheric aerosols. *Nature*, *409*(6821), 695–697. 000166816400037

Jayarathne, T., Stockwell, C. E., Yokelson, R. J., Nakao, S., & Stone, E. A. (2014). Emissions of fine particle fluoride from biomass burning. *Environmental Science & Technology*, *48*(21), 12636–12644. WOS:000344449100020

Jeong, G. Y. (2008). Bulk and single-particle mineralogy of Asian dust and a comparison with its source soils. *Journal of Geophysical Research-Atmospheres*, *113*(D2), 16. WOS:000252825200003. doi:10.1029/2007gl030502

Jeong, G. Y., & Nousiainen, T. (2014). TEM analysis of the internal structures and mineralogy of Asian dust particles and the implications for optical modeling. *Atmospheric Chemistry and Physics*, *14*(14), 7233–7254. WOS:000339934900008

Johnson, K. S., Zuberi, B., Molina, L. T., Molina, M. J., Iedema, M. J., Cowin, J. P., et al. (2005). Processing of soot in an urban environment: Case study from the Mexico City metropolitan area. *Atmospheric Chemistry and Physics*, *5*, 3033–3043. WOS:000233317400001

Johnston, M. V. (2000). Sampling and analysis of individual particles by aerosol mass spectrometry. *Journal of Mass Spectrometry*, *35*(5), 585–595. WOS:000086840900001

Jordanov, N., & Zellner, R. (2006). Investigations of the hygroscopic properties of ammonium sulfate and mixed ammonium sulfate and glutaric acid micro droplets by means of optical levitation and Raman spectroscopy. *Physical Chemistry Chemical Physics*, *8*(23), 2759–2764. WOS:000238130000013

Kaegi, R., & Gasser, P. (2006). Application of the focused ion beam technique in aerosol science: Detailed investigation of selected, airborne particles. *Journal of Microscopy-Oxford*, *224*, 140–145. WOS:000242863700003

Kahnert, M., Nousiainen, T., & Lindqvist, H. (2014). Review: Model particles in atmospheric optics. *Journal of Quantitative Spectroscopy & Radiative Transfer*, *146*, 41–58. WOS: 000339697300004

Kamilli, K. A., Ofner, J., Krause, T., Sattler, T., Schmitt-Kopplin, P., Eitenberger, E., et al. (2016). How salt lakes affect atmospheric new particle formation: A case study in Western Australia. *Science of the Total Environment*, *573*, 985–995. WOS:000390071000095

Kandler, K., Benker, N., Bundke, U., Cuevas, E., Ebert, M., Knippertz, P., et al. (2007). Chemical composition and complex refractive index of Saharan Mineral Dust at Izana, Tenerife (Spain) derived by electron microscopy. *Atmospheric Environment*, *41*(37), 8058–8074. WOS:000251829600006

Kandler, K., Schutz, L., Deutscher, C., Ebert, M., Hofmann, H., Jackel, S., et al. (2009). Size distribution, mass concentration, chemical and mineralogical composition and derived optical parameters of the boundary layer aerosol at Tinfou, Morocco, during SAMUM 2006. *Tellus B: Chemical and Physical Meteorology*, *61*(1), 32–50. WOS:000262514400004

Katrinak, K. A., Rez, P., & Buseck, P. R. (1992). Structural variations in individual carbonaceous particles from an URBAN aerosol. *Environmental Science & Technology*, *26*(10), 1967–1976. WOS:A1992JR07400024

Katrinak, K. A., Rez, P., Perkes, P. R., & Buseck, P. R. (1993). Fractal geometry of carbonaceous aggregates from an urban aerosol. *Environmental Science and Technology*, *27*, 539–547.

Kim, K. H., Kabir, E., & Kabir, S. (2015). A review on the human health impact of airborne particulate matter. *Environment International*, *74*, 136–143. WOS:000346681700017

Kirpes, R. M., Bondy, A. L., Bonanno, D., Moffet, R. C., Wang, B. B., Laskin, A., et al. (2018). Secondary sulfate is internally mixed with sea spray aerosol and organic aerosol in the winter Arctic. *Atmospheric Chemistry and Physics*, *18*(6), 3937–3949. WOS:000427963000001

Kirz, J. (1974). Phase zone plates for X-rays and extreme UV. *Journal of the Optical Society of America*, *64*(3), 301–309. WOS:A1974S532800007

Kneipp, K., Kneipp, H., Itzkan, I., Dasari, R. R., & Feld, M. S. (2002). Surface-enhanced Raman scattering and biophysics. *Journal of Physics: Condensed Matter*, *14*(18), R597–R624. WOS:000175816300003

Knopf, D. A., Koop, T., Luo, B. P., Weers, U. G., & Peter, T. (2002). Homogeneous nucleation of NAD and NAT in liquid stratospheric aerosols: Insufficient to explain denitrification. *Atmospheric Chemistry and Physics*, *2*, 207–214. WOS: 000177879500001

Knopf, D. A., Luo, B. P., Krieger, U. K., & Koop, T. (2003). Thermodynamic dissociation constant of the bisulfate ion from Raman and ion interaction modeling studies of aqueous sulfuric acid at low temperatures. *Journal of Physical Chemistry A*, *107*(21), 4322–4332. WOS:000183065900017

Knopf, D. A., Wang, B., Laskin, A., Moffet, R. C., & Gilles, M. K. (2010). Heterogeneous nucleation of ice on anthropogenic organic particles collected in Mexico City. *Geophysical Research Letters*, *37*, 5. WOS:000278453200004. doi:10.1029/2010gl043362

Knopf, D. A., Wang, P. W., Wong, B. Y., Tomlin, J. M., Veghte, D. P., Lata, N. N., et al. (2023). Physicochemical characterization of free troposphere and marine boundary layer ice-nucleating particles collected by aircraft in the eastern North Atlantic. *Atmospheric Chemistry and Physics*, *23*(15), 8659–8681. WOS:001044291200001

Kohli, R. K., Davis, R. D., & Davies, J. F. (2023). Tutorial: Electrodynamic balance methods for single particle levitation and the physicochemical analysis of aerosol. *Journal of Aerosol Science*, *174*, 28. WOS:001079082900001

Kollensperger, G., Friedbacher, G., Grasserbauer, M., & Dorffner, L. (1997). Investigation of aerosol particles by atomic force microscopy. *Fresenius Journal of Analytical Chemistry*, *358*(1–2), 268–273. WOS:A1997XD21300070

Kollensperger, G., Friedbacher, G., Kotzick, R., Niessner, R., & Grasserbauer, M. (1999). In-situ atomic force microscopy investigation of aerosols exposed to different humidities. *Fresenius Journal of Analytical Chemistry*, *364*(4), 296–304. WOS:000081129800002

Krueger, B. J., Grassian, V. H., Cowin, J. P., & Laskin, A. (2004). Heterogeneous chemistry of individual mineral dust particles from different dust source regions: The importance of particle mineralogy. *Atmospheric Environment*, *38*(36), 6253–6261. WOS:000225010500020

Krueger, B. J., Grassian, V. H., Laskin, A., & Cowin, J. P. (2003). The transformation of solid atmospheric particles into liquid droplets through heterogeneous chemistry: Laboratory insights into the processing of calcium containing mineral dust aerosol in the troposphere. *Geophysical Research Letters*, *30*(3), 4. WOS:000181935400004. doi:10.1029/2002gl016563

Krueger, B. J., Ross, J. L., & Grassian, V. H. (2005). Formation of microcrystals, micropuddles, and other spatial inhomogenieties in surface reactions under ambient conditions: An atomic force microscopy study of water and nitric acid adsorption on MgO(100) and CaCO3(1014). *Langmuir*, *21*(19), 8793–8801. WOS:000231789800037

Laskin, A., Iedema, M. J., Ichkovich, A., Graber, E. R., Taraniuk, I., & Rudich, Y. (2005). Direct observation of completely processed calcium carbonate dust particles. *Faraday Discussions*, *130*, 453–468. WOS:000231804100027

Laskin, A., Laskin, J., & Nizkorodov, S. A. (2012). Mass spectrometric approaches for chemical characterisation of atmospheric aerosols: Critical review of the most recent advances. *Environmental Chemistry*, *9*(3), 163–189. WOS:000305832900002

Laskin, A., Moffet, R. C., & Gilles, M. K. (2019). Chemical imaging of atmospheric particles. *Accounts of Chemical Research*, *52*(12), 3419–3431. WOS:000503910400013

Laskin, A., Wietsma, T. W., Krueger, B. J., & Grassian, V. H. (2005). Heterogeneous chemistry of individual mineral dust particles with nitric acid: A combined CCSEM/EDX, ESEM, and ICP-MS study. *Journal of Geophysical Research-Atmospheres*, *110*(D10), 15. WOS:000229775200002. doi:10.1029/2004jd005206

Laskina, O., Morris, H. S., Grandquist, J. R., Estillore, A. D., Stone, E. A., Grassian, V. H., & Tivanski, A. V. (2015). Substrate-deposited sea spray aerosol particles: Influence of analytical method, substrate, and storage conditions on particle size, phase, and morphology. *Environmental Science & Technology*, *49*(22), 13447–13453. WOS:000365151200041

Laskina, O., Morris, H. S., Grandquist, J. R., Qiu, Z., Stone, E. A., Tivanski, A. V., & Grassian, V. H. (2015). Size matters in the water uptake and hygroscopic growth of atmospherically relevant multicomponent aerosol particles. *Journal of Physical Chemistry A*, *119*(19), 4489–4497. WOS:000354911000022

Laucks, M. L., Roll, G., Schweiger, G., & Davis, E. J. (2000). Physical and chemical (Raman) characterization of bioaerosols—pollen. *Journal of Aerosol Science*, *31*(3), 307–319. WOS:000085609500003

Laurain, A. M. C., & Reid, J. P. (2009). Characterizing internally mixed insoluble organic inclusions in aqueous aerosol droplets and their influence on light absorption. *Journal of Physical Chemistry A*, *113*(25), 7039–7047. WOS:000267205500025

Leaitch, W. R., Macdonald, A. M., Anlauf, K. G., Liu, P. S. K., Toom-Sauntry, D., Li, S. M., et al. (2009). Evidence for Asian dust effects from aerosol plume measurements during INTEX-B 2006 near Whistler, BC. *Atmospheric Chemistry and Physics*, *9*(11), 3523–3546. WOS:000266968200004

Lee, A. K. Y., & Chan, C. K. (2007). Single particle Raman spectroscopy for investigating atmospheric heterogeneous reactions of organic aerosols. *Atmospheric Environment*, *41*(22), 4611–4621. WOS:000247769100005

Lee, H. D., Estillore, A. D., Morris, H. S., Ray, K. K., Alejandro, A., Grassian, V. H., & Tivanski, A. V. (2017). Direct surface tension measurements of individual sub-micrometer particles using atomic force microscopy. *Journal of Physical Chemistry A*, *121*(43), 8296–8305. WOS:000414622800014

Lee, P. K., Yu, S. Y., Chang, H. J., Cho, H. Y., Kang, M. J., & Chae, B. G. (2016). Lead chromate detected as a source of atmospheric Pb and Cr (VI) pollution. *Scientific Reports*, *6*, 10. WOS:000386004800001

Lehmpuhl, D. W., Ramirez-Aguilar, K. A., Michel, A. E., Rowlen, K. L., & Birks, J. W. (1999). Physical and chemical characterization of atmospheric aerosols by atomic force microscopy. *Analytical Chemistry*, *71*(2), 379–383. WOS: 000078108700013

Lew, L. J. N., Ting, M. V., & Preston, T. C. (2018). Determining the size and refractive index of homogeneous spherical aerosol particles using Mie resonance spectroscopy. *Applied Optics*, *57*(16), 4601–4609. WOS:000433959100036

Lewis, C. W., Norris, G. A., Conner, T. L., & Henry, R. C. (2003). Source apportionment of phoenix PM2.5 aerosol with the Unmix receptor model. *Journal of the Air & Waste Management Association*, *53*(3), 325–338. WOS:000181448800007

Li, J., Posfai, M., Hobbs, P. V., & Buseck, P. R. (2003). Individual aerosol particles from biomass burning in southern Africa: 2, compositions and aging of inorganic particles. *Journal of Geophysical Research-Atmospheres*, *108*(D13), 12. WOS:000182141800002. doi:10.1029/2002jd002310

Li, R., Yang, X., Fu, H. B., Hu, Q. Q., Zhang, L. W., & Chen, J. M. (2017). Characterization of typical metal particles during haze episodes in Shanghai, China. *Chemosphere*, *181*, 259–269. WOS:000402343700029

Li, W. J., & Shao, L. Y. (2009a). Observation of nitrate coatings on atmospheric mineral dust particles. *Atmospheric Chemistry and Physics*, *9*(6), 1863–1871. WOS:000264625400002

Li, W. J., & Shao, L. Y. (2009b). Transmission electron microscopy study of aerosol particles from the brown hazes in northern China. *Journal of Geophysical Research-Atmospheres*, *114*, 10. WOS:000265930000002. doi:10.1029/2008jd011285

Li, W. J., & Shao, L. Y. (2010). Mixing and water-soluble characteristics of particulate organic compounds in individual urban aerosol particles. *Journal of Geophysical Research-Atmospheres*, *115*, 9. WOS:000273899300003. doi:10.1029/2009jd012575

Li, W. J., Shao, L. Y., Geng, H., Casuccio, G. S., & Gao, J. (2023). Computer-controlled scanning electron microscope: Methodologies and application scenarios in atmospheric particle research. *Science of the Total Environment*, *885*, 13. WOS:001000566100001

Li, W. J., Shao, L. Y., Zhang, D. Z., Ro, C. U., Hu, M., Bi, X. H., et al. (2016). A review of single aerosol particle studies in the atmosphere of East Asia: Morphology, mixing state, source, and heterogeneous reactions. *Journal of Cleaner Production*, *112*, 1330–1349. WOS:000368206800008

Lieke, K. I., Kristensen, T. B., Korsholm, U. S., Sorensen, J. H., Kandler, K., Weinbruch, S., et al. (2013). Characterization of volcanic ash from the 2011 Grimsvotn eruption by means of single-particle analysis. *Atmospheric Environment*, *79*, 411–420. WOS:000325834700045

Ling, T. Y., & Chan, C. K. (2007). Formation and transformation of metastable double salts from the crystallization of mixed ammonium nitrate and ammonium sulfate particles. *Environmental Science & Technology*, *41*(23), 8077–8083. WOS:000251305500024

Ling, T. Y., & Chan, C. K. (2008). Partial crystallization and deliquescence of particles containing ammonium sulfate and dicarboxylic acids. *Journal of Geophysical Research-Atmospheres*, *113*(D14), 15. WOS:000257798300009. doi:10.1029/2008jd009779

Liu, S., Takahama, S., Russell, L. M., Gilardoni, S., & Baumgardner, D. (2009). Oxygenated organic functional groups and their sources in single and submicron organic particles in MILAGRO 2006 campaign. *Atmospheric Chemistry and Physics*, *9*(18), 6849–6863. WOS:000270131400013

Liu, Y., Cain, J. P., Wang, H., & Laskin, A. (2007). Kinetic study of heterogeneous reaction of deliquesced NaCl particles with gaseous HNO3 using particle-on-substrate stagnation flow reactor approach. *Journal of Physical Chemistry A*, *111*(40), 10026–10043. WOS:000249987100003

Liu, Y., Gibson, E. R., Cain, J. P., Wang, H., Grassian, V. H., & Laskin, A. (2008). Kinetics of heterogeneous reaction of CaCO3 particles with gaseous HNO3 over a wide range of humidity. *Journal of Physical Chemistry A*, *112*(7), 1561–1571. WOS:000253222100024

Liu, Y. C., Liu, C., Ma, J. Z., Ma, Q. X., & He, H. (2010). Structural and hygroscopic changes of soot during heterogeneous reaction with O-3. *Physical Chemistry Chemical Physics*, *12*(36), 10896–10903. WOS:000281613300032

Lohmann, U., & Feichter, J. (2005). Global indirect aerosol effects: A review. *Atmospheric Chemistry and Physics*, *5*(3), 715–737. http://www.atmos-chem-phys.net/5/715/2005

Lu, X. J., Gao, X. Y., Ma, J. B., & Zhang, Y. H. (2019). Investigation on the volatility of ammonium nitrate using optical tweezers. *Spectroscopy and Spectral Analysis*, *39*(5), 1648–1652. WOS:000469899500050

Lv, X. J., Wang, Y., Cai, C., Pang, S. F., Ma, J. B., & Zhang, Y. H. (2018). Investigation of gel formation and volatilization of acetate acid in magnesium acetate droplets by the optical tweezers. *Spectrochimica Acta Part A-Molecular and Biomolecular Spectroscopy*, *200*, 179–185. WOS:000433272500023

Mahrt, F., Alpert, P. A., Dou, J., Gronquist, P., Arroyo, P. C., Ammann, M., et al. (2020). Aging induced changes in ice nucleation activity of combustion aerosol as determined by near edge X-ray absorption fine structure (NEXAFS) spectroscopy. *Environmental Science: Processes & Impacts*, *22*(4), 895–907. WOS:000546453900020

Majestic, B. J., Anbar, A. D., & Herckes, P. (2009). Stable isotopes as a tool to apportion atmospheric iron. *Environmental Science & Technology*, *43*(12), 4327–4333. WOS:000266968500019

Manek, I., Ovchinnikov, Y. B., & Grimm, R. (1998). Generation of a hollow laser beam for atom trapping using an axicon. *Optics Communications*, *147*(1–3), 67–70. WOS: 000073399500014

Maria, S. F., Russell, L. M., Gilles, M. K., & Myneni, S. C. B. (2004). Organic aerosol growth mechanisms and their climate-forcing implications. *Science*, *306*(5703), 1921–1924. WOS:000225695600040

Martin, S. T. (2000). Phase transitions of aqueous atmospheric particles. *Chemical Reviews*, *100*(9), 3403–3453. WOS: 000089384600004

Mazzei, F., D'Alessandro, A., Lucarelli, F., Nava, S., Prati, P., Valli, G., & Vecchi, R. (2008). Characterization of particulate matter sources in an urban environment. *Science of the Total Environment*, *401*(1–3), 81–89. WOS:000257690100009

McGloin, D., Burnham, D. R., Summers, M. D., Rudd, D., Dewar, N., & Anand, S. (2008). Optical manipulation of airborne particles: Techniques and applications. *Faraday Discussions*, *137*, 335–350. WOS:000250173900021

McInnes, L. M., Covert, D. S., Quinn, P. K., & Germani, M. S. (1994). Measurements of chloride depletion and sulfur enrichment in individual sea-salt particles collected from the remote marine boundary-layer. *Journal of Geophysical Research-Atmospheres*, *99*(D4), 8257–8268. WOS:A1994NH26300018. doi:10.1029/93jd03453

McKeown, P. J., Johnston, M. V., & Murphy, D. M. (1991). Online single-particle analysis by laser desorption mass-spectrometry. *Analytical Chemistry*, *63*(18), 2069–2073. WOS:A1991GE85800033

McMurry, P. H., Litchy, M., Huang, P. F., Cai, X. P., Turpin, B. J., Dick, W. D., & Hanson, A. (1996). Elemental composition and morphology of individual particles separated by size and hygroscopicity with the TDMA. *Atmospheric Environment*, *30*(1), 101–108. WOS:A1996TK58200009

Miller, K. A., Siscovick, D. S., Sheppard, L., Shepherd, K., Sullivan, J. H., Anderson, G. L., & Kaufman, J. D. (2007). Long-term exposure to air pollution and incidence of cardiovascular events in women. *New England Journal of Medicine*, *356*(5), 447–458. WOS:000243860700005

Mishra, S. K., & Tripathi, S. N. (2008). Modeling optical properties of mineral dust over the Indian Desert. *Journal of Geophysical Research-Atmospheres*, *113*, 19. WOS:000261472100001. doi:10.1029/2008jd010048

Mitchem, L., Buajarern, J., Ward, A. D., & Reid, J. P. (2006). A strategy for characterizing the mixing state of immiscible aerosol components and the formation of multiphase aerosol particles through coagulation. *Journal of Physical Chemistry B*, *110*(28), 13700–13703. WOS:000239001700006

Mitchem, L., Hopkins, R. J., Buajarern, J., Ward, A. D., & Reid, J. P. (2006). Comparative measurements of aerosol droplet growth. *Chemical Physics Letters*, *432*(1–3), 362–366. WOS:000242520100068

Mitchem, L., & Reid, J. P. (2008). Optical manipulation and characterisation of aerosol particles using a single-beam gradient force optical trap. *Chemical Society Reviews*, *37*(4), 756–769. WOS:000254315700011

Moffet, R., Tivanski, A. V., & Gilles, M. K. (2011). Scanning X-ray transmission microscopy: Applications in atmospheric aerosol research. In R. Signorell & J. P. Reid (Eds.), *Fundamentals and applications in aerosol spectroscopy* (pp. 419–462). Boca Raton, FL: Taylor and Francis Books.

Moffet, R. C., Furutani, H., Rodel, T. C., Henn, T. R., Sprau, P. O., Laskin, A., et al. (2012). Iron speciation and mixing in single aerosol particles from the Asian continental outflow. *Journal of Geophysical Research-Atmospheres*, *117*, 12. WOS:000302528500005. doi:10.1029/2011jd016746

Moffet, R. C., Henn, T. R., Tivanski, A. V., Hopkins, R. J., Desyaterik, Y., Kilcoyne, A. L. D., et al. (2010). Microscopic characterization of carbonaceous aerosol particle aging in the outflow from Mexico City. *Atmospheric Chemistry and Physics*, *10*(3), 961–976. WOS:000274410000008

Morris, H. S., Estillore, A. D., Laskina, O., Grassian, V. H., & Tivanski, A. V. (2016). Quantifying the hygroscopic growth of individual submicrometer particles with atomic force microscopy. *Analytical Chemistry*, *88*(7), 3647–3654. WOS:000373656300033

Morris, H. S., Grassian, V. H., & Tivanski, A. V. (2015). Humidity-dependent surface tension measurements of individual inorganic and organic submicrometre liquid particles. *Chemical Science*, *6*(5), 3242–3247. WOS:000353223100071

Muinonen, K., Nousiainen, T., Lindqvist, H., Munoz, O., & Videen, G. (2009). Light scattering by Gaussian particles with internal inclusions and roughened surfaces using ray optics. *Journal of Quantitative Spectroscopy & Radiative Transfer*, *110*(14–16), 1628–1639. WOS:000268468700036

Muller, A., Rene, M., Behr, H. J., & Kronz, A. (2003). Trace elements and cathodoluminescence of igneous quartz in topaz granites from the Hub Stock (Slavkovsky Les Mts., Czech Republic). *Mineralogy and Petrology*, *79*(3–4), 167–191. WOS:000186904600002

Mund, C., & Zellner, R. (2003). Freezing nucleation of levitated single sulfuric acid/H2O micro-droplets. A combined Raman- and Mie spectroscopic study. *Journal of Molecular Structure*, *661*, 491–500. WOS:000187239400051

Najera, J. J., & Horn, A. B. (2009). Infrared spectroscopic study of the effect of oleic acid on the deliquescence behaviour of ammonium sulfate aerosol particles. *Physical Chemistry Chemical Physics*, *11*(3), 483–494. WOS:000262007000005

Nguyen, H. N., Martinsson, B. G., Wagner, J. B., Carlemalm, E., Ebert, M., Weinbruch, S., et al. (2008). Chemical composition and morphology of individual aerosol particles from a CARIBIC flight at 10 km altitude between 50 degrees N and 30 degrees S. *Journal of Geophysical Research-Atmospheres*, *113*(D23), 12. WOS:000261670100002. doi:10.1029/2008jd009956

Noble, C. A., & Prather, K. A. (2000). Real-time single particle mass spectrometry: A historical review of a quarter century of the chemical analysis of aerosols. *Mass Spectrometry Reviews*, *19*(4), 248–274. WOS:000089044800003

Noziere, B., Kaberer, M., Claeys, M., Allan, J., D'Anna, B., Decesari, S., et al. (2015). The molecular identification of organic compounds in the atmosphere: State of the art and challenges. *Chemical Reviews*, *115*(10), 3919–3983. WOS:000355383900008

O'Brien, R. E., Wang, B. B., Kelly, S. T., Lundt, N., You, Y., Bertram, A. K., et al. (2015). Liquid-liquid phase separation in aerosol particles: Imaging at the nanometer scale. *Environmental Science & Technology*, *49*(8), 4995–5002. WOS:000353610300026

Offroy, M., Moreau, M., Sobanska, S., Milanfar, P., & Duponchel, L. (2015). Pushing back the limits of Raman imaging by coupling super-resolution and chemometrics for aerosols characterization. *Scientific Reports*, *5*, 14. WOS:000358358700001

Ofner, J., Deckert-Gaudig, T., Kamilli, K. A., Held, A., Lohninger, H., Deckert, V., & Lendl, B. (2016). Tip-enhanced Raman spectroscopy of atmospherically relevant aerosol nanoparticles. *Analytical Chemistry*, *88*(19), 9766–9772. WOS:000384842200060

Ofner, J., Kamilli, K. A., Eitenberger, E., Friedbacher, G., Lendl, B., Held, A., & Lohninger, H. (2015). Chemometric analysis of multisensor hyperspectral images of precipitated atmospheric particulate matter. *Analytical Chemistry*, *87*(18), 9413–9420. WOS:000361416000044

Okada, K., Ikegami, M., Zaizen, Y., Tsutsumi, Y., Makino, Y., Jensen, J. B., & Gras, J. L. (2005). Soot particles in the free troposphere over Australia. *Atmospheric Environment*, *39*(28), 5079–5089. WOS:000231885700008

Olson, N. E., Lei, Z. Y., Craig, R. L., Zhang, Y., Chen, Y. Z., Lambe, A. T., et al. (2019). Reactive uptake of isoprene epoxydiols increases the viscosity of the Core of phase-separated aerosol particles. *ACS Earth and Space Chemistry*, *3*(8), 1402–1414. WOS:000492426500003

Olson, N. E., Xiao, Y., Lei, Z. Y., & Ault, A. P. (2020). Simultaneous optical photothermal infrared (O-PTIR) and Raman spectroscopy of submicrometer atmospheric particles. *Analytical Chemistry*, *92*(14), 9932–9939. WOS: 000554986200062

Ono, K., Mizushima, Y., Furuya, M., Kunihisa, R., Tsuchiya, N., Fukuma, T., et al. (2020). Direct measurement of adhesion force of individual aerosol particles by atomic force microscopy. *Atmosphere*, *11*(5), 14. WOS:000541801900084

Ott, E. J. E., Kucinski, T. M., Dawson, J. N., & Freedman, M. A. (2021). Use of transmission electron microscopy for analysis of aerosol particles and strategies for imaging fragile particles. *Analytical Chemistry*, *93*(33), 11347–11356. WOS:000691297400001

Otto, S., Bierwirth, E., Weinzierl, B., Kandler, K., Esselborn, M., Tesche, M., et al. (2009). Solar radiative effects of a Saharan dust plume observed during SAMUM assuming spheroidal model particles. *Tellus B: Chemical and Physical Meteorology*, *61*(1), 270–296. WOS:000262514400020

Parish, C. M., & Russell, P. E. (2007). Scanning cathodoluminescence microscopy. In P. Hawkes (Ed.), *Advances in imaging and electron physics* (Vol. *147*, pp. 1–135). Amsterdam: Elsevier.

Park, K., Kim, J. S., & Park, S. H. (2009). Measurements of hygroscopicity and volatility of atmospheric ultrafine particles during ultrafine particle formation events at urban, industrial, and coastal sites. *Environmental Science & Technology*, *43*(17), 6710–6716. WOS:000269258000046

Paul, W., & Raether, M. (1955). Das elektrische massenfilter. *Zeitschrift Fur Physik*, *140*(3), 262–273. WOS:A1955WT841 00002

Pauley, P. M., Baker, N. L., & Barker, E. H. (1996). An observational study of the "interstate 5" dust storm case. *Bulletin of the American Meteorological Society*, *77*(4), 693–720. https://journals.ametsoc.org/doi/abs/10.1175/1520-0477%281996%29077%3C0693%3AAOSOTD%3E2.0.CO%3B2

Peckhaus, A., Kiselev, A., Wagner, R., Duft, D., & Leisner, T. (2016). Temperature-dependent formation of NaCl dihydrate in levitated NaCl and sea salt aerosol particles. *Journal of Chemical Physics*, *145*(24), 12. WOS:000392174800037

Pekney, N. J., & Davidson, C. I. (2005). Determination of trace elements in ambient aerosol samples. *Analytica Chimica Acta*, *540*(2), 269–277. WOS:000229346900005

Pham, D. Q., O'Brien, R., Fraund, M., Bonanno, D., Laskina, O., Beall, C., et al. (2017). Biological impacts on carbon speciation and morphology of sea spray aerosol. *ACS Earth and Space Chemistry*, *1*(9), 551–561. WOS:000416291100003

Pope, C. A., Burnett, R. T., Thun, M. J., Calle, E. E., Krewski, D., Ito, K., & Thurston, G. D. (2002). Lung cancer, cardiopulmonary mortality, and long-term exposure to fine particulate air pollution. *JAMA: Journal of the American Medical Association*, *287*(9), 1132–1141. 000174186000028

Pope, C. A., Ezzati, M., & Dockery, D. W. (2009). Fine-particulate air pollution and life expectancy in the United States. *New England Journal of Medicine*, *360*(4), 376–386. WOS:000262519900009

Posfai, M., Axisa, D., Tompa, E., Freney, E., Bruintjes, R., & Buseck, P. R. (2013). Interactions of mineral dust with pollution and clouds: An individual-particle TEM study of atmospheric aerosol from Saudi Arabia. *Atmospheric Research*, *122*, 347–361. WOS:000315551500027

Posfai, M., & Buseck, P. R. (2010). Nature and climate effects of individual tropospheric aerosol particles. In R. Jeanloz & K. H. Freeman (Eds.), *Annual review of earth and planetary sciences* (Vol. *38*, pp. 17–43). Palo Alto: Annual Reviews.

Posfai, M., Gelencser, A., Simonics, R., Arato, K., Li, J., Hobbs, P. V., & Buseck, P. R. (2004). Atmospheric tar balls: Particles from biomass and biofuel burning. *Journal of Geophysical Research-Atmospheres*, *109*(D6), 9. WOS:000220622900003. doi:10.1029/2003jd004169

Posfai, M., Simonics, R., Li, J., Hobbs, P. V., & Buseck, P. R. (2003). Individual aerosol particles from biomass burning in southern Africa: 1. Compositions and size distributions of carbonaceous particles. *Journal of Geophysical Research-Atmospheres*, *108*(D13), 13. WOS:000182141800001. doi:10.1029/2002jd002291

Posfai, M., Xu, H. F., Anderson, J. R., & Buseck, P. R. (1998). Wet and dry sizes of atmospheric aerosol particles: An AFM-TEM study. *Geophysical Research Letters*, *25*(11), 1907–1910. WOS:000074033800038. doi:10.1029/98gl01416

Pratt, K. A., & Prather, K. A. (2012a). Mass spectrometry of atmospheric aerosols—Recent developments and applications. Part I: Off-line mass spectrometry techniques. *Mass Spectrometry Reviews*, *31*(1), 1–16. WOS:000298578100001

Pratt, K. A., & Prather, K. A. (2012b). Mass spectrometry of atmospheric aerosols—Recent developments and applications. Part II: On-line mass spectrometry techniques. *Mass Spectrometry Reviews*, *31*(1), 17–48. WOS:000298578100002

Rafferty, A., & Preston, T. C. (2018). Measuring the size and complex refractive index of an aqueous aerosol particle using electromagnetic heating and cavity-enhanced Raman scattering. *Physical Chemistry Chemical Physics*, *20*(25), 17038–17047. WOS:000436571800018

Raman, C. V., & Krishnan, K. S. (1928). A new type of secondary radiation. *Nature*, *121*, 501–502. WOS:000189300700216

Ramirez-Aguilar, K. A., Lehmpuhl, D. W., Michel, A. E., Birks, J. W., & Rowlen, K. L. (1999). Atomic force microscopy for the analysis of environmental particles. *Ultramicroscopy*, *77*(3–4), 187–194. WOS:000081252200009

Ray, K. K., Lee, H. D., Gutierrez, M. A., Chang, F. J., & Tivanski, A. V. (2019). Correlating 3D morphology, phase state, and viscoelastic properties of individual substrate-deposited particles. *Analytical Chemistry*, *91*(12), 7621–7630. WOS:000472682000018

Rindelaub, J. D., Craig, R. L., Nandy, L., Bondy, A. L., Dutcher, C. S., Shepson, P. B., & Ault, A. P. (2016). Direct measurement of pH in individual particles via Raman microspectroscopy and variation in acidity with relative humidity. *Journal of Physical Chemistry A*, *120*(6), 911–917. WOS:000370678600008

Ritchie, N. W. M. (2020). *NIST DTSA-II*. Gaithersburg, MD: National Institute of Standards and Technology. Retrieved from https://www.nist.gov/services-resources/software/nist-dtsa-ii

Rkiouak, L., Tang, M. J., Camp, J. C. J., McGregor, J., Watson, I. M., Cox, R. A., et al. (2014). Optical trapping and Raman spectroscopy of solid particles. *Physical Chemistry Chemical Physics*, *16*(23), 11426–11434. WOS:000336796800040

Ro, C.-U., Osan, J., Szaloki, I., Oh, K. Y., Kim, H., & Van Grieken, R. (2000). Determination of chemical species in individual aerosol particles using ultrathin window EPMA. *Environmental Science & Technology*, *34*(14), 3023–3030. WOS:000088156100023

Ro, C.-U., Osan, J., & van Grieken, R. (1999). Determination of low-Z elements in individual environmental particles using windowless EPMA. *Analytical Chemistry*, *71*(8), 1521–1528. WOS:000079756900007

Rocha-Lima, A., Martins, J. V., Remer, L. A., Krotkov, N. A., Tabacniks, M. H., Ben-Ami, Y., & Artaxo, P. (2014). Optical, microphysical and compositional properties of the Eyjafjallajokull volcanic ash. *Atmospheric Chemistry and Physics*, *14*(19), 10649–10661. WOS:000344164800016

Rosasco, G. J., Etz, E. S., & Cassatt, W. A. (1975). Analysis of discrete fine particles by raman-spectroscopy. *Applied Spectroscopy*, *29*(5), 396–404. WOS:A1975AQ46800005

Rosen, H., & Novakov, T. (1977). Raman-scattering and characterization of atmospheric aerosol-particles. *Nature*, *266*(5604), 708–710. WOS:A1977DC52400034

Rubasinghege, G., & Grassian, V. H. (2013). Role(s) of adsorbed water in the surface chemistry of environmental interfaces. *Chemical Communications*, *49*(30), 3071–3094. WOS: 000316274500001

Russell, L. M., Maria, S. F., & Myneni, S. C. B. (2002). Mapping organic coatings on atmospheric particles. *Geophysical Research Letters*, *29*(16), 4. WOS:000179631600006. doi:10.1029/2002gl014874

Sadezky, A., Muckenhuber, H., Grothe, H., Niessner, R., & Poschl, U. (2005). Raman micro spectroscopy of soot and related carbonaceous materials: Spectral analysis and structural information. *Carbon*, *43*(8), 1731–1742. WOS:000229986400017

Schaffer, M., Wagner, J., Schaffer, B., Schmied, M., & Mulders, H. (2007). Automated three-dimensional X-ray analysis using a dual-beam FIB. *Ultramicroscopy*, *107*(8), 587–597. WOS:000246937000004

Schill, G. P., Genareau, K., & Tolbert, M. A. (2015). Deposition and immersion-mode nucleation of ice by three distinct samples of volcanic ash. *Atmospheric Chemistry and Physics*, *15*(13), 7523–7536. WOS:000357978300023

Schill, G. P., & Tolbert, M. A. (2013). Heterogeneous ice nucleation on phase-separated organic-sulfate particles: Effect of liquid vs. glassy coatings. *Atmospheric Chemistry and Physics*, *13*(9), 4681–4695. WOS:000318941300014

Schulz, F., Commodo, M., Kaiser, K., De Falco, G., Minutolo, P., Meyer, G., et al. (2019). Insights into incipient soot formation by atomic force microscopy. *Proceedings of the Combustion Institute*, *37*(1), 885–892. WOS:000456612200091

Schwartz, R. E., Russell, L. M., Sjostedt, S. J., Vlasenko, A., Slowik, J. G., Abbatt, J. P. D., et al. (2010). Biogenic oxidized organic functional groups in aerosol particles from a mountain forest site and their similarities to laboratory chamber products. *Atmospheric Chemistry and Physics*, *10*(11), 5075–5088. WOS:000278745300005

Schweiger, G. (1991). Raman-scattering on microparticles—Size dependence. *Journal of the Optical Society of America B-Optical Physics*, *8*(8), 1770–1778. WOS:A1991GA28100032

Semeniuk, T. A., Wise, M. E., Martin, S. T., Russell, L. M., & Buseck, P. R. (2007). Hygroscopic behavior of aerosol particles from biomass fires using environmental transmission electron microscopy. *Journal of Atmospheric Chemistry*, *56*(3), 259–273. WOS:000245105900004

Siepka, D., Uzu, G., Stefaniak, E. A., & Sobanska, S. (2018). Combining Raman microspectrometry and chemometrics for determining quantitative molecular composition and mixing state of atmospheric aerosol particles. *Microchemical Journal*, *137*, 119–130. WOS:000423894800018

Skoog, D.A., Holler, F.J., Nieman, T.A. (1998). Principles of Instrumental Analysis, 5th ed., Philadelphia: Harcourt Brace College Publishers.

Small, J. A., & Michael, J. R. (2001). Phase identification of individual crystalline particles by electron backscatter diffraction. *Journal of Microscopy-Oxford*, *201*, 59–69. WOS:000167165700005

Sobanska, S., Falgayrac, G., Laureyns, J., & Bremard, C. (2006). Chemistry at level of individual aerosol particle using multivariate curve resolution of confocal Raman image. *Spectrochimica Acta Part A-Molecular and Biomolecular Spectroscopy*, *64*(5), 1102–1109. WOS:000240093100004

Sobanska, S., Falgayrac, G., Rimetz-Planchon, J., Perdrix, E., Bremard, C., & Barbillat, J. (2014). Resolving the internal structure of individual atmospheric aerosol particle by the combination of atomic force microscopy, ESEM-EDX, Raman and ToF-SIMS imaging. *Microchemical Journal*, *114*, 89–98. WOS:000333495800013

Sobanska, S., Hwang, H., Choel, M., Jung, H. J., Eom, H. J., Kim, H., et al. (2012). Investigation of the chemical mixing state of individual Asian dust particles by the combined use of electron probe X-ray microanalysis and Raman microspectrometry. *Analytical Chemistry*, *84*(7), 3145–3154. WOS:000302829800018

Song, M., Marcolli, C., Krieger, U. K., Zuend, A., & Peter, T. (2012). Liquid-liquid phase separation and morphology of internally mixed dicarboxylic acids/ammonium sulfate/water particles. *Atmospheric Chemistry and Physics*, *12*(5), 2691–2712. WOS:000301547500029

Stevie, F. A., Shane, T. C., Kahora, P. M., Hull, R., Bahnck, D., Kannan, V. C., & David, E. (1995). Applications of focused ion-beams in microelectronics production, design and development. *Surface and Interface Analysis*, *23*(2), 61–68. WOS:A1995QM24200002

Stevie, F. A., Vartuli, C. B., Giannuzzi, L. A., Shofner, T. L., Brown, S. R., Rossie, B., et al. (2001). Application of focused ion beam lift-out specimen preparation to TEM, SEM, STEM, AES and SIMS analysis. *Surface and Interface Analysis*, *31*(5), 345–351. WOS:000169014700001

Sun, Z. L., Duan, F. K., He, K. B., Du, J. J., & Zhu, L. D. (2019). Sulfate-nitrate-ammonium as double salts in PM2.5: Direct observations and implications for haze events. *Science of the Total Environment*, *647*, 204–209. WOS:000447090400022

Takahama, S., Gilardoni, S., & Russell, L. M. (2008). Single-particle oxidation state and morphology of atmospheric iron aerosols. *Journal of Geophysical Research-Atmospheres*, *113*. WOS:000261144300001. doi:10.1029/2008jd009810

Takahama, S., Gilardoni, S., Russell, L. M., & Kilcoyne, A. L. D. (2007). Classification of multiple types of organic carbon composition in atmospheric particles by scanning transmission X-ray microscopy analysis. *Atmospheric Environment*, *41*(40), 9435–9451. WOS:000252355700015

Takahama, S., Liu, S., & Russell, L. M. (2010). Coatings and clusters of carboxylic acids in carbon-containing atmospheric particles from spectromicroscopy and their implications for cloud-nucleating and optical properties. *Journal of Geophysical Research-Atmospheres*, *115*. WOS:000273506700001. doi:10.1029/2009jd012622

Tang, I. N., Fung, K. H., Imre, D. G., & Munkelwitz, H. R. (1995). Phase-transformation and metastability of hygroscopic microparticles. *Aerosol Science and Technology*, *23*(3), 443–453. WOS:A1995RW11800017

Tang, M. J., Chan, C. K., Li, Y. J., Su, H., Ma, Q. X., Wu, Z. J., et al. (2019). A review of experimental techniques for aerosol hygroscopicity studies. *Atmospheric Chemistry and Physics*, *19*(19), 12631–12686. WOS:000489847600001

Thurn, R., & Kiefer, W. (1984). Raman-microsampling technique applying optical levitation by radiation pressure. *Applied Spectroscopy*, *38*(1), 78–83. WOS:A1984SB79400017

Thurn, R., & Kiefer, W. (1985). Structural resonances observed in the Raman-spectra of optically levitated liquid droplets. *Applied Optics*, *24*(10), 1515–1519. WOS:A1985AHJ9500030

Tivanski, A. V., Hopkins, R. J., Tyliszczak, T., & Gilles, M. K. (2007). Oxygenated interface on biomass burn tar balls determined by single particle scanning transmission X-ray microscopy. *Journal of Physical Chemistry A*, *111*(25), 5448–5458. WOS:000247395300013

Tomasevic, M., Vukmirovic, Z., Rajsic, S., Tasic, M., & Stevanovic, B. (2005). Characterization of trace metal particles deposited on some deciduous tree leaves in an urban area. *Chemosphere*, *61*(6), 753–760. WOS:000233189000001

Tóth, A., Hoffer, A., Nyirő-Kósa, I., Pósfai, M., & Gelencsér, A. (2014). Atmospheric tar balls: Aged primary droplets from biomass burning? *Atmospheric Chemistry and Physics*, *14*(13), 6669–6675. https://acp.copernicus.org/articles/14/6669/2014

Trunk, M., Lubben, J. F., Popp, J., Schrader, B., & Kiefer, W. (1997). Investigation of a phase transition in a single optically levitated microdroplet by Raman-Mie scattering. *Applied Optics*, *36*(15), 3305–3309. WOS:A1997WZ08000018

Twomey, S. (1974). Pollution and the planetary albedo. *Atmospheric Environment*, *8*, 1251–1256.

Uchic, M. D., Holzer, L., Inkson, B. J., Principe, E. L., & Munroe, P. (2007). Three-dimensional microstructural characterization using focused ion beam tomography. *MRS Bulletin*, *32*(5), 408–416. WOS:000246770000014

Udisti, R., Becagli, S., Benassai, S., Castellano, E., Fattori, I., Innocenti, M., et al. (2004). Atmosphere-snow interaction by a comparison between aerosol and uppermost snow-layers composition at Dome C, East Antarctica. In J. Jacka (Ed.), *Annals of glaciology* (Vol. *39*, pp. 53–61). Cambridge: International Glaciological Society.

Ueda, S., Osada, K., Hara, K., Yabuki, M., Hashihama, F., & Kanda, J. (2018). Morphological features and mixing states of soot-containing particles in the marine boundary layer over the Indian and southern oceans. *Atmospheric Chemistry and Physics*, *18*(13), 9207–9224. WOS:000437262500002

Vehring, R., Aardahl, C. L., Schweiger, G., & Davis, E. J. (1998). The characterization of fine particles originating from an uncharged aerosol: Size dependence and detection limits for Raman analysis. *Journal of Aerosol Science*, *29*(9), 1045–1061. WOS:000075254100002

Vernooij, M. G. C., Mohr, M., Tzvetkov, G., Zelenay, V., Huthwelker, T., Kaegi, R., et al. (2009). On source identification and alteration of single diesel and wood smoke soot particles in the atmosphere; an x-ray microspectroscopy study. *Environmental Science & Technology*, *43*(14), 5339–5344. WOS:000268138000030

Vester, B. P., Ebert, M., Barnert, E. B., Schneider, J., Kandler, K., Schutz, L., & Weinbruch, S. (2007). Composition and mixing state of the urban background aerosol in the Rhein-Main area (Germany). *Atmospheric Environment*, *41*(29), 6102–6115. WOS:000249752400002

Wang, B. B., Knopf, D. A., China, S., Arey, B. W., Harder, T. H., Gilles, M. K., & Laskin, A. (2016). Direct observation of ice nucleation events on individual atmospheric particles. *Physical Chemistry Chemical Physics*, *18*(43), 29721–29731. WOS:000387773100007

Wang, L., Bai, Z., Pi, Y., Zhai, Y., Zhang, Y., Yan, Z., et al. (2018a). Luminescence characteristics of individual Beijing haze aerosol particles. *Atmospheric Environment*, *190*, 249–255. http://www.sciencedirect.com/science/article/pii/S1352231018304709

Wang, M. J., Zheng, N., Zhu, T., Shang, J., Yu, T., Song, X. J., et al. (2018b). Using X-ray computed tomography and micro-Raman spectrometry to measure individual particle surface area, volume, and morphology towards investigating atmospheric heterogeneous reactions. *Journal of Environmental Sciences-China*, *69*, 23–32. WOS:000439688100005

Wang, W. H., Shao, L. Y., Guo, M. L., Hou, C., Xing, J. P., & Wu, F. (2017). Physicochemical properties of individual airborne particles in Beijing during pollution periods. *Aerosol and Air Quality Research*, *17*(12), 3209–3219. WOS:000423472400023

Wang, Y., Ma, J. B., Zhou, Q., Pang, S. F., & Zhang, Y. H. (2015). Hygroscopicity of mixed glycerol/Mg(NO3)(2)/water droplets affected by the interaction between magnesium ions and glycerol molecules. *Journal of Physical Chemistry B*, *119*(17), 5558–5566. WOS:000353931000024

Warwick, T., Franck, K., Kortright, J. B., Meigs, G., Moronne, M., Myneni, S., et al. (1998). A scanning transmission x-ray microscope for materials science spectromicroscopy at the advanced light source. *Review of Scientific Instruments*, *69*(8), 2964–2973. WOS:000075375000022

Werner, M. L., Nico, P. S., Marcus, M. A., & Anastasio, C. (2007). Use of micro-XANES to speciate chromium in airborne fine particles in the Sacramento Valley. *Environmental Science & Technology*, *41*(14), 4919–4924. WOS:000248363100019

Williams, D. B., & Carter, C. B. (1996a). *Transmission electron microscopy* (Vol. *1* (Vol. Basics)). New York: Plenum.

Williams, D. B., & Carter, C. B. (1996b). *Transmission electron microscopy* (Vol. *3* (Vol. Imaging)). New York: Plenum.

Wise, M. E., Baustian, K. J., Koop, T., Freedman, M. A., Jensen, E. J., & Tolbert, M. A. (2012). Depositional ice nucleation onto crystalline hydrated NaCl particles: A new mechanism for ice formation in the troposphere. *Atmospheric Chemistry and Physics*, *12*(2), 1121–1134. WOS:000300321500033

Wise, M. E., Baustian, K. J., & Tolbert, M. A. (2010). Internally mixed sulfate and organic particles as potential ice nuclei in the tropical tropopause region. *Proceedings of the National Academy of Sciences of the United States of America*, *107*(15), 6693–6698. WOS:000276642100026

Wittmaack, K., & Strigl, N. (2005). Novel approach to identifying supersaturated metastable ambient aerosol particles. *Environmental Science & Technology*, *39*(21), 8177–8184. WOS:000233078000018

Worobiec, A., Potgieter-Vermaak, S., Brooker, A., Darchuk, L., Stefaniak, E., & Van Grieken, R. (2010). Interfaced SEM/EDX and micro-Raman spectrometry for characterisation of heterogeneous environmental particles—fundamental and practical challenges. *Microchemical Journal*, *94*(1), 65–72. WOS:000273238800011

Wu, F. M., Wang, X. W., Jing, B., Zhang, Y. H., & Ge, M. F. (2018). Liquid-liquid phase separation in internally mixed magnesium sulfate/glutaric acid particles. *Atmospheric Environment*, *178*, 286–292. WOS:000428104200031

Wu, G. J., Zhang, X. L., Zhang, C. L., & Xu, T. L. (2016). Mineralogical and morphological properties of individual dust particles in ice cores from the Tibetan Plateau. *Journal of Glaciology*, *62*(231), 46–53. WOS:000376838400005

Xiong, C., & Friedlander, S. K. (2001). Morphological properties of atmospheric aerosol aggregates. *Proceedings of the National Academy of Sciences of the United States of America*, *98*(21), 11851–11856. WOS:000171558900008

Xu, L., Okada, K., Iwasaka, Y., Hara, K., Okuhara, Y., Tsutsumi, Y., & Shi, G. Y. (2001). The composition of individual aerosol particle in the troposphere and stratosphere over Xianghe (39.45 degrees N, 117.0 degrees E), China. *Atmospheric Environment*, *35*(18), 3145–3153. WOS:000169466800008

Zelenyuk, A., & Imre, D. (2009). Beyond single particle mass spectrometry: Multidimensional characterisation of individual aerosol particles. *International Reviews in Physical Chemistry*, *28*(2), 309–358. WOS:000269343700006

Zhang, D. Z., & Iwasaka, Y. (2001). Chlorine deposition on dust particles in marine atmosphere. *Geophysical Research Letters*, *28*(18), 3613–3616. WOS:000171035000049. doi:10.1029/2001gl013333

Zhang, D. Z., Shi, G. Y., Iwasaka, Y., & Hu, M. (2000). Mixture of sulfate and nitrate in coastal atmospheric aerosols: Individual particle studies in Qingdao (36 degrees 04′N, 120 degrees 21′E), China. *Atmospheric Environment*, *34*(17), 2669–2679. WOS:000086852700001

Zhang, J., Liu, L., Xu, L., Lin, Q. H., Zhao, H. J., Wang, Z. B., et al. (2020). Exploring wintertime regional haze in northeast China: Role of coal and biomass burning. *Atmospheric Chemistry and Physics*, *20*(9), 5355–5372. WOS:000531861800001

Zhang, J. X., Aiello, D., & Aker, P. M. (1994). A spectroscopic tour through the liquid aerosol interface—implications for atmospheric chemistry. *Journal of Geophysical Research-Atmospheres*, *99*(D12), 25667–25672. WOS:A1994PY22900027. doi:10.1029/94jd02617

Zhang, Y., Chen, Y. Z., Lambe, A. T., Olson, N. E., Lei, Z. Y., Craig, R. L., et al. (2018). Effect of the aerosol-phase state on secondary organic aerosol formation from the reactive uptake of isoprene-derived epoxydiols (IEPDX). *Environmental Science & Technology Letters*, *5*(3), 167–174. WOS:000427661200007

Zhang, Y. H., & Chan, C. K. (2000). Study of contact ion pairs of supersaturated magnesium sulfate solutions using Raman scattering of levitated single droplets. *Journal of Physical Chemistry A*, *104*(40), 9191–9196. WOS:000090087100018

Zhang, Y. H., & Chan, C. K. (2002). Understanding the hygroscopic properties of supersaturated droplets of metal and ammonium sulfate solutions using Raman spectroscopy. *Journal of Physical Chemistry A*, *106*(2), 285–292. WOS:000173355900011

Zhang, Y. L., Kang, S. C., Li, C. L., Gao, T. G., Cong, Z. Y., Sprenger, M., et al. (2017). Characteristics of black carbon in snow from Laohugou No. 12 glacier on the northern Tibetan Plateau. *Science of the Total Environment*, *607*, 1237–1249. WOS:000408755300124

Zhao, D. F., Song, X. J., Zhu, T., Zhang, Z. F., Liu, Y. J., & Shang, J. (2018). Multiphase oxidation of SO2 by NO2 on CaCO3 particles. *Atmospheric Chemistry and Physics*, *18*(4), 2481–2493. WOS:000425505500008

Zhao, L., Sun, W. Q., Li, X. L., Ye, Z., Huang, J., Zhang, G., & Cai, J. J. (2017). Assessment of particulate emissions from a sinter plant in steelmaking works in China. *Environmental Monitoring and Assessment*, *189*(8), 16. WOS:000405440300003

Zhu, F. F., Yan, F. W., Zhang, Z. H., & Wang, Y. Y. (2017). PM2.5 emission behavior from laboratory-scale combustion of typical municipal solid waste components and their morphological characteristics. *Energy & Fuels*, *31*(9), 10032–10045. WOS:000411771200125

Zimmermann, F., Ebert, M., Worringen, A., Schutz, L., & Weinbruch, S. (2007). Environmental scanning electron microscopy (ESEM) as a new technique to determine the ice nucleation capability of individual atmospheric aerosol particles. *Atmospheric Environment*, *41*(37), 8219–8227. WOS:000251829600018

Zuberi, B., Johnson, K. S., Aleks, G. K., Molina, L. T., & Laskin, A. (2005). Hydrophilic properties of aged soot. *Geophysical Research Letters*, *32*(1), 4. WOS:000226510800003. doi:10.1029/2004gl021496

2

Importance of Microanalysis in Air Quality Studies

Weijun Li, Lei Liu, and Liang Xu

ABSTRACT

Various methods of individual-particle analysis have been widely applied to study the morphology, composition, and mixing state of individual aerosol particles collected in different environments. This chapter summarizes the sources of particles, regional haze formation mechanisms, and particle aging process in highly polluted air and background clean air. In particular, the advantages of microscopic methods allow us to observe complex mixtures of aerosol particles in the highly polluted air. The results substantially improve knowledge of aerosol particles in the atmosphere, including their aging process, hygroscopic, and optical properties.

2.1. INTRODUCTION OF AIR POLLUTION WORLDWIDE

Leaps in anthropogenic activities have led to massive emissions of particulate matter (PM), significantly influencing air quality, climate, and public health. Globally, vehicular traffic is the biggest contributor (25%) to fine particles ($PM_{2.5}$, particles with aerodynamic diameter $\leq 2.5\,\mu m$), followed by domestic fuel burning (20%), natural dust and salt (18%), and industrial activities (15%) (Karagulian et al., 2015). A $10\,\mu g\,m^{-3}$ rise in $PM_{2.5}$ concentration was related to a 4% elevated risk of all-cause mortality (Pope III et al., 2002). It is estimated that 4.28 million deaths were attributed to $PM_{2.5}$ in 2015, and air pollution-related mortality in Asia is the highest due to the large population and poor air quality (Lelieveld et al., 2018).

Asia is one of the largest emission source areas of air pollutants on the Earth (Figure 2.1). In particular, India and China, two populous developing countries, are suffering from severe air pollution issues (Lelieveld, 2017).

Department of Atmospheric Sciences, Zhejiang University, Hangzhou, China

Moreover, large amounts of air pollutants from Asia can be transported to downwind areas as a result of monsoons, causing a global air pollution challenge. Long-range transport of air pollutants from China may affect the Pacific Ocean, North America, and the Arctic (Seinfeld et al., 2004; Heald et al., 2006; Lin et al., 2014). In summer, surface pollutants in the Himalayan-Gangetic Plain and the Sichuan Basin can be pumped up to the upper troposphere-lower stratosphere because of orographically forced deep convection above the Tibetan Plateau, after which the pollutants are distributed globally due to the horizontal wind in the stratosphere (Lau et al., 2018; Yuan et al., 2019a).

During the first quarter of 2013, China experienced severe and persistent haze pollution, affecting more than 1.3 million km^2 and 800 million people (Huang et al., 2014b). Many scientists concentrated on understanding the sources and formation of haze episodes in China and their impacts on human health and climate (Seinfeld et al., 2004; Li et al., 2011b; Guo et al., 2014; Wang et al., 2016; Song et al., 2017; Yao et al., 2018). These results suggest that secondary aerosol species (e.g., sulfates, nitrates, and secondary organic aerosols) were the most important components of haze (Guo et al., 2014), and organic matter was

Microanalysis of Atmospheric Particles: Techniques and Applications, Geophysical Monograph 288, First Edition.
Edited by Joseph M. Conny and Peter R. Buseck.

DOI: 10.1002/9781119554318.ch2

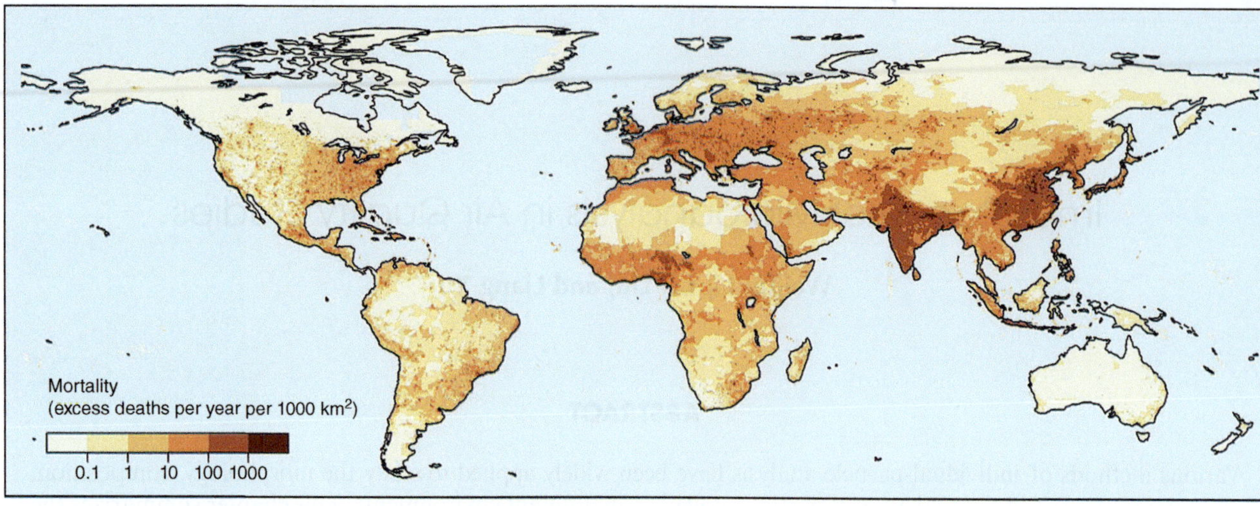

Figure 2.1 Mortality from ambient air pollution in 2015. Lelieveld et al. (2018)/with permission of Elsevier.

dominant in secondary aerosols (Huang et al., 2014b). Most of these researchers applied bulk techniques to acquire the chemical and physical properties of aerosol particles. The prominent air pollution has drawn the attention of the public, scientists, and the government. In recent years, the Chinese government introduced a series of comprehensive policies (e.g., the Air Pollution Prevention and Control Action Plan) to mitigate air pollution (Feng & Liao, 2016). Recent studies have shown that the annual air quality in China has improved significantly year by year since 2013; the decline in annual $PM_{2.5}$ concentration was 39%, 34%, and 26% in Beijing-Tianjin-Hebei, the Yangtze River Delta, and the Pearl River Delta, respectively (Wang et al., 2018).

Bulk techniques are used to analyze numerous aerosol particles simultaneously and obtain the mass concentration of different aerosol species in particles. For example, the aerosol mass spectrometer (AMS) can be used for measurement of nitrate, sulfate, and organics; ion chromatograph (IC) for soluble inorganic ions; gas chromatography-mass spectrometer (GC-MS) for organic matter; and inductively coupled plasma mass spectrometer (ICP-MS) for metallic elements. The information is essential for understanding the formation of haze particles, their source identification, and their impacts on health and climate. However, they cannot provide information about the mixing state and surface properties of aerosol particles (Prather et al., 2008; Pósfai & Buseck, 2010; Ault & Axson, 2016; Li et al., 2016a).

The mixing state of an aerosol particle refers to the distribution of different components in the particle (Prather et al., 2008). Knowledge of the mixing state can be used to infer optical and hygroscopic properties of aerosol particles and to figure out their aging and reaction mechanisms in the atmosphere (Pósfai & Buseck, 2010; Riemer et al., 2019). The aging of atmospheric particles refers to the mechanisms in the atmosphere that can change the physical and chemical properties of aerosol particles, such as coagulation, condensation, and heterogeneous reactions. Because of the limitations of bulk analyses, individual-particle analysis methods were introduced to study the mixing state and surface properties of aerosol particles.

Analytical instruments for individual-particle analysis include the scanning electron microscope (SEM), transmission electron microscope (TEM), atomic force microscope (AFM), aerosol-time-of-flight mass spectrometer/single particle aerosol mass spectrometer (ATOFMS/SPAMS), single-particle soot photometer (SP2), nanometer secondary ion mass spectrometer (NanoSIMS), and scanning transmission X-ray microscopy with near-edge X-ray absorption fine-structure spectroscope (STXM/NEXAFS).

Each instrument brings advantages and limitations. An SEM images the sample surface by scanning it with a high-energy beam of electrons in a raster scan pattern. However, SEM is insufficient for observing particles with diameters less than 100 nm. TEM coupled with energy dispersive X-ray spectrometry (EDS) is capable of analyzing the morphological, physical, crystallographic, and chemical properties of nanosized particles, observing the mixing state and surface of individual particles, and confirming particle phases of some crystalline particles. TEM is a common instrument, but its operation consumes a lot of time and money, so it can be used to analyze only a fraction of the particles collected on the substrate. AFM uses a high-resolution scanning probe microscopy to detect surface morphology, volumes, and viscosities of individual particles, but AFM cannot provide composition information. ATOFMS/SPAMS provides real-time sizes and chemical composition of individual particles online with wide dynamic size range capabilities from 200 nm to 2 mm, but it does not determine the morphology and mixing structures of different aerosol components within individual particles. SP2 can specifically obtain the optical properties, mass and number concentrations, sizes, and mixing state of black carbon (BC) in individual soot-containing aerosol particles by measuring the light scattering and/or incandescence of each particle, but SP2 still cannot provide accurate morphology of soot aggregates (Wang et al., 2021a; Wang et al., 2021b). NanoSIMS can provide high-spatial images of different ion clusters (e.g., C^-, N^-, CN^-, and S^-) within individual particles, showing the mixing states of organics and sulfates or nitrates in individual particles. However, NanoSIMS is an uncommon and extremely expensive instrument, and only a few scientists study individual aerosol particles using this technique. STXM/NEXAFS, utilizing an intense, tightly focused (25 nm) monochromatic X-ray beam to perform chemically resolved imaging, is especially powerful for measuring different light elements (C and N) and metals in individual particles based on differential shell absorption edges. Because this method uses synchrotron radiation-based techniques, it can measure some typical particles with a limited particle number (Li et al., 2016a; Laskin et al., 2019). The above combined techniques for individual particle analysis can efficiently provide the size, composition, morphology, and mixing state of aerosol particles (Li & Shao, 2009a; Moffet & Prather, 2009; Moffet et al., 2010; Adachi et al., 2014; Zhang et al., 2015). Compared with bulk methods, there are some drawbacks of individual-particle analysis methods: the number of particles analyzed by individual-particle methods is limited, the high cost and complicated procedures affect its application, and individual-particle methods cannot quantitatively determine the concentrations of aerosol species. However, characterizations of individual particles can remedy what the bulk methods cannot obtain. Here, we focus on recent advances in the microanalysis of individual particles.

TEM and SEM are two typical electron microscopy methods used in universities and research institutes, where they have been employed to study individual aerosol particles in various environments: urban areas, including Beijing, China (Li et al., 2010b; Niu et al., 2016), Jinan, China (Li et al., 2011b), Jilin city, China (Zhang et al., 2017a), Hong Kong, China (Li et al., 2013a), Incheon, South Korea (Geng et al., 2010a), Mexico City, Mexico (Adachi and Buseck, 2008), and Houston, USA (Laskin et al., 2003); background areas, including Changdao Island, China (Chen et al., 2017), Qinghai–Tibet Plateau, China (Li et al., 2015), and Okinawa Island, Japan (Moffet et al., 2012); mountain sites, including Mt. Lu, China (Li et al., 2013b; Li et al., 2014a) and Mt. Tai, China (Liu et al., 2018); and marine air, including the North Atlantic Ocean (Pósfai et al., 1995; Li et al., 2003a), the Yellow Sea (Li et al., 2017), and the Arctic Ocean (Geng et al., 2010b; Chi et al., 2015). Therefore, there are many findings concerning individual particles in the global air.

Other individual-particle methods have also been used to study individual particles in the atmosphere. For example, Moffet et al. (2013) observed the composition and mixing state of carbonaceous aerosols during transport from urban to background areas in California and found that the mass of organic carbon increased following transport through the combination of SEM and STXM/NEXAFS. Using these two techniques, Laskin et al. (2012) found that organic acids can effectively react with chloride in sea salt aerosols (SSA), causing substantial chloride depletion. Adachi et al. (2016) showed the changes in the mixing states and abundance of light-absorbing carbonaceous particles corresponding to meteorological conditions and air mass sources using TEM and SP2 in Tokyo, Japan; Lee et al. (2017) developed a new AFM technique to directly measure the 3D structure of individual sucrose particles in solid, semisolid, and liquid phases with varying RH. Integration of NanoSIMS and TEM provided evidence to confirm the "Fe acid dissolution" hypothesis: acids formed from anthropogenic gaseous pollutants dissolve iron in individual aerosol particles, enhancing the bioavailable iron input to the oceans (Li et al., 2017).

2.2. IMPORTANCE OF INDIVIDUAL PARTICLE MEASUREMENTS IN POLLUTED AIR

Analytical advances have been motivated by the need to capture precise physical and chemical properties of aerosol particles due to their global-scale impacts on both human health and Earth's climate (Ault and Axson, 2016). Traditional bulk analytical techniques have excellent quantitative capabilities for measuring atmospheric aerosols. However, there have been limitations in understanding the mixing state or physical properties of individual particles (Li et al., 2016a). Knowledge of the mixing state of individual particles is crucial to understanding the hygroscopic and optical properties of particles and provide information about their atmospheric histories (aging and reactions), which is a bridge combining laboratory and field experiments (Pósfai and Buseck, 2010; Li et al., 2016a). In addition, the impacts of aerosols on climate and health are dependent on the properties of the individual particles, such as water uptake by cloud condensation nuclei (CCN) or ice nuclei (IN), light scattering and absorption, and surface heterogeneous reactions with trace gas species

(Jacobson, 2001; Ault et al., 2013a; Ryder et al., 2014; Fierce et al., 2015). Therefore, many kinds of microanalytical techniques are widely used to study the properties of individual particles in air quality studies (Li et al., 2016a). Microscopic and spectroscopic methods have many advantages for probing the physical properties and chemical composition of complex individual particles. The increasing application of microscopy and spectroscopy, particularly for the analysis of individual particles, has facilitated a more thorough understanding of mixing state and the characterization of the particle population in air quality studies (Sobanska et al., 2012; Ault and Axson, 2016). The combination of different microanalytical techniques can obtain the morphology, composition, size, mixing state, and surface properties of individual particles, which can provide direct information on their emission sources and aging processes during transport and further reveal their effects on human health and climate change (Li et al., 2016a).

2.2.1. Microanalysis of Individual Particles from Different Sources

PM pollution has become a major environmental concern worldwide, especially in many developing countries (Van Donkelaar et al., 2010). In polluted regions, air pollution occurs frequently owing to complex emission sources and evolution processes of aerosol particles (Wang et al., 2014; Zheng et al., 2015). Although the secondary formation of aerosol particles contributes significantly to air pollution, intensive emissions of primary particles are the key causes of air pollution (Huang et al., 2014b; Zhang et al., 2017b). Knowledge of primary particles' sources, composition, and evolution processes is of great importance for the effective improvement of air quality (Zhang et al., 2017b). Among the complex emission sources, industrial activities, power plants, biomass and coal burning, vehicular exhaust, and mineral dust are the main anthropogenic sources contributing to air pollution. Biological aerosols and sea salts are abundant, especially in clean atmospheres such as forests, coastal areas, and marine air.

Industrial Activities

Owing to rapid industrialization, industrial activities release large amounts of aerosol particles and gaseous pollutants into the atmosphere. Industrial sources mainly include coal-fired power plants, iron and steelworks, nonferrous smelters, and oil refineries (Li et al., 2016a). Fly-ash (Figure 2.2a) and metal (Figures 2.2b,c) particles are the typical primary particle types emitted from industrial sources (Li et al., 2016a). Fly-ash particles containing Si and Al with minor Fe and Mn are predominantly spherical and nanosized with diameters smaller than 200 nm. Because the elemental composition of fly-ash particles is similar to that of mineral dust particles, they cannot be identified by bulk methods. In contrast, microanalytical techniques such as SEM and TEM can identify these particles based on their composition and morphology. Abundant fly-ash particles were revealed in the brown hazes over northern China using TEM (Li et al., 2011b; Li et al., 2012). By using a dual-beam focused ion beam SEM (FIB-SEM), Chen et al. (2013a) analyzed the modification of morphology, chemical composition, and elemental distribution within individual fly-ash particles during atmospheric transport, measurements that cannot be achieved by bulk methods. Metal particles mainly contain Fe-rich, Zn-rich, Pb-rich, and Mn-rich particles, most of which have sizes smaller than 500 nm, with a median diameter of 200 nm in the haze episodes (Li et al., 2013a). TEM observations reveal that fly-ash and metal particles are frequently internally mixed with secondary components (e.g., sulfates, nitrates, and organics) in polluted air (Li and Shao, 2009a; Li et al., 2013a). Furthermore, internally mixed fly-ash and metal particles can release soluble metal ions during their transport through the acid dissolution process. Li et al. (2017) provided a "smoking gun" for the iron dissolution process with acids in atmospheric particles revealed by using NanoSIMS and scanning transmission electron microscopy (STEM).

Vehicular Exhaust

Vehicular exhaust becomes a main source of air pollution in urban cities. Nearly 50% of the primary particles emitted from vehicular exhausts are soot, with liquid phase inorganic materials and organic matter (OM) coating their surfaces (Figure 2.2d) (Mohankumar & Senthilkumar, 2017). TEM and SEM observations show that soot particles are chain-like aggregates of carbon-bearing spheres with typical diameters ranging from 10 to 100 nm (Buseck et al., 2014). Compared to SEM, TEM better observes the complex morphology of soot particles embedded in secondary aerosols. Because the mixing structures of soot particles can significantly affect light absorption and alter radiative forcing (Jacobson, 2001; Cappa et al., 2012; Wang et al., 2017; Wang et al., 2021a; Wang et al., 2021b), the morphology and mixing structures of soot particles acquired by TEM are of extreme importance to estimate their impacts on climate change (Moffet and Prather, 2009; Adachi et al., 2010; Buseck et al., 2014). Raman spectroscopy is sensitive to graphene layer distribution and therefore is used to characterize carbon materials and their nanostructures (Dresselhaus et al., 2005; Soewono & Rogak, 2011). Ge et al. (2019) characterized soot particles from diesel and gasoline engines using Raman spectroscopy and

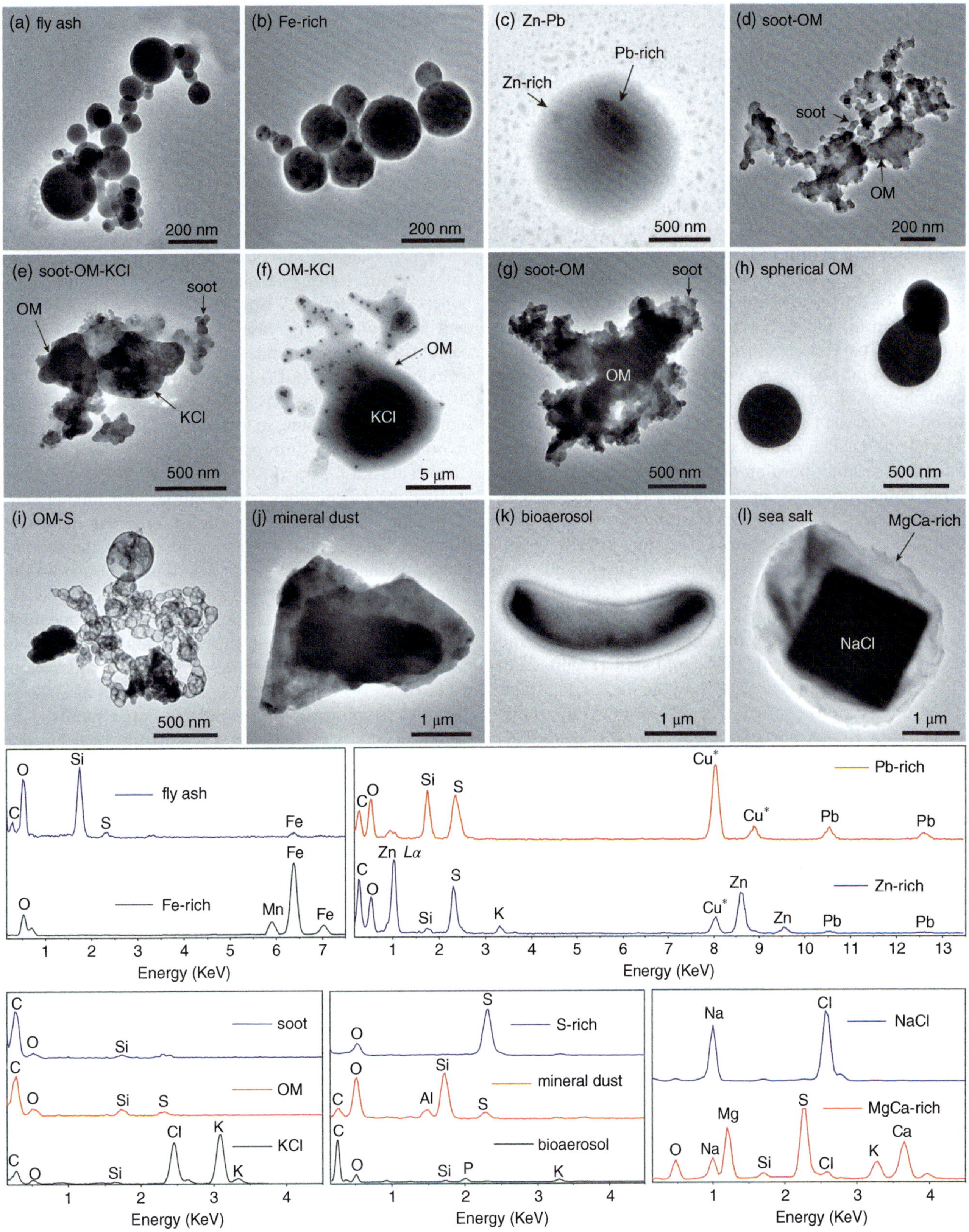

Figure 2.2 TEM images of different types of individual particles emitted from different sources. (a) Fly-ash; (b) Fe-rich; (c) Zn-Pb-rich particles from industrial activities; (d) internally mixed OM and soot (soot-OM) particle from diesel engine exhaust; (e) internal mixture of soot, OM, and KCl (soot-OM-KCl) from the flaming phase of maize straw burning; (f) gel-like OM internally mixed with KCl (OM-KCl) from the smoldering phase of maize straw burning (Liu et al., (2017a) / Springer Nature / CC BY 4.0); (g) soot-OM particle from the flaming phase of low-maturity coal burning; (h) spherical OM particles from the flaming phase of medium-maturity coal burning; (i) internally mixed OM and S-rich particles (OM-S) from the flaming phase of high-maturity coal burning (Zhang et al., 2018a / with permission of American Geophysical Union.); (j) irregular mineral dust particle collected from the atmosphere of the Tengger desert in Northwest China; (k) bioaerosol particle collected from Lesser Khingan Mountains (Li et al., (2020) / with permission of ELSEVIER); (l) sea salt aerosol particle collected from the Arctic (Chi et al., (2015) / Copernicus Publications / CC BY 3.0).

found that diesel soot particles have higher amorphous graphitic carbon than gasoline-derived soot.

Biomass Burning

Wildfire as one type of biomass burning is a widespread and significant global source of gaseous and particulate emissions in the troposphere (Koppmann et al., 2005). Wood and agricultural residues are common energy sources for domestic heating and cooking in many parts of the world. Open burning of agricultural residues during the harvest season in the fields is widespread in East Asia and Southeast Asia (Streets et al., 2003). The contribution of agricultural residual burning is estimated to be 10% of the total mass burned globally (Andreae & Merlet, 2001). The wildfire such as forest fires and grassland and savanna burning occur frequently and make a large contribution to air pollutants (Li et al., 2003b; Streets et al., 2003). Biomass burning can release large amounts of particulate species, including potassium salt (K-rich), primary organic aerosols (POAs), and soot (i.e., BC) (Myers-Pigg et al., 2016). TEM observations show that fresh K-rich particles from wildfire in the flaming phase are mainly KCl with euhedral morphology and are internally mixed with POAs and soot (Figure 2.2e) (Li et al., 2003b; Liu et al., 2017). In the smoldering stage, TEM observations confirm that gel-like POAs internally mixed with K-rich particles are produced (Figure 2.2f) (Liu et al., 2017). During transport in the atmosphere, KCl particles can be converted to K_2SO_4 and KNO_3 with the aging of smoke (Li et al., 2003b). TEM can confirm the transformation of individual K-rich particles by obtaining their morphology, structures, and composition, leading to a determination of their sources and hygroscopicity. SPAMS is an ideal instrument to identify the markers for organics, soot, and K-rich particles. The peak m/z = −59 (i.e., $^{59}C_2H_3O_2^-$) obtained by SPAMS is commonly used as a tracer of biomass combustion particles, as it is a signature of monosaccharides (e.g., levoglucosan) produced from the pyrolysis of cellulose (Pagels et al., 2013). These gel-like POAs can be further aged and become tar balls in the smoke of wildfire and biofuel burning (Pósfai et al., 2004; Adachi et al., 2019). Atmospheric "tar balls" are distinct POA particles that can be readily identified by TEM (Pósfai et al., 2004). TEM analysis shows that tar balls are amorphous, carbonaceous spherules with diameters typically ranging from 30 to 500 nm. EDS spectra indicate that tar balls mainly consist of C with minor O. Electron energy loss spectroscopy (EELS) confirms that tar balls have a non-negligible light-absorbing performance and should be clarified as brown carbon (Pósfai et al., 2004; Alexander et al., 2008).

Residential Coal Burning

Emissions from residential coal burning for cooking and heating are an important contributor to air pollution in developing countries such as China, India, Mongolia, and Nepal (Zhang et al., 2018a; Zhang et al., 2020). Due to low combustion efficiency and lack of emission control devices in residential stoves, emission factors from residential stoves are 100 times higher than those from industrial boilers (Zhang et al., 2008b). Residential coal burning can emit large amounts of gaseous pollutants (e.g., SO_2 and volatile organic compounds) and primary particles, mainly including OM and soot (Zhang et al., 2008b; Zhou et al., 2016). Primary emissions from residential coal burning significantly contribute to the formation of severe haze in northern China during wintertime (Liu et al., 2016; Chen et al., 2017; Li et al., 2018; Liu et al., 2021). TEM can acquire the morphology, composition, and mixing state of primary particles from coals of different maturities (Zhang et al., 2018a). It was found that in the flaming phase, internally mixed OM and soot particles (soot-OM, Figure 2.2g) were mainly emitted from low-maturity coals, OM particles (Figure 2.2h) were emitted from medium-maturity coals, and internally mixed OM and sulfate (S-rich) particles (OM-S, Figure 2.2i) dominated the primary emitted particles of high-maturity coals (Zhang et al., 2018a). Also, TEM observations show that large numbers of spherical OM particles were emitted from coal burning (Figure 2.2h), which are identified as tar balls (Zhang et al., 2018a). The characterization of tar balls by the NanoSIMS displays strong intensities of $^{12}C^{14}N^-$ and $^{12}C^-$ signals but weak intensities of $^{16}O^-$ and $^{32}S^-$ signals (Zhang et al., 2018a). The microscopic observations of individual particles provide direct evidence of the regional light and medium haze formation in northern China in winter, which is mainly attributed to residential coal stoves used for heating and cooking (Chen et al., 2017; Liu et al., 2021).

Dust Sources

Mineral dust, mainly from deserts and arid regions, along with construction activities and resuspended road dust, is one of the most abundant aerosol types in the troposphere and can be transported over long distances (Andreae, 1995; Grini et al., 2005; Prospero et al., 2014). During transport, properties such as particle size, morphology, composition, mixing state, and hygroscopicity undergo modification through atmospheric processing, thus influencing the climate, atmospheric chemistry, and biogeochemical cycles regionally and globally (Sullivan et al., 2007; Shi et al., 2009; Harris et al., 2012; Huang et al., 2014a; Ito and Shi, 2016). SEM and TEM both can observe the irregular shapes of mineral dust particles (Figure 2.2j) and reveal that the sizes of most particles dominate the coarse mode (diameter > 2 μm) (Li and Shao, 2012). Hygroscopicity of mineral dust particles is important for its impacts on climate and the atmospheric environment (Laskin et al., 2005b; Li et al., 2014b). Heterogeneous and/or multiphase reactions during transport can occur on the surface of mineral dust particles, which will modify their composition and mixing state and further change the hygroscopicity and CCN/IN activities (Usher et al., 2003; Li and Shao, 2009b; Crowley et al., 2010; Kulkarni et al., 2015; Tang et al., 2016).

Biological Source

Biological aerosol particles (e.g., bacteria, spores, fungi, viruses, algae, and pollen) are ubiquitous in Earth's atmosphere and constitute a large fraction (15–45%) of aerosol particles, especially in pristine air such as forest, rural, and marine environments. (Pöschl, 2005; Elbert et al., 2007; Hu et al., 2017; May et al., 2018; Li et al., 2020). Biological aerosol particles in the atmosphere can significantly affect climate change, ecosystems, atmospheric processes, and human health (Wittmaack et al., 2005; Després et al., 2012; Pohlker et al., 2012; China et al., 2018). Microanalysis is powerful for characterizing the morphology, size, and mixing states of biological aerosols, which allows one to identify their types and provide insights into the actual state of individual biological particles suspended in the atmosphere (Pósfai et al., 2003; Wittmaack et al., 2005; Martin et al., 2010; Després et al., 2012). Most biological aerosol particles contain more or less phosphorus (P) in individual particles and have unique morphologies identified by electron microscope (e.g., Figure 2.2k) (Pöschl, 2005; Wittmaack et al., 2005). Pósfai et al. (2003) found a few rod-like bacterial particles in marine air using TEM. In the boreal forest of China in summer, Li et al. (2020) found similar rod-like bacterial and fungal particles (Figure 2.2k) with a dominant size range of 1–5 μm with two typical peaks at 1.4 and 3.5 μm. Exposure to biological aerosol particles may lead to adverse health effects, especially for allergy sufferers, immunosuppressed persons, or patients with respiratory diseases (Walser et al., 2015). Moreover, rod-like biological aerosol particles are commonly internally mixed with mineral dust, nanoscale metal, organics, and inorganic salts (Li et al., 2020). Environmental scanning electron microscopy (ESEM) analysis reveals that Amazonian fungal spores can rupture and release fungal fragments that contain hygroscopic sodium salts after exposure to high humidity (China et al., 2016; China et al., 2018). An epifluorescence microscope can be used to count the bacteria number and identify the viability of bacteria collected from the near-surface air (Hu et al., 2017).

Marine Sources

SSA particles can be generated by breaking waves in the marine environment via bubble bursting (Andreae and Rosenfeld, 2008), and they are considered to be the dominant aerosol component in the atmosphere by mass (Liao et al., 2006). Therefore, SSA particles have a significant impact on Earth's radiative forcing directly by scattering light and indirectly by serving as CCN or IN (Haywood et al., 1999; Seinfeld and Pandis, 2006). SSA particles also represent a very large source of halogens in the lower atmosphere (Graedel and Keene, 1995). The major component of SSA particles is sodium chloride (NaCl), which is commonly coated by Mg and Ca salts (Figure 2.2l), but organic species (e.g., long-chain alkanes, alkanols, and acids) have also been identified in SSA particles (Ming & Russell, 2001; Ault et al., 2013b; Chi et al., 2015). Elemental mapping with EDS is a tool to characterize the distribution of the elements within each particle (Conny & Norris, 2011; Ault et al., 2012; Chi et al., 2015). Ault et al. (2013b) did element mapping of SSA particles using SEM-EDS and found most of them were an internal mixture that includes a distinct NaCl cubic core and a ring containing C, O, and Mg. STXM-NEXAFS analysis showed that organic components containing COO^- and C=C groups occur in the ring. Chi et al. (2015) determined the depletion of Cl during the aging of SSA particles through element mapping using STEM. NanoSIMS analysis provided direct evidence of Na_2SO_4 and $NaNO_3$ in fully aged SSA particles. Prather et al. (2013) successfully revealed size-resolved associations between inorganic and organic species within individual sea spray aerosols for sizes from 30 nm to 3 μm by integrating the results from ATOFMS, STXM, and TEM.

2.2.2. Aging of Individual Aerosol Particles and Their Environmental Influences

Particle aging can significantly alter the size, morphology, and mixing structure of individual aerosol particles (Figure 2.3). Detailed knowledge of aging and physicochemical properties (e.g., mixing state, hygroscopicity, and optical properties) of aerosol particles in the atmosphere is useful for understanding their effects on climate and human health.

Mineral Dust

Fresh mineral dust particles can serve as important reactive surfaces for atmospheric heterogeneous reactions. Through surface uptake, alkaline components (e.g., calcite) in mineral dust particles can react with acid gases such as SO_2 and NO_x to form coatings of sulfates and nitrates (Figure 2.4d) (Laskin et al., 2005a; Kojima

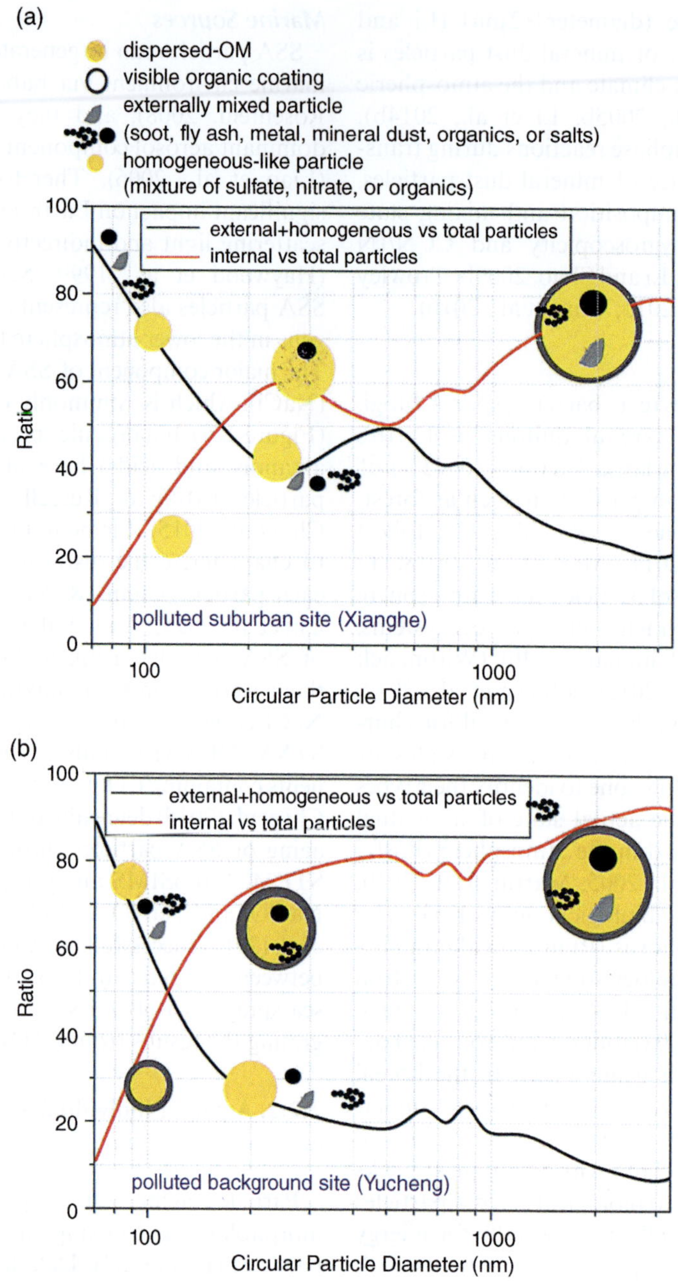

Figure 2.3 Heterogeneously internally mixed particles (internal) and externally mixed particles (external) plus homogeneous-like (homogeneous) particles versus total analyzed particles by particle number at polluted suburban sites (a) and a rural site (b). Li et al. (2016b)/with permission of American Geophysical Union.

et al., 2006; Li & Shao, 2009b). Meanwhile, the coating materials can alter the hygroscopicity of mineral dust particles and change the interaction with water, which in turn influences the reactivity of mineral dust particles with acid gases (Tang et al., 2016).

Regional haze events and dust storms frequently occur in winter and spring in northern China. A mixture of abundant mineral dust particles, SO_2, and NO_X in the air can promote heterogeneous aging reactions of mineral dust particles (Wang et al., 2006). Even worse, under the prevailing westerlies, the polluted dust plume from continental China can be transported over the open sea. Once ocean-originated SSA particles and water vapor are mixed into the polluted dust plume, mineral dust and SSA particles can have competitive reactions with acidic gases (Zhang & Iwasaka, 2001; Tobo et al., 2010).

Additionally, the deposition of mineral dust particles could bring nutrient elements (e.g., Fe, P, and Cu) into the

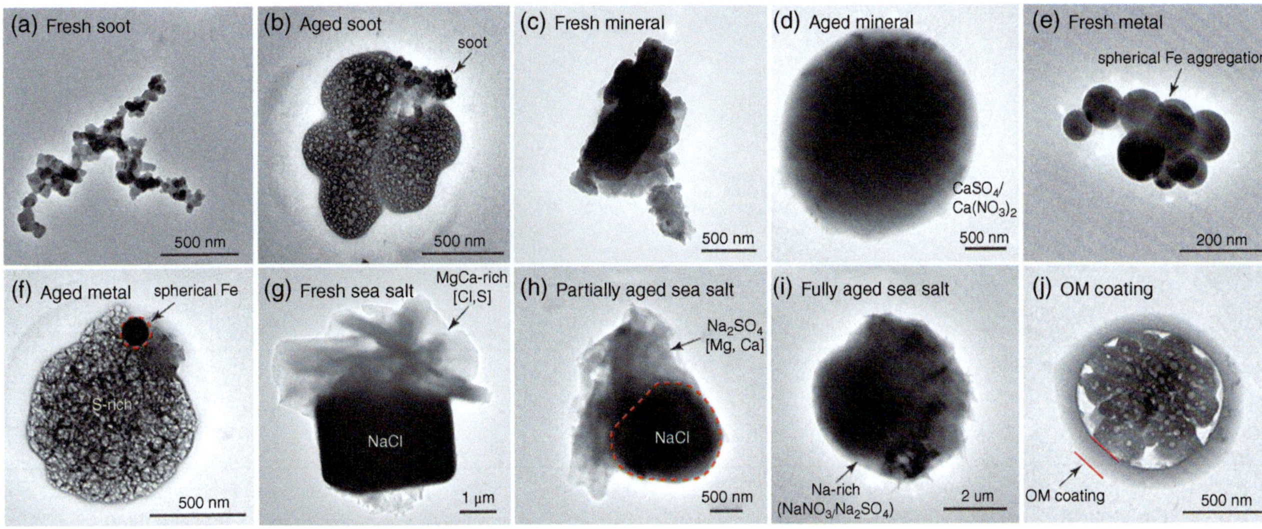

Figure 2.4 TEM images of fresh and aged individual particles. (a) Fresh soot particle; (b) aged soot particle associated with S-rich particle; (c) Irregular fresh mineral dust particle; (d) mixture of mineral and $CaSO_4$ coated with $Ca(NO_3)_2$; (e) One aggregate of Fe; (f) aged Fe particle mixed with S-rich particle; (g) Fresh sea salt aerosol particle containing one square NaCl attached to a mixture of Mg, Ca, S, and Cl; (h) Irregular NaCl mixed with Na_2SO_4; (i) Mixture of $NaNO_3$ and Na_2SO_4; (j) S-rich particle with organic matter (OM) coating.

ocean (Tang et al., 2016). Particularly, aging reactions on the surface of mineral dust particles can dissolve part of the iron (Fe) in the particles and transfer them into a bioavailable form for the growth of phytoplankton in the ocean, eventually resulting in the enhancement in ocean productivity and oceanic carbon uptake (Meskhidze et al., 2005; Mahowald et al., 2010; Li et al., 2017).

Sea Salt Aerosol

SSA serves as the reactive surface in the atmosphere that can react with gaseous SO_2, NO_x, and organic acids, releasing volatile gaseous HCl and causing the depletion of chloride and the enrichment of the corresponding salts in particles (Laskin et al., 2012; Laskin et al., 2016). TEM observation shows that fresh SSA particles normally form square NaCl cores attached to $MgCl_2$ and $CaSO_4$ (Figure 2.4g). Due to chloride depletion, square NaCl particles can form a mixture of NaCl, $NaNO_3$, Na_2SO_4, $Mg(NO_3)_2$, and $MgSO_4$ (Figure 2.4h, i) (Li et al., 2010a; Li et al., 2011a; Chi et al., 2015). Moreover, that SSA may react with organic acids, causing chloride depletion (Laskin et al., 2012; Chi et al., 2015). Reactions of polluted air with SSA could promote ozone formation that deteriorates air quality in densely polluted coastal areas (von Glasow, 2008).

Soot (Black Carbon)

Freshly emitted soot particles are normally fractal, hydrophobic, and externally mixed aggregates (Figure 2.4a). Once soot particles are emitted into the atmosphere and interact with other aerosol components, they undergo structural transformation (Riemer et al., 2004; Adachi & Buseck, 2015). Previous laboratory and in situ studies reported changes in morphology and compactness of soot particles during their aging (Figures 2.4b and 2.5a) and pointed out that the influence of water or other secondary aerosol materials was responsible for the compaction of soot (Ma et al., 2013; China et al., 2015; Peng et al., 2016). The fractal dimension (D_f) is a widely used key parameter to describe the fractal morphology of soot particles (Oh & Sorensen, 1997; Brasil et al., 1999). TEM observation shows that soot particles after long-range transport tend to become more compact, with a larger D_f (Wang et al., 2017). In addition, for soot-bearing particles coated by secondary aerosols, phase changes of the secondary aerosol coating under different RHs can significantly affect the D_f of soot particles (Yuan et al., 2019b).

The variations in morphology and mixing state of soot particles eventually affect their optical properties and radiative forcing (Zhang et al., 2008a; Zhou et al., 2017). The shell of secondary aerosols acts as a lens that can focus light on the soot core and enhance its absorption (Adachi et al., 2010). By assuming an internal mixture of soot core and S-rich shell as an ideal core-shell model, Jacobson (2001) proposed that optical absorption by the S-rich coating on soot particles can be enhanced by a factor of two over externally mixed soot. However, direct in situ measurements in two California regions show that the absorption enhancement of aged soot is only 6% on average at 532 nm (Cappa et al., 2012). The discrepancy in the absorption enhancement of soot particles between

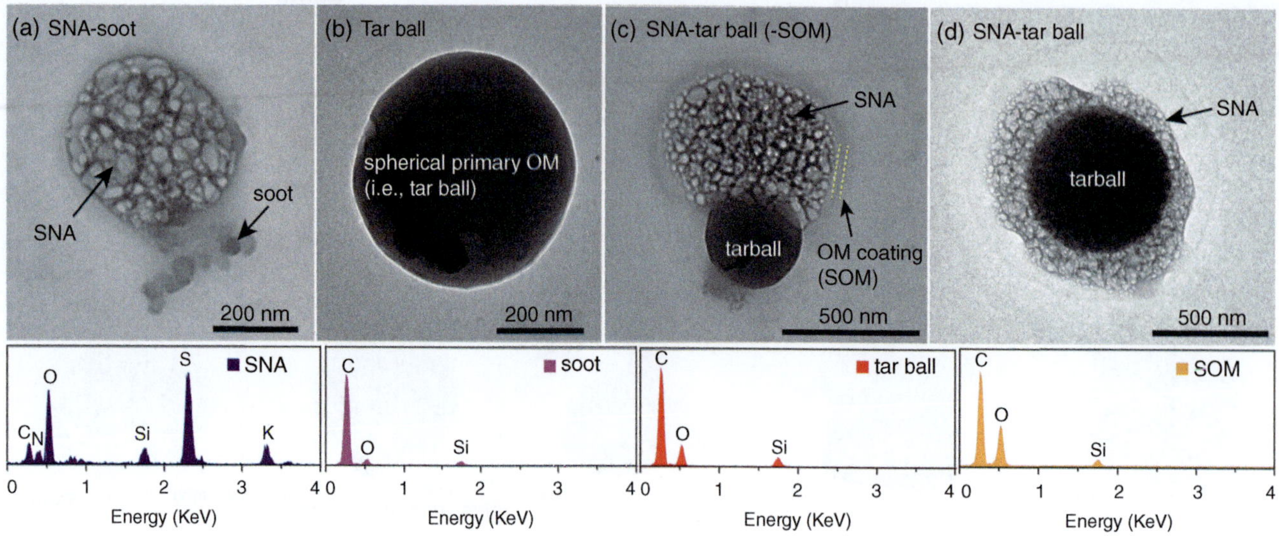

Figure 2.5 Typical transmission electron microscopy (TEM) images of organic aerosols and inorganic aerosols (SNA). (a) Mixture of SNA and soot particles; (b) Bare tar ball; (c) Tar ball partly coated by a mixture of SNA and OM coating (SOM) particles; (d) Tar ball totally coated by SNA particles. Zhang et al. (2021)/with permission of American Geophysical Union.

model estimation and field observation should be attributed to the lack of knowledge about the precise shapes and mixing structures of soot particles in the air (Wang et al., 2021a). Therefore, microscopy-based techniques are a suitable way to observe the aging of soot particles.

Metal and Fly Ash

Metal particles including Fe, Zn, Pb, and Mn are commonly observed in polluted air influenced by heavy industrial or combustion activities; fly-ash particles containing Si, Al, and minor Ca, Mn, and Fe are produced from coal combustion (Li et al., 2016a). Most metal and fly-ash particles are internally mixed with secondary components (i.e., sulfate, nitrate, and organics) (Adachi & Buseck, 2010; Li et al., 2016a).

No sulfate was detected in freshly emitted fly-ash and metal particles (Figure 2.4e); however, a certain amount of sulfate coating was observed in most of the fly-ash and metal particles in ambient air (Li et al., 2017). Fly-ash and metal particles during long-range transport were internally mixed with sulfuric acid, organic acids, and sulfate, leading to the acidic sulfate coating (Figures 2.3 and 2.4f). The acids in the coating could dissolve part of metal components, releasing soluble metal ions into the coating (Li et al., 2013a; Li et al., 2013b). Also, some soluble metal ions in cloud water could alter reaction processes, acting as a catalyst for sulfate production from autooxidation of S(IV) (Rao and Collett, 1998).

Organic Aerosols

OM particles mainly present two different morphologies: spherical morphology with an average shape factor of 0.75 (Figure 2.5b) and OM coating on the surfaces of inorganic aerosols (i.e., sulfates, nitrates, and ammonium [SNA]) particles (Figures 2.4j and 2.5c). Spherical OM particles with high viscosities exist as one type of primary organic particles (Reid et al., 2018; Yuan et al., 2020) and have generally been considered to be tar balls (Pósfai et al., 2004; Adachi and Buseck, 2011; China et al., 2013; Tóth et al., 2014; Liu et al., 2017; Zhang et al., 2018a; Adachi et al., 2019). These studies well documented that tar balls are from fossil fuel combustion emissions and are a good marker for household heating and cooking using biomass and coals (Hand et al., 2005; Girotto et al., 2018; Sedlacek Iii et al., 2018; Zhang et al., 2018b; Li et al., 2019). Once these tar balls are emitted into the atmosphere, they become aged and form internally mixed particles with inorganic aerosol particles (Figure 2.5d). These tar-ball particles have been verified as the primary brown carbon in the atmosphere (Alexander et al., 2008).

OM coatings have been perceived as secondary OM (SOM) aerosols (Li et al., 2016b; Zhang et al., 2017a). The formation of the "OM coating and S-rich core" particle is likely due to liquid-liquid phase separation in the air (You et al., 2012; Li et al., 2021). SEM and STXM/NEXAFS observation of carbonaceous aerosols reveals that the mass of organic carbon on individual particles increased through condensation of secondary organic aerosols in ambient air (Moffet et al., 2013). The NanoSIMS ion intensity map of particles shows a clear OM coating on S-rich particles (Li et al., 2016b; Yu et al., 2019; Li et al., 2021). TEM images can further capture the coating thickness of OM coating on S-rich particles (e.g., Figure 2.5c) (Yu et al., 2019). Previous studies showed that OM coating thickness increased with particle diameter, suggesting that particles with larger sizes underwent more aging than finer particles (Figure 2.3). The thickness of the OM coating on secondary particles can be used to infer the aging degree of particles (Chen et al., 2017; Hou et al., 2018).

2.2.3. Microanalysis of Individual Particles from Urban Areas

Microanalysis has been widely used to study the air quality and trace sources of aerosol particles in urban areas in both developed and developing countries. For example, Chen et al. (2006) conducted an in-depth characterization of individual submicron particles in Lexington, Kentucky, USA, using TEM. They found that S-rich particles constituted the dominant species (45–60%), followed by carbonaceous particles (21–39%), Si-bearing particles (4–11%), and metal oxides (2–8%). The results implicated combustion of fossil fuels as the main source of airborne particles at Lexington. Adachi and Buseck (2008) measured individual particles collected from Mexico City and found that over half of the aerosol particles consisted of internally mixed soot, OM, and S-rich particles. S-rich particles commonly contained K, indicating that biomass and biofuel were major sources of aerosol particles in Mexico City. Regional brown haze occurs frequently in northern China; Li and Shao (2009a) studied the morphology, elemental composition, and mixing state of individual aerosol particles in regional hazes in urban areas of Beijing using TEM. Mineral, complex secondary, organic, soot, fly-ash, and metal (Fe-rich and Zn-bearing) particles were identified. TEM observations revealed that about 70% of these particles are internally mixed, with two or more aerosol components from different sources, including industrial activities, biomass burning, and mineral dust. The TEM analysis revealed the dominance of tar-ball and fly-ash particles from coal burning in the winter of Taiyuan City, a coal-burning region in the eastern China Loess Plateau (Li et al., 2012). Therefore, TEM analysis can not only classify aerosol types and trace their sources but also identify the aging of aerosol particles, such as condensation, dissolution, aqueous chemistry, and coagulation.

2.3. IMPLICATIONS FOR HEALTH EVALUATION

Air quality is of great concern, especially in many developing countries. It is estimated that more than 8.8 million excess deaths occur globally each year due to ambient air pollution (Lelieveld et al., 2020). Epidemiological studies have shown clear correlations between fine particulate matter ($PM_{2.5}$) in polluted air and adverse health effects such as cardiovascular, respiratory, and allergic diseases (Shiraiwa et al., 2012; Liu et al., 2019). Size, number, solubility, and chemical composition of aerosol particles are the main factors determining their adverse effects on human health (Dockery & Pope, 1994; Pöschl, 2005). Elucidating the contributions and interactions of individual components or sources in a multipollutant mixture is a growing research focus because epidemiological studies must account for complex correlations between different pollutants and between the contributions of multiple sources (Costa & Dreher, 1997; West et al., 2016; Huang et al., 2020). In general, toxicological studies have not indicated adverse health effects from sulfate and nitrate (Schlesinger & Cassee, 2003), but one field study in a Chinese city found their concentrations to be closely associated with total and cardiovascular mortality (Cao et al., 2012). This disparity could be explained by understanding the mixing state of individual particles—nanosized soot, OM, and metals, which are toxic components of $PM_{2.5}$ (Dockery & Pope, 1994; Shao et al., 2006; Cao et al., 2012; Kan et al., 2012).

$PM_{2.5}$ has known impacts on human health, and ultrafine particles (d_p < 100 nm) are suspected to be particularly hazardous because of their large surface area and reactive oxygen species (ROS) (Löndahl et al., 2007). Ultrafine particles dominate in most environments by number and are sufficiently small to penetrate the membranes of the respiratory tract and enter the blood circulation or be transported along olfactory nerves into the brain (Oberdörster et al., 2004; Oberdörster et al., 2005; Löndahl et al., 2007; Calderon-Garciduenas et al., 2021). TEM observations reveal that many toxic metal particles (e.g., Fe-rich, Zn-rich, Pb-rich, and Mn-rich) exhibit nanometer sizes and tend to be internally mixed with secondary inorganic components (e.g., sulfates and nitrates) in the atmosphere (Li et al., 2013a; Liu et al., 2018). Microanalysis has shown that the internal mixtures of insoluble metal oxides with low toxicity evolve into toxic metal ions through reactions with acidic aerosols (Baltrusaitis et al., 2012; Oakes et al., 2012; Li et al., 2013a; Li et al., 2013b; Li et al., 2017). For example, Li et al. (2013b) analyzed the element distribution of individual cloud droplet residues internally mixed with metal particles using STEM, with results that aqueous chemical processes in acidic cloud droplets significantly enhance metal solubility. Furthermore, using combined individual-particle analytical techniques, they demonstrated that iron oxides internally mixed with a thick layer of acidic S-rich coating can be converted to soluble Fe sulfate through the acid dissolution process during long-range transport. Therefore, when evaluating the impacts of aerosol particles on human health, we must consider the mixing state of the aerosol particles. Water-soluble metal ions can homogeneously mix with secondary aerosols (Li et al., 2013a) and may induce airway epithelial injury and cytokine gene expression when inhaled and deposited into deeper parts of the human respiratory (Dye et al., 1999). The presence of transition metals—especially iron—in the particles increases the production of reactive oxygen species in vivo, which results in cellular and tissue damage and further initiates and exacerbates inflammation (Hitzfeld et al., 1997; Kadiiska et al., 1997). One study suggests that Zn has robust positive associations with different blood pressures (e.g., diastolic and systolic blood pressure) in the urban atmosphere of China (Wu et al., 2013).

Soot and organic particles (e.g., polycyclic aromatic hydrocarbons [PAHs]) emitted from diesel exhaust, smoke, and incomplete combustion are in the submicron size range that can penetrate deeply into human lungs and have high carcinogenic, mutagenic, and allergenic potentials to human health (Finlayson-Pitts & Pitts Jr, 1999; Lewtas, 2007; Jia et al., 2020). Soot formation is linked to the production of PAHs in flames during combustion (Durlak et al., 1998; Frenklach, 2002), and PAHs reside to a large extent at the surface of soot particles (Pöschl, 2005). Long-lived reactive oxygen intermediate can be produced due to the delocalized π electrons of an aromatic surface that existed in PAHs and soot particles and could cause oxidative stress to human health (Shiraiwa et al., 2012). PAHs have been implicated in inducing lung, skin, and breast cancer by mutations in DNA and in promoting cancer by affecting the proliferative capacity of mutated cells. PAHs also exhibit a wide range of toxic effects, including skin and eye irritation, immunotoxicity, and developmental toxicity (Flowers et al., 2002).

Dry or wet deposition of aerosol particles could further harm the ecosystem on a large scale, including impacts on drinking water, soils, crops, vegetables, and other flora, fauna, and medicinal plants (Mohapatra & Biswal, 2014; Przybysz et al., 2014; Zhang et al., 2017c). Ultimately, aerosol particles could indirectly affect human health through food (Kampa & Castanas, 2008; Gheorghe & Ion, 2011).

2.4. ATMOSPHERIC IMPLICATIONS OF MICROANALYSIS IN AIR QUALITY STUDIES

Microanalysis has been widely used for air quality studies worldwide as mentioned. Here we raise a successful example of how to use microanalysis techniques to understand air pollution and particle transport in the atmosphere. A recent work shows that enormous levels of haze aerosols containing tar balls (over 70% by number) from the North China Plain (NCP) can be rapidly transported into the Yangtze River Delta within 12 hours and further induce regional haze formation in East China during winter (Zhang et al., 2021). Zhang et al. (2021) found that a large number of tar balls emitted from residential coal burning in rural villages of the NCP can be transported thousands of kilometers and influence the air quality of the Yangtze River Delta (Figure 2.6). Tar balls have been generally considered primary brown carbon (Chakrabarty et al., 2010; Tóth et al., 2014; Sedlacek III et al., 2018; Zhang et al., 2018a; Adachi et al., 2019). As a result, abundant

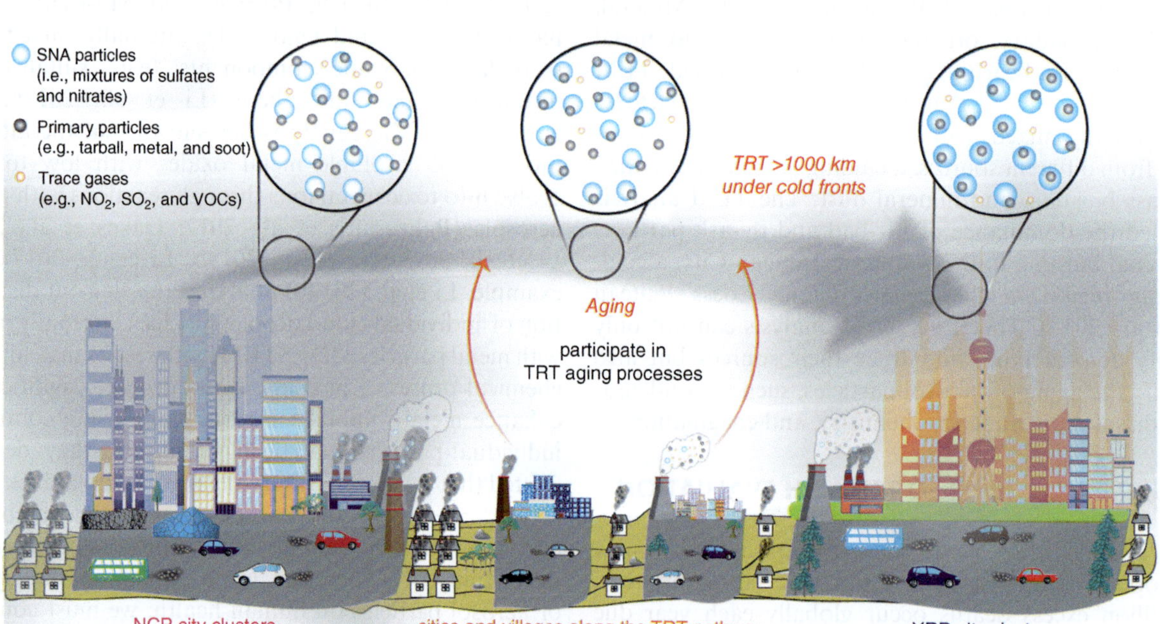

Figure 2.6 A schematic diagram of air pollutants trans-regional transport (TRT) from city clusters in the North China Plain (NCP) to city clusters in the Yangtze River Delta (YRD) under cold fronts during winter. Zhang et al. (2021)/with permission of John Wiley & Sons.

light-absorbing tar balls from the NCP were transported and spread throughout East China following cold fronts (Figure 2.6). These light-absorbing tar balls should be further considered in regional climate models in East China.

In this work, TEM observations not only found large numbers of tar balls but also revealed that many nanosized metal particles mixed with SNA from the NCP were transported into East China (Figure 2.5b, c). Tar balls from residential coal burning contain plentiful carcinogenic PAHs, and metal (e.g., Zn, Pb, Mn, and Cr) particles from heavy industries are typical toxic aerosols for human health (Zhang et al., 2008b; Chen et al., 2013b; Lu et al., 2020). Our study potentially shows that the large population in the city clusters of eastern China could also be exposed to these long-range transported tar balls and metal particles during transregional transport haze events (Figure 2.6).

The aging of haze particles via heterogeneous reactions in transregional transport significantly transformed their mixing structures, such as tar-ball-containing particles changing from a partly-coated state in the NCP to a totally coated state in the Yangtze River Delta (Figure 2.5b–d), which is different from short-range regional transport (Chen et al., 2017). The aging of transregional transport haze particles is important, as it changes these particles' optical properties and activity as potential CCN.

In summary, microanalysis work can provide evidence that transregional transport haze events can carry large numbers of nanosized primary particles (e.g., tar ball, metal, fly ash, soot, and minerals) in addition to trace gases (e.g., SO_2, NO_x, and VOCs) and SNA. This information is the basis for further simulations of regional climate changes and discussing the health effects of aerosol particles.

REFERENCES

Adachi, K., & Buseck, P. R. (2008). Internally mixed soot, sulfates, and organic matter in aerosol particles from Mexico City. *Atmospheric Chemistry and Physics*, *8*(21), 6469–6481.

Adachi, K., & Buseck, P. R. (2010). Hosted and free-floating metal-bearing atmospheric nanoparticles in Mexico City. *Environmental Science & Technology*, *44*(7), 2299–2304.

Adachi, K., & Buseck, P. R. (2011). Atmospheric tar balls from biomass burning in Mexico. *Journal of Geophysical Research*, *116*(D5). doi: 10.1029/2010JD015102

Adachi, K., & Buseck, P. R. (2015). Changes in shape and composition of sea-salt particles upon aging in an urban atmosphere. *Atmospheric Environment*, *100*(0), 1–9.

Adachi, K., Chung, S. H., & Buseck, P. R. (2010). Shapes of soot aerosol particles and implications for their effects on climate. *Journal of Geophysical Research, (Atmospheres)*, *115*(D15), D15206.

Adachi, K., Zaizen, Y., Kajino, M., & Igarashi, Y. (2014). Mixing state of regionally transported soot particles and the coating effect on their size and shape at a mountain site in Japan. *Journal of Geophysical Research: Atmospheres*, *119*(9), 5386–5396.

Adachi, K., Moteki, N., Kondo, Y., & Igarashi, Y. (2016). Mixing states of light-absorbing particles measured using a transmission electron microscope and a single-particle soot photometer in Tokyo, Japan. *Journal of Geophysical Research: Atmospheres*, *121*(15), 9153–9164.

Adachi, K., Sedlacek, A. J., Kleinman, L., Springston, S. R., Wang, J., Chand, D., et al. (2019). Spherical tarball particles form through rapid chemical and physical changes of organic matter in biomass-burning smoke. *Proceedings of the National Academy of Sciences*, *116*(39), 19336–19341.

Alexander, D. T. L., Crozier, P. A., & Anderson, J. R. (2008). Brown carbon spheres in east Asian outflow and their optical properties. *Science*, *321*(5890), 833–836.

Andreae, M. O. (1995). *Climatic effects of changing atmospheric aerosol levels, in future climates of the world: A modeling perspective* (pp. 347–398). Amsterdam: Elsevier.

Andreae, M. O., & Merlet, P. (2001). Emission of trace gases and aerosols from biomass burning. *Global Biogeochemical Cycles*, *15*(4), 955–966.

Andreae, M. O., & Rosenfeld, D. (2008). Aerosol–cloud–precipitation interactions. Part 1. The nature and sources of cloud-active aerosols. *Earth-Science Reviews*, *89*, 13–41.

Ault, A. P., & Axson, J. L. (2016). Atmospheric aerosol chemistry: Spectroscopic and microscopic advances. *Analytical Chemistry*, *89*(1), 430–452.

Ault, A. P., Peters, T. M., Sawvel, E. J., Casuccio, G. S., Willis, R. D., Norris, G. A., & Grassian, V. H. (2012). Single-particle SEM-EDX analysis of Iron-containing coarse particulate matter in an Urban environment: Sources and distribution of Iron within Cleveland, Ohio. *Environmental Science & Technology*, *46*(8), 4331–4339.

Ault, A. P., Guasco, T. L., Ryder, O. S., Baltrusaitis, J., Cuadra-Rodriguez, L. A., Collins, D. B., et al. (2013a). Inside versus outside: Ion redistribution in nitric acid Reacted Sea spray aerosol particles as determined by single particle analysis. *Journal of the American Chemical Society*, *135*(39), 14528–14531.

Ault, A. P., Moffet, R. C., Baltrusaitis, J., Collins, D. B., Ruppel, M. J., Cuadra-Rodriguez, L. A., et al. (2013b). Size-dependent changes in sea spray aerosol composition and properties with different seawater conditions. *Environmental Science & Technology*, *47*(11), 5603–5612.

Baltrusaitis, J., Chen, H., Rubasinghege, G., & Grassian, V. H. (2012). Heterogeneous atmospheric chemistry of Lead oxide particles with nitrogen dioxide increases Lead solubility: Environmental and health implications. *Environmental Science & Technology*, *46*(23), 12806–12813.

Brasil, A. M., Farias, T. L., & Carvalho, M. G. (1999). A recipe for image characterization of fractal-like aggregates. *Journal of Aerosol Science*, *30*(10), 1379–1389.

Buseck, P. R., Adachi, K., Gelencsér, A., Tompa, É., & Pósfai, M. (2014). Ns-soot: A material-based term for strongly light-absorbing carbonaceous particles. *Aerosol Science and Technology*, *48*(7), 777–788.

Calderon-Garciduenas, L., Gonzalez-Maciel, A., Reynoso-Robles, R., Rodriguez-Lopez, J. L., Silva-Pereyra, H. G., Labrada-Delgado, G. J., et al. (2021). Environmental Fe, Ti,

Al, cu, hg, Bi, and Si nanoparticles in the Atrioventricular conduction Axis and the associated Ultrastructural damage in young urbanites: Cardiac arrhythmias caused by anthropogenic, industrial, E-waste, and indoor nanoparticles. *Environmental Science & Technology*, *55*(12), 8203–8214.

Cao, J., Xu, H., Xu, Q., Chen, B., & Kan, H. (2012). Fine particulate matter constituents and cardiopulmonary mortality in a heavily polluted Chinese City. *Environmental Health Perspectives*, *120*(3), 373–378.

Cappa, C. D., Onasch, T. B., Massoli, P., Worsnop, D. R., Bates, T. S., Cross, E. S., et al. (2012). Radiative absorption enhancements due to the mixing state of atmospheric black carbon. *Science*, *337*(6098), 1078–1081.

Chakrabarty, R. K., Moosmüller, H., Chen, L. W. A., Lewis, K., Arnott, W. P., Mazzoleni, C., et al. (2010). Brown carbon in tar balls from smoldering biomass combustion. *Atmospheric Chemistry and Physics*, *10*(13), 6363–6370.

Chen, Y., Shah, N., Huggins, F. E., & Huffman, G. P. (2006). Microanalysis of ambient particles from Lexington, KY, by electron microscopy. *Atmospheric Environment*, *40*(4), 651–663.

Chen, H., Grassian, V. H., Saraf, L. V., & Laskin, A. (2013a). Chemical imaging analysis of environmental particles using the focused ion beam/scanning electron microscopy technique: Microanalysis insights into atmospheric chemistry of fly ash. *Analyst*, *138*(2), 451–460.

Chen, Y., Ebenstein, A., Greenstone, M., & Li, H. (2013b). Evidence on the impact of sustained exposure to air pollution on life expectancy from China's Huai River policy. *Proceedings of the National Academy of Sciences*, *110*(32), 12936–12941.

Chen, S., Xu, L., Zhang, Y., Chen, B., Wang, X., Zhang, X., et al. (2017). Direct observations of organic aerosols in common wintertime hazes in North China: Insights into direct emissions from Chinese residential stoves. *Atmospheric Chemistry and Physics*, *17*(2), 1259–1270.

Chi, J., Li, W., Zhang, D., Zhang, J., Lin, Y., Shen, X., et al. (2015). Sea salt aerosols as a reactive surface for inorganic and organic acidic gases in the Arctic troposphere. *Atmospheric Chemistry and Physics*, *15*(19), 11341–11353.

China, S., Mazzoleni, C., Gorkowski, K., Aiken, A. C., & Dubey, M. K. (2013). Morphology and mixing state of individual freshly emitted wildfire carbonaceous particles. *Nature Communications*, *4*, 2122.

China, S., Scarnato, B., Owen, R. C., Zhang, B., Ampadu, M. T., Kumar, S., et al. (2015). Morphology and mixing state of aged soot particles at a remote marine free troposphere site: Implications for optical properties. *Geophysical Research Letters*, *42*(4), 1243–1250.

China, S., Wang, B. B., Weis, J., Rizzo, L., Brito, J., Cirino, G. G., et al. (2016). Rupturing of biological spores as a source of secondary particles in Amazonia. *Environmental Science & Technology*, *50*(22), 12179–12186.

China, S., Burrows, S. M., Wang, B. B., Harder, T. H., Weis, J., Tanarhte, M., et al. (2018). Fungal spores as a source of sodium salt particles in the Amazon basin. *Nature Communications*, *9*, 4793.

Conny, J. M., & Norris, G. A. (2011). Scanning Electron microanalysis and analytical challenges of mapping elements in Urban atmospheric particles. *Environmental Science & Technology*, *45*(17), 7380–7386.

Costa, D. L., & Dreher, K. L. (1997). Bioavailable transition metals in particulate matter mediate cardiopulmonary injury in healthy and compromised animal models. *Environmental Health Perspectives*, *105*, 1053–1060.

Crowley, J. N., Ammann, M., Cox, R. A., Hynes, R. G., Jenkin, M. E., Mellouki, A., et al. (2010). Evaluated kinetic and photochemical data for atmospheric chemistry: Volume V—heterogeneous reactions on solid substrates. *Atmospheric Chemistry and Physics*, *10*(18), 9059–9223.

Després, V. R., Huffman, J. A., Burrows, S. M., Hoose, C., Safatov, A. S., Buryak, G., et al. (2012). Primary biological aerosol particles in the atmosphere: A review. *Tellus B: Chemical and Physical Meteorology*, *64*, 15598.

Dockery, D. W., & Pope, C. A., III (1994). Acute respiratory effects of particulate air pollution. In G. S. Omenn (Ed.), *Annual review of public health*. Annual Reviews Inc.

Dresselhaus, M. S., Dresselhaus, G., Saito, R., & Jorio, A. (2005). Raman spectroscopy of carbon nanotubes. *Physics Reports*, *409*(2), 47–99.

Durlak, S. K., Biswas, P., Shi, J., & Bernhard, M. J. (1998). Characterization of polycyclic aromatic hydrocarbon particulate and gaseous emissions from polystyrene combustion. *Environmental Science & Technology*, *32*(15), 2301–2307.

Dye, J. A., Adler, K. B., Richards, J. H., & Dreher, K. L. (1999). Role of soluble metals in oil fly ash-induced airway epithelial injury and cytokine gene expression. *American Journal of Physiology-Lung Cellular and Molecular Physiology*, *277*(3), L498–L510.

Elbert, W., Taylor, P. E., Andreae, M. O., & Poschl, U. (2007). Contribution of fungi to primary biogenic aerosols in the atmosphere: Wet and dry discharged spores, carbohydrates, and inorganic ions. *Atmospheric Chemistry and Physics*, *7*(17), 4569–4588.

Feng, L., & Liao, W. (2016). Legislation, plans, and policies for prevention and control of air pollution in China: Achievements, challenges, and improvements. *Journal of Cleaner Production*, *112*, 1549–1558.

Fierce, L., Riemer, N., & Bond, T. C. (2015). Explaining variance in black carbon's aging timescale. *Atmospheric Chemistry and Physics*, *15*(6), 3173–3191.

Finlayson-Pitts, B. J., & Pitts, J. N., Jr. (1999). *Chemistry of the upper and lower atmosphere: Theory, experiments, and applications*. San Diego, California: Academic Press.

Flowers, L., Rieth, S. H., Cogliano, V. J., Foureman, G. L., Hertzberg, R., Hofmann, E. L., et al. (2002). Health assessment of polycyclic aromatic hydrocarbon mixtures: Current practices and future directions. *Polycyclic Aromatic Compounds*, *22*(3–4), 811–821.

Frenklach, M. (2002). Reaction mechanism of soot formation in flames. *Physical Chemistry Chemical Physics*, *4*(11), 2028–2037.

Ge, H. W., Ye, Z. P., & He, R. (2019). Raman spectroscopy of diesel and gasoline engine-out soot using different laser power. *Journal of Environmental Sciences*, *79*, 74–80.

Geng, H., Kang, S., Jung, H.-J., Choël, M., Kim, H., & Ro, C.-U. (2010a). Characterization of individual submicrometer aerosol particles collected in Incheon, Korea, by quantitative transmission electron microscopy energy-dispersive X-ray spectrometry. *Journal of Geophysical Research: Atmospheres, 115*(D15), D15306.

Geng, H., Ryu, J., Jung, H.-J., Chung, H., Ahn, K.-H., & Ro, C.-U. (2010b). Single-particle characterization of summertime Arctic aerosols collected at Ny-Ålesund, Svalbard. *Environmental Science & Technology, 44*(7), 2348–2353.

Gheorghe, I. F., & Ion, B. (2011). The effects of air pollutants on vegetation and the role of vegetation in reducing atmospheric pollution. In M. Khallaf (Ed.), *The impact of air pollution on health, economy, environment and agricultural sources*. InTech.

Girotto, G., China, S., Bhandari, J., Gorkowski, K., Scarnato, B. V., Capek, T., et al. (2018). Fractal-like tar ball aggregates from wildfire smoke. *Environmental Science & Technology Letters, 5*(6), 360–365.

von Glasow, R. (2008). Pollution meets sea salt. *Nature Geoscience, 1*(5), 292–293.

Graedel, T. E., & Keene, W. C. (1995). Tropospheric budget of reactive chlorine. *Global Biogeochemical Cycles, 9*(1), 47–77.

Grini, A., Myhre, G., Zender, C. S., & Isaksen, I. S. (2005). Model simulations of dust sources and transport in the global atmosphere: Effects of soil erodibility and wind speed variability. *Journal of Geophysical Research: Atmospheres, 110*(D2), D02205.

Guo, S., Hu, M., Zamora, M. L., Peng, J., Shang, D., Zheng, J., et al. (2014). Elucidating severe urban haze formation in China. *Proceedings of the National Academy of Sciences, 111*(49), 17373–17378.

Hand, J. L., Malm, W. C., Laskin, A., Day, D., Lee, T., Wang, C., et al. (2005). Optical, physical, and chemical properties of tar balls observed during the Yosemite aerosol characterization study. *Journal of Geophysical Research, 110*(D21). doi: 10.1029/2004JD005728

Harris, E., Sinha, B., Foley, S., Crowley, J. N., Borrmann, S., & Hoppe, P. (2012). Sulfur isotope fractionation during heterogeneous oxidation of SO_2 on mineral dust. *Atmospheric Chemistry and Physics, 12*(11), 4867–4884.

Haywood, J. M., Ramaswamy, V., & Soden, B. J. (1999). Tropospheric aerosol climate forcing in clear-sky satellite observations over the oceans. *Science, 283*(5406), 1299–1303.

Heald, C. L., Jacob, D. J., Park, R. J., Alexander, B., Fairlie, T. D., Yantosca, R. M., & Chu, D. A. (2006). Transpacific transport of Asian anthropogenic aerosols and its impact on surface air quality in the United States. *Journal of Geophysical Research: Atmospheres, 111*(D14), D14310.

Hitzfeld, B., Friedrichs, K. H., Ring, J., & Behrendt, H. (1997). Airborne particulate matter modulates the production of reactive oxygen species in human polymorphonuclear granulocytes. *Toxicology, 120*(3), 185–195.

Hou, C., Shao, L., Hu, W., Zhang, D., Zhao, C., Xing, J., et al. (2018). Characteristics and aging of traffic-derived particles in a highway tunnel at a coastal city in southern China. *Science of the Total Environment, 619-620*, 1385–1393.

Hu, W., Murata, K., Fukuyama, S., Kawai, Y., Oka, E., Uematsu, M., & Zhang, D. Z. (2017). Concentration and viability of airborne Bacteria over the Kuroshio extension region in the northwestern Pacific Ocean: Data from three cruises. *Journal of Geophysical Research: Atmospheres, 122*(23), 12892–12905.

Huang, J. P., Wang, T. H., Wang, W. C., Li, Z. Q., & Yan, H. R. (2014a). Climate effects of dust aerosols over east Asian arid and semiarid regions. *Journal of Geophysical Research: Atmospheres, 119*(19), 11398–11416.

Huang, R.-J., Zhang, Y., Bozzetti, C., Ho, K.-F., Cao, J.-J., Han, Y., et al. (2014b). High secondary aerosol contribution to particulate pollution during haze events in China. *Nature, 514*(7521), 218–222.

Huang, Y., Unger, N., Harper, K., & Heyes, C. (2020). Global climate and human health effects of the gasoline and diesel vehicle fleets. *GeoHealth, 4*(3), e2019GH000240.

Ito, A., & Shi, Z. (2016). Delivery of anthropogenic bioavailable iron from mineral dust and combustion aerosols to the ocean. *Atmospheric Chemistry and Physics, 16*(1), 85–99.

Jacobson, M. Z. (2001). Strong radiative heating due to the mixing state of black carbon in atmospheric aerosols. *Nature, 409*(6821), 695–697.

Jia, H., Li, S., Wu, L., Li, S., Sharma, V. K., & Yan, B. (2020). Cytotoxic free radicals on air-borne soot particles generated by burning wood or low-maturity coals. *Environmental Science & Technology, 54*(9), 5608–5618.

Kadiiska, M. B., Mason, R. P., Dreher, K. L., Costa, D. L., & Ghio, A. J. (1997). In vivo evidence of free radical formation in the rat lung after exposure to an emission source air pollution particle. *Chemical Research in Toxicology, 10*(10), 1104–1108.

Kampa, M., & Castanas, E. (2008). Human health effects of air pollution. *Environmental Pollution, 151*(2), 362–367.

Kan, H., Chen, R., & Tong, S. (2012). Ambient air pollution, climate change, and population health in China. *Environment International, 42*, 10–19.

Karagulian, F., Belis, C. A., Dora, C. F. C., Prüss-Ustün, A. M., Bonjour, S., Adair-Rohani, H., & Amann, M. (2015). Contributions to cities' ambient particulate matter (PM): A systematic review of local source contributions at global level. *Atmospheric Environment, 120*, 475–483.

Kojima, T., Buseck, P. R., Iwasaka, Y., Matsuki, A., & Trochkine, D. (2006). Sulfate-coated dust particles in the free troposphere over Japan. *Atmospheric Research, 82*(3), 698–708.

Koppmann, R., von Czapiewski, K., & Reid, J. S. (2005). A review of biomass burning emissions, part I: Gaseous emissions of carbon monoxide, methane, volatile organic compounds, and nitrogen containing compounds. *Atmospheric Chemistry and Physics Discussions, 2005*, 10455–10516.

Kulkarni, G., Zhang, K., Zhao, C., Nandasiri, M., Shutthanandan, V., Liu, X. H., et al. (2015). Ice formation on nitric acid-coated dust particles: Laboratory and modeling studies. *Journal of Geophysical Research: Atmospheres, 120*(15), 7682–7698.

Laskin, A., Iedema, M. J., & Cowin, J. P. (2003). Time-resolved aerosol collector for CCSEM/EDX single-particle analysis. *Aerosol Science and Technology, 37*(3), 246–260.

Laskin, A., Iedema, M. J., Ichkovich, A., Graber, E. R., Taraniuk, I., & Rudich, Y. (2005a). Direct observation of completely processed calcium carbonate dust particles. *Faraday Discussions*, *130*, 453–468.

Laskin, A., Wietsma, T. W., Krueger, B. J., & Grassian, V. H. (2005b). Heterogeneous chemistry of individual mineral dust particles with nitric acid: A combined CCSEM/EDX, ESEM, and ICP-MS study. *Journal of Geophysical Research: Atmospheres*, *110*(D10), D10208.

Laskin, A., Moffet, R. C., Gilles, M. K., Fast, J. D., Zaveri, R. A., Wang, B., et al. (2012). Tropospheric chemistry of internally mixed sea salt and organic particles: Surprising reactivity of NaCl with weak organic acids. *Journal of Geophysical Research: Atmospheres*, *117*(D15), D15302.

Laskin, A., Gilles, M. K., Knopf, D. A., Wang, B., & China, S. (2016). Progress in the analysis of complex atmospheric particles. *Annual Review of Analytical Chemistry*, *9*(1), 117–143.

Laskin, A., Moffet, R. C., & Gilles, M. K. (2019). Chemical imaging of atmospheric particles. *Accounts of Chemical Research*, *52*(12), 3419–3431.

Lau, W. K. M., Yuan, C., & Li, Z. (2018). Origin, maintenance and variability of the Asian Tropopause aerosol layer (ATAL): The roles of monsoon dynamics. *Scientific Reports*, *8*(1), 3960.

Lee, H. D., Ray, K. K., & Tivanski, A. V. (2017). Solid, semi-solid, and liquid phase states of individual submicrometer particles directly probed using atomic force microscopy. *Analytical Chemistry*, *89*(23), 12720–12726.

Lelieveld, J. (2017). Clean air in the Anthropocene. *Faraday Discussions*, *200*, 693–703.

Lelieveld, J., Haines, A., & Pozzer, A. (2018). Age-dependent health risk from ambient air pollution: A modelling and data analysis of childhood mortality in middle-income and low-income countries. *The Lancet Planetary Health*, *2*(7), e292–e300.

Lelieveld, J., Pozzer, A., Pöschl, U., Fnais, M., Haines, A., & Münzel, T. (2020). Loss of life expectancy from air pollution compared to other risk factors: A worldwide perspective. *Cardiovascular Research*, *116*(11), 1910–1917.

Lewtas, J. (2007). Air pollution combustion emissions: Characterization of causative agents and mechanisms associated with cancer, reproductive, and cardiovascular effects. *Mutation Research/Reviews in Mutation Research*, *636*(1–3), 95–133.

Li, J., Anderson, J. R., & Buseck, P. R. (2003a). TEM study of aerosol particles from clean and polluted marine boundary layers over the North Atlantic. *Journal of Geophysical Research: Atmospheres*, *108*(D6), 4189.

Li, J., Pósfai, M., Hobbs, P. V., & Buseck, P. R. (2003b). Individual aerosol particles from biomass burning in southern Africa: 2, compositions and aging of inorganic particles. *Journal of Geophysical Research: Atmospheres*, *108*(D13), 8484.

Li, W., & Shao, L. (2009a). Transmission electron microscopy study of aerosol particles from the brown hazes in northern China. *Journal of Geophysical Research: Atmospheres*, *114*(D9), D09302.

Li, W. J., & Shao, L. Y. (2009b). Observation of nitrate coatings on atmospheric mineral dust particles. *Atmospheric Chemistry and Physics*, *9*(6), 1863–1871.

Li, W., & Shao, L. (2012). Chemical modification of dust particles during different dust storm episodes. *Aerosol and Air Quality Research*, *12*(6), 1095–1104.

Li, W., Shao, L., Wang, Z., Shen, R., Yang, S., & Tang, U. (2010a). Size, composition, and mixing state of individual aerosol particles in a South China coastal city. *Journal of Environmental Sciences*, *22*(4), 561–569.

Li, W. J., Shao, L. Y., & Buseck, P. R. (2010b). Haze types in Beijing and the influence of agricultural biomass burning. *Atmospheric Chemistry and Physics*, *10*(17), 8119–8130.

Li, W., Shao, L., Shen, R., Yang, S., Wang, Z., & Tang, U. (2011a). Internally mixed sea salt, soot, and sulfates at Macao, a Coastal City in South China. *Journal of the Air & Waste Management Association*, *61*(11), 1166–1173.

Li, W., Zhou, S., Wang, X., Xu, Z., Yuan, C., Yu, Y., et al. (2011b). Integrated evaluation of aerosols from regional brown hazes over northern China in winter: Concentrations, sources, transformation, and mixing states. *Journal of Geophysical Research: Atmospheres*, *116*(D9), D09301.

Li, W., Shi, Z., Zhang, D., Zhang, X., Li, P., Feng, Q., et al. (2012). Haze particles over a coal-burning region in the China loess plateau in winter: Three flight missions in December 2010. *Journal of Geophysical Research: Atmospheres*, *117*(D12), D12306.

Li, W., Wang, T., Zhou, S., Lee, S., Huang, Y., Gao, Y., & Wang, W. (2013a). Microscopic observation of metal-containing particles from Chinese continental outflow observed from a non-industrial site. *Environmental Science & Technology*, *47*(16), 9124–9131.

Li, W., Wang, Y., Collett, J. L., Chen, J., Zhang, X., Wang, Z., & Wang, W. (2013b). Microscopic evaluation of trace metals in cloud droplets in an acid precipitation region. *Environmental Science & Technology*, *47*(9), 4172–4180.

Li, W., Chi, J., Shi, Z., Wang, X., Chen, B., Wang, Y., et al. (2014a). Composition and hygroscopicity of aerosol particles at Mt. Lu in South China: Implications for acid precipitation. *Atmospheric Environment*, *94*, 626–636.

Li, W., Shao, L., Shi, Z., Chen, J., Yang, L., Yuan, Q., et al. (2014b). Mixing state and hygroscopicity of dust and haze particles before leaving Asian continent. *Journal of Geophysical Research: Atmospheres*, *119*(2), 1044–1059.

Li, W. J., Chen, S. R., Xu, Y. S., Guo, X. C., Sun, Y. L., Yang, X. Y., et al. (2015). Mixing state and sources of submicron regional background aerosols in the northern Qinghai–Tibet plateau and the influence of biomass burning. *Atmospheric Chemistry and Physics*, *15*(23), 13365–13376.

Li, W., Shao, L., Zhang, D., Ro, C.-U., Hu, M., Bi, X., et al. (2016a). A review of single aerosol particle studies in the atmosphere of East Asia: Morphology, mixing state, source, and heterogeneous reactions. *Journal of Cleaner Production*, *112*, 1330–1349.

Li, W., Sun, J., Xu, L., Shi, Z., Riemer, N., Sun, Y., et al. (2016b). A conceptual framework for mixing structures in individual aerosol particles. *Journal of Geophysical Research: Atmospheres*, *121*(22), 13784–13798.

Li, W., Xu, L., Liu, X., Zhang, J., Lin, Y., Yao, X., et al. (2017). Air pollution–aerosol interactions produce more bioavailable iron for ocean ecosystems. *Science Advances*, *3*(3), e1601749.

Li, X., Wu, J. R., Elser, M., Feng, T., Cao, J. J., El-Haddad, I., et al. (2018). Contributions of residential coal combustion to the air quality in Beijing-Tianjin-Hebei (BTH), China: A case study. *Atmospheric Chemistry and Physics*, *18*(14), 10675–10691.

Li, C., He, Q., Schade, J., Passig, J., Zimmermann, R., Meidan, D., et al. (2019). Dynamic changes in optical and chemical properties of tar ball aerosols by atmospheric photochemical aging. *Atmospheric Chemistry and Physics*, *19*(1), 139–163.

Li, W., Liu, L., Xu, L., Zhang, J., Yuan, Q., Ding, X., et al. (2020). Overview of primary biological aerosol particles from a Chinese boreal forest: Insight into morphology, size, and mixing state at microscopic scale. *Science of the Total Environment*, *719*, 137520.

Li, W., Liu, L., Zhang, J., Xu, L., Wang, Y., Sun, Y., & Shi, Z. (2021). Microscopic evidence for phase separation of organic species and inorganic salts in fine ambient aerosol particles. *Environmental Science & Technology*, *55*(4), 2234–2242.

Liao, H., Chen, W.-T., & Seinfeld, J. H. (2006). Role of climate change in global predictions of future tropospheric ozone and aerosols. *Journal of Geophysical Research: Atmospheres*, *111*(D12), D12304.

Lin, J., Pan, D., Davis, S. J., Zhang, Q., He, K., Wang, C., et al. (2014). China's international trade and air pollution in the United States. *Proceedings of the National Academy of Sciences*, *111*(5), 1736–1741.

Liu, J., Mauzerall, D. L., Chen, Q., Zhang, Q., Song, Y., Peng, W., et al. (2016). Air pollutant emissions from Chinese households: A major and underappreciated ambient pollution source. *Proceedings of the National Academy of Sciences of the United States of America*, *113*(28), 7756–7761.

Liu, L., Kong, S., Zhang, Y., Wang, Y., Xu, L., Yan, Q., et al. (2017). Morphology, composition, and mixing state of primary particles from combustion sources—crop residue, wood, and solid waste. *Scientific Reports*, *7*, 5047.

Liu, L., Zhang, J., Xu, L., Yuan, Q., Huang, D., Chen, J., et al. (2018). Cloud scavenging of anthropogenic refractory particles at a mountain site in North China. *Atmospheric Chemistry and Physics*, *18*(19), 14681–14693.

Liu, C., Chen, R., Sera, F., Vicedo-Cabrera, A. M., Guo, Y., Tong, S., et al. (2019). Ambient particulate air pollution and daily mortality in 652 cities. *New England Journal of Medicine*, *381*(8), 705–715.

Liu, L., Zhang, J., Zhang, Y., Wang, Y., Xu, L., Yuan, Q., et al. (2021). Persistent residential burning-related primary organic particles during wintertime hazes in North China: Insights into their aging and optical changes. *Atmospheric Chemistry and Physics*, *21*(3), 2251–2265.

Löndahl, J., Massling, A., Pagels, J., Swietlicki, E., Vaclavik, E., & Loft, S. (2007). Size-resolved respiratory-tract deposition of fine and ultrafine hydrophobic and hygroscopic aerosol particles during rest and exercise. *Inhalation Toxicology*, *19*(2), 109–116.

Lu, D., Luo, Q., Chen, R., Zhuansun, Y., Jiang, J., Wang, W., et al. (2020). Chemical multi-fingerprinting of exogenous ultrafine particles in human serum and pleural effusion. *Nature Communications*, *11*, 2567.

Ma, X., Zangmeister, C. D., Gigault, J., Mulholland, G. W., & Zachariah, M. R. (2013). Soot aggregate restructuring during water processing. *Journal of Aerosol Science*, *66*, 209–219.

Mahowald, N. M., Kloster, S., Engelstaedter, S., Moore, J. K., Mukhopadhyay, S., McConnell, J. R., et al. (2010). Observed 20th century desert dust variability: Impact on climate and biogeochemistry. *Atmospheric Chemistry and Physics*, *10*(22), 10875–10893.

Martin, S. T., Andreae, M. O., Artaxo, P., Baumgardner, D., Chen, Q., Goldstein, A. H., et al. (2010). Sources and properties of Amazonian aerosol particles. *Reviews of Geophysics*, *48*, RG2002.

May, N. W., Olson, N. E., Panas, M., Axson, J. L., Tirella, P. S., Kirpes, R. M., et al. (2018). Aerosol emissions from Great Lakes harmful algal blooms. *Environmental Science & Technology*, *52*(2), 397–405.

Meskhidze, N., Chameides, W., & Nenes, A. (2005). Dust and pollution: A recipe for enhanced ocean fertilization? *Journal of Geophysical Research: Atmospheres*, *110*(D3), D03301.

Ming, Y., & Russell, L. M. (2001). Predicted hygroscopic growth of sea salt aerosol. *Journal of Geophysical Research: Atmospheres*, *106*(D22), 28259–28274.

Moffet, R. C., & Prather, K. A. (2009). In-situ measurements of the mixing state and optical properties of soot with implications for radiative forcing estimates. *Proceedings of the National Academy of Sciences*, *106*(29), 11872–11877.

Moffet, R. C., Henn, T. R., Tivanski, A. V., Hopkins, R. J., Desyaterik, Y., Kilcoyne, A. L. D., et al. (2010). Microscopic characterization of carbonaceous aerosol particle aging in the outflow from Mexico City. *Atmospheric Chemistry and Physics*, *10*(3), 961–976.

Moffet, R. C., Furutani, H., Rödel, T. C., Henn, T. R., Sprau, P. O., Laskin, A., et al. (2012). Iron speciation and mixing in single aerosol particles from the Asian continental outflow. *Journal of Geophysical Research: Atmospheres*, *117*(D7), D07204.

Moffet, R. C., Rödel, T., Kelly, S. T., Yu, X.-Y., Carroll, G., Fast, J., et al. (2013). Spectro-microscopic measurements of carbonaceous aerosol aging in Central California. *Atmospheric Chemistry and Physics*, *13*(20), 10445–10459.

Mohankumar, S., & Senthilkumar, P. (2017). Particulate matter formation and its control methodologies for diesel engine: A comprehensive review. *Renewable and Sustainable Energy Reviews*, *80*, 1227–1238.

Mohapatra, K., & Biswal, S. (2014). Effect of particulate matter (PM) on plants, climate, ecosystem and human health. *International Journal of Advanced Technology in Engineering and Science*, *2*(4), 2348–7550.

Myers-Pigg, A. N., Griffin, R. J., Louchouarn, P., Norwood, M. J., Sterne, A., & Cevik, B. K. (2016). Signatures of biomass burning aerosols in the plume of a saltmarsh wildfire in South Texas. *Environmental Science & Technology*, *50*(17), 9308–9314.

Niu, H., Hu, W., Zhang, D., Wu, Z., Guo, S., Pian, W., et al. (2016). Variations of fine particle physicochemical properties during a heavy haze episode in the winter of Beijing. *Science of the Total Environment*, *571*, 103–109.

Oakes, M., Weber, R. J., Lai, B., Russell, A., & Ingall, E. D. (2012). Characterization of iron speciation in urban and rural single particles using XANES spectroscopy and micro X-ray fluorescence measurements: Investigating the relationship between speciation and fractional iron solubility. *Atmospheric Chemistry and Physics*, *12*(2), 745–756.

Oberdörster, G., Sharp, Z., Atudorei, V., Elder, A., Gelein, R., Kreyling, W., & Cox, C. (2004). Translocation of inhaled ultrafine particles to the brain. *Inhalation Toxicology*, *16*(6–7), 437–445.

Oberdörster, G., Oberdörster, E., & Oberdörster, J. (2005). Nanotoxicology: An emerging discipline evolving from studies of ultrafine particles. *Environmental Health Perspectives*, *113*(7), 823–839.

Oh, C., & Sorensen, C. M. (1997). The effect of overlap between monomers on the determination of fractal cluster morphology. *Journal of Colloid and Interface Science*, *193*(1), 17–25.

Pagels, J., Dutcher, D. D., Stolzenburg, M. R., McMurry, P. H., Gälli, M. E., & Gross, D. S. (2013). Fine-particle emissions from solid biofuel combustion studied with single-particle mass spectrometry: Identification of markers for organics, soot, and ash components. *Journal of Geophysical Research: Atmospheres*, *118*(2), 859–870.

Peng, J., Hu, M., Guo, S., Du, Z., Zheng, J., Shang, D., et al. (2016). Markedly enhanced absorption and direct radiative forcing of black carbon under polluted urban environments. *Proceedings of the National Academy of Sciences*, *113*(16), 4266–4271.

Pohlker, C., Wiedemann, K. T., Sinha, B., Shiraiwa, M., Gunthe, S. S., Smith, M., et al. (2012). Biogenic potassium salt particles as seeds for secondary organic aerosol in the Amazon. *Science*, *337*(6098), 1075–1078.

Pope, C. A., III, Burnett, R. T., Thun, M. J., Calle, E. E., Krewski, D., Ito, K., & Thurston, G. D. (2002). Lung cancer, cardiopulmonary mortality, and long-term exposure to fine particulate air pollution. *JAMA*, *287*(9), 1132–1141.

Pöschl, U. (2005). Atmospheric aerosols: Composition, transformation, climate and health effects. *Angewandte Chemie-International Edition*, *44*(46), 7520–7540.

Pósfai, M., & Buseck, P. R. (2010). Nature and climate effects of individual tropospheric aerosol particles. *Annual Review of Earth and Planetary Sciences*, *38*(1), 17–43.

Pósfai, M., Anderson, J. R., Buseck, P. R., & Sievering, H. (1995). Compositional variations of sea-salt-mode aerosol particles from the North Atlantic. *Journal of Geophysical Research: Atmospheres*, *100*(D11), 23063–23074.

Pósfai, M., Li, J., Anderson, J. R., & Buseck, P. R. (2003). Aerosol bacteria over the Southern Ocean during ACE-1. *Atmospheric Research*, *66*(4), 231–240.

Pósfai, M., Gelencsér, A., Simonics, R., Arató, K., Li, J., Hobbs, P. V., & Buseck, P. R. (2004). Atmospheric tar balls: Particles from biomass and biofuel burning. *Journal of Geophysical Research: Atmospheres*, *109*(D6), D06213.

Prather, K. A., Hatch, C. D., & Grassian, V. H. (2008). Analysis of atmospheric aerosols. *Annual Review of Analytical Chemistry*, *1*(1), 485–514.

Prather, K. A., Bertram, T. H., Grassian, V. H., Deane, G. B., Stokes, M. D., DeMott, P. J., et al. (2013). Bringing the ocean into the laboratory to probe the chemical complexity of sea spray aerosol. *Proceedings of the National Academy of Sciences*, *110*(19), 7550–7555.

Prospero, J. M., Collard, F. X., Molinie, J., & Jeannot, A. (2014). Characterizing the annual cycle of African dust transport to the Caribbean Basin and South America and its impact on the environment and air quality. *Global Biogeochemical Cycles*, *28*(7), 757–773.

Przybysz, A., Saebo, A., Hanslin, H. M., & Gawronski, S. W. (2014). Accumulation of particulate matter and trace elements on vegetation as affected by pollution level, rainfall and the passage of time. *Science of the Total Environment*, *481*, 360–369.

Rao, X., & Collett, J. L. (1998). The drop size-dependence of Iron and manganese concentrations in clouds and fogs: Implications for sulfate production. *Journal of Atmospheric Chemistry*, *30*(2), 273–289.

Reid, J. P., Bertram, A. K., Topping, D. O., Laskin, A., Martin, S. T., Petters, M. D., et al. (2018). The viscosity of atmospherically relevant organic particles. *Nature Communications*, *9*, 956.

Riemer, N., Vogel, H., & Vogel, B. (2004). Soot aging time scales in polluted regions during day and night. *Atmospheric Chemistry and Physics*, *4*(7), 1885–1893.

Riemer, N., Ault, A. P., West, M., Craig, R. L., & Curtis, J. H. (2019). Aerosol mixing state: Measurements, modeling, and impacts. *Reviews of Geophysics*, *57*(2), 187–249.

Ryder, O. S., Ault, A. P., Cahill, J. F., Guasco, T. L., Riedel, T. P., Cuadra-Rodriguez, L. A., et al. (2014). On the role of particle inorganic mixing state in the reactive uptake of N2O5 to ambient aerosol particles. *Environmental Science & Technology*, *48*(3), 1618–1627.

Schlesinger, R. B., & Cassee, F. (2003). Atmospheric secondary inorganic particulate matter: The toxicological perspective as a basis for health effects risk assessment. *Inhalation Toxicology*, *15*(3), 197–235.

Sedlacek, A. J., III, Buseck, P. R., Adachi, K., Onasch, T. B., Springston, S. R., & Kleinman, L. (2018). Formation and evolution of tar balls from northwestern US wildfires. *Atmospheric Chemistry and Physics*, *18*(15), 11289–11301.

Seinfeld, J. H., & Pandis, S. N. (2006). *Atmospheric chemistry and physics: From air pollution to climate change*. John Wiley & Sons.

Seinfeld, J. H., Carmichael, G. R., Arimoto, R., Conant, W. C., Brechtel, F. J., Bates, T. S., et al. (2004). ACE-ASIA: Regional climatic and atmospheric chemical effects of Asian dust and pollution. *Bulletin of the American Meteorological Society*, *85*(3), 367–380.

Shao, L., Shi, Z., Jones, T. P., Li, J., Whittaker, A. G., & BéruBé, K. A. (2006). Bioreactivity of particulate matter in Beijing air: Results from plasmid DNA assay. *Science of the Total Environment*, *367*(1), 261–272.

Shi, Z. B., Krom, M. D., Bonneville, S., Baker, A. R., Jickells, T. D., & Benning, L. G. (2009). Formation of Iron nanoparticles and increase in Iron reactivity in mineral dust during simulated cloud processing. *Environmental Science & Technology*, *43*(17), 6592–6596.

Shiraiwa, M., Selzle, K., & Pöschl, U. (2012). Hazardous components and health effects of atmospheric aerosol particles: Reactive oxygen species, soot, polycyclic aromatic compounds and allergenic proteins. *Free Radical Research*, *46*(8), 927–939.

Sobanska, S., Hwang, H., Choël, M., Jung, H.-J., Eom, H.-J., Kim, H., et al. (2012). Investigation of the chemical mixing state of individual Asian dust particles by the combined use of Electron probe X-ray microanalysis and Raman microspectrometry. *Analytical Chemistry*, *84*(7), 3145–3154.

Soewono, A., & Rogak, S. (2011). Morphology and Raman spectra of engine-emitted particulates. *Aerosol Science and Technology*, *45*(10), 1206–1216.

Song, C., He, J., Wu, L., Jin, T., Chen, X., Li, R., et al. (2017). Health burden attributable to ambient PM2.5 in China. *Environmental Pollution*, *223*, 575–586.

Streets, D. G., Yarber, K. F., Woo, J. H., & Carmichael, G. R. (2003). Biomass burning in Asia: Annual and seasonal estimates and atmospheric emissions. *Global Biogeochemical Cycles*, *17*(4), 1099.

Sullivan, R. C., Guazzotti, S. A., Sodeman, D. A., & Prather, K. A. (2007). Direct observations of the atmospheric processing of Asian mineral dust. *Atmospheric Chemistry and Physics*, *7*, 1213–1236.

Tang, M., Cziczo, D. J., & Grassian, V. H. (2016). Interactions of water with mineral dust aerosol: Water adsorption, Hygroscopicity, cloud condensation, and ice nucleation. *Chemical Reviews*, *116*(7), 4205–4259.

Tobo, Y., Zhang, D., Matsuki, A., & Iwasaka, Y. (2010). Asian dust particles converted into aqueous droplets under remote marine atmospheric conditions. *Proceedings of the National Academy of Sciences*, *107*(42), 17905–17910.

Tóth, A., Hoffer, A., Nyirő-Kósa, I., Pósfai, M., & Gelencsér, A. (2014). Atmospheric tar balls: Aged primary droplets from biomass burning? *Atmospheric Chemistry and Physics*, *14*(13), 6669–6675.

Usher, C. R., Michel, A. E., & Grassian, V. H. (2003). Reactions on mineral dust. *Chemical Reviews*, *103*(12), 4883–4939.

Van Donkelaar, A., Martin, R. V., Brauer, M., Kahn, R., Levy, R., Verduzco, C., & Villeneuve, P. J. (2010). Global estimates of ambient fine particulate matter concentrations from satellite-based aerosol optical depth: Development and application. *Environmental Health Perspectives*, *118*(6), 847–855.

Walser, S. M., Gerstner, D. G., Brenner, B., Bünger, J., Eikmann, T., Janssen, B., et al. (2015). Evaluation of exposure–response relationships for health effects of microbial bioaerosols – A systematic review. *International Journal of Hygiene and Environmental Health*, *218*(7), 577–589.

Wang, Y., Zhuang, G., Sun, Y., & An, Z. (2006). The variation of characteristics and formation mechanisms of aerosols in dust, haze, and clear days in Beijing. *Atmospheric Environment*, *40*(34), 6579–6591.

Wang, Y. S., Yao, L., Wang, L. L., Liu, Z. R., Ji, D. S., Tang, G. Q., et al. (2014). Mechanism for the formation of the January 2013 heavy haze pollution episode over central and eastern China. *Science China-Earth Sciences*, *57*(1), 14–25.

Wang, G., Zhang, R., Gomez, M. E., Yang, L., Levy Zamora, M., Hu, M., et al. (2016). Persistent sulfate formation from London fog to Chinese haze. *Proceedings of the National Academy of Sciences*, *113*(48), 13630–13635.

Wang, Y., Liu, F., He, C., Bi, L., Cheng, T., Wang, Z., et al. (2017). Fractal dimensions and mixing structures of soot particles during atmospheric processing. *Environmental Science & Technology Letters*, *4*(11), 487–493.

Wang, J., Lei, Y., & Ning, M. (2018). Chinese model for improving air quality: An assessment of action plan of air pollution prevention and control (in Chinese). *Environmental Protection*, *46*(2), 7–11.

Wang, Y., Li, W., Huang, J., Liu, L., Pang, Y., He, C., et al. (2021a). Nonlinear enhancement of Radiative absorption by black carbon in response to particle mixing structure. *Geophysical Research Letters*, *48*(24), e2021GL096437.

Wang, Y., Pang, Y., Huang, J., Bi, L., Che, H., Zhang, X., & Li, W. (2021b). Constructing shapes and mixing structures of black carbon particles with applications to optical calculations. *Journal of Geophysical Research: Atmospheres*, *126*(10), e2021JD034620.

West, J. J., Cohen, A., Dentener, F., Brunekreef, B., Zhu, T., Armstrong, B., et al. (2016). What we breathe impacts our health: Improving understanding of the link between air pollution and health. *Environmental Science & Technology*, *50*(10), 4895–4904.

Wittmaack, K., Wehnes, H., Heinzmann, U., & Agerer, R. (2005). An overview on bioaerosols viewed by scanning electron microscopy. *Science of the Total Environment*, *346*(1–3), 244–255.

Wu, S. W., Deng, F. R., Huang, J., Wang, H. Y., Shima, M., Wang, X., et al. (2013). Blood pressure changes and chemical constituents of particulate air pollution: Results from the healthy volunteer natural relocation (HVNR) study. *Environmental Health Perspectives*, *121*(1), 66–72.

Yao, L., Garmash, O., Bianchi, F., Zheng, J., Yan, C., Kontkanen, J., et al. (2018). Atmospheric new particle formation from sulfuric acid and amines in a Chinese megacity. *Science*, *361*(6399), 278–281.

You, Y., Renbaum-Wolff, L., Carreras-Sospedra, M., Hanna, S. J., Hiranuma, N., Kamal, S., et al. (2012). Images reveal that atmospheric particles can undergo liquid–liquid phase separations. *Proceedings of the National Academy of Sciences*, *109*(33), 13188–13193.

Yu, H., Li, W., Zhang, Y., Tunved, P., Dall'Osto, M., Shen, X., et al. (2019). Organic coating on sulfate and soot particles during late summer in the Svalbard archipelago. *Atmospheric Chemistry and Physics*, *19*(15), 10433–10446.

Yuan, C., Lau, W. K. M., Li, Z., & Cribb, M. (2019a). Relationship between Asian monsoon strength and transport of surface aerosols to the Asian Tropopause aerosol layer (ATAL): Interannual variability and decadal changes. *Atmospheric Chemistry and Physics*, *19*(3), 1901–1913.

Yuan, Q., Xu, J., Wang, Y., Zhang, X., Pang, Y., Liu, L., et al. (2019b). Mixing state and fractal dimension of soot particles at a remote site in the southeastern Tibetan plateau. *Environmental Science & Technology*, *53*(14), 8227–8234.

Yuan, Q., Xu, J., Liu, L., Zhang, A., Liu, Y., Zhang, J., et al. (2020). Evidence for large amounts of Brown carbonaceous Tarballs in the Himalayan atmosphere. *Environmental Science & Technology Letters*.

Zhang, D. Z., & Iwasaka, Y. (2001). Chlorine deposition on dust particles in marine atmosphere. *Geophysical Research Letters*, *28*(18), 3613–3616.

Zhang, R., Khalizov, A. F., Pagels, J., Zhang, D., Xue, H., & McMurry, P. H. (2008a). Variability in morphology, hygroscopicity, and optical properties of soot aerosols during atmospheric processing. *Proceedings of the National Academy of Sciences*, *105*(30), 10291–10296.

Zhang, Y., Schauer, J. J., Zhang, Y., Zeng, L., Wei, Y., Liu, Y., & Shao, M. (2008b). Characteristics of particulate carbon emissions from real-world Chinese coal combustion. *Environmental Science & Technology*, *42*(14), 5068–5073.

Zhang, G., Han, B., Bi, X., Dai, S., Huang, W., Chen, D., et al. (2015). Characteristics of individual particles in the atmosphere of Guangzhou by single particle mass spectrometry. *Atmospheric Research*, *153*, 286–295.

Zhang, J., Liu, L., Wang, Y., Ren, Y., Wang, X., Shi, Z., et al. (2017a). Chemical composition, source, and process of urban aerosols during winter haze formation in Northeast China. *Environmental Pollution*, *231*, 357–366.

Zhang, R., Liu, C., Zhou, G., Sun, J., Liu, N., Hsu, P.-C., et al. (2017b). Morphology and property investigation of primary particulate matter particles from different sources. *Nano Research*.

Zhang, T., Bai, Y. X., Hong, X. L., Sun, L. W., & Liu, Y. J. (2017c). Particulate matter and heavy metal deposition on the leaves of Euonymus japonicus during the east Asian monsoon in Beijing, China. *PLoS One*, *12*(6), 16.

Zhang, Y., Yuan, Q., Huang, D., Kong, S., Zhang, J., Wang, X., et al. (2018a). Direct observations of fine primary particles from residential coal burning: Insights into their morphology, composition, and hygroscopicity. *Journal of Geophysical Research: Atmospheres*, *123*(22), 12964–12979.

Zhang, Y., Yuan, Q., Huang, D., Kong, S., Zhang, J., Wang, X., et al. (2018b). Direct observations of fine primary particles from residential coal burning: Insights into their morphology, composition, and Hygroscopicity. *Journal of Geophysical Research: Atmospheres*, *123*(22), 12,964–912,979.

Zhang, J., Liu, L., Xu, L., Lin, Q., Zhao, H., Wang, Z., et al. (2020). Exploring wintertime regional haze in Northeast China: Role of coal and biomass burning. *Atmospheric Chemistry and Physics*, *20*(9), 5355–5372.

Zhang, J., Yuan, Q., Liu, L., Wang, Y., Zhang, Y., Xu, L., et al. (2021). Trans-regional transport of haze particles from the North China plain to Yangtze River Delta during winter. *Journal of Geophysical Research: Atmospheres*, *126*(8), e2020JD033778.

Zheng, G. J., Duan, F. K., Su, H., Ma, Y. L., Cheng, Y., Zheng, B., et al. (2015). Exploring the severe winter haze in Beijing: The impact of synoptic weather, regional transport and heterogeneous reactions. *Atmospheric Chemistry and Physics*, *15*(6), 2969–2983.

Zhou, W., Jiang, J., Duan, L., & Hao, J. (2016). Evolution of submicrometer organic aerosols during a complete residential coal combustion process. *Environmental Science & Technology*, *50*(14), 7861–7869.

Zhou, C., Zhang, H., Zhao, S., & Li, J. (2017). Simulated effects of internal mixing of anthropogenic aerosols on the aerosol–radiation interaction and global temperature. *International Journal of Climatology*, *37*(S1), 972–986.

3

Importance of Microanalysis in Climate Studies of Atmospheric Aerosol Particles

Cynthia H. Twohy[1,2]

ABSTRACT

Atmospheric aerosol particles impact climate through direct interactions with shortwave and longwave radiation. They also affect climate indirectly through multiple types of interactions with clouds. The magnitude and sign of these effects depend on small-scale characteristics of individual particles, such as size, shape, and composition. These are properties that can be measured well with microanalysis techniques. A variety of studies linking aerosol particles to climate impacts using microanalysis methods such as electron microscopy, scanning transmission X-ray microscopy, Raman microspectroscopy, and atomic force microscopy are discussed.

3.1. INTRODUCTION

Aerosol particles impact the Earth's climate directly because they scatter solar radiation back to space and also absorb shortwave and longwave radiation within the atmosphere. These are called "direct" aerosol effects. The optical properties, or shortwave scattering and absorption characteristics, vary greatly depending on aerosol type. Sulfates, sea spray, and most organic particles predominately scatter sunlight back into space, producing a cooling effect. In contrast, black and brown carbon (Andreae & Gelencsér, 2006) absorb radiation across a broad wavelength range from the ultraviolet to the infrared (Liu et al., 2020), for a warming effect. The single-scattering albedo, or ratio of scattering to total extinction of light by particles, is influenced by the amount of light-absorbing material, such as black or brown carbon, in particles. Thus, chemical composition is of first-order importance in understanding climate impacts of aerosols. Either cooling or warming can dominate regionally; the net global effect of this direct interaction of aerosol particles with radiation is cooling, in contrast to, but not sufficient to counteract global warming due to, increasing greenhouse gases (Arias et al., 2021).

In addition to composition, other properties of aerosols that impact climate are their size and shape. Particle size directly impacts the amount of radiation that is scattered or absorbed for particles of a certain composition. Particle shape is also important. Although optical properties of aerosol particles are typically modeled using Mie theory that assumes they are spherical, many particle types, such as black carbon, mineral dust, and nonhydrated sea spray, are irregularly shaped. This affects their angular scattering properties (e.g., Jaggard et al., 1981) and therefore the fraction of light that they scatter back into space. Particle size and shape also affect the transport and sedimentation rates of particles, which influence their lifetime within the atmosphere.

[1] *NorthWest Research Associates, Redmond, WA, USA*

[2] *Scripps Institution of Oceanography, University of California at San Diego, San Diego, CA, USA*

Microanalysis of Atmospheric Particles: Techniques and Applications, Geophysical Monograph 288, First Edition.
Edited by Joseph M. Conny and Peter R. Buseck.

DOI: 10.1002/9781119554318.ch3

Interactions of aerosol particles with clouds are an important influence on climate because particles act as cloud condensation nuclei (CCN) and ice-nucleating particles (INPs). Because clouds are ubiquitous in the atmosphere, and cloud droplets and ice crystals are much larger than most aerosol particles, these "indirect" aerosol effects potentially have a large radiative impact. The largest uncertainties in predicting a future climate (Boucher et al., 2013) are due to the complexity of aerosol/cloud interactions. Direct and indirect climate effects of aerosol particles can also create feedbacks that influence aerosol sources and chemical and biological processes associated with them (Figure 3.1) (Pöschl, 2005).

In the case of CCN, which form cloud droplets, cloud-nucleating efficiency is determined by particle hygroscopicity and size (and, to a lesser extent, surface tension), in addition to atmospheric conditions. Factors influencing how effectively particles act as INPs are more complex (Kanji et al., 2017) and are determined not only by temperature, relative humidity, and nucleation mode but also by particle composition, structure, and size. Some biological entities can also form ice, and these are often identifiable by microscopy and associated methods.

Thus, aerosol composition, size, and shape are all important in determining their direct and indirect climate impact, and an accurate assessment of these properties is needed for input into regional and global climate models. These properties are all amenable to measurement via microanalysis techniques, some of which are discussed in this chapter. Indeed, these techniques can reveal characteristics of particles, such as aspect ratio, structure, and mixing state, that are not generally measurable by bulk or other single-particle methods. For example, these techniques are critical for assessing aerosol types like mineral dust that can have complex and varying morphologies (Kandler et al., 2009; Mishra & Tripathi, 2008).

In the following section, studies using various microanalysis techniques to examine aerosol climatic impacts are discussed. These include scanning electron microscopy (SEM) and transmission electron microscopy (TEM) and their counterparts under environmental conditions (ESEM and ETEM, respectively), scanning transmission X-ray microscopy (STXM), energy-dispersive X-ray spectrometry (EDS), electron energy-loss spectrometry (EELS), and near-edge X-ray absorption fine-structure spectroscopy (NEXAFS). Also included are a few publications describing the use of Raman microspectroscopy, selected-area electron diffraction (SAED), and atomic force microscopy (AFM). These methods are described in detail in other chapters; here, we focus on their use for studies relevant to climate.

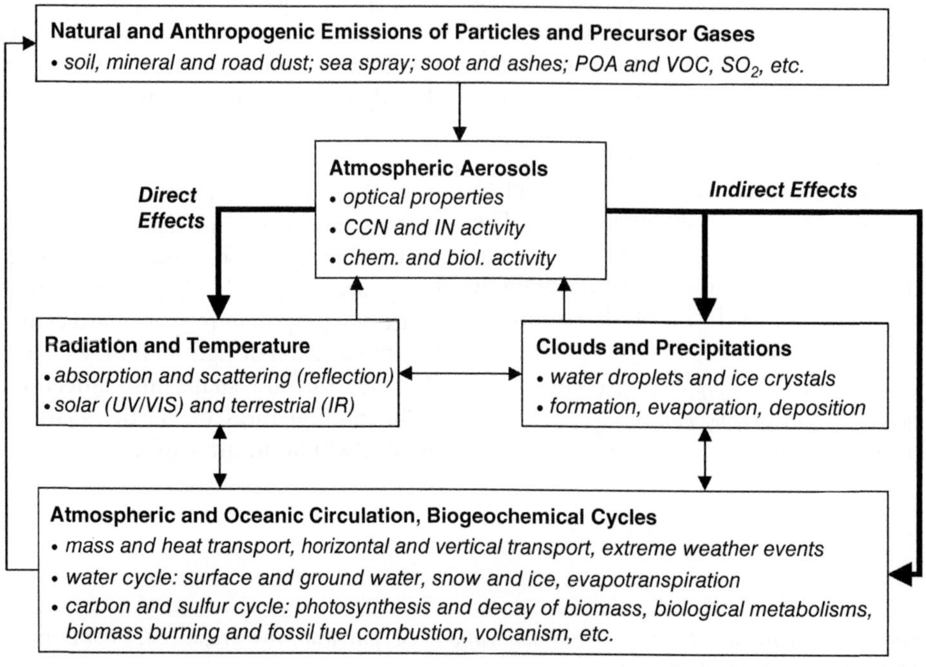

Figure 3.1 Direct and indirect aerosol effects and major feedback loops in the climate system. Pöschl (2005)/with permission of John Wiley & Sons.

3.2. CLIMATE STUDIES USING MICROANALYSIS TECHNIQUES

3.2.1. Aerosol Optical Properties

The most common microanalysis techniques utilized to determine the size and shape of atmospheric particles are SEM and TEM (Pósfai & Buseck, 2010). Together with the relative abundance of different particle compositions, as analyzed via particle counting and EDS and other techniques, their optical properties and net radiative impact can be derived. Electron microscopy techniques can also be used to aid development and understanding of new real-time instruments, such as those measuring the scattering properties of different types of nonspherical dust (Glen & Brooks, 2013), and in studying new particle formation from gaseous precursors (O'Dowd et al., 2002; Twohy et al., 2002).

McInnes et al. (1997) found that in the marine boundary layer, sea-salt particles dominated the coarse mode, whereas sulfates formed the majority of the fine mode number, with some contribution from organics. European continental influence on the marine atmosphere was shown to enhance the amount of sulfate and industrial material (Bruynseels et al., 1988). An early microanalysis study by Anderson et al. (1996) reported that African dust could be transported across the Atlantic and found in the marine boundary layer over Bermuda. Zhang et al. (1993) found that continental aerosol particles were externally mixed, with primarily inorganic particles being more hygroscopic and primarily carbonaceous ones less hygroscopic. The great diversity of compositions of aerosols transported westward from Asia was documented by Arimoto et al. (2006). The complex refractive index for continental rural and urban aerosol particles was measured by Ebert et al. (2004) via SEM (Figure 3.2) and X-ray analysis. TEM/EDS has also been used to analyze particle size, shape, and composition in the upper troposphere/lower stratosphere (Sheridan et al., 1994).

More recent studies have focused on the organic carbon content of aerosol particles and how this affects radiative forcing. Secondary organic aerosol (SOA), namely that produced through gas-to-particle conversion, can be important in increasing aerosol absorption or scattering cross-section, as well as increasing particle hygroscopicity and cloud interactions (Shrivastava et al., 2017). SEM has been used by Moffet et al. (2010) and others to identify organic coatings on sulfate salts, likely produced through secondary processes. Maria et al. (2004) used STXM/NEXAFS (Figure 3.3) to help show that atmospheric aerosols had slower oxidation rates and were less hygroscopic than previously thought, making them less likely to be removed by clouds and leading to a stronger net cooling. Russell et al. (2002) and Takahama et al. (2007) utilized STXM/NEXAFS to measure the composition of organic coatings that may change optical properties and hygroscopicity. Takahama et al. (2007) developed a classification scheme for 14 different atmospheric particle types with organic components. STXM/NEXAFS was also used to measure the organic volume fraction of individual particles (Fraund et al., 2019). The mixing of organic with inorganic material during atmospheric transport was studied with SEM/EDS and AFM by Kirpes et al. (2020).

Black carbon (soot) particles absorb light and therefore have an opposite effect on climate than aerosol particles that primarily scatter light. This climate effect varies with mixing state, however. For example, whether black carbon is internally or externally mixed with light-scattering material makes a large difference in its climate effect, with absorption being substantially enhanced for internal mixtures (Jacobson, 2000; Schnaiter et al., 2005). Microanalysis methods are well suited to determining particle heterogeneity and mixing state and have demonstrated that black carbon is in fact often internally mixed with other particle types in the atmosphere, particularly after aging (Adachi et al., 2010; Hasegawa & Ohta, 2002; Pósfai et al., 1999; Pósfai et al., 2003).

Some types of organic carbon, termed brown carbon, can also absorb ultraviolet and visible wavelengths and be important for climate (Laskin et al., 2015). The refractive indices of laboratory-generated organic acid particles, including humic-like substances, were measured by Laskina et al. (2014), with supporting SEM measurements. Yu et al. (2019) used a variety of microanalysis methods to measure organic coatings on particles in the Arctic, and together with Mie theory, showed that brown carbon coatings would produce a substantial increase in aerosol light absorption. Tar balls, a unique type of viscous brown carbon derived from biomass burning, have been identified via SEM (Hand et al., 2005) and STXM/NEXAFS (Tivanski et al., 2007). Tar balls were determined via TEM/EELS to be strongly absorbing (Alexander et al., 2008) and, as discussed in the following section, may also be active in nucleating ice crystals under some conditions.

Microanalysis techniques have been used successfully to test the validity of Mie theory, which assumes spherical particles in radiative transfer equations. For example, Mishchenko et al. (1995) found that due to the nonsphericity of dust particles, errors of 100% or more in satellite measurements of aerosol optical thickness were possible. Dust particles exposed to different organic compounds exhibited altered optical properties, appearing more spherical (Alexander et al., 2015). van Poppel et al. (2005) used TEM and tomography to measure the three-dimensional properties of fractal soot and found that large errors occurred in calculating its lifetime and

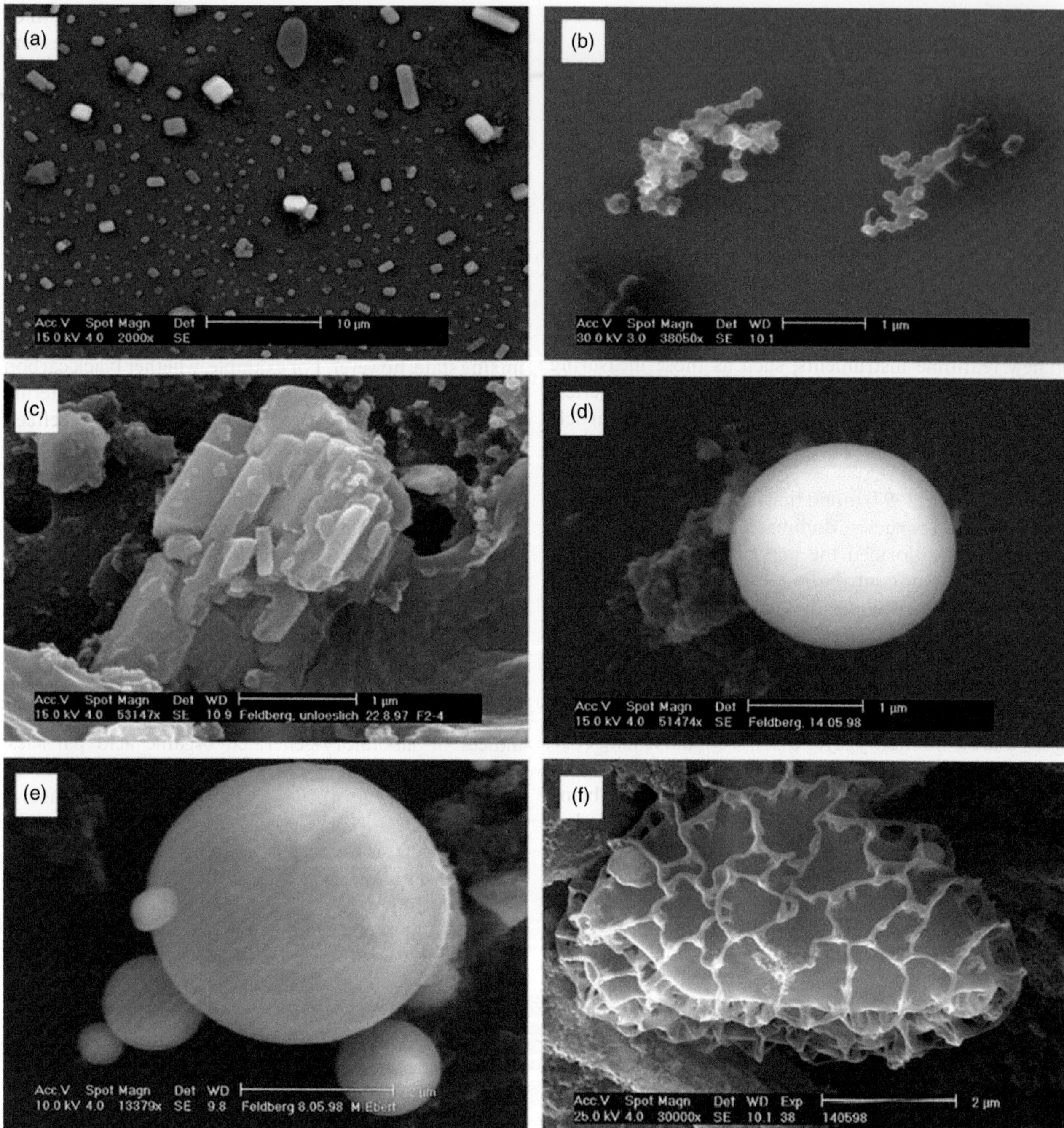

Figure 3.2 Typical secondary electron images of particles from different groups: (a) ammonium sulfates, (b) soot, (c) aluminosilicate particle (mineral dust), (d) aluminosilicate particle (fly ash), (e) iron oxide (fly ash), and (f) biological particle (fungi spore). Ebert et al. (2004)/with permission of Elsevier.

radiative properties if a spherical shape was assumed. Adachi et al. (2011) used ETEM to show that not all deliquesced particles were spherical; sulfate particles internally mixed with organic matter remained irregular. For these particles larger than 1.0 µm, Mie theory's assumption of spherical shapes could underestimate light-scattering efficiencies by ~50%.

Microanalysis techniques have also been used to measure particles in ice cores and snow to examine past and future climate. For example, SEM-EDX was used to

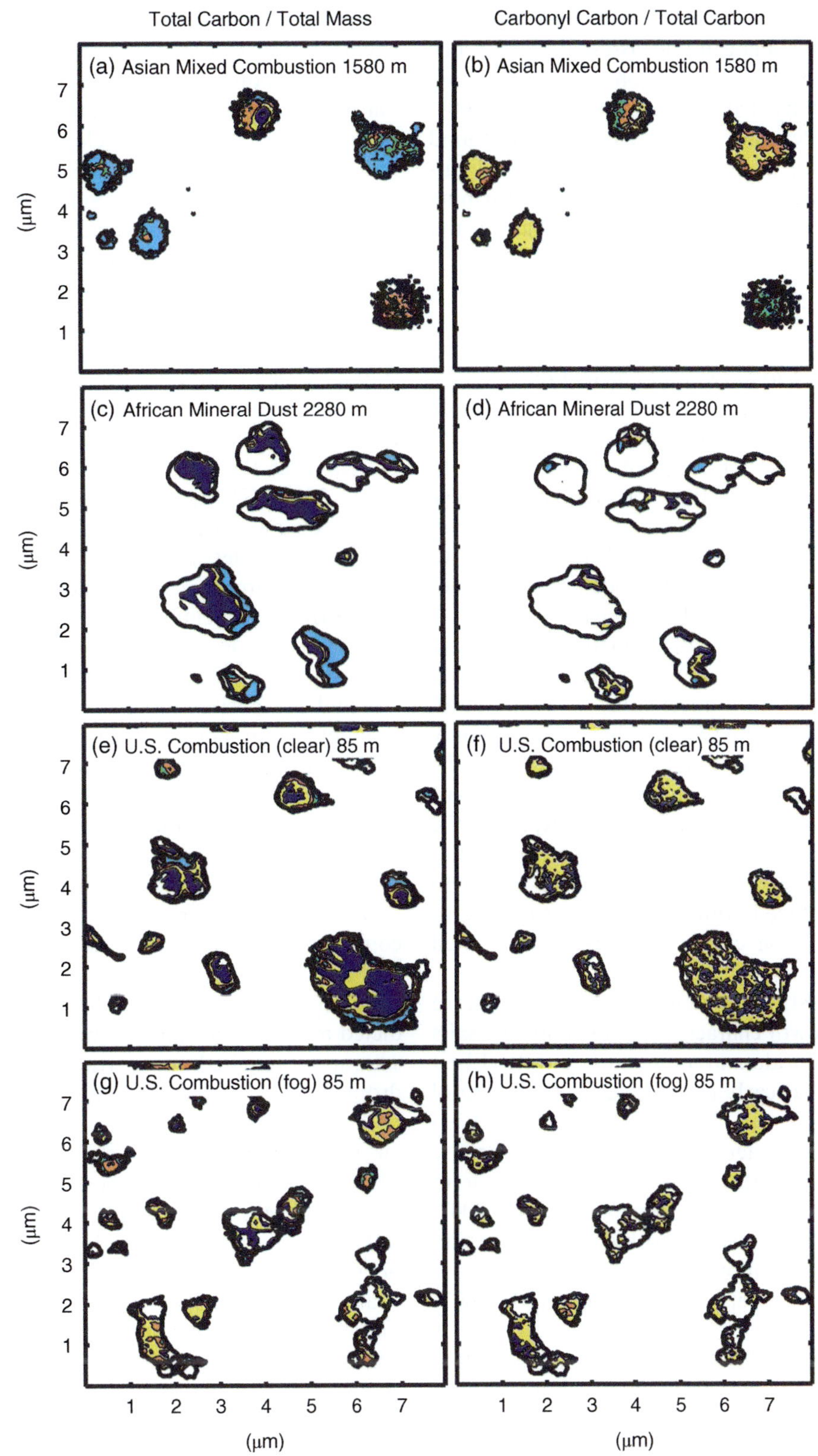

Figure 3.3 Two-dimensional spectrally resolved maps of organic composition for particles representative of two different aerosol types. The total carbon–to–total mass ratios are plotted in (a) and (c) demonstrating ratios of 0.02–0.2 (dark blue), 0.2–0.4 (yellow), 0.4–0.6 (orange), 0.6–0.8 (green), and >0.8 (light blue). The carbonyl-to-total carbon ratio is plotted in (b) and (d), with ratios of 0.02–0.08 (dark blue), 0.08–0.16 (yellow), 0.16–0.24 (orange), 0.24–0.32 (green), and >0.32 (light blue). Maria et al. (2004) with permission of American Association for the Advancement of Science.

study the type and characteristics of dust from ice cores on the Tibetan Plateau and assess its long-range transport and potential effect on glacier melting (Wu et al., 2016). Dong et al. (2014) and Cong et al. (2009) collected aerosols deposited in snow in Asia and found substantial influence from Asian dust and soot, light-absorbing components that could enhance warming and glacial melting (Sarangi et al., 2020; Warren and Wiscombe, 1980).

3.2.2. Impacts on Clouds

Studies Relevant to Cloud Droplet Nucleation

As detailed in the previous section, early studies used microanalysis techniques to characterize the dominant types of aerosol particles in different environments. In evaluating their cloud-nucleating potential, studies have focused on their size, composition, and affinity to take on water. Before or in conjunction with X-ray methods to supplement SEM/TEM, microscope grids were sometimes treated with chemicals to identify non-sea-salt sulfates as a major type of hygroscopic atmospheric particles and CCN (Bigg, 1973; Bigg et al., 1974; Parungo et al., 1992; Parungo et al., 1993). In a novel study by Pósfai et al. (1998), AFM and TEM/SAED were both used to study the amount of volatile material on particles over the Atlantic, and the authors concluded that organic coatings were likely present. AFM has also explored the changes in aerosol structure under different humidities (Köllensperger et al., 1999; Ramirez-Aguilar et al., 1999) and has been combined with infrared spectroscopy to provide information on chemical composition as well (Bondy et al., 2017).

In later studies, ESEM has been used to directly observe the hygroscopic growth characteristics of inorganic salts such as sodium chloride and ammonium nitrate (Ebert et al., 2002) as well as carbonaceous agricultural aerosol particles (Hiranuma et al., 2008). ETEM has also been used to examine how aerosol particles change phase and take up water at relative humidities relative to cloud formation (Freney et al., 2009; Semeniuk et al., 2007; Wise et al., 2007). In one study, both deliquescence and efflorescence (crystallization) humidities and structure were documented (Wise et al., 2005; Wise et al., 2007) (Figure 3.4).

SEM and TEM with X-ray spectrometry of residual particles extracted directly from cloud droplets have been used to show that a variety of different individual particle types can serve as CCN. Noone et al. (2000) and Hopkins et al. (2008) used SEM, electron probe X-ray analysis, and other techniques to show that cloud droplet residuals off the coast of California were composed primarily of sea salt and partially neutralized sulfates with lesser amounts of organic material. Studies show that sodium-based salts often dominate in clean marine clouds, whereas sulfates and organic particles become more common in clouds with anthropogenic influence (Bruynseels et al., 1988; Twohy et al., 2013; Twohy & Anderson, 2008). Sulfur-based CCN are also produced through gas-to-particle conversion from marine phytoplankton and can be important in upwelling regions or near Antarctica in the summertime (Bigg, 1980; Parungo et al., 1986; Twohy et al., 2021b).

Mineral dust particles internally mixed with sea salt or sulfates were found to be effective CCN and giant CCN (Levin et al., 2005); however, uncoated Saharan dust particles can act as CCN when the Saharan air layer is present over the Atlantic Ocean (Twohy, 2015; Twohy et al., 2009). TEM, in conjunction with single-particle microscopy, was used to identify salts from dry lake beds as cloud droplet residues (Pratt et al., 2010). The increasing hygroscopicity of calcium carbonate dust as it transforms to calcium nitrate with atmospheric aging was studied via Raman spectroscopy by Liu et al. (2008). Electron microscopy has been used to determine the size and composition of giant aerosol particles that are thought to be important in directly initiating supercooled drizzle drops that can be important in precipitation formation (Lasher-Trapp et al., 2008). TEM has also been used to study the fractal properties and droplet activation properties of soot aggregates when exposed to ozone for various periods of time (Wu et al., 2020).

Over time, the potential importance of organic compounds in influencing the properties of CCN and INPs has been recognized. Organic coatings on inorganic material, as observed via STXM/NEXAFS, were predicted to have a substantial impact on the critical supersaturation required for particle activation into cloud droplets (Takahama et al., 2010). Pöschl et al. (2010) used both SEM/EDS and AFM to show that SOA can be an important CCN source over the Amazon rainforest. Marine emissions can also be a source of organic CCN (Gantt & Meskhidze, 2013), and often, organic and sulfates are internally mixed, with the proportion of organics increasing with decreasing particle size (Saliba et al., 2020). Bacterial activity can dramatically decrease the hygroscopicity of sea-spray aerosol through the addition of organic matter (Collins et al., 2013), and cryogenic TEM has been used to show the complex mixing states of salts and organics in sea-spray aerosol (Patterson et al., 2016).

In polluted air near India, nitrogenated organic particles were also identified as CCN using TEM with EELS (Twohy et al., 2005). TEM has also been used to examine

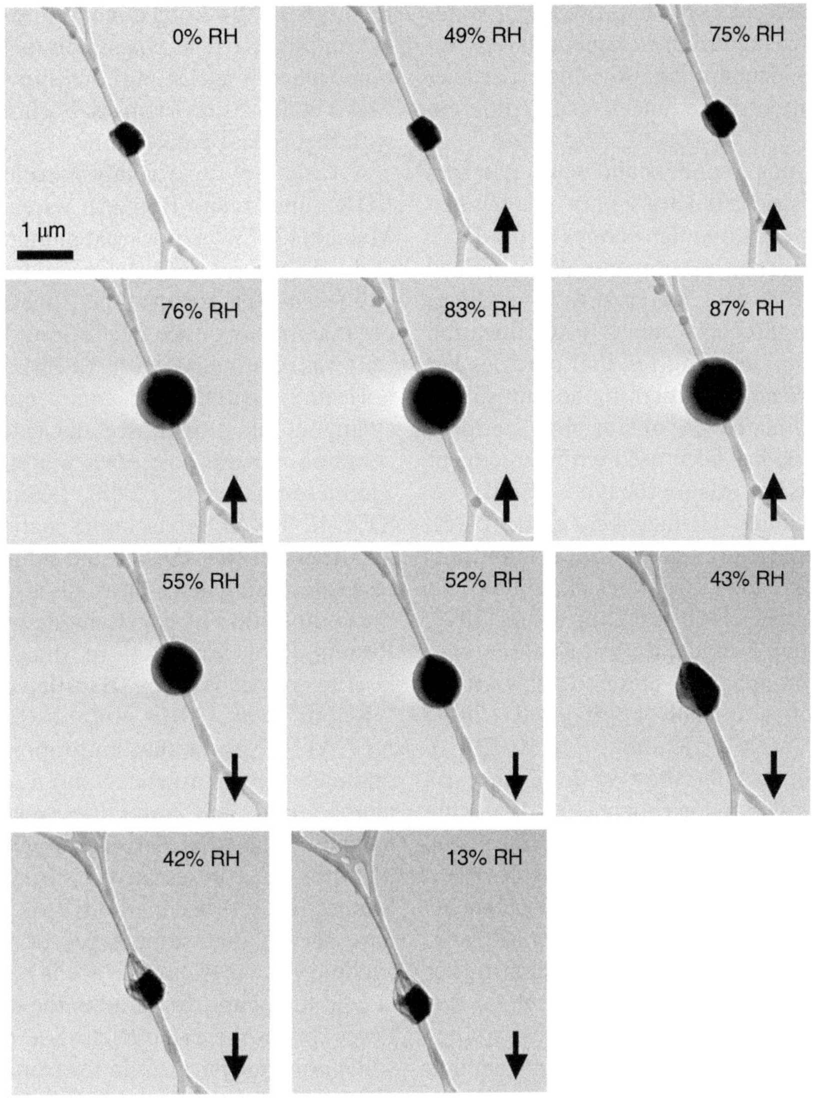

Figure 3.4 ETEM images of an NaCl particle as the RH is increased from 0% to 87% and then decreased from 87% to 13% at ~279 K. The up arrows indicate increasing water vapor pressure, and the down arrows indicate decreasing water vapor pressure in the environmental cell. Humidification experiments have also been used with in situ STXM to study mass-based hygroscopic properties of 15 complex aerosol types (Piens et al., 2016). Wise et al. (2005)/with permission of Taylor and Francis.

biomass-burning particles and their cloud-nucleating potential (Li et al., 2003). Even predominately organic biomass-burning particles can act as CCN, as some organics are hygroscopic, and internally mixed inorganic components can render them more so (Semeniuk et al., 2007; Twohy et al., 2021a).

Studies Relevant to Ice Nucleation

Microanalysis has been used for many decades to examine the types of particles that nucleate ice in clouds, using both laboratory experiments and in situ collection methods (Cziczo et al., 2017). Understanding these INPs and their sources is extremely important to climate because clouds containing mostly ice, typically at colder, higher altitudes in the atmosphere, can have substantially different effects on climate than those containing mostly droplets. In addition, precipitation in many cloud types is formed through ice-phase processes. Although there are many different types of INPs that vary with the nucleation process and environmental conditions, particles that act as INPs are usually much lower in concentration than those that act as CCN. These factors make them difficult to measure accurately. In recent years, improved microanalysis methods and instrumentation in general have contributed to greater understanding of which particles nucleate ice in the atmosphere under many different conditions.

Early studies attempting to understand which particles act as ice nuclei analyzed particles within precipitated snow, sometimes using electron microscopy (Isono, 1955; Kumai, 1961). Cloud chamber experiments also examined nuclei in laboratory-generated ice crystals (Rucklidge, 1965). In a more direct approach, recent INP identification studies have focused on analyzing residual particles left when ice crystals are collected in situ from the atmosphere and immediately sublimated to obtain their residual material. This technique has been used with subsequent EM and X-ray analysis to measure the types of particles within natural cirrus clouds (Heintzenberg et al., 1996), aircraft contrails (Petzold et al., 1998; Twohy & Gandrud, 1998), orographic ice clouds (Targino et al., 2006), and thunderstorm anvil cirrus clouds (Prenni et al., 2007; Twohy and Poellot, 2005). Residual ice particles are sometimes collected from mountaintop sites and analyzed via microanalysis techniques as well (Ebert et al., 2011; Eriksen Hammer et al., 2018; Kamphus et al., 2010). These studies together usually show a dominance of mineral dust (both silicates and carbonates) and metallic oxides in ice residuals, indicating that heterogeneous nucleation is an important process in the atmosphere.

Ice crystals can also be grown within an instrument such as a continuous-flow diffusion chamber (Rogers, 1993) and collected for later analysis of residual nuclei via microanalysis techniques (Cornwell et al., 2019; Kreidenweis et al., 1998; Kulkarni, 2018). These studies also often show a dominance of mineral dust and metallic components acting as heterogeneous ice nucleators (Richardson et al., 2007). For example, Chen et al. (1998) found that heterogeneous INPs in the upper troposphere/lower stratosphere included metallic, crustal, and carbonaceous material and were deficient in sulfate particle types, and Rogers et al. (2001) and Prenni et al. (2009a) found similar results for the Arctic mid-troposphere in springtime. China et al. (2017) used electron microscopy to study the effects of airmass history and atmospheric aging on INPs in the free troposphere over the Atlantic. Dust, sea salt, organics, and soot were all identified as INPs. Due to the potential loss of volatile material in post-collection microscopy analysis, others have used real-time mass spectroscopy to measure the composition of ice crystal residuals (e.g., Cziczo et al., 2004). Cziczo et al. (2013) used single-particle mass spectrometry together with EM in a variety of locations to show that most cirrus cloud residual particles were mineral dust and metallic particles, implying that heterogeneous nucleation was the dominant ice formation mechanism in many environments. Some field studies also have shown that INP types shift from mineral and metallic in nature to hydrophilic particle types such as sulfates under colder, homogeneous nucleating conditions (DeMott et al., 2003; Froyd et al., 2010; Twohy & Poellot, 2005), in agreement with theoretical expectations.

A variety of microanalysis techniques (AFM, SEM-EDX, and micro-Raman) were used by Iwata and Matsuki (2018) to show that aluminosilicate mineral dust and calcium-rich particles were dominant ice-nucleating entities in Asian dust and to examine the effects of organic matter on ice nucleation. A novel ice-nucleation cell was combined with ESEM to study the spatial nucleation patterns of one mineral type, kaolinite (Wang et al., 2016). Ice nucleation on a variety of common mineral dust types was studied with ESEM by Zimmermann et al. (2008). Leaitch et al. (2009) used STXM/NEXAFS to examine coatings on Asian dust and hypothesized that these could inhibit ice nucleation as a result. Raman spectroscopy has also been used to analyze the composition of coarse-mode particles nucleating ice. Results were compared to those from single-particle laser mass spectroscopy (Baustian et al., 2012).

Knopf et al. (2010) utilized SEM/EDX and STXM/NEXAFS to show that anthropogenic aerosol particles dominated by organics can play a role in ice nucleation at temperatures and humidities relative to cirrus cloud formation and that particles used as laboratory proxies may not be representative of true urban environments. Studies using other methods (e.g., Murray et al., 2010) have shown that some types of organic particles can become viscous enough to act as heterogeneous ice nuclei at cold temperatures found in the upper troposphere and lower stratosphere. Ice-nucleation mechanisms on mixed sulfate/organic particles in the tropical tropopause region were studied via optical microscopy coupled with Raman microscopy by Wise et al. (2010). The potential of black carbon and organic biomass-burning particles to nucleate ice has been examined in a number of studies. Electron microscopy techniques were used by McCluskey et al. (2014) to show that black carbon, organics, and mineral particles from biomass fires can all act as INPs under some conditions (Levin et al., 2014). A small proportion of biomass-burning particles, including tar balls and fine ash particles, were identified via scanning TEM and other techniques to be INPs (Adachi et al., 2022; Barry et al., 2021). Between their ability to absorb light directly and act as CCN and sometimes INPs, biomass-burning particles exert complex effects that must be taken together to determine their net effect on climate in different regions (e.g., Jacobson, 2014).

Laboratory studies have shown that some primary biological particles act to nucleate ice at much warmer temperatures than nucleators such as mineral dust (Möhler et al., 2007), but the importance of biological particles to the global INP population in the atmosphere is still under debate. Matthias-Maser and Jaenicke (1994) used SEM/EDX to identify biological particles >0.4 μm diameter in the atmospheric aerosol and found that they could represent a high number fraction under the post-frontal clean conditions studied. Biological types were found to dominate the INPs and temperatures warmer than −27 °C in the Amazon during the wet season when biomass burning was limited (Prenni et al., 2009b). Tobo et al. (2013) and others have identified fungal spores as INPs. Organic soils and plant matter have also been identified as ice-nucleating agents at relatively warm temperatures (Knopf et al., 2021; Suski et al., 2018; Testa et al., 2021; Tobo et al., 2014).

Various types of primary marine organic material have been identified via microanalysis techniques (Ault et al., 2013a; Ault et al., 2013b; Hawkins & Russell, 2010) (Figure 3.5), and some have been identified as ice-nucleating agents (McCluskey et al., 2018a). Microanalysis techniques have been used to investigate microfilms from biological activity in rivers (Lawrence et al., 2003) (Figure 3.6) and oceans (McCluskey et al., 2018b; Wilson et al., 2015), some of which have been shown to have ice-nucleating activity. Marine organic material was found to be prevalent in INPs collected over the Southern Ocean (Saliba et al., 2020; Twohy et al., 2021b). In addition, biological particles also have been observed in cloud residuals in some locations (e.g., Creamean et al., 2013; Pratt et al., 2009). However, a large-scale field and modeling study suggested that biological INPs comprise a small fraction of particles forming ice relative to mineral dust/metals throughout the troposphere on a global scale (Cziczo et al., 2013).

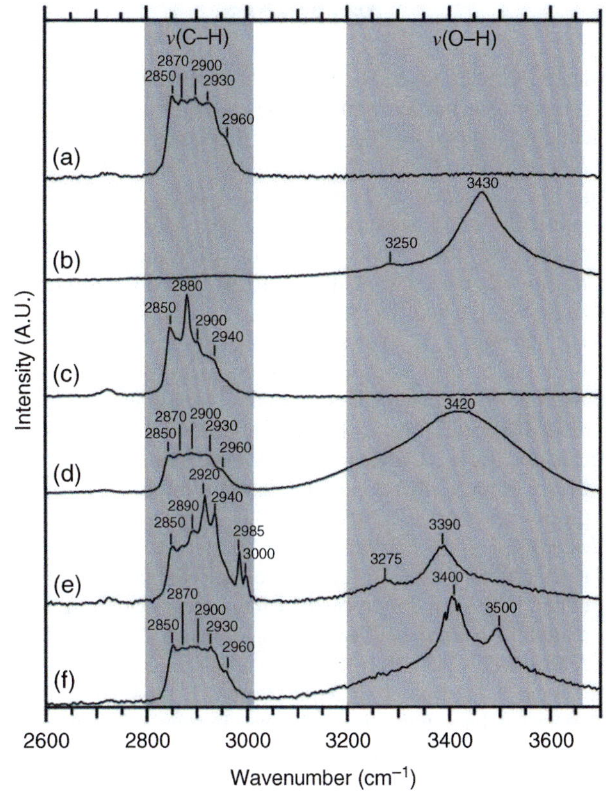

Figure 3.5 Raman spectra of six different sea-spray aerosol particles in the C—H and O—H stretching regions, with variability attributed mostly to the presence of different organic components in and on the particles. (a–c) Particles were collected before addition of bacteria and phytoplankton to the sample tank; (d–f) particles were collected after addition. Ault et al. (2013a)/Royal Society of Chemistry/CC-BY-3.0.

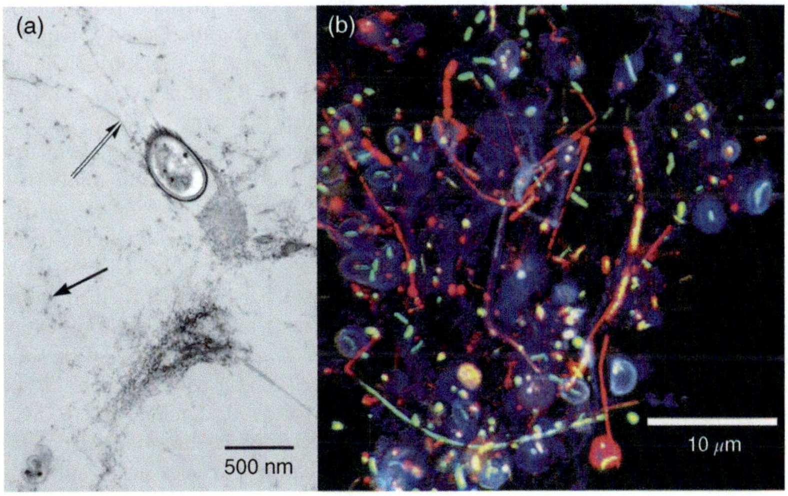

Figure 3.6 (a) TEM image showing a bacterium held within a biofilm by a porous (water-rich) matrix of fibrillar extracellular polymeric substances ("fibrils"); fibrils (double arrow) cross-connect bacteria and other colloids. The single arrow identifies nanoscale agglomerations on parts of some fibrils. (b) Confocal laser scanning microscopy (CLSM) images after staining, illustrating the distribution of nucleic acids (green), lipid (red), and polysaccharide (blue) regions. Lawrence et al. (2003)/with permission of American Society For Microbiology.

REFERENCES

Adachi, K., Chung, S. H., & Buseck, P. R. (2010). Shapes of soot aerosol particles and implications for their effects on climate. *Journal of Geophysical Research – Atmospheres, 115*(D15). doi: 10.1029/2009JD012868

Adachi, K., Freney, E. J., & Buseck, P. R. (2011). Shapes of internally mixed hygroscopic aerosol particles after deliquescence, and their effect on light scattering. *Geophysical Research Letters, 38*(13). doi: 10.1029/2011GL047540

Adachi, K., Dibb, J. E., Scheuer, E., Katich, J. M., Schwarz, J. P., Perring, A. E., et al. (2022). Fine ash-bearing particles as a major aerosol component in biomass burning smoke. *Journal of Geophysical Research – Atmospheres, 127*(2), e2021JD035657. doi: 10.1029/2021JD035657

Alexander, D. T., Crozier, P. A., & Anderson, J. R. (2008). Brown carbon spheres in east Asian outflow and their optical properties. *Science, 321*(5890), 833–836.

Alexander, J. M., Grassian, V. H., Young, M. A., & Kleiber, P. D. (2015). Optical properties of selected components of mineral dust aerosol processed with organic acids and humic material. *Journal of Geophysical Research – Atmospheres, 120*(6), 2437–2452. doi: 10.1002/2014JD022782

Anderson, J. R., Buseck, P. R., Patterson, T. L., & Arimoto, R. (1996). Characterization of the Bermuda tropospheric aerosol by combined individual-particle and bulk-aerosol analysis. *Atmospheric Environment, 30*(2), 319–338.

Andreae, M. O., & Gelencsér, A. (2006). Black carbon or brown carbon? The nature of light-absorbing carbonaceous aerosols. *Atmospheric Chemistry and Physics, 6*(10), 3131–3148.

Arias, P. A., Belluouin, N., Coppola, E., Jones, R. G., Krinner, G., Marotzke, J., et al. (2021). *Climate change 2021: the physical science basis* (contribution of Working Group I to the Sixth Assessment Report of the Intergovernmental Panel on Climate Change). Cambridge,UK: Cambridge University Press.

Arimoto, R., Kim, Y. J., Kim, Y. P., Quinn, P. K., Bates, T. S., Anderson, T. L., et al. (2006). Characterization of Asian dust during ACE-Asia. *Global and Planetary Change, 52*(1), 23–56.

Ault, A. P., Zhao, D., Ebben, C. J., Tauber, M. J., Geiger, F. M., Prather, K. A., & Grassian, V. H. (2013a). Raman microspectroscopy and vibrational sum frequency generation spectroscopy as probes of the bulk and surface compositions of size-resolved sea spray aerosol particles. *Physical Chemistry Chemical Physics, 15*(17), 6206–6214.

Ault, A. P., Moffet, R. C., Baltrusaitis, J., Collins, D. B., Ruppel, M. J., Cuadra-Rodriguez, L. A., et al. (2013b). Size-dependent changes in sea spray aerosol composition and properties with different seawater conditions. *Environmental Science & Technology, 47*(11), 5603–5612.

Barry, K. R., Hill, T. C. J., Levin, E. J. T., Twohy, C. H., Moore, K. A., Weller, Z. D., et al. (2021). Observations of ice nucleating particles in the free troposphere from Western US wildfires. *Journal of Geophysical Research – Atmospheres, 126*(3), e2020JD033752.

Baustian, K. J., Cziczo, D. J., Wise, M. E., Pratt, K. A., Kulkarni, G., Hallar, A. G., & Tolbert, M. A. (2012). Importance of aerosol composition, mixing state, and morphology for heterogeneous ice nucleation: A combined field and laboratory approach. *Journal of Geophysical Research – Atmospheres, 117*(D6). doi: 10.1029/2011JD016784

Bigg, E. K. (1973). Ice nucleus concentrations in remote areas. *Journal of the Atmospheric Sciences, 30*(6), 1153–1157.

Bigg, E. K. (1980). Comparison of aerosol at four baseline atmospheric monitoring stations. *Journal of Applied Meteorology and Climatology, 19*(5), 521–533.

Bigg, E. K., Ono, A., & Williams, J. A. (1974). Chemical tests for individual submicron aerosol particles. *Atmospheric Environment (1967), 8*(1), 1–13.

Bondy, A. L., Kirpes, R. M., Merzel, R. L., Pratt, K. A., Banaszak Holl, M. M., & Ault, A. P. (2017). Atomic force microscopy-infrared spectroscopy of individual atmospheric aerosol particles: Subdiffraction limit vibrational spectroscopy and morphological analysis. *Analytical Chemistry, 89*(17), 8594–8598.

Boucher, O., Randall, D., Artaxo, P., Bretherton, C., Feingold, G., Forster, P., et al. (2013). Clouds and aerosols. In T. F. Stocker, D. Qin, G.-K. Plattner, M. Tignor, S. K. Allen, J. Boschung, et al. (Eds.), *Climate change 2013: The physical science basis. Contribution of working group I to the fifth assessment report of the intergovernmental panel on climate change*. Cambridge, United Kingdom and New York, NY, USA: Midgley Cambridge University Press.

Bruynseels, F., Storms, H., Van Grieken, R., & Van Der Auwera, L. (1988). Characterization of North Sea aerosols by individual particle analyses. *Atmospheric Environment (1967), 22*(11), 2593–2602.

Chen, Y., Kreidenweis, S. M., McInnes, L. M., Rogers, D. C., & DeMott, P. J. (1998). Single particle analyses of ice nucleating aerosols in the upper troposphere and lower stratosphere. *Geophysical Research Letters, 25*(9), 1391–1394. doi: 10.1029/97GL03261

China, S., Alpert, P. A., Zhang, B., Schum, S., Dzepina, K., Wright, K., et al. (2017). Ice cloud formation potential by free tropospheric particles from long-range transport over the northern Atlantic Ocean. *Journal of Geophysical Research – Atmospheres, 122*(5), 3065–3079. doi: 10.1002/2016JD025817

Collins, D. B., Ault, A. P., Moffet, R. C., Ruppel, M. J., Cuadra-Rodriguez, L. A., Guasco, T. L., et al. (2013). Impact of marine biogeochemistry on the chemical mixing state and cloud forming ability of nascent sea spray aerosol. *Journal of Geophysical Research – Atmospheres, 118*(15), 8553–8565. doi: 10.1002/jgrd.50598

Cong, Z., Kang, S., & Qin, D. (2009). Seasonal features of aerosol particles recorded in snow from Mt. Qomolangma (Everest) and their environmental implications. *Journal of Environmental Sciences, 21*(7), 914–919.

Cornwell, G. C., McCluskey, C. S., Levin, E. J. T., Suski, K. J., DeMott, P. J., Kreidenweis, S. M., & Prather, K. A. (2019). Direct online mass spectrometry measurements of ice nucleating particles at a California coastal site. *Journal of Geophysical Research – Atmospheres, 124*(22), 12157–12172. doi: 10.1029/2019JD030466

Creamean, J. M., Suski, K. J., Rosenfeld, D., Cazorla, A., DeMott, P. J., Sullivan, R. C., et al. (2013). Dust and biological aerosols from the Sahara and Asia influence precipitation in the western U. S. *Science*, *339*(6127), 1572–1578.

Cziczo, D. J., Murphy, D. M., Hudson, P. K., & Thomson, D. S. (2004). Single particle measurements of the chemical composition of cirrus ice residue during CRYSTAL-FACE. *Journal of Geophysical Research – Atmospheres*, *109*(D4). n/a–n/a

Cziczo, D. J., Froyd, K. D., Hoose, C., Jensen, E. J., Diao, M., Zondlo, M. A., et al. (2013). Clarifying the dominant sources and mechanisms of cirrus cloud formation. *Science*, *340*(6138), 1320–1324.

Cziczo, D. J., Ladino, L., Boose, Y., Kanji, Z. A., Kupiszewski, P., Lance, S., et al. (2017). Measurements of ice nucleating particles and ice residuals. *Meteorological Monographs*, *58*, 8.1–8.13. doi: 10.1029/2003JD004032

DeMott, P. J., Cziczo, D. J., Prenni, A. J., Murphy, D. M., Kreidenweis, S. M., Thomson, D. S., et al. (2003). Measurements of the concentration and composition of nuclei for cirrus formation. *Proceedings of the National Academy of Sciences*, *100*(25), 14655.

Dong, Z., Qin, D., Kang, S., Ren, J., Chen, J., Cui, X., et al. (2014). Physicochemical characteristics and sources of atmospheric dust deposition in snow packs on the glaciers of western Qilian Mountains, China. *Tellus B*, *66*(1), 20956.

Ebert, M., Inerle-Hof, M., & Weinbruch, S. (2002). Environmental scanning electron microscopy as a new technique to determine the hygroscopic behaviour of individual aerosol particles. *Atmospheric Environment*, *36*(39), 5909–5916.

Ebert, M., Weinbruch, S., Hoffmann, P., & Ortner, H. M. (2004). The chemical composition and complex refractive index of rural and urban influenced aerosols determined by individual particle analysis. *Atmospheric Environment*, *38*(38), 6531–6545.

Ebert, M., Worringen, A., Benker, N., Mertes, S., Weingartner, E., & Weinbruch, S. (2011). Chemical composition and mixing-state of ice residuals sampled within mixed phase clouds. *Atmospheric Chemistry and Physics*, *11*(6), 2805–2816.

Eriksen Hammer, S., Mertes, S., Schneider, J., Ebert, M., Kandler, K., & Weinbruch, S. (2018). Composition of ice particle residuals in mixed-phase clouds at Jungfraujoch (Switzerland): Enrichment and depletion of particle groups relative to total aerosol. *Atmospheric Chemistry and Physics*, *18*(19), 13987–14003.

Fraund, M., Park, T., Yao, L., Bonanno, D., Pham, D. Q., & Moffet, R. C. (2019). Quantitative capabilities of STXM to measure spatially resolved organic volume fractions of mixed organic/inorganic particles. *Atmospheric Measurement Techniques*, *12*(3), 1619–1633.

Freney, E. J., Martin, S. T., & Buseck, P. R. (2009). Deliquescence and efflorescence of potassium salts relevant to biomass-burning aerosol particles. *Aerosol Science and Technology*, *43*(8), 799–807.

Froyd, K. D., Murphy, D. M., Lawson, P., Baumgardner, D., & Herman, R. L. (2010). Aerosols that form subvisible cirrus at the tropical tropopause. *Atmospheric Chemistry and Physics*, *10*(1), 209–218.

Gantt, B., & Meskhidze, N. (2013). The physical and chemical characteristics of marine primary organic aerosol: A review. *Atmospheric Chemistry and Physics*, *13*(8), 3979–3996.

Glen, A., & Brooks, S. D. (2013). A new method for measuring optical scattering properties of atmospherically relevant dusts using the cloud and aerosol spectrometer with polarization (CASPOL). *Atmospheric Chemistry and Physics*, *13*(3), 1345–1356.

Hand, J. L., Malm, W. C., Laskin, A., Day, D., Lee, T., Wang, C., et al. (2005). Optical, physical, and chemical properties of tar balls observed during the Yosemite aerosol characterization study. *Journal of Geophysical Research – Atmospheres*, *110*(D21). doi: 10.1029/2004JD005728

Hasegawa, S., & Ohta, S. (2002). Some measurements of the mixing state of soot-containing particles at urban and non-urban sites. *Atmospheric Environment*, *36*(24), 3899–3908.

Hawkins, L. N., & Russell, L. M. (2010). Polysaccharides, proteins, and phytoplankton fragments: Four chemically distinct types of marine primary organic aerosol classified by single particle spectromicroscopy. *Advances in Meteorology*, *2010*, 14.

Heintzenberg, J., Okada, K., & Ström, J. (1996). On the composition of non-volatile material in upper tropospheric aerosols and cirrus crystals. *Atmospheric Research*, *41*(1), 81–88.

Hiranuma, N., Brooks, S. D., Auvermann, B. W., & Littleton, R. (2008). Using environmental scanning electron microscopy to determine the hygroscopic properties of agricultural aerosols. *Atmospheric Environment*, *42*(9), 1983–1994.

Hopkins, R. J., Desyaterik, Y., Tivanski, A. V., Zaveri, R. A., Berkowitz, C. M., Tyliszczak, T., et al. (2008). Chemical speciation of sulfur in marine cloud droplets and particles: Analysis of individual particles from the marine boundary layer over the California current. *Journal of Geophysical Research–Atmospheres*, *113*(D4). doi: 10.1029/2007JD008954

Isono, K. (1955). On ice-crystal nuclei and other substances found in snow crystals. *Journal of Atmospheric Sciences*, *12*(5), 456–462.

Iwata, A., & Matsuki, A. (2018). Characterization of individual ice residual particles by the single droplet freezing method: A case study in the Asian dust outflow region. *Atmospheric Chemistry and Physics*, *18*(3), 1785–1804.

Jacobson, M. Z. (2000). A physically-based treatment of elemental carbon optics: Implications for global direct forcing of aerosols. *Geophysical Research Letters*, *27*(2), 217–220. doi: 10.1029/1999GL010968

Jacobson, M. Z. (2014). Effects of biomass burning on climate, accounting for heat and moisture fluxes, black and brown carbon, and cloud absorption effects. *Journal of Geophysical Research – Atmospheres*, *119*(14), 8980–9002. doi: 10.1002/2014JD021861

Jaggard, D. L., Hill, C., Shorthill, R. W., Stuart, D., Glantz, M., Rosswog, F., et al. (1981). Light scattering from particles of regular and irregular shape. *Atmospheric Environment (1967)*, *15*(12), 2511–2519.

Kamphus, M., Ettner-Mahl, M., Klimach, T., Drewnick, F., Keller, L., Cziczo, D. J., et al. (2010). Chemical composition of ambient aerosol, ice residues and cloud droplet residues in mixed-phase clouds: Single particle analysis during the cloud and aerosol characterization experiment (CLACE 6). *Atmospheric Chemistry and Physics*, *10*(16), 8077–8095.

Kandler, K., SchÜTz, L., Deutscher, C., Ebert, M., Hofmann, H., JÄCkel, S., et al. (2009). Size distribution, mass concentration, chemical and mineralogical composition and derived optical parameters of the boundary layer aerosol at Tinfou, Morocco, during SAMUM 2006. *Tellus B*, *61*(1), 32–50.

Kanji, Z. A., Ladino, L. A., Wex, H., Boose, Y., Burkert-Kohn, M., Cziczo, D. J., & Krämer, M. (2017). Overview of ice nucleating particles. *Meteorological Monographs*, *58*.

Kirpes, R. M., Rodriguez, B., Kim, S., China, S., Laskin, A., Park, K., et al. (2020). Emerging investigator series: Influence of marine emissions and atmospheric processing on individual particle composition of summertime Arctic aerosol over the Bering Strait and Chukchi Sea. *Environmental Science: Processes & Impacts*, *22*(5), 1201–1213.

Knopf, D. A., Wang, B., Laskin, A., Moffet, R. C., & Gilles, M. K. (2010). Heterogeneous nucleation of ice on anthropogenic organic particles collected in Mexico City. *Geophysical Research Letters*, *37*(11). doi: 10.1029/2010GL043362

Knopf, D. A., Barry, K. R., Brubaker, T. A., Jahl, L. G., Jankowski, K. A., Li, J., et al. (2021). Aerosol–ice formation closure: A southern Great Plains field campaign. *Bulletin of the American Meteorological Society*, *102*(10), E1952–E1971.

Köllensperger, G., Friedbacher, G., Kotzick, R., Niessner, R., & Grasserbauer, M. (1999). In-situ atomic force microscopy investigation of aerosols exposed to different humidities. *Fresenius' Journal of Analytical Chemistry*, *364*(4), 296–304.

Kreidenweis, S. M., Chen, Y., Rogers, D. C., & DeMott, P. J. (1998). Isolating and identifying atmospheric ice-nucleating aerosols: A new technique. *Atmospheric Research*, *46*(3), 263–278.

Kulkarni, G. (2018). Immersion freezing of Total ambient aerosols and ice residuals. *Atmosphere*, *9*(2).

Kumai, M. (1961). Snow crystals and the identification of the nuclei in the northern United States of America. *Journal of Meteorology*, *18*(2), 139–150.

Lasher-Trapp, S., Anderson-Bereznicki, S., Shackelford, A., Twohy, C. H., & Hudson, J. G. (2008). An investigation of the influence of droplet number concentration and Giant aerosol particles upon Supercooled large drop formation in wintertime Stratiform clouds. *Journal of Applied Meteorology and Climatology*, *47*(10), 2659–2678.

Laskin, A., Laskin, J., & Nizkorodov, S. A. (2015). Chemistry of atmospheric Brown carbon. *Chemical Reviews*, *115*(10), 4335–4382.

Laskina, O., Young, M. A., Kleiber, P. D., & Grassian, V. H. (2014). Infrared optical constants of organic aerosols: Organic acids and model humic-like substances (HULIS). *Aerosol Science and Technology*, *48*(6), 630–637.

Lawrence, J. R., Swerhone, G. D. W., Leppard, G. G., Araki, T., Zhang, X., West, M. M., & Hitchcock, A. P. (2003). Scanning transmission X-ray, laser scanning, and transmission electron microscopy mapping of the exopolymeric matrix of microbial biofilms. *Applied and Environmental Microbiology*, *69*(9), 5543–5554.

Leaitch, W. R., Macdonald, A. M., Anlauf, K. G., Liu, P. S. K., Toom-Sauntry, D., Li, S. M., et al. (2009). Evidence for Asian dust effects from aerosol plume measurements during INTEX-B 2006 near whistler, BC. *Atmospheric Chemistry and Physics*, *9*(11), 3523–3546.

Levin, Z., Teller, A., Ganor, E., & Yin, Y. (2005). On the interactions of mineral dust, sea-salt particles, and clouds: A measurement and modeling study from the Mediterranean Israeli dust experiment campaign. *Journal of Geophysical Research*, *110*(D20). doi: 10.1029/2005jd005810

Levin, E. J. T., McMeeking, G. R., DeMott, P. J., McCluskey, C. S., Stockwell, C. E., Yokelson, R. J., & Kreidenweis, S. M. (2014). A new method to determine the number concentrations of refractory black carbon ice nucleating particles. *Aerosol Science and Technology*, *48*(12), 1264–1275.

Li, J., Pósfai, M., Hobbs, P. V., & Buseck, P. R. (2003). Individual aerosol particles from biomass burning in southern Africa: 2, compositions and aging of inorganic particles. *Journal of Geophysical Research – Atmospheres*, *108*(D13). doi: 10.1029/2002JD002310

Liu, Y. J., Zhu, T., Zhao, D. F., & Zhang, Z. F. (2008). Investigation of the hygroscopic properties of Ca(NO3)2 and internally mixed Ca(NO3)2/CaCO3 particles by micro-Raman spectrometry. *Atmospheric Chemistry and Physics*, *8*(23), 7205–7215.

Liu, F., Yon, J., Fuentes, A., Lobo, P., Smallwood, G. J., & Corbin, J. C. (2020). Review of recent literature on the light absorption properties of black carbon: Refractive index, mass absorption cross section, and absorption function. *Aerosol Science and Technology*, *54*(1), 33–51.

Maria, S. F., Russell, L. M., Gilles, M. K., & Myneni, S. C. B. (2004). Organic aerosol growth mechanisms and their climate-forcing implications. *Science*, *306*(5703), 1921.

Matthias-Maser, S., & Jaenicke, R. (1994). Examination of atmospheric bioaerosol particles with radii > 0.2 μm. *Journal of Aerosol Science*, *25*(8), 1605–1613.

McCluskey, C. S., DeMott, P. J., Prenni, A. J., Levin, E. J. T., McMeeking, G. R., Sullivan, A. P., et al. (2014). Characteristics of atmospheric ice nucleating particles associated with biomass burning in the US: Prescribed burns and wildfires. *Journal of Geophysical Research – Atmospheres*, *119*(17), 10458–10470.

McCluskey, C. S., Hill, T. C. J., Sultana, C. M., Laskina, O., Trueblood, J., Santander, M. V., et al. (2018a). A Mesocosm double feature: Insights into the chemical makeup of marine ice nucleating particles. *Journal of the Atmospheric Sciences*, *75*(7), 2405–2423.

McCluskey, C. S., Hill, T. C. J., Humphries, R. S., Rauker, A. M., Moreau, S., Strutton, P. G., et al. (2018b). Observations of ice nucleating particles over Southern Ocean waters. *Geophysical Research Letters*, *45*(21), 11,989–911,997. doi: 10.1002/2014JD021980

McInnes, L., Covert, D., & Baker, B. (1997). The number of sea-salt, sulfate, and carbonaceous particles in the marine atmosphere. *Tellus B*, *49*(3), 300–313.

Mishchenko, M. I., Lacis, A. A., Carlson, B. E., & Travis, L. D. (1995). Nonsphericity of dust-like tropospheric aerosols: Implications for aerosol remote sensing and climate modeling. *Geophysical Research Letters*, *22*(9), 1077–1080. doi: 10.1029/95GL00798

Mishra, S. K., & Tripathi, S. N. (2008). Modeling optical properties of mineral dust over the Indian Desert. *Journal of Geophysical Research – Atmospheres*, *113*(D23). doi: 10.1029/2008JD010048

Moffet, R. C., Henn, T. R., Tivanski, A. V., Hopkins, R. J., Desyaterik, Y., Kilcoyne, A. L. D., et al. (2010). Microscopic characterization of carbonaceous aerosol particle aging in the outflow from Mexico City. *Atmospheric Chemistry and Physics*, *10*(3), 961–976.

Möhler, O., DeMott, P. J., Vali, G., & Levin, Z. (2007). Microbiology and atmospheric processes: The role of biological particles in cloud physics. *Biogeosciences*, *4*(6), 1059–1071.

Murray, B. J., Wilson, T. W., Dobbie, S., Cui, Z., Al-Jumur, S. M. R. K., Möhler, O., et al. (2010). Heterogeneous nucleation of ice particles on glassy aerosols under cirrus conditions. *Nature Geoscience*, *3*(4), 233–237.

Noone, K. J., Öström, E., Ferek, R. J., Garrett, T., Hobbs, P. V., Johnson, D. W., et al. (2000). A case study of ships forming and not forming tracks in moderately polluted clouds. *Journal of the Atmospheric Sciences*, *57*(16), 2729–2747.

O'Dowd, C. D., Hämeri, K., Mäkelä, J. M., Pirjola, L., Kulmala, M., Jennings, S. G., et al. (2002). A dedicated study of new particle formation and fate in the coastal environment (PARFORCE): Overview of objectives and achievements. *Journal of Geophysical Research – Atmospheres*, *107*(D19). doi: 10.1029/2001JD000555

Parungo, F. P., Nagamoto, C. T., Rosinski, J., & Haagenson, P. L. (1986). A study of marine aerosols over the Pacific Ocean. *Journal of Atmospheric Chemistry*, *4*(2), 199–226.

Parungo, F., Kopcewicz, B., Nagamoto, C., Schnell, R., Sheridan, P., Zhu, C., & Harris, J. (1992). Aerosol particles in the Kuwait oil fire plumes: Their morphology, size distribution, chemical composition, transport, and potential effect on climate. *Journal of Geophysical Research – Atmospheres*, *97*(D14), 15867–15882. doi: 10.1029/92JD01223

Parungo, F., Rosinski, J., Wu, M. L. C., Nagamoto, C. T., Minyu, Z., & Ni, Z. (1993). Variation of sulfate aerosol concentrations over the western Pacific and their effect on clouds, radiation and precipitatio. *Acta Oceanologica Sinica*, *4*, 521–534.

Patterson, J. P., Collins, D. B., Michaud, J. M., Axson, J. L., Sultana, C. M., Moser, T., et al. (2016). Sea spray aerosol structure and composition using cryogenic transmission Electron microscopy. *ACS Central Science*, *2*(1), 40–47.

Petzold, A., Ström, J., Ohlsson, S., & Schröder, F. P. (1998). Elemental composition and morphology of ice-crystal residual particles in cirrus clouds and contrails. *Atmospheric Research*, *49*(1), 21–34.

Piens, D. S., Kelly, S. T., Harder, T. H., Petters, M. D., O'Brien, R. E., Wang, B., et al. (2016). Measuring mass-based Hygroscopicity of atmospheric particles through in situ imaging. *Environmental Science & Technology*, *50*(10), 5172–5180.

van Poppel, L. H., Friedrich, H., Spinsby, J., Chung, S. H., Seinfeld, J. H., & Buseck, P. R. (2005). Electron tomography of nanoparticle clusters: Implications for atmospheric lifetimes and radiative forcing of soot. *Geophysical Research Letters*, *32*(24).

Pöschl, U. (2005). Atmospheric aerosols: Composition, transformation, climate and health. *Angewandte Chemie International Edition*, *44*(46), 7520–7540. doi: 10.1029/2005GL024461effects

Pöschl, U., Martin, S. T., Sinha, B., Chen, Q., Gunthe, S. S., Huffman, J. A., et al. (2010). Rainforest aerosols as biogenic nuclei of clouds and precipitation in the Amazon. *Science*, *329*(5998), 1513.

Pósfai, M., & Buseck, P. R. (2010). Nature and climate effects of individual tropospheric aerosol particles. *Annual Review of Earth and Planetary Sciences*, *38*(1), 17–43.

Pósfai, M., Xu, H., Anderson, J. R., & Buseck, P. R. (1998). Wet and dry sizes of atmospheric aerosol particles: An AFM-TEM study. *Geophysical Research Letters*, *25*(11), 1907–1910. doi: 10.1029/98GL01416

Pósfai, M., Anderson, J. R., Buseck, P. R., & Sievering, H. (1999). Soot and sulfate aerosol particles in the remote marine troposphere. *Journal of Geophysical Research – Atmospheres*, *104*(D17), 21685–21693. doi: 10.1029/1999JD900208

Pósfai, M., Simonics, R., Li, J., Hobbs, P. V., & Buseck, P. R. (2003). Individual aerosol particles from biomass burning in southern Africa: 1. Compositions and size distributions of carbonaceous particles. *Journal of Geophysical Research – Atmospheres*, *108*(D13). doi: 10.1029/2002JD002291

Pratt, K. A., DeMott, P. J., French, J. R., Wang, Z., Westphal, D. L., Heymsfield, A. J., et al. (2009). In situ detection of biological particles in cloud ice-crystals. *Nature Geoscience*, *2*(6), 397–400.

Pratt, K. A., Twohy, C. H., Murphy, S. M., Moffet, R. C., Heymsfield, A. J., Gaston, C. J., et al. (2010). Observation of playa salts as nuclei in orographic wave clouds. *Journal of Geophysical Research-Atmospheres*, *115*. doi: 10.1029/2009JD013150

Prenni, A. J., DeMott, P. J., Twohy, C., Poellot, M. R., Kreidenweis, S. M., Rogers, D. C., et al. (2007). Examinations of ice formation processes in Florida cumuli using ice nuclei measurements of anvil ice crystal particle residues. *Journal of Geophysical Research-Atmospheres*, *112*(D10). doi: 10.1029/2006jd007549

Prenni, A. J., Demott, P. J., Rogers, D. C., Kreidenweis, S. M., McFarquhar, G. M., Zhang, G., & Poellot, M. R. (2009a). Ice nuclei characteristics from M-PACE and their relation to ice formation in clouds. *Tellus B*, *61*(2), 436–448.

Prenni, A. J., Petters, M. D., Kreidenweis, S. M., Heald, C. L., Martin, S. T., Artaxo, P., et al. (2009b). Relative roles of biogenic emissions and Saharan dust as ice nuclei in the Amazon basin. *Nature Geoscience*, *2*(6), 402–405.

Ramirez-Aguilar, K. A., Lehmpuhl, D. W., Michel, A. E., Birks, J. W., & Rowlen, K. L. (1999). Atomic force microscopy for the analysis of environmental particles. *Ultramicroscopy*, *77*(3), 187–194.

Richardson, M. S., DeMott, P. J., Kreidenweis, S. M., Cziczo, D. J., Dunlea, E. J., Jimenez, J. L., et al. (2007). Measurements of heterogeneous ice nuclei in the western United States in springtime and their relation to aerosol characteristics. *Journal of Geophysical Research – Atmospheres*, *112*(D2). doi: 10.1029/2006JD007500

Rogers, D. C. (1993). Measurements of natural ice nuclei with a continuous flow diffusion chamber. *Atmospheric Research*, *29*(3), 209–228.

Rogers, D. C., DeMott, P. J., & Kreidenweis, S. M. (2001). Airborne measurements of tropospheric ice-nucleating aerosol particles in the Arctic spring. *Journal of Geophysical Research – Atmospheres*, *106*(D14), 15053–15063. doi: 10.1029/2000JD900790

Rucklidge, J. (1965). The examination by Electron microscope of ice crystal nuclei from cloud chamber experiments. *Journal of Atmospheric Sciences*, *22*(3), 301–308.

Russell, L. M., Maria, S. F., & Myneni, S. C. B. (2002). Mapping organic coatings on atmospheric particles. *Geophysical Research Letters*, *29*(16). doi: 10.1029/2002GL014874

Saliba, G., Sanchez, K. J., Russell, L. M., Twohy, C. H., Roberts, G. C., Lewis, S., et al. (2020). Organic composition of three different size ranges of aerosol particles over the Southern Ocean. *Aerosol Science and Technology*, *55*(3), 268–288.

Sarangi, C., Qian, Y., Rittger, K., Ruby Leung, L., Chand, D., Bormann, K. J., & Painter, T. H. (2020). Dust dominates high-altitude snow darkening and melt over high-mountain Asia. *Nature Climate Change*, *10*(11), 1045–1051.

Schnaiter, M., Linke, C., Möhler, O., Naumann, K. H., Saathoff, H., Wagner, R., et al. (2005). Absorption amplification of black carbon internally mixed with secondary organic aerosol. *Journal of Geophysical Research – Atmospheres*, *110*(D19). doi: 10.1029/2005JD006046

Semeniuk, T. A., Wise, M. E., Martin, S. T., Russell, L. M., & Buseck, P. R. (2007). Hygroscopic behavior of aerosol particles from biomass fires using environmental transmission electron microscopy. *Journal of Atmospheric Chemistry*, *56*(3), 259–273.

Sheridan, P. J., Brock, C. A., & Wilson, J. C. (1994). Aerosol particles in the upper troposphere and lower stratosphere: Elemental composition and morphology of individual particles in northern midlatitudes. *Geophysical Research Letters*, *21*(23), 2587–2590. doi: 10.1029/94GL01387

Shrivastava, M., Cappa, C. D., Fan, J., Goldstein, A. H., Guenther, A. B., Jimenez, J. L., et al. (2017). Recent advances in understanding secondary organic aerosol: Implications for global climate forcing. *Reviews of Geophysics*, *55*(2), 509–559.

Suski, K. J., Hill, T. C. J., Levin, E. J. T., Miller, A., DeMott, P. J., & Kreidenweis, S. M. (2018). Agricultural harvesting emissions of ice-nucleating particles. *Atmospheric Chemistry and Physics*, *18*(18), 13755–13771.

Takahama, S., Gilardoni, S., Russell, L. M., & Kilcoyne, A. L. D. (2007). Classification of multiple types of organic carbon composition in atmospheric particles by scanning transmission X-ray microscopy analysis. *Atmospheric Environment*, *41*(40), 9435–9451.

Takahama, S., Liu, S., & Russell, L. M. (2010). Coatings and clusters of carboxylic acids in carbon-containing atmospheric particles from spectromicroscopy and their implications for cloud-nucleating and optical properties. *Journal of Geophysical Research – Atmospheres*, *115*(D1). doi: 10.1029/2009JD012622

Targino, A. C., Krejci, R., Noone, K. J., & Glantz, P. (2006). Single particle analysis of ice crystal residuals observed in orographic wave clouds over Scandinavia during INTACC experiment. *Atmospheric Chemistry and Physics*, *6*(7), 1977–1990.

Testa, B., Hill, T. C. J., Marsden, N. A., Barry, K. R., Hume, C. C., Bian, Q., et al. (2021). Ice nucleating particle connections to regional Argentinian land surface emissions and weather during the cloud, aerosol, and complex terrain interactions experiment. *Journal of Geophysical Research – Atmospheres*, *126*(23), e2021JD035186.

Tivanski, A. V., Hopkins, R. J., Tyliszczak, T., & Gilles, M. K. (2007). Oxygenated Interface on biomass burn tar balls determined by single particle scanning transmission X-ray microscopy. *The Journal of Physical Chemistry A*, *111*(25), 5448–5458.

Tobo, Y., Prenni, A. J., DeMott, P. J., Huffman, J. A., McCluskey, C. S., Tian, G., et al. (2013). Biological aerosol particles as a key determinant of ice nuclei populations in a forest ecosystem. *Journal of Geophysical Research – Atmospheres*, *118*(17), 10,100–110,110. doi: 10.1002/jgrd.50801

Tobo, Y., DeMott, P. J., Hill, T. C. J., Prenni, A. J., Swoboda-Colberg, N. G., Franc, G. D., & Kreidenweis, S. M. (2014). Organic matter matters for ice nuclei of agricultural soil origin. *Atmospheric Chemistry and Physics*, *14*(16), 8521–8531.

Twohy, C. H. (2015). Measurements of Saharan dust in convective clouds over the tropical eastern Atlantic Ocean. *Journal of the Atmospheric Sciences*, *72*(1), 75–81.

Twohy, C. H., & Anderson, J. R. (2008). Droplet nuclei in non-precipitating clouds: Composition and size matter. *Environmental Research Letters*, *3*(4).

Twohy, C. H., & Gandrud, B. W. (1998). Electron microscope analysis of residual particles from aircraft contrails. *Geophysical Research Letters*, *25*(9), 1359–1362.

Twohy, C. H., & Poellot, M. R. (2005). Chemical characteristics of ice residual nuclei in anvil cirrus clouds: Evidence for homogeneous and heterogeneous ice formation. *Atmospheric Chemistry and Physics*, *5*, 2289–2297.

Twohy, C. H., Clement, C. F., Gandrud, B. W., Weinheimer, A. J., Campos, T. L., Baumgardner, D., et al. (2002). Deep convection as a source of new particles in the midlatitude upper troposphere. *Journal of Geophysical Research-Atmospheres*, *107*(D21). doi: 10.1029/2001jd000323

Twohy, C. H., Anderson, J. R., & Crozier, P. A. (2005). Nitrogenated organic aerosols as cloud condensation nuclei. *Geophysical Research Letters*, *32*(19). doi: 10.1029/2005gl023605

Twohy, C. H., Kreidenweis, S. M., Eidhammer, T., Browell, E. V., Heymsfield, A. J., Bansemer, A. R., et al. (2009). Saharan dust particles nucleate droplets in eastern Atlantic clouds. *Geophysical Research Letters*, 36. doi: 10.1029/2008gl035846

Twohy, C. H., Anderson, J. R., Toohey, D. W., Andrejczuk, M., Adams, A., Lytle, M., et al. (2013). Impacts of aerosol particles on the microphysical and radiative properties of stratocumulus clouds over the Southeast Pacific Ocean. *Atmospheric Chemistry and Physics*, *13*(5), 2541–2562.

Twohy, C. H., Toohey, D. W., Levin, E. J. T., DeMott, P. J., Rainwater, B., Garofalo, L. A., et al. (2021a). Biomass burning smoke and its influence on clouds over the Western U. S. *Geophysical Research Letters*, *48*(15), e2021GL094224.

Twohy, C. H., DeMott, P. J., Russell, L. M., Toohey, D. W., Rainwater, B., Geiss, R., et al. (2021b). Cloud-nucleating particles over the Southern Ocean in a changing climate. *Earth's Future*, *9*(3), e2020EF001673.

Wang, B., Knopf, D. A., China, S., Arey, B. W., Harder, T. H., Gilles, M. K., & Laskin, A. (2016). Direct observation of ice nucleation events on individual atmospheric particles. *Physical Chemistry Chemical Physics*, *18*(43), 29721–29731.

Warren, S. G., & Wiscombe, W. J. (1980). A model for the spectral albedo of snow. II: Snow containing atmospheric aerosols. *Journal of Atmospheric Sciences*, *37*(12), 2734–2745.

Wilson, T. W., Ladino, L. A., Alpert, P. A., Breckels, M. N., Brooks, I. M., Browse, J., et al. (2015). A marine biogenic source of atmospheric ice-nucleating particles. *Nature*, *525*(7568), 234–238.

Wise, M. E., Biskos, G., Martin, S. T., Russell, L. M., & Buseck, P. R. (2005). Phase transitions of single salt particles studied using a transmission Electron microscope with an environmental cell. *Aerosol Science and Technology*, *39*(9), 849–856.

Wise, M. E., Semeniuk, T. A., Bruintjes, R., Martin, S. T., Russell, L. M., & Buseck, P. R. (2007). Hygroscopic behavior of NaCl-bearing natural aerosol particles using environmental transmission electron microscopy. *Journal of Geophysical Research – Atmospheres*, *112*(D10). doi: 10.1029/2006JD007678

Wise, M. E., Baustian, K. J., & Tolbert, M. A. (2010). Internally mixed sulfate and organic particles as potential ice nuclei in the tropical tropopause region. *Proceedings of the National Academy of Sciences*, *107*(15), 6693.

Wu, G., Zhang, X., Zhang, C., & Xu, T. (2016). Mineralogical and morphological properties of individual dust particles in ice cores from the Tibetan plateau. *Journal of Glaciology*, *62*(231), 46–53.

Wu, J., Faccinetto, A., Grimonprez, S., Batut, S., Yon, J., Desgroux, P., & Petitprez, D. (2020). Influence of the dry aerosol particle size distribution and morphology on the cloud condensation nuclei activation. An experimental and theoretical investigation. *Atmospheric Chemistry and Physics*, *20*(7), 4209–4225.

Yu, H., Li, W., Zhang, Y., Tunved, P., Dall'Osto, M., Shen, X., et al. (2019). Organic coating on sulfate and soot particles during late summer in the Svalbard archipelago. *Atmospheric Chemistry and Physics*, *19*(15), 10433–10446.

Zhang, X. Q., McMurray, P. H., Hering, S. V., & Casuccio, G. S. (1993). Mixing characteristics and water content of submicron aerosols measured in Los Angeles and at the grand canyon. *Atmospheric Environment. Part A. General Topics*, *27*(10), 1593–1607.

Zimmermann, F., Weinbruch, S., Schütz, L., Hofmann, H., Ebert, M., Kandler, K., & Worringen, A. (2008). Ice nucleation properties of the most abundant mineral dust phases. *Journal of Geophysical Research – Atmospheres*, *113*(D23). doi: 10.1029/2008JD010655

Part II
Microanalytical Techniques

4

Particle Population Analysis by Automated Scanning Electron Microscopy

Gary S. Casuccio

ABSTRACT

Automated scanning electron microscopy (SEM) analysis of particle populations associated with atmospheric particles has evolved from an exploratory research tool to a mature analytical technology that can now be considered integral to the characterization of atmospheric particulate matter. The ability to investigate air masses based on the morphology and composition of thousands of individual particles provides environmental scientists with detailed information on particle characteristics which can be used to identify sources of particulate matter and their impact on the environment. This chapter provides an overview of the evolution of automated SEM and its applications from an atmospheric particle analysis perspective.

4.1. INTRODUCTION

The scanning electron microscope (SEM) has historically been used to provide valuable information on the characteristics associated with atmospheric particles; however, its impact in environmental studies was hindered in some respects due to the amount of time and associated costs required to perform a statistical particle analysis on particle populations. To compensate for the time and tedium of a particle analysis performed in a manual fashion, development of automated approaches to the analysis of individual particles using an SEM was pursued in the 1970s, with the goal of utilizing computers to control the SEM and replace the analyst where the eye/brain is weak (routine/repetitive data collection) and to assist the analyst where the eye/brain is best (data interpretation).

The analysis of atmospheric particles using automated SEM techniques, often referred to as computer controlled scanning electron microscopy (CCSEM), has evolved from exploratory evaluations at a few academic and industrial research centers to routine and research-related studies at industrial facilities, governmental agencies, universities, and contract research organizations. Over its evolution, CCSEM has become a mature analytical technique used to obtain size, shape, and elemental composition, as well as, a digital image of thousands of individual particles in a time-efficient manner. The ability to provide particle classification data based on morphology and elemental composition associated with individual particles provides an additional dimension related to the characterization of atmospheric particulate matter and complements data obtained using bulk and other individual particle analysis techniques.

Most of the environmental applications utilizing CCSEM have focused on the statistical evaluation of particulate matter populations associated with air masses and the identification and quantification of particulate species to assist in the evaluation of source impacts. CCSEM has also been used to characterize emissions from point and fugitive sources, assess health effects from exposure to particles in epidemiological studies, provide insight into particle transport, and impact of particles on climate.

In this chapter, an overview of CCSEM is provided, focusing on the evolution of the technology and applications related to the analysis of atmospheric particles.

RJ Lee Group, Pittsburgh, PA, USA

Microanalysis of Atmospheric Particles: Techniques and Applications, Geophysical Monograph 288, First Edition.
Edited by Joseph M. Conny and Peter R. Buseck.

DOI: 10.1002/9781119554318.ch4

4.2. THE EVOLUTION OF CCSEM

Automation of any new machine, instrument, or process typically evolves over time in a step-by-step process; each step represents a useful incremental advancement that serves well until the next step arrives. CCSEM is no different; digital beam control, X-ray acquisition, image processing, image/data storage, computing processing speed, and other components of CCSEM all developed incrementally and along parallel paths. Today's CCSEM systems represent a confluence of these paths and the accumulated vision of the pioneers and contributors along the way.

The first commercial SEMs in the mid-to-late 1960s were completely analog; there was no control aside from a human operator pressing buttons and turning knobs. Digital technology did exist elsewhere in the form of mainframe computers and the earliest minicomputers, and early solid-state analog-to-digital converters (A/Ds) were available to convert analog signals into digital data. The latent possibilities embodied by these building blocks were recognized by Eugene White and a group of researchers at the Materials Research Laboratory at Pennsylvania State University (PSU). Their approach, referred to as the CESEMI (computer evaluations of SEM images) (McMillan et al., 1969), used an A/D to access the secondary and backscattered electron signals during the analog scan and convert those signals into digital words (*pixels,* as they are known today) proportional to the analog voltage, which itself is proportional to the intensity of the electron signal. As the scan progressed, the digital data generated was written directly onto magnetic tape for storage. Each scan produced up to 275 digital words (pixels) per line with up to 1,000 lines per image.

Having captured the digital data on magnetic tape, that media would then be loaded to an IBM 360/67 mainframe computer, where a software program would be used to read image data from the tape, apply a digital threshold to determine which pixels were on particles, combine adjacent pixels into data structures representing particle entities, and compute size and shape metrics for each particle. This was all accomplished with less than 8 K of program memory. The software even saved a representation of the digital data as arrays of asterisks—this latter function allowed the data to be printed out on a teletype printer so that researchers could visualize the digitized spatial information in a human-readable form.

Strictly speaking, the CESEMI system was not computer controlled. It was, however, a powerful step toward a CCSEM system, made more significant because it was the first step. The fact that researchers used it for numerous studies, including analyzing lunar dust collected by the Apollo space program (Gorz et al., 1971), speaks eloquently of its value to contemporary science.

Meanwhile, researchers were busy on another piece of technology, one that would allow CCSEM to not only detect and measure particles but also quantify the elements present. In 1968, building on existing nuclear spectroscopy technology, the first practical—albeit experimental—solid-state energy dispersive X-ray spectroscopy (EDS or EDX) system for SEM was constructed (Fitzgerald et al., 1968), consisting of a lithium-drifted silicon diode (Si[Li]) detector, amplifiers, and a multichannel analyzer (MCA) that converted X-ray energy measurements into an X-ray spectrum. The 600 eV resolution of this pioneering SEM/EDS device would evolve into the sub-130 eV detectors of today. Following this breakthrough, which used an ORTEC diode and amplifier chain along with a Nuclear Data MCA, new vendors such as Kevex, Princeton Gamma Tech, and Tracor Northern entered the SEM/EDS marketplace in the 1970s with commercially available detectors, MCAs, and, eventually, full X-ray analysis systems.

Integration of the concepts of EDS with particle recognition resulted in what we today call *active control.* Active control was the next step in the process of developing a CCSEM system, and digital beam control was the required innovation. Rather than the passive approach implemented in CESEMI, this would permit the electron beam to be moved under computer control in sync with the digitization of the electron signals. One of the pioneers in this area was Lemont Scientific, an offshoot of the PSU group that developed CESEMI. Digital-to-analog conversion (DAC) electronics translated digital *x*/*y* positions into the analog signals required to drive an SEM's scan coils, thus allowing a computer to calculate and move the beam to any desired location under software control while monitoring the electron signal at each position. With this capability, the digital scan generator could perform real-time measurements of individual particles by "drawing" chords in various directions and measuring the distance over which the signal remained within a predetermined grayscale threshold value. It also could vary the point spacing—wider spacing while searching for particles to optimize speed, and narrower spacing when measuring the size of a particle to increase precision of the measurement.

At this time, minicomputers that had appropriate processing and storage capabilities emerged as smaller, lower-cost alternatives to mainframe computers. These minicomputers proved to be the ideal tools for the tasks associated with SEM automation. The Lemont Scientific digital beam control systems were built around the Interdata 6/16 minicomputer, whereas Tracor Northern's NS-880—the first truly integrated EDS analyzer—used the Digital Equipment Corporation PDP-11 series of minicomputers. The NS-880 also introduced full software programmability, with Tracor's cryptic but powerful FLEXTRAN programming language, in the early 1970s (Schamber, 2009). With an architecture that interfaced all devices to the PDP-11, researchers were able to write software that controlled all aspects of X-ray collection and had direct access to spectral memory and a rich set of software libraries for control and data manipulation. By the mid-1970s, the TN-1310 Modular Electron Column Automation hardware added digital beam control capability to Tracor Northern EDS analyzers for a more fully integrated CCSEM platform.

One example of a highly functioning early CCSEM system was developed at the United States Steel Research Center (U.S. Steel Research) during the mid-1970s through a collaboration with Lemont Scientific and Tracor Northern. The Lemont Scientific collaboration focused on development of the digital beam control technology, whereas the Tracor Northern collaboration involved a research and development venture directed at extending the application of online computer control of electron microscopes (Lee & Fisher, 1980). The initial U.S. Steel Research CCSEM system integrated an ETEC Autoscan SEM with the Tracor Northern NS-880 X-ray analyzer and Lemont Scientific digital beam control, along with peripherals such as a dual 8-in. floppy drive, LSI ADM-3A video display terminal, and teletype printer. This system incorporated a "homemade" backscattered electron detector following the approach used by the researchers at PSU on the CESEMI system (Lebiedzik, 1975). Figure 4.1 provides a photograph of an early version of a CCSEM system developed at U.S. Steel Research.

The FLEXTRAN-based automation software was flexible enough to analyze inclusions in steel as well as environmental air particulates. The analysis, typically performed using an accelerating voltage of 20 kV, moved the electron beam over the sample in discrete steps while monitoring the resultant secondary or backscattered electron signals. Once a coordinate was reached where the signal was above the threshold level, the size of the particle was measured using a series of measurements (rotated chord approach; Casuccio et al. (1983)), and the average, maximum, and minimum diameters were recorded on computer media.

After sizing was completed, the electron beam was positioned on the centroid of the particle, and collection of characteristic X-rays was initiated (~3–5 seconds). The energy and abundance of the X-rays emitted by the particle were displayed as a spectrum of X-ray counts versus energy, and the minicomputer processed the X-ray data to identify elements present in the resultant spectrum. If the peak for an element was determined to be above background, the net X-ray counts for that element were obtained by subtracting out background counts. The elements were then further processed to obtain their relative concentrations expressed as normalized percentages. Once the spectrum analysis was completed and elements identified, a density was assigned to the particle based on the elemental composition, and its volume was estimated by assuming that the particle was in the shape of a prolate spheroid. Particle mass could then be calculated by multiplying the assigned particle density by its volume. Particles were classified into species according to their elemental composition (and shape, for spherical particles which was based on aspect ratio) using rules developed through examination of the results obtained from the CCSEM analysis of reference materials. The classification, along with the size and elemental composition of individual particles, were compiled into printable number, area, volume, and mass-distribution tables. The data was stored in a data file on magnetic media for archive and future review.

The systems at U.S. Steel Research in the early 1980s were capable of characterizing about 500 particles

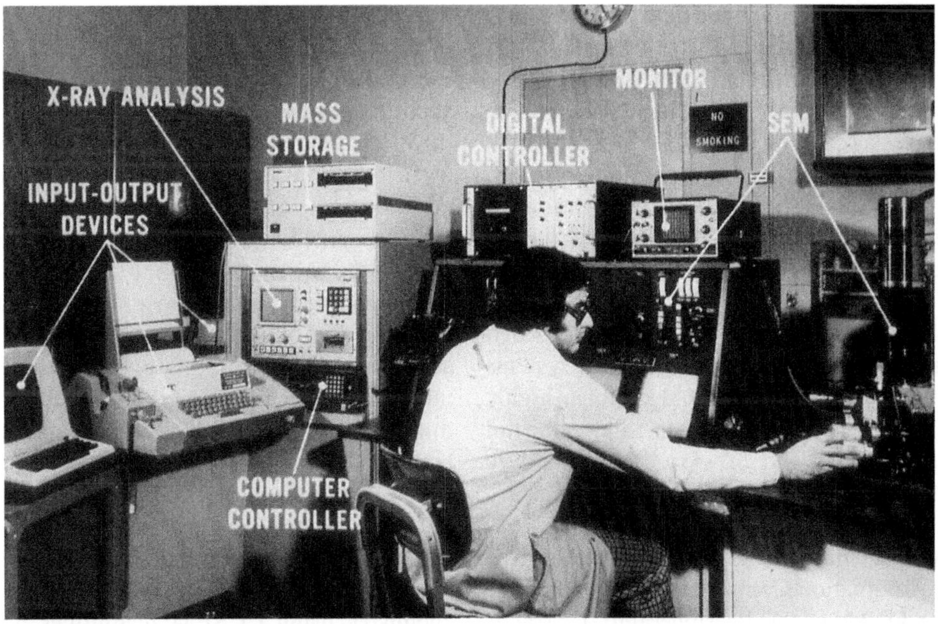

Figure 4.1 Early version of a CCSEM system developed at U.S. Steel Research capable of measuring size and obtaining elemental composition on hundreds of particles per hour. *Source:* Lee and Fisher (1980).

per hour on multiple SEM platforms, including both ETEC and JEOL SEMs using the NS-880 X-ray analysis systems and its descendants, including Tracor Northern's TN-2000, TN-4000, and TN-5500 EDS systems (McCarthy et al., 1982). These early CCSEM systems, although limited compared to today's systems, provided a major step forward in terms of particle characterization throughput. To put this into perspective, the time it would have taken to acquire and process size and elemental composition data on 500 particles using manual SEM methods would have been on the order of weeks or longer. In recent discussions with Dr. Richard Lee, he recalled that it took 6 months to analyze and process 8,000 inclusions in steel samples using the SEM in a manual manner. The early CCSEM systems at U.S. Steel were able to characterize that many particles as well as summarize the data within a few days (personal communication).

There were many factors that contributed to a manual analysis taking a significant amount of time. Particle sizing using manual techniques was typically done by measuring the length and width of a particle using a ruler on the CRT screen or estimating the size by eye using the scale bar as a reference. Elemental concentrations were initially obtained by taking a Polaroid photo of the EDS screen and measuring the peak heights using a ruler. As EDS technology evolved, the concentrations of elements in a particle were determined by defining a *region of interest* (ROI) on the energy scale for a specific element. The computer would record the X-ray counts in the ROIs from all of the elements of interest, and then estimates of the elemental concentrations could be determined. Although these EDS systems were able to provide an output summarizing the concentrations more efficiently, the elemental concentrations, along with the particle size data, would then need to be recorded in a notebook by hand and summarized. Also worth noting is that because manual SEM data needed to be acquired, recorded, and processed by hand, it was much more prone to errors as compared to a CCSEM approach.

It should be realized that both manual and automated SEM analysis are valuable in the characterization of individual particles. Some of the strengths and limitations of each technique were noted by Schamber and van Beek (2006).

One of the last pieces required for fully unattended automation was stage control. The CCSEM systems of the late 1970s and early 1980s could perform highly automated particle detection, measurement, and elemental characterization within a field of view, but to analyze multiple fields, an operator was required to turn the *x*/*y* stage knobs and restart the analysis after each field was completed. By the mid-1980s, the TN-5600 PAC (Programmable Automation Controller) became available as another FLEXTRAN-enabled peripheral to the Tracor Northern X-ray analyzer. This, along with the existing beam control and EDS capabilities, enabled a fully unattended analysis. Although a manual setup step was required at the beginning of an analysis to focus the SEM and set up the particle detection threshold, once the automated analysis started, it could run without human intervention for many hours.

4.3. EARLY APPLICATIONS OF CCSEM IN THE CHARACTERIZATION OF ATMOSPHERIC PARTICLES

Similar to the evolution of CCSEM hardware and software, analysis of atmospheric particles using CCSEM has evolved over time. The earliest attempt to characterize atmospheric aerosols using CCSEM techniques was noted in a publication by Byers and co-workers (1971). Objectives of the Byers study included evaluating the feasibility of adapting the SEM for quantitative analysis of the size of atmospheric dust particles as well as developing a suitable atmospheric sampling technique for direct analysis of particles in the SEM. To this end, particles were collected over a 1-hour period on a tube lined with aluminum foil using an electrostatic precipitator (Bendix Model 959). Samples were prepared for direct SEM examination by removing a ~1 cm^2 section of the foil and gluing it to a sample block using Duco cement. The analysis was performed using an accelerating voltage of 25 kV at a magnification of 3,000X in a manner described previously for the CESEMI system. Results were presented comparing the automated SEM size measurements to those obtained using manual SEM and showed that the automated SEM size distribution results were slightly less than those reported for the manual SEM analysis. Byers noted a limitation related to the analysis of air particulate samples using the SEM: volatile particles cannot be handled readily in the SEM due to the high vacuum required for the analysis. Advantages of the SEM technique were noted as better resolution and depth of field as compared to light microscopic techniques. In addition, the automation enables thousands of particles to be counted, making it more reliable than manual counting. He also noted that the SEM has the potential of being able to link atmospheric particles to a specific source based on size, shape, and chemical composition; advance the understanding of air ecology; and yield more extensive information for interpretation of epidemiological data. Finally, Byers described this work as being a first step toward a more complete characterization of suspended particles and noted that work was underway to provide improved automated analysis based on shape characteristics and chemical composition.

Characterization of atmospheric particles incorporating chemistry using CCSEM techniques was described by Lee et al. (1979). Lee and colleagues briefly described the analysis process and particle classification scheme and reported for the sample analyzed that particles greater than 15 μm accounted for about 50% of the sample mass and were composed of clays, quartz, and carbon types. They noted that improved methods for detection of carbon particles needed to be developed, and:

> The development of automated SEM techniques requires a costly and time-consuming program. But if the ability to identify individual sources of airborne particles, and to obtain reliable, reproducible data on their size, shape, and composition can contribute to economical control strategies, it will justify the effort. (Lee et al., 1979)

Additional papers on aerosol applications of CCSEM incorporating EDS analysis were provided by Lee and Kelly (1980) and Lee and Fisher (1980). These papers provided information on the effects of pixel spacing on detection of particles; creation of object vectors based on size, shape, and elemental composition collected per particle; and guidance on sample preparation. Lee also provided, what may be the first, quantitative data summary reported for an air particulate CCSEM analysis, which is provided in Table 4.1.

It is interesting to note that although EDS technology at that time was limited to detecting elements with atomic numbers greater than 11 (sodium), the Lee papers nevertheless provide values for carbon particles. The "carbon" results were based on backscattered-electron detection of a particle that had low or no detectible X-ray counts and was assumed to be carbonaceous.

Kelly and co-workers (1980) discussed characterization of fine particles as well as issues associated with preparation of samples collected on glass fiber, cellulose, and polytetrafluoroethylene (PTFE) filters for CCSEM analysis. Preparation for CCSEM at this time typically required redeposition of the particles from the as-collected filter onto a membrane filter (typically a Nuclepore polycarbonate filter). A brief overview of this process was discussed in a report by Casuccio and Janocko (1981a, 1981b). Kelly also provided guidance on the area that should be analyzed to have statistical significance; detection of particles based on various signals; and applications in air quality studies, including source characterization, particle transport and deposition, and source apportionment.

In summary, the pioneers of CCSEM discussed in this section and others helped to open the door to the development of automated SEM technology and methodologies for the analysis of atmospheric particles. Their efforts, along with concurrent advances in SEM, EDS and computer technology, set the stage to develop CCSEM software and systems on a commercial basis. The first achievement in this area was the development of the Tracor Northern ADEM (automated digital electron microscope) in 1986, which was the first all-digital SEM that allowed scanning for both manual SEM and CCSEM to be controlled by a single set of digital electronics. The ADEM represented the first SEM designed from a CCSEM perspective and incorporated a five-axis automated stage, an integrated EDS system, and an airlock compartment to enable continuous throughput of samples. However, the ADEM did not have a long run from a commercial perspective due, in part perhaps, to the technology being somewhat ahead of its time; the system was built around a minicomputer when the world was beginning to move toward personal computers for analytical instrumentation, and an SEM/EDS system built and sold by one company was viewed as a disruptive technology by other manufacturers in the industry. It is only relatively recently that industry has started to appreciate the benefits of an integrated system with both EDS and beam control capabilities. Although the ADEM had a relatively short commercial life, it opened the door to a totally automated SEM system and led to the creation of the Personal SEM (PSEM).

Table 4.1 Summary of Particle Type and Size Distribution Results from CCSEM Analysis of an Air Particulate Sample

	Size Distribution (μm)				
Type	0.2–2.5	2.5–5.0	5–15	>15	Total
Fly ash	94	5	1	0	85
Quartz	79	15	5	0	83
Ca-rich	72	16	12	0	68
Fe-rich	95	4	1	0	52
Ca-S	99	1	0	0	120
Clay	85	11	4	0	182
Ca-Si	43	40	16	1	70
Carbon	96	4	1	0	75

Values reported under the size distribution bins are the percentages of particles in each size range. Total column indicates the number of particles classified for each particle type (735 particles analyzed). *Source:* Lee and Kelly (1980).

4.4. THE IMPACT OF THE PERSONAL COMPUTER

In addition to the introduction of the ADEM, the 1980s heralded the introduction of the microcomputer. Just as minicomputers had earlier emerged as low-cost alternatives to mainframes, microcomputers—or personal computers (PCs), as they came to be known—created a new niche that was not only far less expensive than minicomputers but also far more user-friendly. Over a relatively short period of time, the PC replaced the minicomputer as a viable platform for instrument control and changed the paradigm in the way CCSEM data was acquired, processed, and evaluated. Importantly, the PC would address significant limitations related to the lack of capabilities to effectively review, manage, and report CCSEM data that was identified as far back as

1970 (in a paper not published until 1972), where the creators of the CESEMI system lamented the task of transforming large amounts of raw data into useful information:

> One of the problems facing us at the time of writing is what to do with the mass of data generated by the CESEMI analysis. There appears to be little precedent for digesting so much data generated on a particle by particle basis. (White et al., 1972)

Further, the power of microscopy historically had been based on the ability to ascertain information from visual inspection of a sample. Unfortunately, the acquisition of digital images for review during offline interpretation of the data was not possible with the initial CCSEM systems, which was a significant limitation to the automated analysis approach. Therefore, it is not surprising that during the rise of PCs in the 1980s, researchers began focusing on the possibilities of PC-based data-reduction and image processing/display to address the limitations of the contemporary CCSEM systems. Research conducted at RJ Lee Group (RJLG) in the mid-1980s focused on downloading data from a minicomputer-based Tracor Northern TN-5500 EDS analyzer to a desktop PC. As a result of this effort, technology was developed to permit collection of digital particle images during the CCSEM analysis, which, along with the X-ray spectra, numeric size/shape data, and elemental concentrations data, were transferred onto PC media after the analysis was completed. This enabled the review of the data at offline PC workstations. This work was integral in developing the concept of *microimaging,* which refers to the collection of low-pixel resolution digital particle images in an automated manner during a CCSEM analysis (Schwoeble et al., 1988). Transferring the microimages and associated data to a PC was accomplished by the development of a particle-image storage system (Mershon et al., 1988). Once in the PC, data could be processed and visualized using software developed to save/share data sets using a variety of removable disc storage systems (Bernoulli Box, WORM drives, ZIP drives) for offline visual review on a then state-of-the-art IBM 386 PC (Henderson et al., 1989).

The first application of CCSEM incorporating microimage technology related to the characterization of atmospheric particles was performed in a study commissioned by the Electric Power Research Institute (EPRI) that focused on emissions from a coal-fired power plant (Casuccio et al., 1988). In this study, particulate emissions collected in the power plant stack were compared with particulate matter upwind and downwind of the stack. Samples were collected in the plume from the stack using a helicopter particle-collector system specially developed for the project by Dr. A.R. McFarland and Mr. C. Ortiz of Texas A&M University; upwind of the plant using a small, fixed-wing aircraft; and within the stack at stack temperature using stack dilution sampling techniques. The CCSEM particle data obtained from these samples were grouped into classes based on elemental composition and morphology that was based, in part, on visual review of the microimages. Examples of results from that study are provided in Tables 4.2 and 4.3. Results showed a strong correlation between the diluted stack and downwind plume samples.

Table 4.2 Comparison of CCSEM Particle Type Results for Diluted Stack, Downwind Plume, and Upwind Background Samples (wt%)

Particle type	Diluted stack	Downwind plume	Upwind background
Fly ash	29	28	1
Crustal	15	23	53
Sulfur-rich	24	20	3
Carbon-rich	3	6	7
Unknown	29	23	36

Source: Adapted from Casuccio et al., (1988).

An example of classifying particle data using microimages is provided in Table 4.3 for particles identified as fly ash, which shows a strong correlation with the rounded and fused morphological characteristics of the fly ash component for the diluted stack and the downwind plume sample.

Although this data retrieval and review software would be slow by today's standards, it was a significant advance over previous approaches and represented the first PC-based, offline data management system for the evaluation of microimages and particle data from CCSEM analyses.

Building upon the proliferation of the PC, the PSEM was introduced in 1992 (Microscopy Today, 1992); it was the first SEM to utilize a PC to control the operation of the SEM via a single set of digital electronics. This SEM was designed specifically to support the requirements of CCSEM applications and as such introduced a new paradigm related to the traditional SEM design approach. During the 1990s and continuing to this day, all SEMs have evolved to digital platforms, most of which offer an application programming interface (API) that allows third-party software to control the scanning system and all aspects of SEM operation.

Table 4.3 Comparison of CCSEM Results Incorporating Shape Classifications for Fly Ash Associated with Diluted Stack, Downwind Plume, and Upwind Background Samples (wt%)

Fly ash particle types	Diluted stack	Downwind plume	Upwind background
Round	12	10	1
Fused	15	13	0
Agglomerated	1	4	0

Source: Adapted from Casuccio et al., (1988).

4.5. PARTICLE CLASSIFICATION OF CCSEM DATA

Classification of the particles analyzed with CCSEM was initially, and still is to a large degree, based on hierarchical rules using elemental concentrations and shape (e.g., aspect ratio) which group particles having similar characteristics (Casuccio & Janocko, 1983; Conny et al., 2019; Hu et al., 2022; McIntyre et al., 1980; Peters et al., 2016; Willis et al., 2002). Although this approach is well-suited to group particles into particle classes, a limitation of this approach is that it requires experience to get well-defined particle sorting. Therefore, researchers have explored using various cluster analysis techniques as an alternative to rule-based sorting.

Kim and Hopke (1987) investigated the use of various hierarchical and nonhierarchical cluster analysis methods to define particle classes for auto emissions, and results suggested that it was possible to identify groups of particles that characterize emissions from highways.

Wienke et al. (1994) and Hopke and Xie (1994) applied adaptive resonance artificial neural network (ART-2a) approaches to cluster airborne particles based on compositional data and particle shape. Wienke experimented initially with a small set of SEM images to investigate the potential of formulating meaningful particle classes based on particle shape and then expanded the study to include elemental composition using source data from natural and industrial emissions sources. It was noted that the ART-2a approach can be used for supervised and unsupervised classifications, and results from the study demonstrated the potential for future online classification of microimages combined with quantitative individual particle composition. Hopke and Xie's work involving ART-2a concluded that a more accurate particle classification can be obtained by combining the chemical compositions and a shape index and that this approach had a high potential value in particle classification because the particle class library can be dynamically updated, which aids in the identification of new sources of airborne particles.

Ault et al. (2012) evaluated the physicochemical properties obtained from CCSEM analysis of coarse-mode, iron-containing particles collected using passive samplers. The CCSEM data was clustered using a k-means algorithm (MATLAB, MathWorks, Inc.), which was noted as being as effective as ART-2a while being simpler and faster to run. Results of the study indicated that spherical iron particles that were representative of anthropogenic sources decreased strongly with distance from the source(s) and demonstrated that CCSEM analysis is a powerful tool to study spatial and temporal gradients in components of coarse particulate matter.

Song et al. (1999) explored the combined use of two neural networks to apportion airborne crustal (soil) particles. ART-2a was first used to partition particles into homogeneous classes based on the elemental composition data. The mass fractions of each particle type class were then calculated to develop source profiles for the crustal dust sources. A back-propagation (BP) neural network was then applied to build the model between the mass fractions of different particle types and the mass contributions. The model was tested using three physical crustal source samples and artificial ambient samples that were prepared by randomly mixing particles from the three source samples. Song noted that satisfactory predictions for the mass contributions of the three sources to the ambient samples were obtained, indicating that the method is a promising tool for the source apportionment of chemically similar soil samples.

Kindratenko et al. (1994) explored using fractal dimensions to identify particles originating from different sources that have similar compositions, such as fly ash and soil dust, to evaluate the applicability of fractals for morphological metrics related to shapes of individual aerosol particles analyzed with CCSEM. Evaluation of SEM images from spherically shaped fly ash, irregularly shaped fly ash, and soil dust indicated that spherical-shaped fly ash had no fractality, whereas the irregularly shaped fly ash could be characterized by textural and structural fractal dimensions. Soil dust particles could only be characterized by a textural fractal dimension. The authors concluded that the fractal dimensions, both textural and structural, can be used to apportion soil dust and fly ash particles.

4.6. PRECISION, ACCURACY, AND REPRODUCIBILITY OF CCSEM DATA

As CCSEM evolved, there were, and still are, questions related to the accuracy of the individual particle data. Although the strengths and limitations of the CCSEM analysis have been reported in the scientific literature, it is worth reviewing in this chapter. Especially because the technology has improved significantly since its inception in terms of both hardware and software.

Whereas the precision associated with a CCSEM analysis is one of its strengths and has been well documented in the literature, it is more difficult to determine the accuracy of CCSEM data for atmospheric particles. In general, CCSEM results for inorganic particles will be of higher accuracy than for organic particles such as carbonaceous species, which are more difficult to detect on a carbon substrate (e.g., polycarbonate track-etched filter) typically used in environmental studies. Furthermore, particle types that are present in major or minor concentrations would typically be more accurate than those present at trace levels. The size of the particles is also a factor when evaluating accuracy, as larger particles are easier to detect and measure than smaller particles.

Johnson et al. (1981) were the first to investigate the quality of CCSEM data for air quality samples. To this end, they developed a methodology to estimate a sample's bulk elemental composition by aggregating the elemental data from the analysis of individual particles, which could then be compared to data obtained from bulk elemental X-ray fluorescence (XRF) analysis. Johnson's evaluation was based on the analysis of particles collected on the coarse-fraction PTFE filters collected with a dichotomous sampler. The filters selected for CCSEM analysis had mass loadings that ranged from 50 to 100 $\mu g\,cm^{-2}$, which he described as being ideal for the CCSEM analysis, and 500–1,000 particles were analyzed per filter. To calculate the weight of a particle, the volume was computed and multiplied by the density assigned to specific particle classes, and the elemental composition of the particle was expressed on a weight percent basis using a defined stoichiometry to account for the weight of the missing light elements (C, N, O, and H). Johnson concluded that when the elemental mass loadings were normalized, good agreement of major elements (>5 wt.%) by the two methods was observed and noted that CCSEM analysis had the potential to be a powerful tool for source apportionment of atmospheric particulate matter.

Casuccio and Janocko (1983), Casuccio, Janocko, Dattner et al. (1983), and Casucco, Janocko, Lee et al. (1983) built on the work of Johnson and reported on the reproducibility of the analysis as well as the precision and accuracy of CCSEM data obtained from analysis of ambient particulate matter collected on total suspended particulate (TSP) glass fiber and cellulose filters in a study investigating the impact of a lead smelter. Similar to the Johnson approach, CCSEM elemental results for individual particles were processed into a "bulk" sample elemental composition and were compared with bulk XRF, proton-induced X-ray emission (PIXE), atomic absorption (AA), and ion chromatography (IC) data, and in general showed good agreement.

van Dyck et al. (1984) analyzed spherical glass particles (NBS Standard #610) with diameters ranging from 2.5 to 14 μm, utilized atomic number, absorption, and fluorescence (ZAF) correction of the X-ray spectral data with different particle models, and produced results comparable to certified concentrations. Germani and Buseck (1991) investigated the precision of CCSEM to characterize mineral particles 1 μm and greater in diameter. Samples evaluated in the study were collected on polycarbonate filters, which were noted as being well-suited for CCSEM analysis. Results indicated that it was possible to identify major particle types associated with mineral particles as small as 1 μm in diameter with a precision of +/−10% based on replicate analysis. Results obtained from a chemical mass balance (CMB) between individual particle analyses and comparison with bulk chemical compositions of standard mineral powders indicated a precision of better than +/−15% for most elements.

Mamane et al. (2001) utilized CCSEM to characterize particles collected on a coarse-fraction polycarbonate filter using a dichotomous sampler and evaluated several parameters, including the number of particles per field (*N*/*F*), average particle diameter (D_{ave}), average mass loading per field M_{ave}, and average particle composition. Subsets of the data were also evaluated to assess the stability of the CCSEM analysis over time. The number of particles needed to obtain representative results was also investigated. It was noted that the comparisons demonstrated excellent stability of CCSEM over an 8-hour analysis session and that the physical properties represented by *N*/*F*, D_{ave}, and M_{ave} were well characterized by analyzing approximately 400 particles. It was also noted that the average elemental composition and major chemical class abundances converged to within a few percent of their final values after analyzing about 1,000 particles, but for many applications analysis of several hundred particles may be sufficient. Offline review of the CCSEM data identified errors with the analysis, including missed particles, overlapping particles and substrate contrast artifacts, sizing errors, and heterogeneous particles, but it was noted that most of these errors could be corrected or eliminated using the offline review of the data or avoided by maintaining a proper particle loading on the filter.

Although the previous studies provide insight into the ability of CCSEM to provide "quantitative" data, there has been limited information on the accuracy of the CCSEM particle class data. Given the complexity of addressing the issue of accuracy related to particle classes without access to standards representative of atmospheric aerosols, another approach to evaluate accuracy is by comparing CCSEM results with data obtained from Federal Reference Method (FRM) and Federal Equivalent Method (FEM) samplers for particulate matter. For example, Edgerton et al. (2009) used CCSEM to speciate particles on coarse-fraction polycarbonate filters collected using a dichotomous sampler to augment bulk analytical analysis. Results showed that a reasonable mass balance was obtained by combining bulk analytical data with the CCSEM carbon results. For example, analysis of selected samples indicated that measurements of sulfate, nitrate, ammonium, and major metal oxides obtained from bulk analysis accounted for about 35% of the $PM_{10-2.5}$ mass, whereas the CCSEM analysis indicated that carbon particles accounted for approximately 70% of the mass. Overall, CCSEM was able to account for 83 ± 25% of the $PM_{10-2.5}$ mass while speciating the carbon component into several classes, including pollen/spores, vegetative detritus, carbon chain agglomerates (soot), and "other."

Coz et al. (2010) used the combination of the CCSEM analysis (ASPEX PSEM) followed by manual evaluation of particles in a field emission SEM (FE-SEM; FEI Sirion 400) to obtain estimates of primary biogenic organic aerosols (PBOA) associated with $PM_{2.5}$. As part of this study, $PM_{2.5}$ mass concentrations as determined by CCSEM were compared to a FRM sampler, and results compared well (R^2 =0.98 and slope = 0.98). CCSEM results were also compared to organic carbon/elemental carbon (OC/EC) results and showed good correlation (R^2 =0.92 and slope = 1.1).

Wagner and Leith (2001) developed a technique for the measurement of PM_{10}, $PM_{2.5}$, and $PM_{10-2.5}$ concentrations based on the collection of ambient particles using a passive sampler (referred to as the UNC-Passive Aerosol Sampler) and subsequent individual particle analysis. The methodology estimates ambient mass size distributions ($\mu g\,m^{-3}$) using CCSEM data and an empirical deposition velocity model to convert the deposited particles to an average concentration over the sampling duration. Although this method involves potential variability due to both the passive collection and the CCSEM analysis, results have been shown to correlate well with concentrations obtained from FRM and FEM monitors. For example, Wagner and Casuccio (2014) reported good agreement for concentrations compared to colocated FRM samplers, as shown in Figure 4.2. In this figure, results from the analysis of three colocated UNC-passive aerosol samplers collected over a 7-day period were compared to the average of seven 24-hour samples collected with FRM PM_{10} and $PM_{2.5}$ samplers. The analyses of the passive samples were performed at

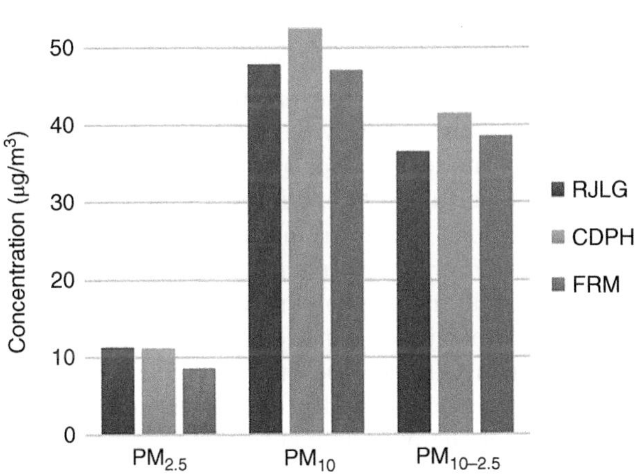

Figure 4.2 Comparison of ambient particle concentrations based on the CCSEM analysis of UNC-Passive Aerosol Samplers to Federal Reference Method gravimetry results. *Source:* Wagner and Casuccio (2014).

the California Department of Public Health (CDPH) and RJLG. The samples were collected in the Phoenix, AZ, area as part of an EPA study.

A more challenging evaluation using the passive sampling technology developed by Wagner was conducted by Sawvel et al. (2015). This work, performed as part of an EPA air quality study in Cleveland, OH, focused on assessing spatial heterogeneity and compositional components of $PM_{10-2.5}$ in a metropolitan area with substantial industrial activity. Results from the study are presented as normalized concentration maps in Figure 4.3 for $PM_{10-2.5}$ and various components of the coarse particulate, including Si/Al particles, Ca-rich particles, and Fe-oxide particles. The plots enable visualization of the spatial heterogeneity with dark regions indicating concentrations greater than the mean and the light regions indicating those less than the mean. The plots also show substantial heterogeneity with elevated concentrations of $PM_{10-2.5}$, Ca-rich, and Fe-oxide particles observed in the industrial region within the Cleveland metropolitan area. The Si/Al-rich class, representative primarily of crustal material (soil), was less heterogeneous compared to the other compositional classes. A quantitative measure of spatial heterogeneity, coefficient of divergence (COD), is provided in Figure 4.3 for $PM_{10-2.5}$ and selected particle types. A COD of zero indicates no difference between site concentrations (homogeneous), whereas a COD greater than 0.2 is considered to indicate substantial spatial heterogeneity. Percent spatial heterogeneity (SH%), which refers to the uneven distribution of various concentrations of each species within an area, varied from 0% for the Si/Al-rich to 100% for Fe-rich and for $PM_{10-2.5}$ class.

As part of this study, Peters et al. (2016) evaluated the precision and accuracy of the sampling and analysis methodology. Overall precision expressed as root-mean-square coefficient of variation (CV_{RMS}) was found to improve with increasing concentrations (37% for all samples; 19% for $PM_{10-2.5}$ >10 $\mu g\,m^{-3}$ and 10% for $PM_{10-2.5}$ > 15 $\mu g\,m^{-3}$). The improvement in precision with increasing concentration is primarily related to having more particles available for analysis (better analytical statistics) and leads to the recommendation to sample for a sufficient time to obtain a particle loading of at least 30 particles mm^{-2}. With respect to accuracy, the linear regression of $PM_{10-2.5}$ concentrations as measured with the passive sampler to that measured with a colocated FRM dichotomous sampler exhibited an intercept not statistically different than zero (p =0.46), and the slope was not statistically different from unity (p =0.92).

Castillo et al. (2019) performed a study evaluating the use of the Wagner passive sampler for measurement of $PM_{2.5}$ in urban areas (Boston, New York, and San Francisco). As part of this study, an improved velocity

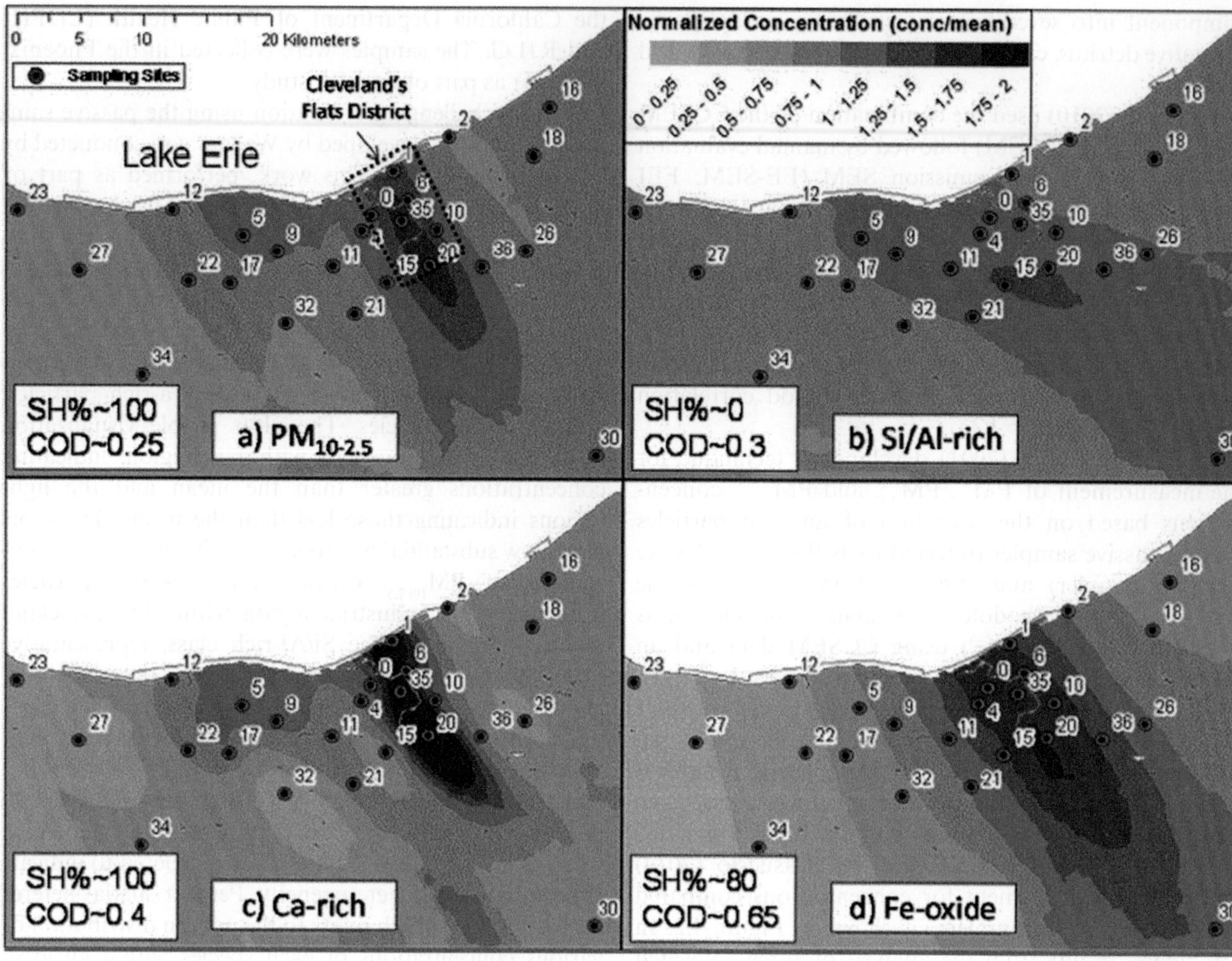

Figure 4.3 Maps of normalized concentration for (a) $PM_{10-2.5}$, (b) Si/Al-rich, (c) Ca-rich, and (d) Fe-oxide. The normalized concentration scale bar in panel (b) applies to all maps. Dots with numbers represent sample locations and identification numbers. *Source:* Sawvel et al., (2015).

deposition model was developed to better account for submicron particles, and improvements in the detection and measurement of carbonaceous particles were made by performing the analysis using an SEM accelerating voltage of 5kV. Overall, good precision was observed with a single sample error coefficient of variation of 14.2%. Comparison of the passive sampler results with colocated FRM filter and FEM continuous monitor results were, on average, accurate to within 13.6% and −1%, respectively.

4.7. APPLICATIONS TO RECEPTOR MODELING

Identifying sources of atmospheric particles is an area where individual particle analysis is valuable because the combination of morphological information with elemental compositional associations provides speciation not obtained using bulk analytical techniques. In many instances, an individual particle can be associated with a specific source or source type (McCrone and Delly, 1973; Edgerton et al., 2009). The capability of resolving source impact can be enhanced by incorporating CCSEM data into receptor models. Casuccio et al. (1981a, 1982, 1983) and Janocko et al. (1982) utilized CCSEM data in a custom-designed receptor model referred to as the *fingerprint/ratio receptor model.* The fingerprint/ratio receptor model was based on the concepts of Friedlander (1973) and apportioned particles using a tracer(s), which is defined as a particle type that is unique to a specific source. Once a tracer was identified on the receptor filter, the balance of the particle types constituting the source signature was tested against the receptor particle types for best fit and apportioned accordingly using a least squares approach.

Kim and Hopke (1988) also investigated the potential of using CCSEM data in receptor models, focusing on a particle class balance (PCB) approach to source attribution using CCSEM data obtained in the El Paso Quantitative Microscopy Study (Casuccio and Janocko, 1983). The authors noted the advantages of the particle-by-particle approach as incorporated in the PCB as compared to the CMB, including (a) integration of size and shape parameters as well as chemical composition; (b) large numbers of homogeneous particle classes provide potentially better resolution in a PCB analysis than the limited number of elemental concentrations can yield in a CMB analysis; and (c) similar sources such as various minerals and soils can be used in PCB source profiles without eliminating any source due to collinearity. A potential limitation of the PCB approach was noted as being related to the problem associated with volatility of some organic particles when exposed to the high vacuum in the SEM and/or extended exposure to the electron beam in the SEM, resulting in modification or loss of the particle.

Additional investigations incorporating SEM data into receptor models were performed by Dzubay and Mamane (1989), where they explored the feasibility of increasing the resolution of the CMB method by using SEM data from the analysis of coarse particles (2.5–10 μm aerodynamic diameter) to complement bulk XRF data. For example, the SEM can be used to provide information on carbonaceous components and morphological information can be used to distinguish fly ash from soil particles. The SEM data was obtained by analyzing 200–300 particles per sample using manual SEM techniques. Data from the SEM analysis for coal fly ash and botanical particles were included with the XRF data in the CMB calculations and compared to CMB results, incorporating only XRF data. The authors noted that knowledge of the contribution of coal fly ash can be valuable in developing cost-effective strategies for meeting air quality standards, which could not be determined by XRF data alone. The authors also noted that the costs associated with a manual SEM analysis limited its application in environmental studies, but CCSEM systems could improve the speed of the analysis resulting in lower costs as well as better precision of the data. Although this work was done using manual SEM analysis, it provided the framework for incorporation of SEM data in the CMB model.

Casuccio et al. (1996, 1998) presented data based on CCSEM analysis of source and receptor samples to apportion particulate matter using the CMB receptor model to specific sources within a steel plant. Source samples were collected and analyzed by CCSEM from point and fugitive emission sources at the steel plant to establish source signatures. Receptor samples evaluated in the study consisted of fine (<2.5 μm) and coarse ($PM_{10-2.5}$) samples collected with a dichotomous sampler, PM_{10} collected with a Hi-Vol sampler, and PM_{10} collected using an ACCU sampler in tandem with a Rupprecht and Patashnick Tapered Element Oscillating Microbalance (TEOM) continuous PM_{10} monitor. The ACCU sampler was programmed to collect a sample on a 0.4 μm polycarbonate filter when PM_{10} concentrations exceeded 150 $\mu g\,m^{-3}$. Data obtained from the analysis of these samples were evaluated in the CMB receptor model. As an example, Table 4.4 provides a summary of CMB source apportionment results for an ACCU sample evaluated in the study. The sample was collected over a 9-hour period (1:00 a.m. to 10:00 a.m.) when ambient concentrations ranged from about 170 to 500 $\mu g\,m^{-3}$. CMB output illustrating the apportionment of the primary sources impacting the receptor presented in Table 4.4 shows that the sampler was impacted primarily from emissions originating from the blast furnace (BFRV-2; 45 wt.%), sinter plant fugitive emissions (SPFUG; 33 wt.%), and sulfur–calcium particles (S/Ca; 10 wt.%). This combination of sources had the best model diagnostics (i.e., percent mass = 88.9; R-square = 0.88; Chi-square = 0.99).

In addition to the CMB receptor model, the positive matrix factorization (PMF) receptor model (Norris & Duvall, 2014) can be applied to CCSEM data to assist in the identification of the source types and apportion their contribution to the receptor. Unlike the CMB receptor model, the PMF does not require the collection and analysis of source samples to establish source signatures. The PMF is a multivariate factor analysis tool that decomposes matrixes of speciated receptor sample data and the associated uncertainties into a factor matrix (source profiles) and a contribution matrix (source contributions). The factor profiles are interpreted by the user to identify source types using experience and/or other information (e.g., emission inventory data).

To illustrate the application of CCSEM data in the PMF, a collection of PM_{10} receptor samples used to evaluate the impact of steel plant emissions on ambient air quality with the CMB receptor model (see Table 4.4) was also applied in the PMF model. The source profiles identified by the deconvolution of the receptor particle type data in the PMF were compared to the CCSEM data of samples collected directly from the process sources at the plant, and the results revealed that all but one of the source profiles identified by the PMF were relatively consistent with the source sample CCSEM data, as shown in Figure 4.4.

Table 4.5 presents an example of the source apportionment of PM_{10} on June 1st comparing the PMF and CMB results. The PMF model further revealed that the normalized source contributions at the receptors correlated with the CMB findings. As an example, Table 4.5 shows the estimated contribution to the PM_{10} mass concentration from each source profile on a receptor sample (i.e., the June 1ACCU sample). The CMB model apportioned

Table 4.4 CMB Source Apportionment Results Based on CCSEM Analysis of TEOM ACCU Sample Collected Over a 9-Hour Period when PM_{10} Concentrations Exceeded 150 $\mu g\,m^{-3}$

SOURCE CONTRIBUTION ESTIMATES -	SITE: SUMMIT		DATE: 3/13/96		CMB7 33889
SAMPLE DURATION	<1	START HOUR	AM	SIZE:	ACCU AM PM_{-10}
R SQUARE	0.88	PERCENT MASS	88.9		
CHI-SQUARE	0.99	DF	17		

SOURCE *	TYPE	SCE (%)	STD ERR	TSTAT
5	SPFUG	0.3327	0.0661	5.0363
19	BFRV-2	0.4520	0.0657	6.8753
22	S/CA	0.1045	0.0277	3.7691

SPECIES CONCENTRATIONS -	SITE: SUMMIT		DATE: 3/13/96	CMB7 33889
SAMPLE DURATION	<1	START HOUR	AM	Size: ACCU AM PM_{-10}
R SQUARE	0.88	PERCENT MASS	88.9	
CHI-SQUARE	0.99	DF	17	

SPECIES		I	MEAS		CALC		RATIO C/M		RATIO R/U
1	MASS	T	1.00000+–	0.15000	0.88919+–	0.07908	0.89+–	0.16	–0.7
2	Fe-ric	*	0.22500+–	0.02500	0.23409+–	0.01980	1.04+–	0.15	0.3
3	Fe(Zn)	*	0.00600<	0.01700	0.00000<	0.00057	0.00<	0.10	–0.4
4	Zn/Cl	*	0.00000<	0.00200	0.00000<	0.00057	0.00<	0.00	0.0
5	Zn-bea	*	0.00200<	0.00700	0.00000<	0.00057	0.00<	0.29	–0.3
6	Fe(Mg)	*	0.01500<	0.01700	0.00889<	0.00698	0.59<	0.82	–0.3
7	Ca/Mg	*	0.01900+–	0.01800	0.02928+–	0.00974	1.54+–	1.55	0.5
8	Ca/Si/	*	0.01100<	0.01500	0.01200<	0.00888	1.09<	1.69	0.1
9	Mg/S	*	0.01000<	0.01900	0.00542<	0.00457	0.54<	1.13	–0.2
10	Mg-bea	*	0.02400+–	0.01900	0.01465+–	0.01304	0.61+–	0.73	–0.4
11	Ca/Fe	*	0.13400+–	0.02100	0.07001+–	0.01094	0.52+–	0.12	–2.7
12	K/Cl	*	0.00000<	0.00200	0.00000<	0.00151	0.00<	0.00	0.0
13	S/K	*	0.12700+–	0.02500	0.13966+–	0.02080	1.10+–	0.27	0.4
14	S/Ca	*	0.11700+–	0.02400	0.11700+–	0.01322	1.00+–	0.23	0.0
15	Ca-ric	*	0.07600+–	0.02400	0.11224+–	0.01066	1.48+–	0.49	1.4
16	Ca/Si	*	0.03500+–	0.01800	0.03227+–	0.00712	0.92+–	0.52	–0.1
17	Crusta	*	0.08100+–	0.02700	0.04309+–	0.01592	0.53+–	0.26	–1.2
18	Si/S	*	0.00000<	0.00200	0.00000<	0.00151	0.00<	0.00	0.0
19	S-bear	*	0.05300+–	0.02200	0.00528+–	0.00963	0.10+–	0.19	–2.0
20	Mn-bea	*	0.01500<	0.02000	0.03613<	0.01738	2.4K	3.41	0.8
21	Misc.	*	0.05000+–	0.02400	0.02812+–	0.01535	0.56+–	0.41	–0.8

Results indicate that the most significant sources were associated with a steel plant blast furnace and sinter fugitive emissions.

47.9% of the mass at the ambient PM_{10} monitor to the SPFUG at the steel mill, whereas 11.7% of the mass was attributed to stack emissions from the blast furnace. The mass fractions obtained by the PMF model estimated 35.7% sinter plant fugitive dust and 11.7% blast furnace stack emissions. Note that the sulfur–calcium particle class was not included in the PMF apportionment results because it was not identified in any of the steel plant source samples. It is assumed that the sulfur–calcium class was related to the reaction of calcium particles with sulfur in the atmosphere. In the CMB model, sulfur–calcium was treated as its own source.

Further, in personal communications with Adriana Pietrodangelo (National Research Council of Italy - Institute of Atmospheric Pollution Research (CNR-IIA)), she has explored the potential of using the PMF to sort individual particles into particle species. To evaluate the potential of this approach, RJLG utilized the particle-by-particle EDS data from the sinter plant fugitive dust source in the steel plant emission study as input in the PMF to classify particles. Particle classifications (factors) and the estimated weight percent contribution of each factor were generally comparable to that obtained with user-defined rules for the same data set, as shown in

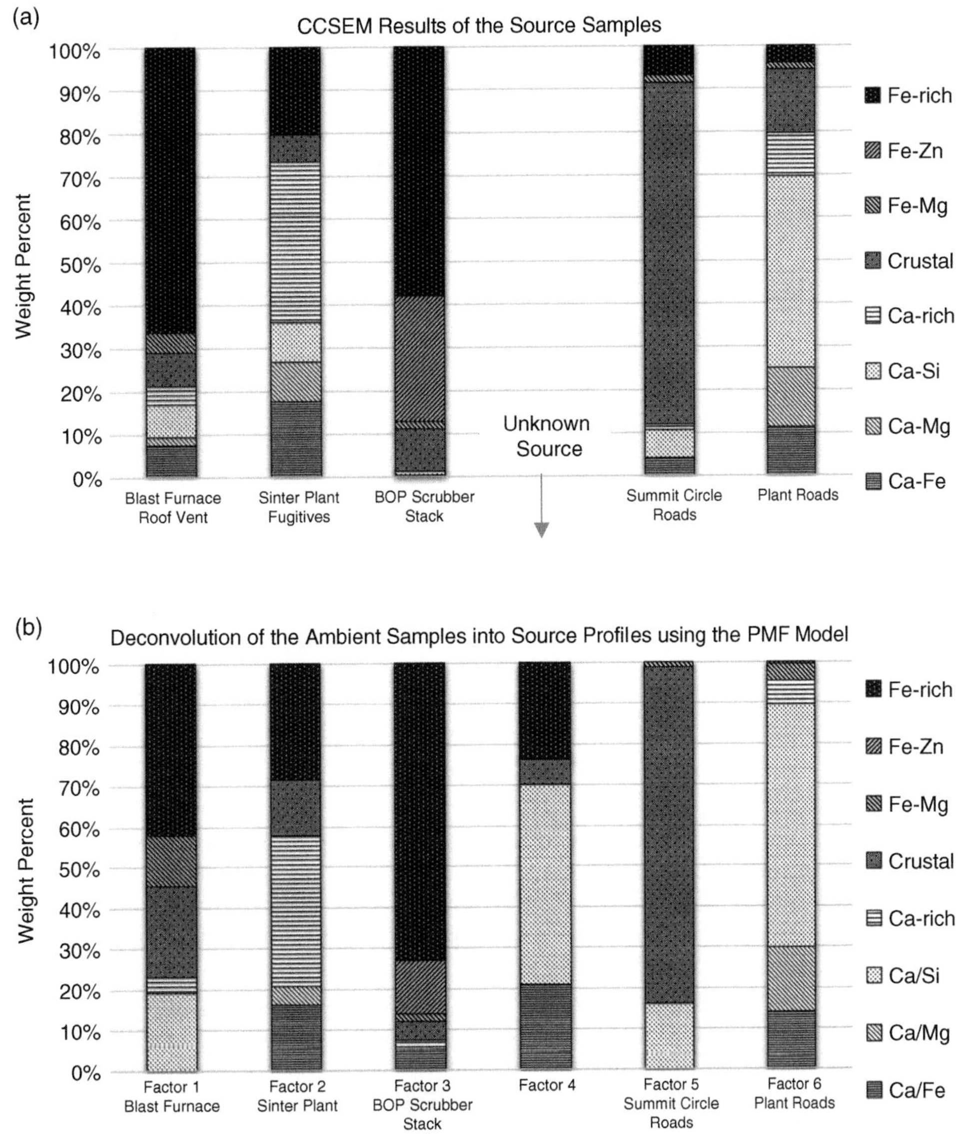

Figure 4.4 Comparison of CCSEM results of source samples collected at a steel plant (a) to the source profiles identified by the deconvolution of the ambient data in the PMF (b). Five of the six source profiles obtained by the PMF were relatively consistent with CCSEM source signatures. Factor 4 is a profile primarily associated with Ca/Si (slag), Ca/Fe, and Fe-rich particles that was not associated with any of the source samples.

Table 4.5 Comparison of CMB and PMF Receptor Model Results Both Applying the Particle Type CCSEM Data Collected at the Receptor on the Same Day (June 1 PM_{10} ACCU Sample)

	Apportionment results (wt.%)	
Source	CMB	PMF
Sinter plant	47.9	35.7
Road dust	13.4	29.8
Blast furnace	11.7	11.7
Sulfur–calcium	14.2	NA
Unknown source	12.8	22.8

Figure 4.5, suggesting the PMF has the potential to classify particles into homogenous particle types. Differences observed in the classification of the mixture of calcium, iron, and iron-calcium particle types are likely due to the challenge of defining the boundaries between elements, which can lead to variations in classification.

4.8. CCSEM ANALYSIS OF ATMOSPHERIC PARTICLES IN THE TWENTY-FIRST CENTURY

The establishment of the National Ambient Air Quality Standards (NAAQS) $PM_{2.5}$ standard in 1997 set the direction for ambient air quality in the twenty-first

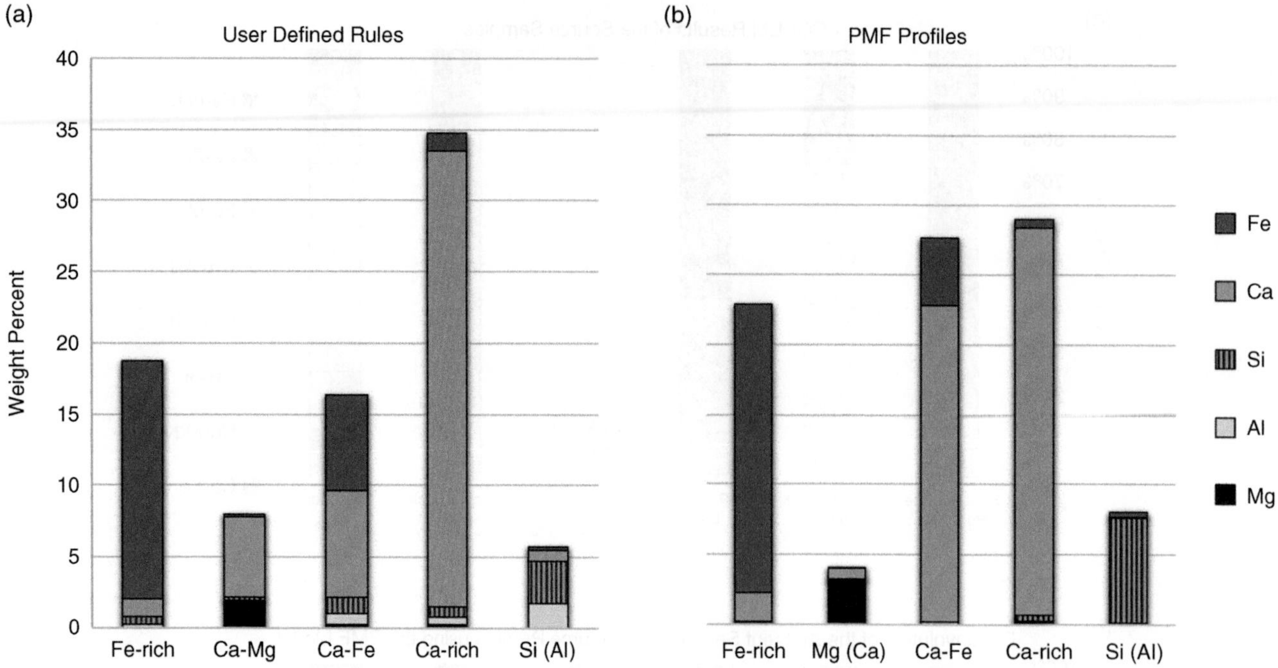

Figure 4.5 CCSEM user-defined rules (a) for the sinter plant fugitive source sample compared to PMF factors (b). The estimated weight percent contribution of each factor was obtained from applying individual particle EDS data in the PMF model.

century. Research and environmental studies related to air studies began to focus more on fine particulate matter and its impact on human health. Particles in the $PM_{2.5}$ size range, especially those in the sub-micrometer range, are more difficult to detect and characterize using SEMs with a tungsten emission source. Fortunately, the twenty-first century introduced significant advances in SEM and sampling instrumentation, computer technology, and software that greatly enhanced CCSEM analysis of atmospheric aerosols. With respect to SEM instrumentation, the FE-SEM, which operates under a higher vacuum than a tungsten-source SEM and produces a more focused electron beam resulting in image spatial resolution under 5 nm, has evolved to the point where it is being used routinely for applications requiring the high-resolution SEM imaging needed for sub-micrometer aerosol characterization (Goldstein et al., 2003). With respect to EDS technology, silicon drift detectors (SDDs) with digital pulse-processing systems and Peltier semiconductor coolers have replaced high-maintenance liquid-nitrogen cooling systems associated with Si[Li] EDS detector technology, with energy resolution under 130 eV now commonplace. EDS systems incorporating SDD detector systems have been shown to have much better X-ray throughput than Si[Li] EDS detectors with accuracy and precision equivalent to that of a high spectral resolution wavelength-dispersive spectrometer used on an electron probe microanalyzer platform (Newbury & Richie, 2015). In addition to SEM and EDS improvements, computer performance has generally been doubling approximately every 2 years since the 1960s, and software for CCSEM applications has taken advantage of the expanding PC capabilities.

As CCSEM technology advanced, applications related to atmospheric aerosols continued to become more sophisticated with evaluations of the physicochemical properties associated with atmospheric particulate matter such as size distribution, chemical composition, mixing state, and morphology, which can vary significantly as a function of space and time. CCSEM analysis in the sub-micrometer range was documented by Laskin and Cowin (2001) using an FE-SEM (LEO Gemini 982) equipped with a Pentafet 6901 EDS spectrometer and DiskInspector hardware and software (Oxford Instruments) for automated particle analysis. Their study was performed on particles collected on a transmission electron microscopy (TEM) grid with an ultrathin carbon film. It was noted that particles collected on this substrate can be observed with superior contrast in the SEM and that low-atomic-number elements can be optimized using a mixed backscattered and secondary electron signal. Further, the use of a thin carbon film substrate supported by a TEM grid yields very low background carbon fluorescence from the substrate compared to a carbon filter and provides quantitative EDS analysis results of low-atomic-number elements (oxygen, nitrogen, and particularly carbon).

Coz et al. (2008) evaluated the relationship between individual particle morphology of $PM_{2.5}$ and aerosol acidity. The CCSEM analysis, incorporating EDS, was performed at an accelerating voltage of 20 kV using a tungsten-source SEM (ASPEX PSEM) to characterize particles greater than 0.2 μm. High-resolution SEM analysis was performed in a manual fashion at an accelerating voltage of 5 kV with a working distance of 3 mm using a FE-SEM (FEI Sirion 400). This analysis was mainly focused on the particle solid or liquid appearance in the size ranges below 1 μm. Morphological parameters of the high-resolution images were measured with programmed macros in Aphelion (ADCIS). Results from this evaluation were grouped into five morphological groups: particles with low shape parameters that presented signs of being in a liquid state when they were deposited on the filter; soot-like; particles with geometric form (mostly spherical, quadrangular/cubic and hexagonal), some of which were surrounded by a liquid layer; minerals and biogenic; and a mixed type. The CCSEM results provided morphological differentiation of samples taken on days with high acidity from days with lower acidity. The high-resolution FE-SEM images were used to examine the morphological properties in more detail and link shape and possible liquid state of the particles. The results showed that particle morphology, size, and sulfur content were different in samples associated with high aerosol acidity from those samples compared to neutral conditions.

With respect to evaluating particle characteristics over time, Laskin et al. (2002a) discussed the development of a sampling device that could collect up to 151 samples on TEM grids with a thin film. During sampling, each thin film substrate can be exposed individually over a time as short as 1 minute. In the Texas 2000 Air Quality Study, nearly 3,000 samples were obtained over a 1-month period. Particles analyzed using CCSEM techniques were grouped into sulfur-bearing and non-sulfur-bearing classes in size ranges from 0.2 to 0.7 μm and from 0.7 to 2.5 μm. The sulfur-bearing class contained particles classified as ammonium sulfate-like particles, carbon-sulfur bearing and silicon-sulfur bearing, and sulfur-bearing particles. The non-sulfur-bearing class contained carbonaceous particles and other non-sulfur-bearing particles. Plots of the results showed the variation of the particle species over time. Laskin et al. (2002b) also used this sampling technology to show the evolution of sea salt particles over time with conversion to sodium nitrate during daylight hours.

Hu et al. (2022) used CCSEM to evaluate the physicochemical characteristics of dust particles from the Taklimakan Desert sand dunes and Gobi (the two types of surfaces constituting the desert) as part of a study to evaluate particle variation during long-distance transport. In the study, 43,222 particles were characterized using IntelliSEM software (RJLG) to control a FE-SEM (Tescan MAIA3) equipped with two 60 mm² SDDs (Bruker). Results included number and mineral class distributions associated with various minerals comprising the Taklimakan sand dunes and Gobi samples.

Investigations of the impact of emissions from wildfires were evaluated by Sparks and Wagner (2021). Study results indicated that $PM_{2.5}$ concentrations at three sites during the wildfires ranged from 46 to 59 $\mu g\, m^{-3}$ and were dominated by submicron *tar balls*. Concentrations of $PM_{2.5}$ during non-wildfire periods ranged from 20 to 27 $\mu g\, m^{-3}$. It was reported that performing the analysis at an accelerating voltage of 5 kV resulted in better detection of the tar balls as compared to an analysis performed at 20 kV. It was also noted that automated CCSEM peak detection algorithms could not always identify very small X-ray peaks even when subjectively present.

With respect to CCSEM software, it is now possible to detect and process particle compositional inhomogeneities, which was outlined by Shah et al. (2018) in a study involving the characterization of inclusions in steel. This approach is now being applied to atmospheric particles. Also, CCSEM software has focused more on efficient management and transfer of the data for offline review and data mining, which enables researchers who are not microscopists to have access to the SEM data for review in a manner similar to being on the SEM. An example of a CCSEM data management/review software tool is shown in Figure 4.6 (IntelliSEM EPAS 2.0 software).

Research exploring the use of machine learning (ML) and computer vision (CV) to classify CCSEM data was performed by Gao et al. (2020). In this study, inclusions in steel were evaluated based solely on digital images acquired during CCSEM analyses using the ASCAT (Automated Steel Cleanliness Analysis Tool). In this work, a publicly available supervised convolutional neural network (CNN) incorporating Visual Geometry Group (VGG16) was utilized to recognize image features, develop numerical representations of these features, and then use these representations for classifications (inclusion or non-inclusion). Digital images acquired during the ASCAT analysis were used for (a) training: images and their classifications were used as inputs to the CNN algorithm to optimize the model parameters so predictions match ground truth classification; (b) validation: a subset of training data was used as a check of parameter optimization; and (c) testing: a set of images independent from those used in training and validation were classified by CNN, and the classifications were compared to the ground truth. The overall accuracy of the model predictions of the test data was 98%, which indicates that a ML/CV approach could be used to predict whether a feature detected was an inclusion with high accuracy. Applying the ML/CV approach to air quality samples,

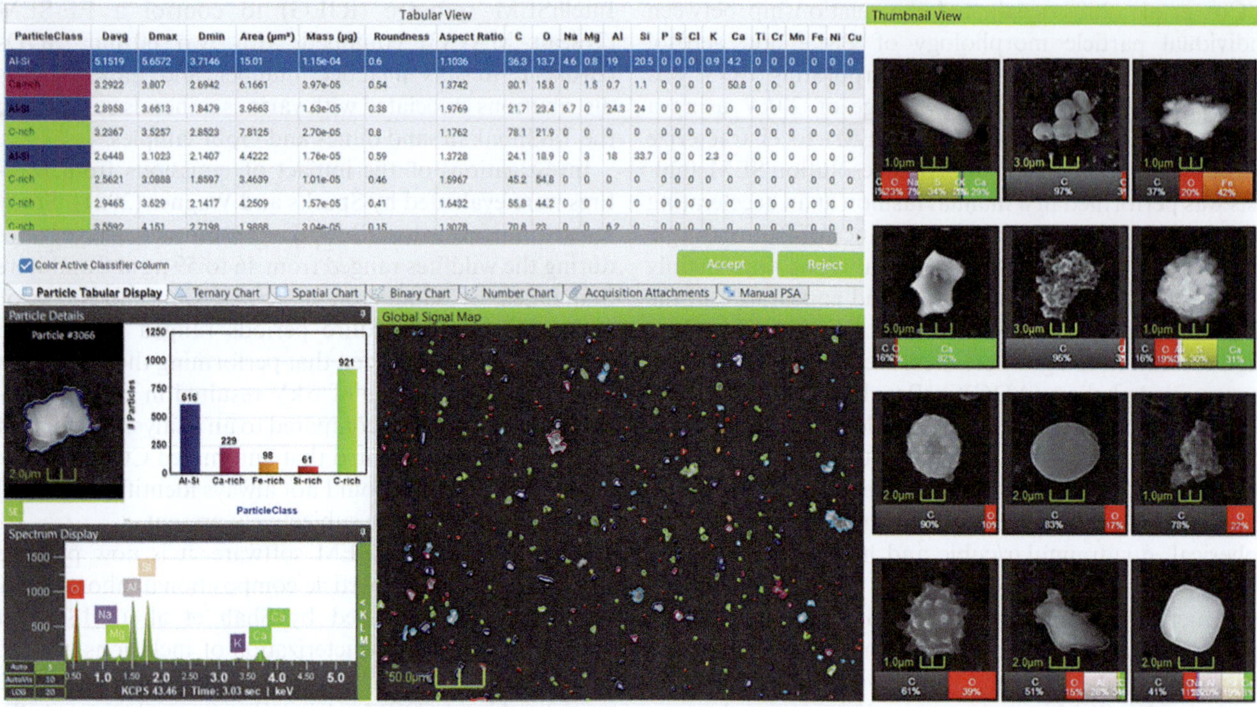

Figure 4.6 CCSEM data management/review software showing some of the ways the individual particle data can be reviewed offline: Tabular View with size, shape, and elemental composition data associated with individual particles; Global Signal Map showing particle distribution associated with particle types (color coded); Thumbnail View showing digital microimages collected during the IntelliSEM analysis with elemental composition bar; Particle Details showing microimage and elemental spectrum along with the number of particles in each particle class. The layout can be modified by the user to suit personal preferences during review of the data.

a CNN model may be able to discern "real" particles from substrate artifacts.

4.9. CONCLUSIONS

CCSEM has evolved greatly since its inception in the 1970s, and the original goals to replace the analyst where the eye/brain is weak (routine/repetitive data collection) and assist the analyst where the eye/brain is best (data interpretation) have to a large degree been met. In addition to improvements in software capabilities, CCSEM has benefited from advancements in SEM, EDS, and computer technology and will continue to benefit from advancements in these areas in the future as the technologies advance. For example, today's SEMs incorporating a field emission source have orders of magnitude better image resolution than the original tungsten-source SEMs and offer the opportunity to perform a CCSEM analysis on particles even in the nanosize range. As an example, Figure 4.7 provides results from a CCSEM analysis using IntelliSEM software of National Institute of Standards and Technology (NIST) 30 nm (RM 8012) and 60 nm (RM 8013) gold nanoparticle reference materials.

EDS technology has advanced significantly with the introduction of SDDs, which are capable of processing X-rays at rates greater than an order of magnitude faster than previous generation X-ray detectors, as well as better spatial resolution provided by FE-SEMs for submicrometer particles. Today's PCs with high-performance multicore CPUs and large amounts of RAM and hard disk space, and network/Internet/Cloud connectivity, have much greater capabilities than the early CCSEM systems. As PCs became more powerful in terms of computing power, memory, and peripherals, CCSEM became less constrained by file size and could process, store, and manage data and images efficiently. CCSEM software has also advanced significantly over the years to not only controlling the electron beam, moving the stage, and collecting/processing EDS spectra and images, but also detecting areas of interest for the analysis, performing X-ray calibration checks prior to the analysis, automatically adjusting the particle threshold detection setting and focus, and performing quality assurance checks during the analysis including beam stability monitoring. Further, the ability to review CCSEM data offline in a manner similar to being on the SEM permits researchers to have access to the individual particle data in an

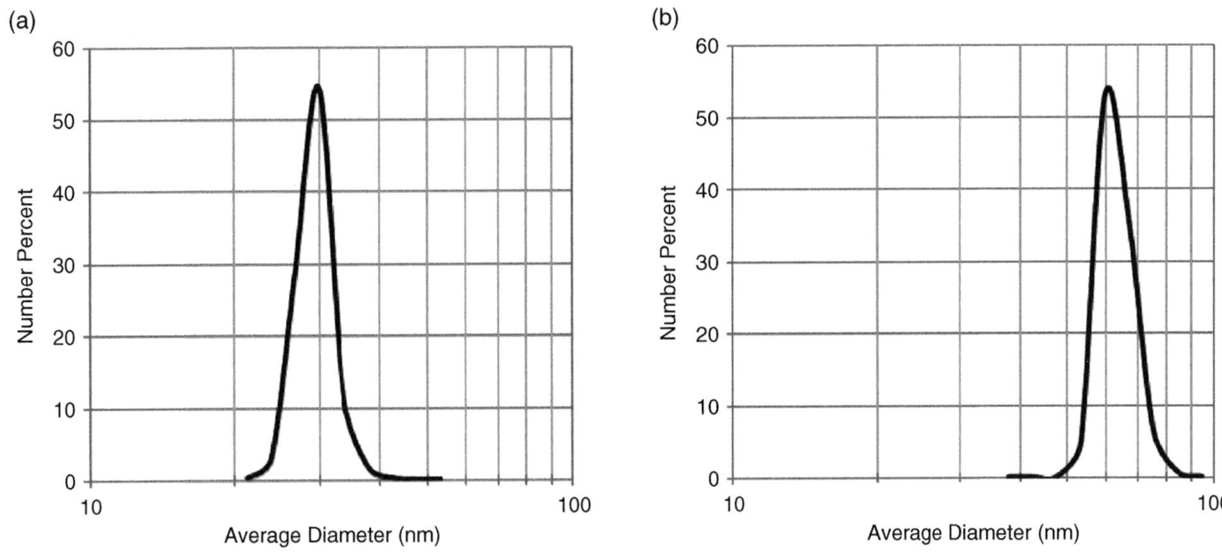

Figure 4.7 Particle size distributions measured by CCSEM of NIST nanoparticle reference materials comprised of spherical gold particles with nominal diameters of 30 nm (a) and 60 nm (b). Plots courtesy Traci Lersch, RJ Lee Group.

efficient manner. Taken as a whole, CCSEM has evolved to become a powerful individual particle analysis tool for the statistical analysis of particle populations associated with atmospheric aerosols.

However, as with all analytical techniques, CCSEM is not without limitations. Many of the limitations have been noted in this chapter, but one of the most significant limitations is that as the technology has improved, the speed of the analysis has increased resulting in increasing amounts of data collected during the analysis. Although this is a good thing from the particle population perspective, the amount of information collected related to size/shape metrics and elemental composition in addition to collection of digital images of individual particles has increased as well and is now getting to the point where review and interpretation of the data can take significantly more time than the analysis itself—echoing the lament of White et al. (1972), the designers of the original CESEMI system back in the early 1970s. Fortunately, as has happened many times in the evolution of CCSEM, technological advancements provide a means to minimize or eliminate the limitations. In this instance, exciting opportunities exist to use machine learning and computer vision to address many of the current limitations associated with the CCSEM analysis, including false positives associated with substrate artifacts, identifying particles that are in contact or are in the form of a "flake" resulting in overestimating of particle size and mass, the ability to perform online grouping of particles in a more consistent and reproducible manner, providing better estimates of carbon subtraction when analyzing fine particles on a carbon substrate, extracting detailed morphological information from the digital image in near real-time, and associating the analyzed particle with a source in an automated fashion. Further, machine learning/computer vision may be able to assist in providing information on volatile particles that dissipate under the electron beam. Should this come to fruition, CCSEM may be able to do more than just assist where the eye/brain is best, and in the process, CCSEM may become a more routine analytical tool utilized in the analysis of atmospheric particles to evaluate the impact and effects of particles on the environment, climate, and human health.

ACKNOWLEDGMENTS

The author wishes to acknowledge the early pioneers of CCSEM, including the researchers at Pennsylvania State University (Eugene W. White, Ralph E. McMillan, John W. Davis, and R. Lee Byers), who explored the potential of integrating computers with scanning electron microscopes; and the researchers at US Steel Research, in particular Richard J. Lee and James F. Kelly, as well as David L. Johnson at the State University of New York, who did the initial investigations on atmospheric particles using automated techniques incorporating EDS. The author also wishes to acknowledge Henry Lentz, who contributed to the section on the evolution of CCSEM, and Traci Lersch, who contributed to the section on applications to receptor modeling.

AVAILABILITY STATEMENT

Data and other information discussed in this chapter are available from the author upon request.

REFERENCES

Ault, A. P., Peters, T. M., Sawvel, E. J., Casuccio, G. S., Willis, R. D., Norris, G. A., & Grassian, V. H. (2012). Single-particle SEM-EDX analysis of iron-containing coarse particulate matter in an urban environment: Sources and distribution of iron within Cleveland, Ohio. *Environmental Science & Technology*, *46*, 4331–4339.

Byers, R. L., Davis, J. W., White, E. W., & McMillan, R. E. (1971). A computerized method for size characterization of atmospheric aerosols by scanning electron microscope. *Environmental Science & Technology*, *5*(6), 517–521.

Castillo, M. D., Wagner, J., Casuccio, G. S., West, R. R., Freedman, F. R., Eisl, H. M., et al. (2019). Field testing of a low-cost passive aerosol sampler for long-term measurement of ambient PM2.5 concentrations and particle composition. *Atmospheric Environment*, *216*, 116905.

Casuccio, G. S. & Janocko, P. B. (1981a). Quantifying blast furnace cast house contribution to TSP using a unique receptor model. Paper presented at the 74th Annual Meeting of the Air Pollution Control Association, Philadelphia, Pennsylvania, June 21–26.

Casuccio G. S. & Janocko, P. B. (1981b). Quantifying the contribution of total suspended particulates from the Shenango Incorporated Neville Island Plant to Avalon (report prepared for Shenango Incorporated).

Casuccio, G. S., Janocko, P. B. & Lucas, J. H. (1982). A comparison of the particulate matter emission inventory to dispersion modeling and receptor modeling results at the coke and iron division of Shenango Incorporated. Kansas City, MO: Emission Inventories & Air Quality Management.

Casuccio, G. S. & Janocko, P. B. (1983). Identification of the sources of total suspended particulates and particulate lead in the El Paso area by quantitative microscopic analysis (Final Report, Volumes I and II). Energy Technology Consultants.

Casuccio, G. S., Janocko, P. B., Dattner, S. L., Mgebroff, J. S. & Zalar, J. L. (1983). Measurement of ambient hi-vol filters by microscopic and bulk chemistry methods. Paper presented at the 76th Annual Meeting of the Air Pollution Control Association.

Casuccio, G. S., Janocko, P. B., Lee, R. J., Kelly, J. F., Dattner, S. L., & Mgebroff, J. S. (1983). The use of computer-controlled scanning electron microscopy in environmental studies. *Journal of the Air Pollution Control Association*, *33*(10), 937–943.

Casuccio, G. S., Schwoeble, A. J., Henderson, B. C., Lee, R. J., Hopke, P. K. & Sverdrup, G. M. (1988). The use of CCSEM/microimaging to study source/receptor relationships. *Transactions – Receptor Models in Air Resources Management, Air & Waste Management Association* (February).

Casuccio, G. S., Lersch, T. L., Simmons, L. & Lambert, L. (1996). Apportionment of PM10 at the WSX plant using a combination of meteorological and microscopic analyses. Paper presented at the Air & Waste Management Association, 89th Annual Meeting & Exhibition, Nashville, TN, June 23–28.

Casuccio, G. S. & Lersch, T. L. (1998). Source apportionment of particulate matter collected during PM10 exceedance days in Weirton, West Virginia using CCSEM data and the CMB7 receptor model (Report prepared for Weirton Steel).

Conny, J. M., Willis, R. D., & Ortiz-Montalvo, D. L. (2019). Analysis and optical modeling of individual heterogeneous Asian dust particles collected at Mauna Loa observatory. *Journal of Geophysical Research: Atmospheres*, 124. https://doi.org/10.1029/2018JD029387

Coz, E., Artinano, B., Clark, L. M., Hernandez, M., Robinson, A. L., Casuccio, G. S., et al. (2010). Characterization of fine primary biogenic organic aerosol in an urban area in the northeastern United States. *Atmospheric Environment*, *44*, 3952–3962.

Coz, E., Artinano, B., Robinson, A. L., Casuccio, G. S., Learch, T. L., & Pandis, S. N. (2008). Individual particle morphology and acidity. *Aerosol Science and Technology*, *42*, 224–232.

Dzubay, T. G., & Mamane, Y. (1989). Use of electron microscopy data in receptor models for PM-10. *Atmospheric Environment*, *23*(2), 467–476.

Edgerton, E. S., Casuccio, G. S., Saylor, R. D., Lersch, T. L., Hartsell, B. E., Jansen, J. J., & Hansen, D. A. (2009). Measurements of OC and EC in coarse particulate matter in the southeastern United States. *Journal of Air & Waste Management Association*, *59*, 78–90.

Fitzgerald, R., Keil, K., & Heinrich, K. (1968). Solid-state energy-dispersion spectrometer for electron-microprobe X-ray analysis. *Science*, *159*(3814), 528–530.

Friedlander, S. K. (1973). Chemical element balances and identification of air pollution sources. *Environmental Science & Technology*, *7*, 235.

Gao, N., Abdulsalam, M., Potter, M., Casuccio, G., Holm, E., & Webler, B. (2020). *Inclusions classification by computer vision and machine learning*. Iron & Steel Technology.

Germani, M. S., & Buseck, P. R. (1991). Automated scanning electron microscopy for atmospheric particle analysis. *Analytical Chemistry*, *63*, 2232–2237.

Goldstein, J. I., Newbury, D. E., Joy, D. C., Lyman, C. E., Echlin, P., Lifshin, E., et al. (2003). *Scanning electron microscopy and X-ray microanalysis* (3rd ed.). Springer.

Gorz, H., White, E. W., Roy, R., & Johnson, G. G., Jr. (1971). Particle size and shape distributions of lunar fines by CESEMI. In *Proceedings of the second lunar science conference* (Vol. *3*, pp. 2021–2025). The M.I.T. Press.

Henderson, B. C., Stewart, I. M., & Casuccio, G. S. (1989). Rapid acquisition/storage of electron microscope images. *American Laboratory*, 49–52.

Hopke, P. K., & Xie, Y. (1994). Airborne particle classification with a combination of chemical composition and shape index utilizing an adaptive resonance artificial neural network. *Environmental Science & Technology*, *28*(11).

Hu, T., Wu, F., Song, Y., Liu, S., Duan, J., Zhu, Y., et al. (2022). Morphology and mineralogical composition of sandblasting dust particles from the Taklimakan Desert. *Science of the Total Environment*, *834*, 155315.

Janocko, P. B., Casuccio, G. S., Dattner, S. L., Johnson, D. L., & Crutcher, E. R. (1982). The El Paso Airshed: source apportionment using complementary analyses and receptor models, in receptor models applied to contemporary pollution problems. In S. L. Dattner & P. K. Hopke (Eds.), *Proceedings no. SP-48, Air Pollution Control Association* (pp. 249–265). Pittsburgh, PA.

Johnson, D. L., McIntyre, B. L., Fortmann, R., Stevens, R. K., & Hanna, R. B. (1981). A chemical element comparison of individual particle analysis and bulk analysis methods. *Scanning Microscopy*, *1*(1), 469–476.

Kelly, J. F., Lee, R. J., & Lentz, S. (1980). Automated characterization of fine particulates. *Scanning Electron Microscopy*, *1*, 311–322.

Kim, D. S., & Hopke, P. K. (1987). Multivariate analysis of CCSEM auto emission data. *Science of the Total Environment*, *59*, 141–155.

Kim, D. S., & Hopke, P. K. (1988). Source apportionment of the El Paso aerosol by particle class balance analysis. *Aerosol Science and Technology*, *9*(3), 221–235.

Kindratenko, V. V., van Espen, P. J. M., Treiger, B. A., & Van Grieken, R. E. (1994). Fractal dimensional classification of aerosol particles by computer-controlled scanning electron microscopy. *Environmental Science and Technology*, *28*, 2197–2202.

Laskin, A., & Cowin, J. P. (2001). Automated single-particle SEM/EDX analysis of submicrometer particles down to 0.1μm. *Analytical Chemistry*, *73*(5).

Laskin, A., Iedema, M. J., & Cowin, J. P. (2002a). Time-resolved aerosol collector for CCSEM/EDX single-particle analysis. *Aerosol Science and Technology*, *37*(3), 246–260.

Laskin, A., Iedema, M. J., & Cowin, J. P. (2002b). Quantitative time-resolved monitoring of nitrate formation in sea salt particles using a CCSEM/EDX single particle analysis. *Environmental Science & Technology*, *36*(23), 4948–4955.

Lebiedzik, J. (1975). *Multiple electron detector method for quantitative microtopographic characterization in the SEM* (Doctoral dissertation). Pennsylvania State University.

Lee, R. J., Fasiska, E. J., Janocko, P., McFarland, D., & Penkala, S. (1979). Electron-beam particle analysis. *Industrial Research/Development.*

Lee, R.J. & Fisher, R.M. (1980). Quantitative characterization of particulates by scanning and high voltage electron microscopy (National Bureau of Standards Special Publication 533, From the Proceedings of the Particle Analysis Session of the 13th Annual Conference of the Microbeam Analysis Society, Ann Arbor, MI, June 22, 1978).

Lee, R. J., & Kelly, J. F. (1980). Overview of SEM-based automated image analysis. *Scanning Electron Microscopy*, *1*, 303–310.

Mamane, Y., Willis, R., & Conner, T. (2001). Evaluation of computer-controlled scanning electron microscopy applied to an ambient urban aerosol sample. *Aerosol Science and Technology*, *34*, 97–107.

McCarthy, J. J., Fisher, R. M., & Lee, R. J. (1982). Applications of computers in electron microscopy. *Ultramicroscopy*, *8*(3), 351–360.

McCrone, W.C. & Delly, J.G. (1973). The Particle Atlas, 2nd Edition, Volume III, Ann Arbor Science Publishers Inc.

McIntyre, B.L., Minnis, M.N. & Johnson, D.L. (1980). An individual particle analysis method and its application to atmospheric transport. *Proceedings of the Symposium on Intermediate Range Transport Processes and Technology Assessment, DOE & Oak Ridge National Laboratory*, 435–444.

McMillan, R.E., Johnson, Jr., G.G., & White, E.W. (1969). Computer processing of binary maps of SEM images, Second Annual Scanning Electron Microscopy Symposium, IIT Research Institute, Chicago.

Mershon, W. J., Powers, Jr., W. H., Lentz, H. P. & Boughman, G.D. (1988). Application of microimaging to computer-controlled scanning electron microscopy. Paper presented at the Joint Meeting of the Electron Microscopy Society of America, Microbeam Analysis Society, Microscopical Society of Canada, Milwaukee, Wisconsin, August 7–12.

New and/or interesting at EMSA/MAS/MSC Conference '92. (1992, September). *Microscopy Today* (Issue 7): 6–7.

Newbury, D. E., & Richie, N. W. M. (2015). Performing elemental microanalysis with high accuracy and high precision by scanning electron microscopy/silicon drift detector energy-dispersive X-ray spectrometry (SEM/SDD-EDS). *Journal of Materials Science*, *50*, 493–518.

Norris, G. & Duvall, R. (2014). *EPA Positive Matrix Factorization (PMF) 5.0 fundamentals and user guide*. Retrieved from http://www.epa.gov/sites/default/files/2015-02/documents/pmf_5.0_user_guide.pdf

Peters, T. M., Sawvel, E. J., Willis, R., West, R. R., & Casuccio, G. S. (2016). Performance of passive samplers analyzed by computer-controlled scanning electron microscopy to measure PM10-2.5. *Environmental Science & Technology*, *50*(14), 7581–7589.

Sawvel, E. J., Willis, R., West, R. R., Casuccio, G. S., Norris, G., Numar, N., et al. (2015). Passive sampling to capture the spatial variability of coarse particles by composition in Cleveland, OH. *Atmospheric Environment*, *105*, 61–69.

Schamber, F. H. (2009). 35 years of EDS software. *Microscopy and Microanalysis*, *15*, 491–504.

Schamber, F. H., & van Beek, C. G. (2006). A different kind of microscopy: Analyzing features with an automated electron beam. *Microscopy and Microanalysis*, *12*(S02), 1782–1783.

Schwoeble, A. J., Dalley, A. M., Henderson, B. C., & Casuccio, G. S. (1988). Computer-controlled SEM and microimaging of fine particles. *Journal of Metals*, *40*(8), 11–14.

Shah, M. N., Story, S. R., Potter, M. S., Casuccio, G. S. & Lentz, H. P. (2018). Detection, measurement and characterization of inclusions using automated SEM techniques, part 2: multiple threshold analysis. American Iron and Steel Technology (AIST) Proceedings.

Song, X., Hadjiiski, L., Hopke, P. K., Ashbaugh, L. L., Carvacho, O., Casuccio, G. S., & Schlaegle, S. (1999). Source apportionment of soil samples by the combination of two neural networks based on computer-controlled scanning electron microscopy. *J. Air & Waste Management Association*, *49*, 773–783.

Sparks, T. L., & Wagner, J. (2021, 2021). Composition of particulate matter during a wildfire smoke episode in an urban area. *Aerosol Science and Technology*, *55*(6), 734–747.

van Dyck, P., Storms, H., & van Grieken, R. (1984). Automated quantitative electron microprobe analysis of particulate material. *Journal de Physique Colloques*, *45*(C2). C2-781-C2-784

Wagner, J., & Casuccio, G. (2014). Spectral imaging and passive sampling to investigate particle sources in urban desert regions. *Environmental Science: Processes & Impacts*, *16*(7), 1745–1753.

Wagner, J., & Leith, D. (2001). Passive aerosol sampler. I: Principle of operation. *Aerosol Science and Technology*, *34*, 186–192.

White, E. W., Mayberry, K., & Johnson, G. G., Jr. (1972). Computer analysis of Multi-Channel SEM and X-ray images from fine particles. In *Pattern recognition* (Vol. *4*, pp. 173–193). Pergamon Press.

Wienke, D., Xie, Y., & Hopke, P. K. (1994). An adaptive resonance theory based artificial neural network (ART-2a) for rapid identification of airborne particle shapes from their scanning electron microscopy images. *Chemometrics and Intelligent Laboratory Systems*, *25*, 367–387.

Willis, R. D., Blanchard, F. T. & Conner, T. L. (2002). Guidelines for the application of SEM/EDX analytical techniques to particulate matter samples (EPA-600/R-02-070). North Carolina: Research Triangle Park.

5

Characterization of Atmospheric Particles by Electron Energy Loss Spectrometry

Karine Deboudt

ABSTRACT

Among the microanalysis techniques capable of providing information on atmospheric particles with a view to elucidating their impact on climate change and air quality, electron energy loss spectrometry (EELS) is particularly useful and suited to fine particles. It explores the loss of energy due to inelastic scattering when an electron beam passes through a thin sample. Depending on the inelastic interaction considered, and therefore the region of the electron energy loss spectrum, different information can be obtained: the elemental composition of atmospheric particles by TEM-EELS, elemental maps and tomography using energy filters by EF-TEM, oxidation state and bonding information by exploring the fine structure of nearby edges (ELNES), and the optical properties of particles deduced from low-loss spectrometry (plasmon). For each of these techniques, the principle will first be briefly presented, with references to the literature if more detailed information is required. Then, a few applications to atmospheric particles will be developed to illustrate their performance and limitations.

5.1. INTRODUCTION

When an electron beam crosses a thin specimen, it loses energy due to inelastic scattering. The energy distribution measurement of electrons that have interacted with the specimen is referred to as electron energy loss spectrometry (EELS). Several kinds of inelastic interaction can be observed and quantified on the electron energy loss spectrum (Egerton, 2009), as illustrated in Figure 5.1. The high peak at zero is the *zero-loss peak* corresponding to electrons that did not interact with the sample or interacted with elastic interaction only. The *low-loss* region, from 0 to about 50 eV, is constituted by this zero-loss peak and the plasmon peak caused by the oscillation of the weakly bound electrons in the conductance or valence band. This plasmon peak provides some information about the specimen thickness or dielectric properties. Then, the fast decay of the spectral intensity requires a gain change to observe the *core-loss* region (from 50 to about 2,000 eV), where the ionization edges are superimposed on the background. These edges corresponding to the ionization of the atom's inner shells are observed at energies typical of the atomic number of the ionized atoms, and their shape leads to chemical information. When the fine structure around the edge provides information on the valence state or bonding of the ionized atom, the post-edge fine structure gives information on its atomic environment as a radial distribution function or symmetry. These two spectrometric domains (Figure 5.1) correspond to energy loss near-edge structure spectroscopy (ELNES) and extended energy loss fine structure spectroscopy (EXELFS), respectively.

EELS typically operates with high-energy electrons, typically 100–300 keV, as used in a transmission electron microscope (TEM), and it was initially based on relatively

Laboratory of Physical Chemistry of the Atmosphere, University of the Littoral Opal Coast, Dunkirk, France

Microanalysis of Atmospheric Particles: Techniques and Applications, Geophysical Monograph 288, First Edition.
Edited by Joseph M. Conny and Peter R. Buseck.

DOI: 10.1002/9781119554318.ch5

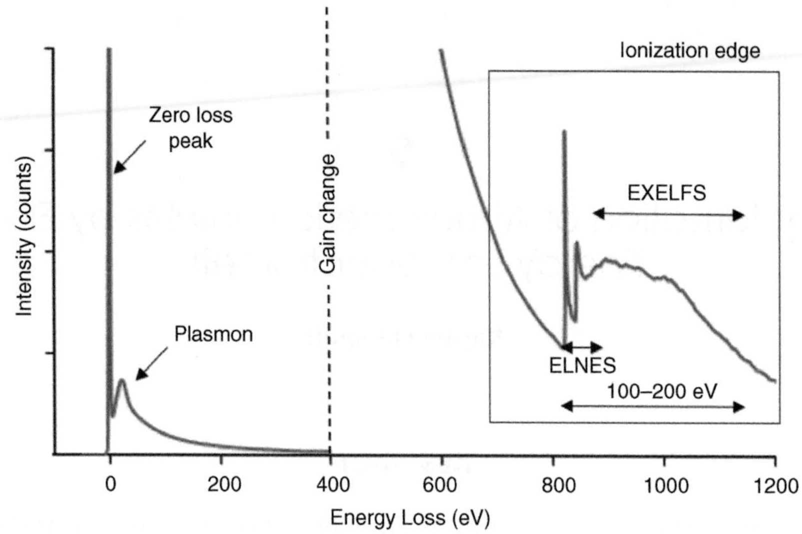

Figure 5.1 Schematic view of an electron energy loss spectrum.

basic instrumentation. Several types of imaging spectrometers were built, but the two most popular commercial instruments are the Gatan Imaging Filter (GIF) and the Zeiss Omega Filter (Williams & Carter, 2009). To observe EELS spectra with the GIF system, a simple magnetic prism spectrometer, which bends the trajectories of transmitted electrons according to their energy, is interfaced below the view screen of a TEM. After the bend, an energy-selecting slit and a set of quadrupole and sextupole lenses focus the beam on a charge-coupled device (CCD) camera, providing the spectra from a nanometer-scale area. The Omega filter is an in-column design composed of four magnetic sectors, giving a typical omega shape to the electron trajectory. An energy-dispersed diffraction pattern is formed at the exit of the filter, and the required energy range is selected by an aperture. The electron beam of the microscope can be in parallel mode (TEM-EELS) or scanning mode (STEM-EELS), an EELS spectrum being acquired for each image pixel in this last case.

The energy resolution is clearly a key factor in EELS. When the spectrometer is not a limiting factor, this resolution is determined by the gun of the microscope. Thermionic sources (W or LaB_6) lead to higher resolution than 1 eV, which is sufficient for ionization loss spectrometry, but an energy resolution width in the 100 meV range obtained by a FEG is needed to resolve vibrational or electronic modes in ELNES or EXELFS. This type of instrument is nevertheless available at most academic electron microscopy centers. In addition, the instrumental progress made over time in EELS with the development of monochromators and in electron microscopy with aberration correctors results today in more sophisticated instruments. The most powerful among them achieve an atomic spatial resolution joined to an ultra-high-energy resolution (width in the 10 meV range), opening new spectroscopic signatures for appropriate samples (Gloter et al., 2017). However, this resolution is obtained for very high-end instruments that are generally not accessible to the average user and available only under specific conditions.

The practical limit to the excellent spatial resolution of EELS is set by radiation damage of the specimen. The inelastic collisions that give all the useful signals can also cause specimen damage under the electron beam and in the vacuum of the microscope (Egerton, 2012), especially for nonconductive samples. Radiolysis, heating, and/or knock-on displacements can change the nanostructure or chemistry of the thin specimen, which is no longer representative of its parent material. Low-electron-dose microscopy techniques, often combined with specimen cooling (cryo-microscopy) and low-noise CCD cameras, are common strategies to limit this damage (Patterson et al., 2016), but that is not always sufficient (Xhoffer et al., 1995). As a consequence, the interpretation of images or spectra can sometimes be difficult.

Moreover, specimen thickness should be on the order of the mean free path of electrons for a single scattering event to avoid multiple inelastic scattering associated with additional peaks in the energy-loss spectrum. As an example, the total inelastic mean free path with a beam energy of 200 keV is 160 nm for amorphous carbon, and it decreases inversely with material density (Egerton, 2011), so specimens <100 nm should be used. For samples consisting of heavier elements (corresponding to lower mean free paths for inelastic scattering), even finer specimens would be preferable to avoid multiple scattering events blurring the spectrum. In the case of atmospheric particles, ultrafine particles are particularly adapted to EELS spectrometry and can be directly sampled on TEM grids. For larger particles and, notably, micrometer particles from the coarse mode highly abundant in the atmosphere, a specific preparation, including ion-beam milling by focused ion-beam (FIB) techniques, is often necessary, such as the one proposed by Conny et al. (2014).

In conclusion, it is essential to use adequate sample preparation, choose good experimental conditions for spectral or image acquisition, and take the necessary precautions when observing potentially unstable atmospheric particles to obtain reliable conclusions. A detailed description of the physics of EELS and its applications can be found in Egerton (2011), and Williams and Carter (2009) have developed the practical aspects more precisely. In this chapter, the different ways of exploring EELS in atmospheric particle studies will be investigated. We will discuss in particular (a) elemental composition by TEM-EELS, (b) elemental maps and tomography by energy-filtered TEM (EF-TEM), (c) oxidation state and bonding information by exploring ELNES, and (d) particle optical properties deduced from low-loss spectrometry.

5.2. ELEMENTAL COMPOSITION BY TEM-EELS

5.2.1. Principles

When the electron beam passes through the sample, some of the primary electrons transfer enough energy to sample atoms that core-shell electrons can be ejected from the attractive field of the nucleus, and these atoms are ionized. These primary electrons, which have lost part of their energy by ionization loss, are then collected and constitute the electron energy loss spectrum. This spectrum is composed of the ionization edges from K, L, M, etc., shell electrons described by a specific nomenclature (Figure 5.2). Every element has its own set of ionization edges (K, L, M, etc.) at specific energy-loss values, which can be used for elemental identification.

In practice, the transmitted electrons are filtered in the EELS spectrometer within an energy slit whose width limits the energy range of the energy-loss spectrum. The wider the slit, the wider the recorded spectral range, but the more degraded the spectral resolution is.

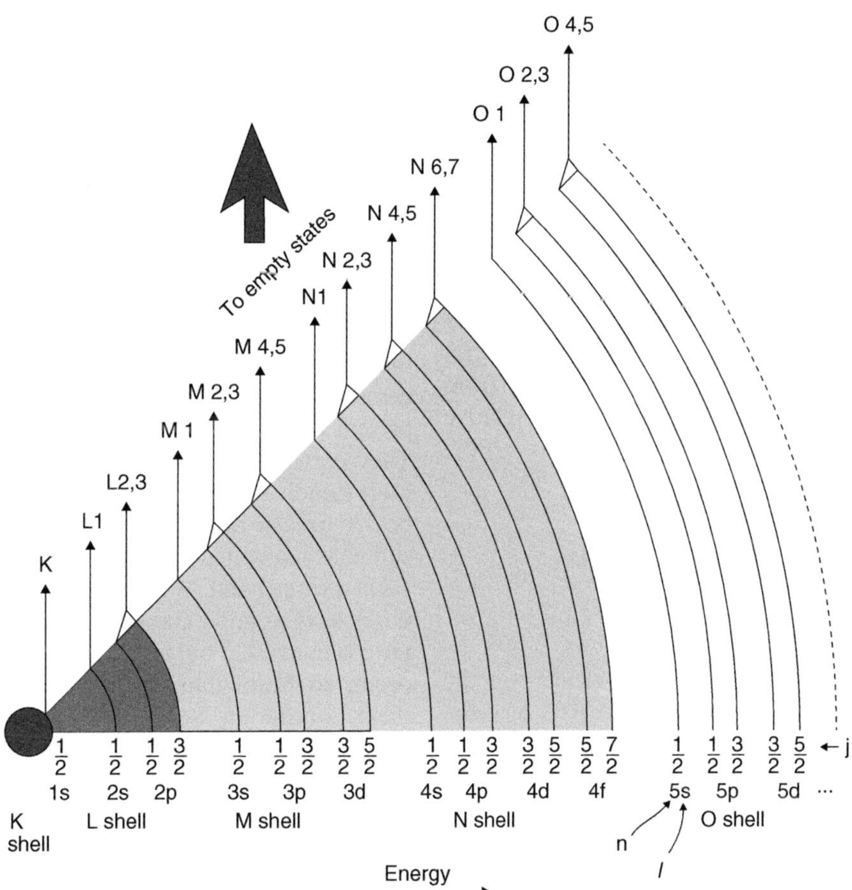

Figure 5.2 Terminology for the edges in the EELS spectrum due to core-shell ionization. Adapted from Williams and Carter (2009).

The great sensitivity of EELS to light elements such as carbon, nitrogen, and oxygen explained the initial interest in the study of atmospheric aerosols in the 1990s, notably carbonaceous particles. This advantage allowed us to overcome the difficulties of energy-dispersive X-ray spectrometry (EDX) to quantify these light elements at that time. Moreover, EELS offers the realistic possibility to obtain compositional analysis on individual particles a few nanometers in diameter. However, edge intensity being approximately inversely proportional to the square of energy loss, a limitation is the very low count rate observed at high energies corresponding to EELS edges for heavy elements (as an example, the Pb M edge is at 2484 eV). Thus, this difficulty in identifying and quantifying light and heavy elements from the same EELS spectrum often leads to the use of EELS as a complement to EDX.

Prior to inner-shell analysis, spectra of the low-loss region (0–100 eV) have to be recorded to determine the relative thickness (t/λ) of each specimen according to Equation 5.1:

$$t/\lambda = \log I_0/I_t \tag{5.1}$$

where I_0 is the intensity of elastically scattered transmitted electrons (zero peak), I_t is the intensity of inelastically scattered transmitted electrons (approximately the plasmon peak), t is the absolute thickness, and λ is the mean free path. An ideal relative thickness lower than 0.5 ensures that no multiscattering correction has to be applied to the spectra. Then, to obtain the atomic ratios simply (Mavrocordatos & Fortin, 2002), the background is properly extrapolated over at least 25 eV prior to the edge and over 100 eV according to a power law function and subtracted from the studied spectra. The integral of each ionization edge is then measured directly from the spectra or fitted with a model to obtain the relative quantification with Equation 5.2 (Egerton, 2009; Williams & Carter, 2009)

$$\frac{N_{Fe}}{N_O} = \frac{\sigma_O(\beta,\Delta)}{\sigma_{Fe}(\beta,\Delta)} \times \frac{\int_E^{E+\Delta} I_{Fe}(\beta,\Delta)}{\int_E^{E+\Delta} I_O(\beta,\Delta)} \tag{5.2}$$

where N is the number of atoms of an element in the analyzed volume (Fe or O in Equation 5.2), σ (β, Δ) is the partial cross-section of the element integrated with an acceptance angle β and over an energy range Δ, and I (β, Δ) is the integral of the intensity of the absorption edge signal I with an acceptance angle β, over an energy range Δ. The partial ionization cross-sections are commonly calculated using a theoretical approach (Egerton & Malac, 2005) because their experimental determination would require standards with the same thickness and the same bonding characteristics as the unknown specimen.

To obtain an absolute quantification, the total intensity of the spectrum has to be deduced from the low-loss spectrum, which must be done with care considering its high intensity (saturation risk of the diode array or afterglow of the scintillator). In practice, the quantification based on elemental ratios is always used for applications in environmental science. It leads to an accuracy of 10–20%, which is sufficient for atmospheric particles, considering the high natural variability of their composition and the nature of the conclusions that will result from this quantification. As an example, it is possible to distinguish kaolinite from smectite or illite by the Si/Al ratios in clay particles (Laribi et al., 2007).

5.2.2. Applications to Atmospheric Particles

Maynard and Brown (1992) were among the first to suggest the use of EELS for the identification and relative elemental quantification of nanometer-size atmospheric particles, notably demonstrating its ability to quantify light elements such as oxygen in 20 nm particles of iron oxides. Qualitative and quantitative data on ultrafine particles directly collected on TEM grids were available down to atomic concentrations of a few percent, although the presented results were limited by the long spectral acquisition time (about 30 s) due to the buildup of carbon contamination of the specimen (Maynard, 1995). However, their comparison with EDX analysis for nanometer-sized particles showed EELS to give higher count rates, leading to higher-quality data over short analysis periods for quantitative analysis. EDX was shown to be preferable to EELS for the detection and analysis of high-atomic-mass elements such as lead. Xhoffer et al. (1995) tried to evaluate the impact of beam damage on the elemental quantification of standard sub-micrometer aerosols by cryo-EELS using three compounds commonly present in the atmosphere: NaCl, $(NH_4)_2SO_4$, and KNO_3. They concluded that the volatile elements are released more rapidly than the others, resulting in a loss of nitrogen before oxygen before chlorine, potassium, sodium, and sulfur, as illustrated in Figure 5.3.

On the one hand, long counting times and high electron doses lead to good statistics because the signal-to-noise ratio is increased; but on the other hand, mass losses can occur, so beam damage becomes more important for these conditions. So, EELS is not fully adapted for quantitative determination of atmospheric particles and notably for inorganic salt particles suffering from beam damage such as $(NH_4)_2SO_4$ and KNO_3, and for which EELS should be restricted to qualitative purposes (Xhoffer et al., 1995). All this explains why the quantitative analysis of atmospheric particles was finally limited to refractory nanometer particles, essentially primary particles, such as in the study of welding fume particles indicating an increasing manganese content with particle

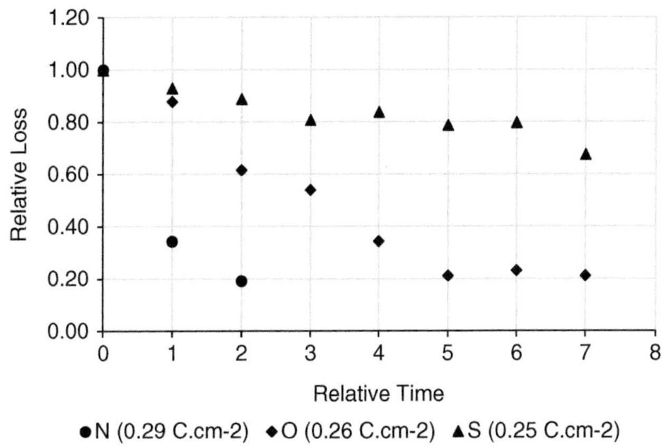

Figure 5.3 Relative loss of nitrogen, oxygen, and sulfur evaluated from EELS spectra of $(NH_4)_2SO_4$ particles as a function of time (relative time in units of 5 min) and collected at a nearly constant electron dose. Xhoffer et al. (1995)/with permission of ELSEVIER.

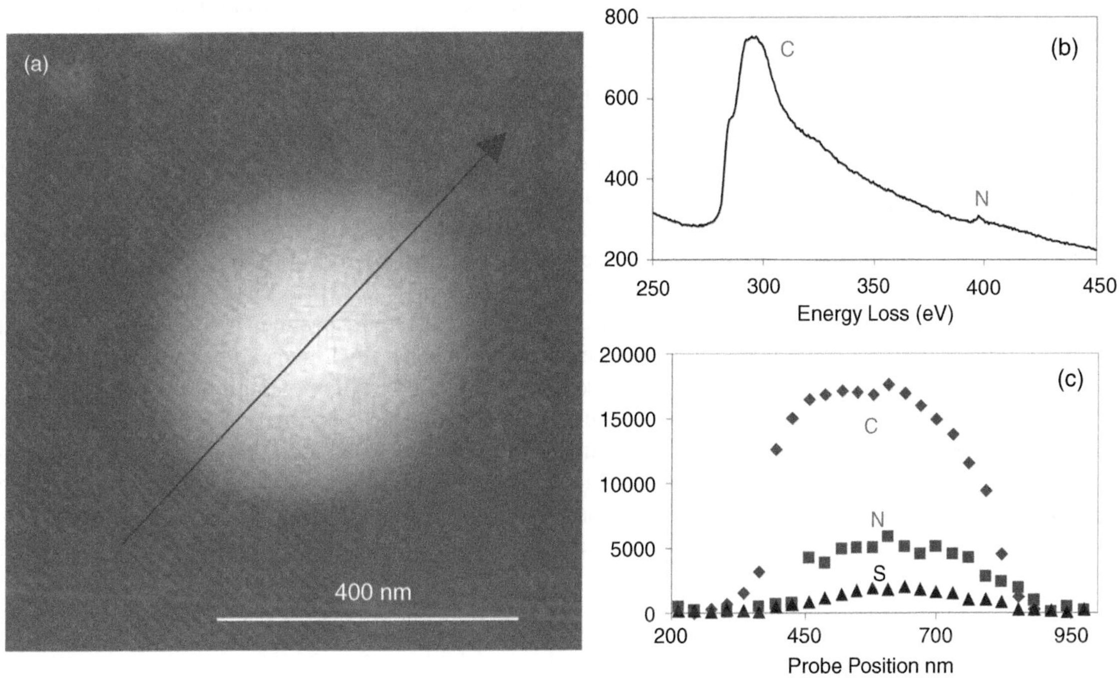

Figure 5.4 Detailed analysis of organic droplet residual particle. (a) Annular dark-field electron micrograph. Diagonal arrow shows the path followed by the electron probe during the acquisition of spatially resolved EELS spectra. (b) Typical energy-loss spectra recorded from organic particles showing C and N signals (intensity in arbitrary units). (c) Elemental profiles obtained from the line scan across particles showing distribution of C, N, and S in relative units. Twohy et al. (2005)/with permission of John Wiley & Sons.

size (Richman et al., 2011). In most atmospheric particle studies, EELS is used for its excellent capacity to detect light elements at the nanoscale, such as the detection of carbon and oxygen in diesel exhaust particles (Stoehr et al., 2015). It also provides information on the distribution of light elements (N and S) that can be carried on the surface of urban carbonaceous particles (Paoletti et al., 2002) or homogeneously distributed inside particles, as illustrated by the EELS line scan in Figure 5.4 for a nitrogenated organic particle (Twohy et al., 2005).

In this last case, the scan was acquired with a dwell time of 1 s per pixel to minimize the beam damage. Although concentration ratios are calculated, no crucial conclusions are deduced from this quantification except that among C, N, and S, nitrogen is the second most abundant element of this organic particle.

In summary, although individual particle analysis by TEM-EELS is time-consuming, it is particularly adapted to study the elemental composition of ultrafine particles considering its excellent spatial resolution. Identification and elemental quantification by TEM-EELS is notably recommended for atmospheric aerosols containing light elements such as carbonaceous particles, supported by EDX when high-atomic-mass elements are expected.

5.3. ELEMENTAL MAPS AND TOMOGRAPHY BY EF-TEM

5.3.1. Principles

EF-TEM is a powerful tool combining chemical information from EELS spectra with spatial information at the nanoscale through qualitative or quantitative elemental maps. It allows for the creation of a TEM image from electrons with specific energy losses. For that, EF-TEM imaging is acquired at specific loss energies by flooding the sample with primary electrons and filtering the transmitted electrons with an energy slit whose width determines the spectral energy window. EF-TEM images are not directly interpretable as chemical maps due to the high background of an EELS spectrum, which is not specific to elements. The most common technique used for data processing is the "three-window" technique, resulting in quantitative maps, but the "jump-ratio" technique leading to qualitative maps can also be used (Verbeeck et al., 2004). For the three-window technique, three images made up of electrons coming from three different energy regions in the spectrum are used: two pre-edge images to estimate the nonspecific background in each pixel of the image, and the post-edge image containing the ionization edge of interest and leading to the elemental map after background subtraction. Because the brightness in maps is proportional to the elemental concentrations, these maps give a good indication of the relative concentrations. In the case of particulate samples, the effect of varying sample thickness (as opposed to thin films) should be taken into account when interpreting the resulting maps, particularly on quantification. Indeed, the signal intensity will be lower in areas where the particle is thin (edges), and there may also be no signal in areas where the particle is very thick. Thus, the variations in thickness inherent to the particles can confuse the results.

In general, a spatial resolution of 1 nm or better is achieved for realistic parameters and a wide range of energy losses, but specimen drift during map acquisition can also make these experiments far from trivial, inducing false quantification results, especially near discontinuities in the sample.

Advanced EF-TEM applications like imaging spectroscopy technique and the spectrum imaging technique have also been developed, resulting in an (x,y,E) data cube, as schematically presented in Figure 5.5.

For the imaging spectroscopy technique, a large set of energy-filtered images at different energy losses is recorded by EF-TEM, whereas for the spectrum imaging technique, an EELS spectrum is acquired at each pixel of the x–y specimen plane scanned by the electron beam (EF-STEM). In this last case, the spectral resolution is better, allowing access to the fine structure of ionization edges, which is powerful for studying valency or bonding effects in different specimen areas. However, the exposure times are generally longer, so beam damage and or image drift can make its deployment more difficult.

In a more sophisticated way, EF-TEM tomography can also be used to reconstruct three-dimensional element-specific images using the tilt axis of the microscope to acquire an EF-TEM tilted series for reconstruction (Mobus et al., 2003). It can be particularly

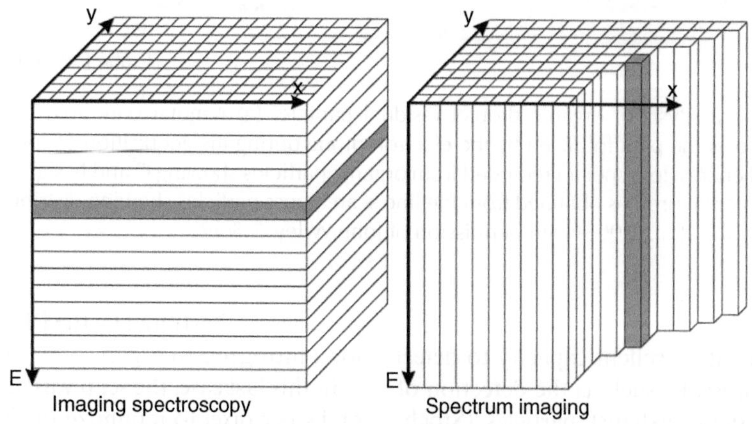

Figure 5.5 Schematic view of the two different ways to experimentally capture part of the three-dimensional data cube, where x–y is a specimen plane and E is the electron energy loss. Verbeeck et al. (2004)/with permission of ELSEVIER.

powerful for providing information on the internal structure of studied specimens.

5.3.2. Applications to Atmospheric Particles

Among atmospheric particles, tar balls from biomass and biofuel burning are often studied by EF-TEM (Hand et al., 2005; Posfai et al., 2003; Posfai et al., 2004) because they do not volatilize under the vacuum of TEM and do not visibly change when exposed to the electron beam. EELS maps of these tar balls can reveal their internal structure, which sometimes includes a core-shell structure, as observed in Figure 5.6a: although the distributions of C and possibly N in the tar balls are rather homogeneous, the concentration of O is strongly higher in the 30 nm outer layer of the particle, probably resulting from atmospheric processing.

This technique has also allowed us to demonstrate that smoke plumes can also contain soot more or less enriched in sulfur (Németh et al., 2015) or organic particles with inorganic inclusions whose composition depends on the combustion source and the age of the smoke, such as the KCl inclusions observed in Figure 5.6.

Thus, EF-TEM can be a helpful tool for the identification and internal structure analysis of nanometer-scale atmospheric particles such as carbonaceous particles, but it has also provided a fast and convenient way to map particles containing sulfur or transition metals in urban (Chen et al., 2005b), desert (Deboudt et al., 2012), and industrial environments (Marris et al., 2013). This EELS mapping has also demonstrated its utility for laboratory experiments, such as in the study of carbon and oxygen redistribution in seasalt particles after heterogeneous reactions with nitric acid (Ault et al., 2013).

In summary, EF-TEM is an interesting technique resulting in elemental maps at the nanoscale. It notably leads to the distinction between homogeneous atmospheric particles and core/shell particles or particles

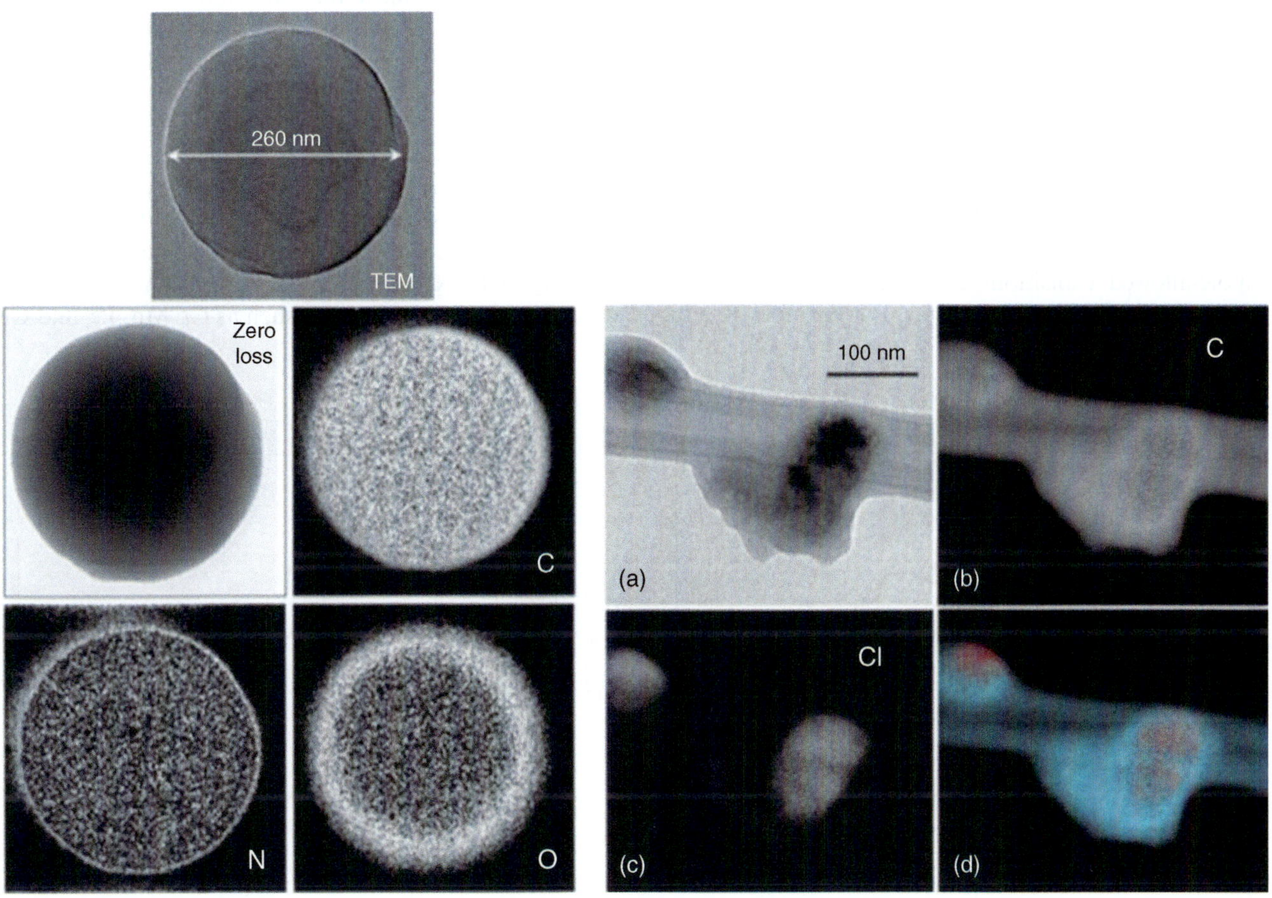

Figure 5.6 TEM image and EELS maps. (Left) A tar ball particle with a bright-field image and the zero loss, carbon, nitrogen, and oxygen maps. Hand et al. (2005)/with permission of American Geophysical Union. (Right) Two KCl-bearing organic particles with (a) bright-field image, (b) carbon map, (c) chloride map, and (d) composite image showing the distribution of C (white/blue) and Cl (black/red). Posfai et al. (2003)/with permission of American Geophysical Union.

containing inclusions. For atmospheric particles that are particularly stable under the electron beam, a 3D reconstruction by EF-TEM tomography can provide additional information on their internal structure.

5.4. OXIDATION STATE AND BONDING INFORMATION BY EXPLORING NEAR-EDGE FINE STRUCTURE

5.4.1. Principles

The intense variations at about 30 eV from the onset of an ionization edge (post-edge), called the *near-edge fine structure* (also known as ELNES), contain very useful information from molecular and/or solid-state effects. According to atomic theory, K-edges have a triangular (sawtooth) shape, whereas L-, M-, and higher-order edges are rounded or sharply peaked at the edge threshold (displaying so-called white lines), notably for transition metals (Egerton, 2011). Apart from a slowly varying matrix-element term, the core-loss intensity is proportional to the density of empty states at an energy above the Fermi level. More specifically, this density is local because inner-shell excitation takes place close to the center of a particular species of atom. Thus, the fine structure is dissimilar for edges that represent different sites of the same element (e.g., C—H and C—Cl sites in a chlorinated hydrocarbon) or different elements in a chemical compound. Moreover, this density of empty states is symmetry-projected because dipole-allowed transitions dominate the edge intensity, and a dipole selection rule works. For example, K-edges represent the excitation of 1 s electrons and the final states of p-character.

As an application example, carbon structural variations in carbonaceous compounds can be determined by the relative amounts of σ^* and π^* bondings from the C K-edge of energy loss spectra (Katrinak et al., 1992). The 1 s transition to the π^* conduction band occurs at 285.5 eV, whereas the 1 s transition to the σ^* conduction band occurs at about 290 eV or higher, depending on the bond distance. Consequently, spectra with higher π^*/σ^* ratios indicate more π bonding, so a structure with a greater graphitic character, as illustrated in Figure 5.7.

Moreover, the σ^* peak is tall and sharp with a characteristic fine structure of the shape just after the peak when structural ordering occurs as in the graphite spectrum, whereas amorphous carbon gives a broad peak with a smooth shape.

Another useful application of ELNES is the change of the $L_{2,3}$ edge of transition metals according to their oxidation. This ionization edge shows characteristic intense white lines due to the transition of excited 2p core-electrons into unoccupied d-orbitals. The white-line intensity ratio L_2/L_3, the normalized total white-line intensity in reference to the intensity due to transitions into the continuum state, and the energy shift of this L-ionization edges vary with d-electron occupancy (Egerton, 2011; Williams & Carter, 2009). Consequently, these characteristics can be investigated to measure the oxidation state of transition metals with a nanoscale spatial resolution. This capacity of ELNES has been largely used in material sciences for various compounds such as Cr, Mn, Fe, and Cu

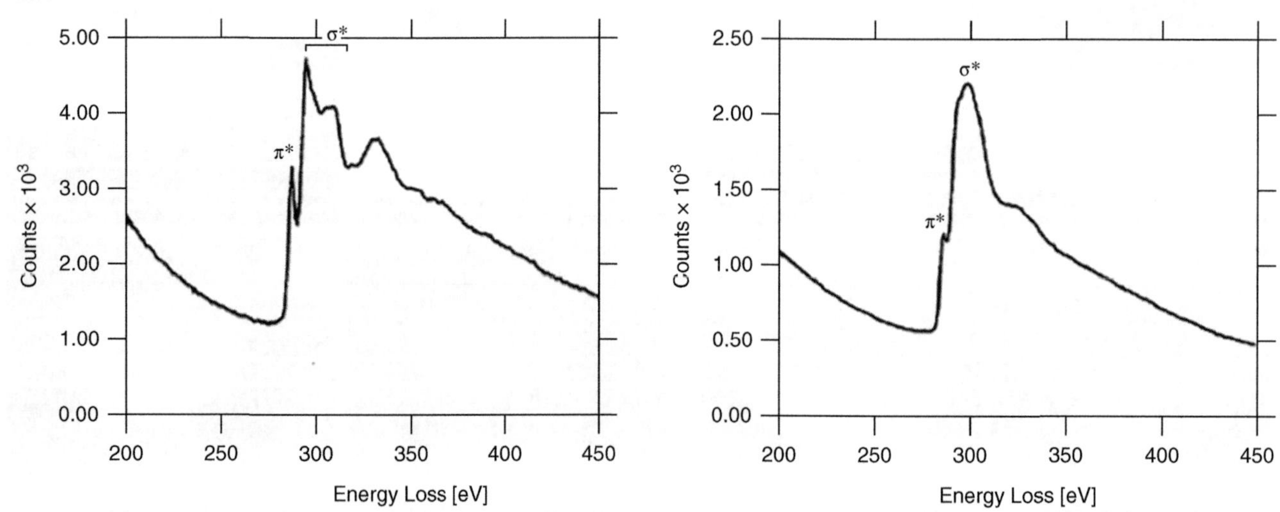

Figure 5.7 Carbon K-edge for (left) a Ticonderoga graphite spectrum with a large σ^* peak indicating a fully ordered structure, $\pi^*/\sigma^* = 0.61$; and (right) an evaporated carbon film spectrum with a broad configurations typical of an amorphous material, $\pi^*/\sigma^* = 0.41$. Katrinak et al. (1992)/with permission of the American Chemical Society.

oxides or minerals (Cave et al., 2006; Daulton & Little, 2006; Laffont & Gibot, 2010; Loomer et al., 2007). In practice, care is first required during the spectral acquisition of specimens and standards of different oxidation states because beam damage can modify the valence, notably for Fe and Mn (Garvie et al., 2004; Livi et al., 2012). The useful information is then extracted from these spectra by analyzing them with more or less sophisticated curve-fitting methods, which can result in a relative standard deviation of about 5% considering the nominal values of oxidation states for standards (Calvert et al., 2005; Garvie & Buseck, 1998; Kurata & Colliex, 1993). The use of valence state information provided by the white lines and the near-edge fine structure can also be explored in valence state mapping (Wang et al., 2000) or in a line scan to elucidate, as an example, the core/shell distribution of metal oxidation states in synthetic nanoparticles (Estradé et al., 2012).

5.4.2. Applications to Atmospheric Particles

The ability of ELNES to derive information on carbon bonding has played an important role in the characterization of atmospheric carbonaceous particles such as soot or black carbon, tar balls, and other organic particles (Buffat, 1999), notably due to their different impacts on local and global climate. All these particles exhibit quite different morphologies and microstructure. Whether from biomass burning, fossil fuel combustion, or any other source, ELNES reveals that soot particles have a fractal morphology and a high proportion of graphitic areas (high π^*/σ^* ratio), whereas tar balls and organic particles are amorphous particles (low π^*/σ^* ratio) that are rather compact and spherical (Chen et al., 2006, 2004, 2005a; Deboudt et al., 2010; Katrinak et al., 1992). The presence of nitrogenous species in these carbonaceous particles is also often revealed by the N-edge at about 401 eV in the EELS spectra, not far from the C—K edge at about 286 eV (Adachi & Buseck, 2011; Chen et al., 2005b).

Of course, the conclusions drawn from all these works must be placed in a context where not all carbonaceous compounds are stable under the electron beam. As an example, by comparison with the near-edge X-ray absorption fine spectra (NEXAFS), the signature of aliphatics and other organic species detectable at energies ranging from 286 to 289 eV is completely missing in the EELS spectra of diesel soot particles (Braun et al., 2005), probably due to the alteration of samples under the highly energetic beam of the microscope. In fact, the EELS spectra actually give a reliable signature of the graphitic part and stable amorphous solid part of atmospheric particles.

A good knowledge of the oxidation state of transition metals in atmospheric particles is essential because, on the one hand, they can act as catalysts via redox processes in atmospheric reactions, leading to the generation of reactive oxygen species; and on the other hand, the toxicity of most of them is highly dependent on this oxidation state (Borch et al., 2010; Cwiertny et al., 2008; Deguillaume et al., 2005). Iron and manganese are commonly present in atmospheric aerosols, so they have often been studied by EELS. Whether at the roadside in urban areas (Calvert et al., 2004; O'Day et al., 2022; Pattammattel et al., 2021; Sanderson et al., 2016) or in an industrial environment near metallurgical complexes (Marris et al., 2013), studies of individual atmospheric particles by ELNES reveal that iron is mainly present in its Fe^{3+} form and manganese as Mn^{2+}, although a certain diversity of oxidation state was observed from one particle to another. In addition, the excellent spatial resolution of the ELNES also made it possible to account for variability in the degree of oxidation at the particle scale itself, as shown in Figure 5.8. The line acquisition from pixel A to pixel B has demonstrated the variation of Fe valence inside a Fe micro-aggregate (diameter ≈ 20 nm) of a typical Fe-containing aluminosilicate particle collected in North-Western Africa (Figure 5.8).

In summary, the implementation of ELNES analyses is easy and inexpensive to investigate the local chemistry of carbon or transition metals in atmospheric aerosols, compared notably to synchrotron techniques; and although it has a lower energy resolution, it provides relevant results with the advantage of a high spatial resolution. In particular, ELNES has been used extensively to study the relative amounts of σ^* and π^* bondings in carbonaceous particles and the oxidation states of transition metals in dust particles.

5.5. PARTICLE OPTICAL PROPERTIES DEDUCED FROM LOW-LOSS SPECTROMETRY (PLASMON)

5.5.1. Principles

A typical low-energy-loss spectrum contains an elastic or zero-loss peak and plasmon peaks due to a collective effect of electrons in the conductance or valence band of the specimen. Indeed, a transmitted electron produces in its wake an oscillation of the electron density of outer-shell electrons (conduction electrons in a metal, valence electrons in a semiconductor or insulator) weakly bound to atoms but coupled to each other by electrostatic forces (Egerton, 2011). This plasmon wake produces a backward force on the transmitted electron, leading it to lose an average energy and resulting in a plasmon peak centered around an energy $E_{p,}$ which is not element-specific and replicates at multiples of that energy as sample thickness increases. This plasmon peak is thus commonly used to measure the local thickness of the specimen. The shape

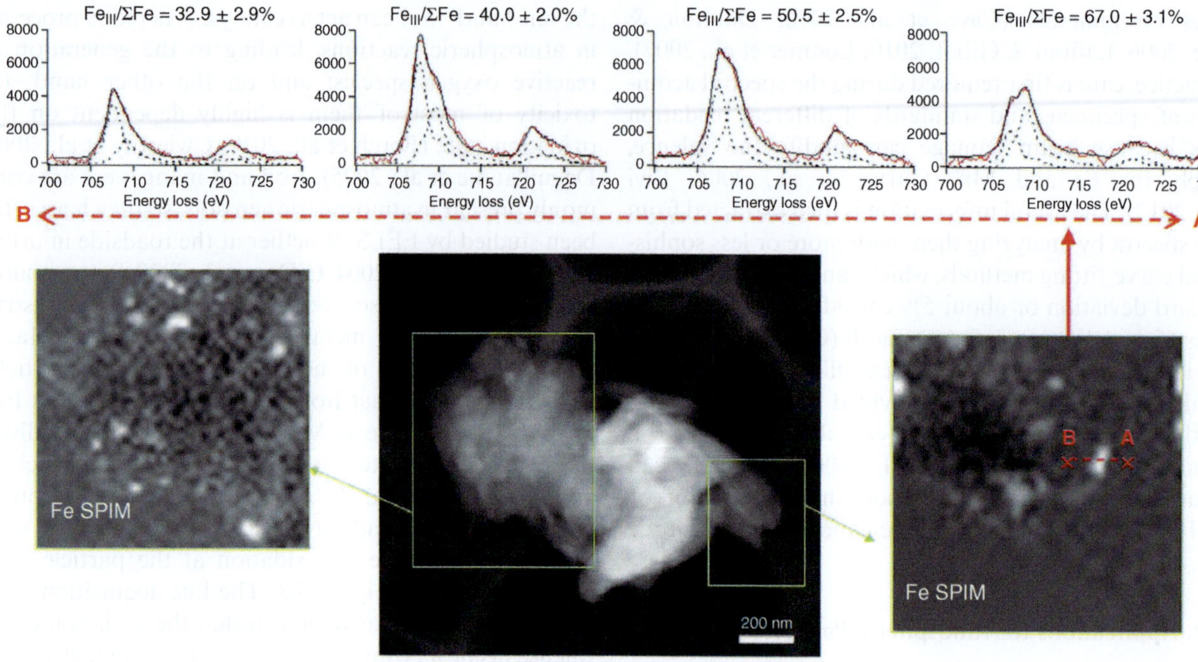

Figure 5.8 Dark-field picture and EELS Fe spectrum images (Fe SPIM) of a typical Fe-containing aluminosilicate particle (Fe SPIM = 1 EELS Fe spectra acquired per pixel). At the top, the extraction of some EELS Fe L2,3 edges of a line acquisition from pixel A to pixel B demonstrates the variation of Fe valence inside the Fe micro-aggregate (diameter≈20 nm) presented in the right Fe SPIM. Deboudt et al. (2012)/with permission of John Wiley & Sons.

of the low-loss spectrum can also help to identify specific phases or features in the TEM image through a "fingerprinting" process by comparison with the spectra of known specimens (Erni et al., 2005; Williams & Carter, 2009). In addition, because plasmon generation is caused by the electric field of the incident electron, a similar effect can be created by an electromagnetic wave, and the plasmon peak can be related to the dielectric properties of the specimen or its permittivity (ε). In practice, the first step in spectral processing to calculate this permittivity is the removal of the zero-loss peak from the inelastic part of the spectrum by a fitting procedure. Then, because a single scattering spectrum is needed, the next step is to remove the plural scattering intensity by Fourier-logarithmic deconvolution. Finally, a Kramer–Kronig analysis allows extraction of the real part of the dielectric constant (ε_1) from the imaginary part (ε_2) (Egerton, 2011). The dielectric function ($\varepsilon(E) = \varepsilon 1 + i\varepsilon 2$, where E is the energy loss) can then be transformed into the refractive index ($N = n - ik = \sqrt{\varepsilon}$). Considering that the energy loss is $E = hc/\lambda$ (with Planck's constant h, optical wavelength λ, and speed of light c), to accurately determine the optical properties of the specimen in the visible regime (from λ = 400 to 700 nm), the low-loss spectrum needs to be processed for energy loss from 1.8 to 3.1 eV with a spectral resolution of some 100 meV.

5.5.2. Applications to Atmospheric Particles

Due to the requirements in terms of spectral resolution and the technical difficulties of working in the low-loss region, the only application of low-loss spectroscopy in the field of atmospheric sciences to our knowledge is the determination of the optical properties of carbonaceous aerosols (Alexander et al., 2008; Mathai et al., 2023; Zhu et al., 2014). Light-absorbing carbonaceous aerosols are believed to be among the most important emissions contributing to global climate warming, but the quantification of their climatic effects is still under debate due to large uncertainties in assessing their radiative forcing (IPCC, 2013). There are several types of carbonaceous atmospheric aerosols with different optical properties, including black carbon, essentially composed of light-absorbing graphitic carbon, and brown carbon, associated with a high fraction of amorphous carbon, which has a significant absorbing component at short wavelengths. To improve the accuracy of direct forcing models, it is necessary, among other things, to have reliable information on the refractive index of atmospheric particles. Nevertheless, there is a very wide range of refractive indices for carbonaceous particles depending on their source, combustion conditions, and atmospheric processes involved during their aging; hence, there is an

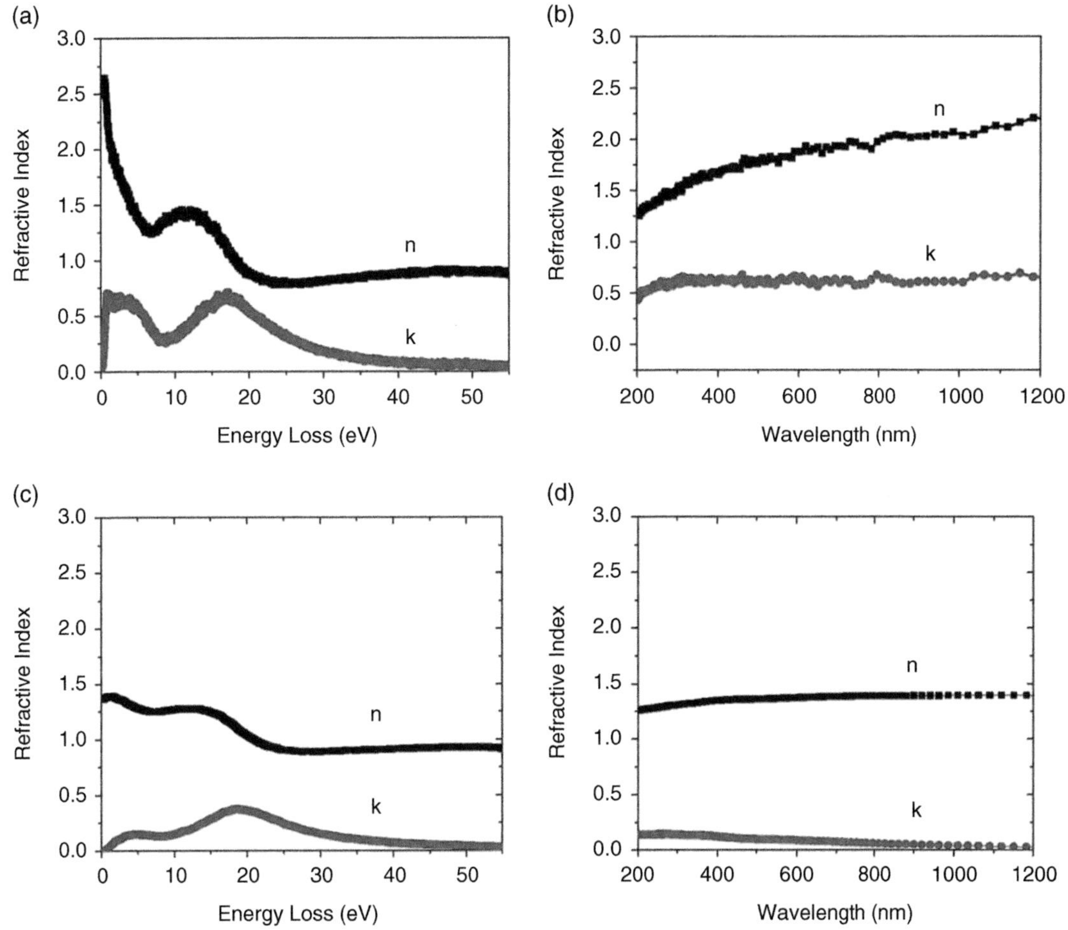

Figure 5.9 The refractive index (n – ik) calculated from the plasmon peak for typical graphitic carbon particles in the range of (a) 0–55 eV and converted in the (b) optical wavelength range of 200–1,200 nm (corresponding to about 6.2–1.0 eV), and for typical amorphous carbon particles (c) and (d), respectively. Zhu et al. (2014)/with permission of Oxford University Press.

interest in determining them precisely. It is difficult to determine these indices from in situ light-optical measurements because the optical signal is averaged over all aerosol components. Hence, the use of low-loss spectrometry at the particle scale is a good alternative. The refractive index ($n - ik$) obtained by low-loss spectrometry for carbonaceous particles collected from East Asia is presented in Figure 5.9. For graphitic carbon particles, the imaginary part of the refractive index k is relatively high and fairly constant in the wavelength range 200–1200 nm, which is characteristic of black carbon. For amorphous carbon particles, k is lower than for black carbon, indicating the lower absorption properties for these particles. In addition, k decreases with higher wavelengths, which confirms that this type of carbonaceous particle appears visually brown.

In summary, the major interest of low-loss spectrometry in the study of atmospheric particles is to provide information on their dielectric properties and in particular their refractive index, a key parameter in the quantification of their climate impact

5.6. CONCLUSION

In conclusion, electron energy loss spectrometry associated with transmission electron microscopy can be an easy-to-use technique for accessing chemical information at the nanoscale. Once it has been ensured that the sample is stable under the electron beam for the duration of the data acquisition, it leads to helpful information on the micro-characteristics of atmospheric particles resulting from their origin or their atmospheric reactivity. Depending on the spectral domain considered and the data acquisition method used, it notably makes it possible to

- Identify and quantify the elements that make up atmospheric particles by TEM-EELS, particularly light elements.

- Map the spatial distribution of elements within nanoparticles by EF-TEM and provide information on their internal microstructure.
- Determine the relative amounts of σ* and π* bondings in carbonaceous particles or the oxidation state of transition metals in dust particles by ELNES.
- Deduce the refractive index of individual particles from low-loss or plasmon spectrometry, particularly useful for the processing of in situ atmospheric optical measurements.

It also makes sense to use synchrotron X-ray absorption spectroscopy (XAS) techniques as a complement to EELS, which, although having lower spatial resolution, generally generates less beam damage. As an example, where the high energy of the primary electron beam allows the contribution of refractory materials to be collected with EELS, near-edge X-ray absorption fine structure spectroscopy (NEXAFS) can provide information on the associated sensitive compounds, such as organic materials contained in soot particles (Braun et al., 2005). Moreover, although EELS provides high spatial resolution imaging of individual atmospheric particles, obtaining quantitative chemical or phase information can be difficult because it requires statistical averages of many measurements. The use of a combination of bulk and spatially resolved spectrometry and microscopy methods can then be a very good approach, as realized as an example by Pattammattel et al. (2021) in the study of iron speciation in urban aerosols by EELS and XAS spectroscopies.

It should be kept in mind that this chapter does not aim to develop the best of EELS but to present what it is possible to measure with this technique when applied to atmospheric science. Of course, considering technological advances, EELS has the potential to provide accurate elemental and structural information at the nanoscale, especially for very high-end instruments and under specific conditions, as long as the samples are suitable. This is probably the limiting factor of EELS applications in atmospheric science: samples are not the clean, reproducible synthetic materials that microscopists particularly appreciate. Atmospheric particles are very heterogeneous, both at the particle scale and at the particle population scale, generally composed of both inorganic and organic materials, refractory or sensitive to the observation conditions. They are therefore not good candidates for testing the latest advances in microspectrometry, which are therefore not developed in this chapter. Of course, EELS is a manual technique that is time-consuming and can only be used on a limited number of atmospheric particles. So, it requires further investigations on a large number of particles using analytical tools such as automated SEM-EDX to ensure that the particles studied by EELS are representative and that the conclusions that can be drawn from them are robust. But in these conditions, it is a powerful tool providing valuable information on the chemistry of atmospheric particles, relevant notably to toxicological, epidemiological, climatic, and source-apportionment studies.

REFERENCES

Adachi, K., & Buseck, P. R. (2011). Atmospheric tar balls from biomass burning in Mexico. *Journal of Geophysical Research, 116*, D05204. doi: 10.1029/2010JD015102

Alexander, D. T. L., Crozier, P. A., & Anderson, J. R. (2008). Brown carbon spheres in east Asian outflow and their optical properties. *Science, 321*, 833–836. doi: 10.1126/science.1155296

Ault, A. P., Guasco, T. L., Ryder, O. S., Baltrusaitis, J., Cuadra-Rodriguez, L. A., Collins, D. B., et al. (2013). Inside versus outside: Ion redistribution in nitric acid Reacted Sea spray aerosol particles as determined by single particle analysis. *Journal of the American Chemical Society, 135*, 14528–14531. doi: 10.1021/ja407117x

Borch, T., Kretzschmar, R., Kappler, A., Van Cappellen, P., Ginder-Vogel, M., Voegelin, A., & Campbell, K. (2010). Biogeochemical redox processes and their impact on contaminant dynamics. *Environmental Science & Technology, 44*, 15–23. doi.org/10.1021/es9026248

Braun, A., Huggins, F. E., Shah, N., Chen, Y., Wirick, S., Mun, S. B., et al. (2005). Advantages of soft X-ray absorption over TEM-EELS for solid carbon studies - a comparative study on diesel soot with EELS and NEXAFS. *Carbon, 43*, 117–124. doi: 10.1016/j.carbon.2004.08.029

Buffat, P. A. (1999). Electron microscopy for the characterisation of atmospheric particles. *Analusis, 27*, 340–346. doi: 10.1051/analusis:1999270340

Calvert, C. C., Brown, A. P., Brydson, R., Lingard, J. J. N., Tomlin, A. S., & Smith, S. (2004). Characterisation of the <500 nm fraction of airborne particulates using TEM, EDX and EELS. In *Electron Microscopy and Analysis 2003, Proceedings of the Institute of Physics Electron Microscopy and Analysis Group Conference, 3-5 September 2003* (pp. 323–326). IOP. doi: 10.1201/9781482269130

Calvert, C. C., Brown, A., & Brydson, R. (2005). Determination of the local chemistry of iron in inorganic and organic materials. *Journal of Electron Spectroscopy and Related Phenomena, 143*, 173–187. doi: 10.1016/j.elspec.2004.03.012

Cave, L., Al, T., Loomer, D., Cogswell, S., & Weaver, L. (2006). A STEM/EELS method for mapping iron valence ratios in oxide minerals. *Micron, 37*, 301–309. doi: 10.1016/j.micron.2005.10.006

Chen, Y., Shah, N., Huggins, F. E., & Huffman, G. P. (2004). Investigation of the microcharacteristics of PM2.5 in residual oil Fly ash by analytical transmission Electron microscopy. *Environmental Science & Technology, 38*, 6553–6560. doi: 10.1021/es049872h

Chen, Y. Z., Shah, N., Braun, A., Huggins, F. E., & Huffman, G. P. (2005a). Electron microscopy investigation of carbonaceous particulate matter generated by combustion of fossil fuels. *Energy & Fuels, 19*, 1644–1651. doi: 10.1021/ef049736y

Chen, Y. Z., Shah, N., Huggins, F. E., & Huffman, G. P. (2005b). Characterization of ambient airborne particles by energy-filtered transmission electron microscopy. *Aerosol Science and Technology*, *39*, 509–518. doi: 10.1080/027868291001402

Chen, Y., Shah, N., Huggins, F. E., & Huffman, G. P. (2006). Microanalysis of ambient particles from Lexington, KY, by electron microscopy. *Atmospheric Environment*, *40*, 651–663. doi: 10.1016/j.atmosenv.2005.09.036

Conny, J. M., Collins, S. M., & Herzing, A. A. (2014). Qualitative multiplatform microanalysis of individual heterogeneous atmospheric particles from high-volume air samples. *Analytical Chemistry*, *86*, 9709–9716. doi: 10.1021/ac5022612

Cwiertny, D. M., Young, M. A., & Grassian, V. H. (2008). Chemistry and photochemistry of mineral dust aerosol. *Annual Review of Physical Chemistry*, *59*, 27–51. doi: 10.1146/annurev.physchem.59.032607.093630

Daulton, T., & Little, B. (2006). Determination of chromium valence over the range Cr(0)-Cr(VI) by electron energy loss spectroscopy. *Ultramicroscopy*, *106*, 561–573. doi: 10.1016/j.ultramic.2006.02.005

Deboudt, K., Flament, P., Choel, M., Gloter, A., Sobanska, S., & Colliex, C. (2010). Mixing state of aerosols and direct observation of carbonaceous and marine coatings on African dust by individual particle analysis. *Journal of Geophysical Research: Atmospheres*, *115*. doi: 10.1029/2010JD013921

Deboudt, K., Gloter, A., Mussi, A., & Flament, P. (2012). Red-ox speciation and mixing state of iron in individual African dust particles. *Journal of Geophysical Research: Atmospheres*, *117*. doi: 10.1029/2011JD017298

Deguillaume, L., Leriche, M., Desboeufs, K., Mailhot, G., George, C., & Chaumerliac, N. (2005). Transition metals in atmospheric liquid phases: Sources, reactivity, and sensitive parameters. *Chemical Reviews*, *105*, 3388–3431. doi: 10.1021/cr040649c

Egerton, R. F. (2009). Electron energy loss spectroscopy in the TEM. *Reports on Progress in Physics*, *72*, 1–25.

Egerton, R. F. (2011). *Energy-loss spectroscopy in the Electron microscope* (3rd ed.). New-York: Springer.

Egerton, R. F. (2012). Mechanisms of radiation damage in beam-sensitive specimens, for TEM accelerating voltages between 10 and 300 kV. *Microscopy Research and Technique*, *75*, 1550–1556. doi: 10.1002/jemt.22099

Egerton, R. F., & Malac, M. (2005). EELS in the TEM. *Journal of Electron Spectroscopy and Related Phenomena*, *143*, 43–50. doi: 10.1016/j.elspec.2003.12.009

Erni, R., Browning, N. D., Dai, Z. R., & Bradley, J. P. (2005). Analysis of extraterrestrial particles using monochromated electron energy-loss spectroscopy. *Micron*, *36*, 369–379. doi: 10.1016/j.micron.2004.12.011

Estradé, S., Yedra, L., López-Ortega, A., Estrader, M., Salazar-Alvarez, G., Baró, M. D., et al. (2012). Distinguishing the core from the shell in MnOx/MnOy and FeOx/MnOx core/shell nanoparticles through quantitative electron energy loss spectroscopy (EELS) analysis. *Micron*, *43*, 30–36. doi: 10.1016/j.micron.2011.04.002

Garvie, L., & Buseck, P. (1998). Ratios of ferrous to ferric iron from nanometre-sized areas in minerals. *Nature*, *396*, 667–670. doi: 10.1038/25334

Garvie, L. A., Zega, T. J., Rez, P., & Buseck, P. R. (2004). Nanometer-scale measurements of Fe3+/ΣFe by electron energy-loss spectroscopy: A cautionary note. *American Mineralogist*, *89*, 1610–1616. doi: 10.2138/am-2004-11-1204

Gloter, A., Badjeck, V., Bocher, L., Brun, N., March, K., Marinova, M., et al. (2017). Atomically resolved mapping of EELS fine structures. *Materials Science in Semiconductor Processing*, Advanced transmission electron microscopy for semiconductor and materials science, *65*, 2–17. doi: 10.1016/j.mssp.2016.07.006

Hand, J. L., Malm, W. C., Laskin, A., Day, D., Lee, T., Wang, C., et al. (2005). Optical, physical, and chemical properties of tar balls observed during the Yosemite aerosol characterization study. *Journal of Geophysical Research: Atmospheres*, *110*. doi: 10.1029/2004JD005728

IPCC (2013). In T. F. Stocker, D. Qin, G.-K. Plattner, M. Tignor, S. K. Allen, J. Boschung, et al. (Eds.), *Climate change 2013: The physical science basis. Contribution of working group I to the fifth assessment report of the intergovernmental panel on climate change*. Cambridge, United Kingdom and New York, NY, USA: Cambridge University Press.

Katrinak, K. A., Rez, P., & Buseck, P. R. (1992). Structural variations in individual carbonaceous particles from an urban aerosol. *Environmental Science & Technology*, *26*, 1967–1976. doi: 10.1021/es00034a014

Kurata, H., & Colliex, C. (1993). Electron-energy-loss core-edge structures in manganese oxides. *Physical Review B*, *48*, 2102. doi: 10.1103/PhysRevB.48.2102

Laffont, L., & Gibot, P. (2010). High resolution electron energy loss spectroscopy of manganese oxides: Application to Mn3O4 nanoparticles. *Materials Characterization*, *61*, 1268–1273. doi: 10.1016/j.matchar.2010.09.001

Laribi, S., Jouffrey, B., & Fleureauc, J. M. (2007). Experimental electron energy loss spectroscopy of clays. *European Physical Journal Applied Physics*, *39*, 257265. doi: 10.1051/epjap:2007115

Livi, K. J. T., Lafferty, B., Zhu, M., Zhang, S., Gaillot, A.-C., & Sparks, D. L. (2012). Electron energy-loss safe-dose limits for manganese valence measurements in environmentally relevant manganese oxides. *Environmental Science & Technology*, *46*, 970–976. doi: 10.1021/es203516h

Loomer, D. B., Al, T. A., Weaver, L., & Cogswell, S. (2007). Manganese valence imaging in Mn minerals at the nanoscale using STEM-EELS. *American Mineralogist*, *92*, 72–79. doi: 10.2138/am.2007.2252

Marris, H., Deboudt, K., Flament, P., Grobéty, B., & Gieré, R. (2013). Fe and Mn oxidation states by TEM-EELS in fine-particle emissions from a Fe-Mn alloy making plant. *Environmental Science & Technology*, *47*, 10832–10840. doi: 10.1021/es400368s

Mathai, S., Veghte, D., Kovarik, L., Mazzoleni, C., Tseng, K.-P., Bucci, S., et al. (2023). Optical properties of individual tar balls in the free troposphere. *Environmental Science and Technology*, *57*(44), 16834–16842. doi: 10.1021/acs.est.3c03498

Mavrocordatos, D., & Fortin, D. (2002). Quantitative characterization of biotic iron oxides by analytical electron microscopy. *American Mineralogist*, *87*, 940–946. doi: 10.2138/am-2002-0717

Maynard, A. D. (1995). The application of Electron-energy-loss spectroscopy to the analysis of ultrafine aerosol-particles. *Journal of Aerosol Science*, *26*, 757–777. doi: 10.1016/0021-8502(95)00006-X

Maynard, A. D., & Brown, L. (1992). Electron energy loss spectroscopy of ultrafine aerosol particles in the scanning transmission electron microscope. *Journal of Aerosol Science*, Proceedings of the 1992 European Aerosol Conference, *23*, 433–436. doi: 10.1016/0021-8502(92)90442-X

Mobus, G., Doole, R. C., & Inkson, B. J. (2003). Spectroscopic electron tomography. *Ultramicroscopy*, *96*, 433–451. doi: 10.1016/s0304-3991(03)00106-2

Németh, Z., Pósfai, M., Nyirő-Kósa, I., Aalto, P., Kulmala, M., & Salma, I. (2015). Images and properties of individual nucleated particles. *Atmospheric Environment*, *123*, 166–170. doi: 10.1016/j.atmosenv.2015.10.051

O'Day, P. A., Pattammattel, A., Aronstein, P., Leppert, V. J., & Jay Forman, H. (2022). Iron speciation in Respirable particulate matter and implications for human health. *Environmental Science & Technology*, *56*(11), 7006–7016. doi: 10.1021/acs.est.1c06962

Paoletti, L., De Berardis, B., & Diociaiuti, M. (2002). Physicochemical characterisation of the inhalable particulate matter (PM10) in an urban area: An analysis of the seasonal trend. *Science of the Total Environment*, *292*, 265–275. doi: 10.1016/S0048-9697(01)01134-2

Pattammattel, A., Leppert, V. J., Aronstein, P., Robinson, M., Mousavi, A., Sioutas, C., et al. (2021). Iron speciation in particulate matter (PM2.5) from urban Los Angeles using spectro-microscopy methods. *Atmospheric Environment*, *245*, 117988. doi: 10.1016/j.atmosenv.2020.117988

Patterson, J. P., Collins, D. B., Michaud, J. M., Axson, J. L., Sultana, C. M., Moser, T., et al. (2016). Sea spray aerosol structure and composition using cryogenic transmission Electron microscopy. *ACS Central Science*, *2*, 40–47. doi: 10.1021/acscentsci.5b00344

Posfai, M., Simonics, R., Li, J., Hobbs, P. V., & Buseck, P. R. (2003). Individual aerosol particles from biomass burning in southern Africa: 1. Compositions and size distributions of carbonaceous particles. *Journal of Geophysical Research: Atmospheres*, *108*, NIL_75–NIL_87. doi: 10.1029/2002JD002291

Posfai, M., Gelencser, A., Simonics, R., Arato, K., Li, J., Hobbs, P. V., & Buseck, P. R. (2004). Atmospheric tar balls: Particles from biomass and biofuel burning. *Journal of Geophysical Research: Atmospheres*, *109*. doi: 10.1029/2003JD004169

Richman, J., Livi, K., & Geyh, A. (2011). A scanning transmission electron microscopy method for determination of manganese composition in welding fume as a function of primary particle size. *Journal of Aerosol Science*, *42*, 408–418. doi: 10.1016/j.jaerosci.2011.03.004

Sanderson, P., Su, S. S., Chang, I. T. H., Delgado Saborit, J. M., Kepaptsoglou, D. M., Weber, R. J. M., & Harrison, R. M. (2016). Characterisation of iron-rich atmospheric submicrometre particles in the roadside environment. *Atmospheric Environment*, *140*, 167–175. doi: 10.1016/j.atmosenv.2016.05.040

Stoehr, L. C., Madl, P., Boyles, M. S. P., Zauner, R., Wimmer, M., Wiegand, H., et al. (2015). Enhanced deposition by electrostatic field-assistance aggravating diesel exhaust aerosol toxicity for human lung cells. *Environmental Science & Technology*, *49*, 8721–8730. doi: 10.1021/acs.est.5b02503

Twohy, C. H., Anderson, J. R., & Crozier, P. A. (2005). Nitrogenated organic aerosols as cloud condensation nuclei. *Geophysical Research Letters*, *32*, L19805. doi: 10.1029/2005GL023605

Verbeeck, J., Van Dyck, D., & Van Tendeloo, G. (2004). Energy-filtered transmission electron microscopy: An overview. *Spectrochimica Acta Part B: Atomic Spectroscopy*, *59*, 1529–1534. doi: 10.1016/j.sab.2004.03.020

Wang, Z. L., Bentley, J., & Evans, N. D. (2000). Valence state mapping of cobalt and manganese using near-edge fine structures. *Micron*, *31*, 355–362. doi: 10.1016/S0968-4328(99)00114-6

Williams, D. B., & Carter, C. B. (2009). *Transmission Electron microscopy - a textbook for materials science* (3rd ed.). Springer.

Xhoffer, C., Jacob, W., Buseck, P. R., & Vangrieken, R. (1995). Problems in quantitatively analyzing individual salt aerosol-particles using Electron-energy-loss spectroscopy. *Spectrochimica Acta Part B: Atomic Spectroscopy*, *50*, 1281–1292. doi: 10.1016/0584-8547(95)01338-F

Zhu, J., Crozier, P. A., Ercius, P., & Anderson, J. R. (2014). Derivation of optical properties of carbonaceous aerosols by monochromated electron energy-loss spectroscopy. *Microscopy and Microanalysis*, *20*, 748–759. doi: 10.1017/S143192761400049X

6

Raman Microspectroscopy of Particles

Sophie Sobanska, Sébastien Bonhommeau, Arnaud Desmedt, and David Talaga

ABSTRACT

Confocal Raman microspectroscopy (RMS) is a powerful technique for the chemical analysis of atmospheric particles, whether homogeneous or heterogeneous. It can determine the chemical composition and state of single particles as well as their acidity, with a spatial resolution close to 1 µm³. After introducing the basics of RMS and Raman imaging and their various applications in aerosol science, this chapter will be mainly devoted to highlighting the advantages and limitations of RMS for single-particle analysis. Data processing of Raman spectra and images using chemometric analysis methods is discussed, as these statistical approaches become increasingly important when large amounts of data, generated using the popular automated acquisition mode for Raman measurements, must be processed. The relevance of surface and tip-enhanced Raman spectroscopy to identify inorganic and organic compounds inside atmospheric particles is also presented. Finally, the chapter concludes by describing the advantages of coupling RMS over other single-particle analysis techniques.

6.1. INTRODUCTION

Raman microspectroscopy (RMS) is a single-particle analysis (SPA) technique based on vibrational spectroscopy. RMS is a laser-based, nondestructive technique developed in the 1970s (Delhaye & Dhamelincourt, 1975) consisting of a coupling between an optical microscope and a Raman spectrometer. It is an offline technique that can provide an optical image and a local analysis of a microsized heterogeneous object without any specific sample preparation. The ability of Raman spectrometers to employ visible laser excitation and high-quality microscope objectives results in a diffraction-limited laser spot about 1 µm in size. RMS was used for the first time in the late 1970s for characterizing atmospheric particles: more precisely, inorganic and organic compounds within single solid, semisolid, or liquid particles, with convenient 1 µm³ spatial resolution (Rosen & Novakov, 1977; Delhaye et al., 1979). Indeed, collected particles with sizes ranging between 1 and 100 µm can be analyzed in RMS, which is complementary to microscopic techniques used for SPA—mainly electron microscopy-based techniques (reviewed in Ault & Axson, 2017). RMS can operate at ambient pressure and thus ambient relative humidity (RH), which avoids a potential loss of water and/or semivolatile compounds. The growing interest in the characterization of chemical composition (identification of chemical species) and chemical heterogeneity within individual particles (Sobanska et al., 2006; Ivleva et al., 2007; Song et al., 2010; Ivleva et al., 2013; Jentzsch et al., 2012; Ault and Axson, 2017; Wang et al., 2021, to cite a few), including bioaerosols (reviewed in An et al., 2023 and Huang et al., 2024), has led to increasing use of RMS in atmospheric

Institute of Molecular Sciences, University of Bordeaux, CNRS, UMR 5255, Talence, France

Microanalysis of Atmospheric Particles: Techniques and Applications, Geophysical Monograph 288, First Edition.
Edited by Joseph M. Conny and Peter R. Buseck.

DOI: 10.1002/9781119554318.ch6

chemistry in the last decade. The development of imaging and automation techniques supported by chemometric data processing methods (Windig et al., 2002) allowed for the extraction of detailed chemical distributions, even below the classical diffraction-limited spatial resolution, from Raman images (Offroy et al., 2010, 2015).

Beyond the identification of chemical species within collected particle samples, RMS is a powerful technique to follow the chemical reactivity of model particles during laboratory experiments (Liang et al., 2022). By exploiting specially designed sample holders and flow reactor assemblies, water uptake by particles and their subsequent phase transformations and ice nucleation can be quantified (Schill & Tolbert, 2014; Laskina et al., 2015; Seng et al., 2018; Mael et al., 2019; Ishizaka et al., 2020; DeMott et al., 2023, as examples). The acid–base properties and hygroscopicity of particles, as well as their chemical reactivity, have been successfully monitored using RMS (Ault, 2020; Ault et al., 2013, 2014; Chang et al., 2020; Rindelaub et al., 2016; Wu et al., 2020; Wu & Ro, 2020, among others). However, precautions must be taken for the analysis of objects of micrometer size using optical spectroscopy techniques. The spatial resolution—that is, lateral and depth resolutions—is one of the crucial parameters to consider for a proper interpretation of Raman spectra, especially when the description of the full volume of a single heterogeneous particle is probed. Because atmospheric and model particles are necessarily collected on substrates before RMS analysis, the final spectrum intensity, and thus the adequate characterization of compounds, may be affected by the substrate contribution (Godoi et al., 2006; Zhou et al., 2014). Recently, some works have reported online automated measurements for semicontinuous analysis of atmospheric aerosols (Craig et al., 2017; Doughty & Hill, 2017, 2020a, 2020b) that open a perspective for more extensive RMS-based studies of aerosols.

This chapter aims to explain the main theoretical basis of the Raman phenomenon and RMS measurements for analysis of atmospheric and model particles to better understand the advantages and limitations of RMS.

The principle of Raman spectroscopy is first briefly described. Because microanalysis is suitable for the analysis of single particles with sizes smaller than 10 μm, microanalysis and imaging techniques are presented with a particular emphasis on the spatial resolution offered by the confocality of RMS and the implied limitations. The combination of RMS with chemometrics is briefly discussed to show its potential for the analysis and processing of large spectral datasets originating from automated measurements and imaging. Finally, the relevance of surface and tip-enhanced Raman spectroscopies for the analysis of atmospheric particles is presented.

6.2. PRINCIPLES OF RAMAN SPECTROSCOPY

Raman scattering spectroscopy, like infrared (IR) absorption spectroscopy—widespread in research laboratories—is a technique giving access to the vibrational (and rotational, in the case of gases) states of a molecule. A fundamental difference between these two techniques lies in the radiation–matter interaction: Raman spectroscopy is a technique based on the scattering of an electromagnetic wave, whereas IR spectroscopy is due to the absorption of an electromagnetic wave. Therefore, these spectroscopic techniques allow the analysis of vibrational spectra through different physical quantities. Variation of the molecular dipole moment is probed by IR absorption spectroscopy, and variation of the molecular polarizability is probed by Raman spectroscopy. Because of this matter–radiation interaction difference, a complementarity between these two spectroscopic techniques exists in terms of selection rules (i.e., vibrational modes activity). In this section, the focus will be on Raman scattering techniques; the reader is referred to many existing textbooks for IR absorption techniques.

Discovered in 1928, Raman spectroscopy expanded in the 1960s with the development of laser sources, allowing the analysis of small samples with reduced acquisition time. In the 1970s, the coupling of Raman spectrometers with (confocal) microscopes yielded significant improvements in the resulting spatial resolution (reaching about 1 μm^3) and the expansion of micro-Raman imaging. Over the last decade, technological innovation in laser diodes and detectors has allowed the development of miniaturized and mobile Raman (micro)spectrometers, useful for field campaigns, for instance. In addition, thanks to the high power provided by pulsed lasers and the tunability of modern laser sources, multiple nonlinear and/or resonant phenomena can also be studied in laboratories (stimulated Raman, hyper-Raman, coherent anti-Stokes Raman scattering, etc.). In the present description of the Raman spectroscopy principle, only the case of spontaneous Raman scattering will be treated.

Raman spectroscopy experiments consist of detecting the light scattered by a sample illuminated with a coherent monochromatic light source (laser) of frequency ν_0. A large fraction of the scattered light (defined by its frequency ν_s) possesses the same wavelength as the incident beam. This is Rayleigh scattering or elastic light scattering, for which $h\nu_s = h\nu_0$. A weak fraction of the light (10^{-6}–10^{-9}) is scattered, with a wavelength different from the incident one. This is Raman scattering or inelastic light scattering, for which $h\nu_s \neq h\nu_0$. When the scattered photon energy is lower than the incident photon energy (i.e., $h\nu_s < h\nu_0$), Stokes Raman scattering happens.

When the scattered photon energy is higher than the incident photon energy (i.e., $h\nu_s > h\nu_0$), anti-Stokes Raman scattering occurs (Figure 6.1). Nonresonant (Rayleigh and Raman) scattering processes are commonly described by considering virtual states in between ground and excited electronic states of the molecule (Figure 6.1). Each photon energy difference $|h\upsilon_s - h\upsilon_0|$ then corresponds to molecular vibrational energy, resulting in a Raman band associated with a molecular vibrational mode.

Light-scattering processes determining the impact of an oscillating electric field $\vec{E}$ (V/m) on molecular vibrational frequencies can be described from classical mechanics considerations. The electric field of frequency ν_0 (Hz) of an incident monochromatic beam is

$$\vec{E} = \vec{E}_0.\cos\left(2\pi\nu_0 t\right) \tag{6.1}$$

For a diatomic molecule behaving as a harmonic oscillator, the associated vibrational stretching frequency ν can be described by introducing a normal mode coordinate Q as

$$Q = Q_0.\cos\left(2\pi\nu t\right) \tag{6.2}$$

Just as a polarizable atom becomes an oscillating dipole when submitted to an electric field, a molecule under an applied electric field experiences an induced dipole moment $\vec{P}$ (C m) due to the displacement of the electronic "cloud" with respect to the nuclei in the laboratory frame (XYZ):

$$\vec{P} = \bar{\bar{\alpha}} \cdot \vec{E} \text{ or } \begin{pmatrix} P_X \\ P_Y \\ P_Z \end{pmatrix} = \begin{pmatrix} \alpha_{XX} & \alpha_{XY} & \alpha_{XZ} \\ \alpha_{YX} & \alpha_{YY} & \alpha_{YZ} \\ \alpha_{ZX} & \alpha_{ZY} & \alpha_{ZZ} \end{pmatrix} \begin{pmatrix} E_X \\ E_Y \\ E_Z \end{pmatrix} \tag{6.3}$$

where $\bar{\bar{\alpha}}$ (C m^2 V^{-1}) is the molecular polarizability tensor. Elements of this (second-rank) tensor depend on molecular symmetry and vary according to the considered vibrational mode. Using the first-order Taylor series expansion around the equilibrium position (Q = 0), $\bar{\bar{\alpha}}$ can be written as

$$\bar{\bar{\alpha}} = \bar{\bar{\alpha}}_0 + \bar{\bar{\alpha}}'.\,Q \text{ where } \bar{\bar{\alpha}}' = \overline{\overline{\left(\frac{\partial\alpha}{\partial Q}\right)_{Q=0}}} \tag{6.4}$$

where $\bar{\bar{\alpha}}'$ is the differential polarizability tensor. It should be noted that if $\bar{\bar{\alpha}}' = \overline{\overline{\left(\frac{\partial\alpha}{\partial Q}\right)_{Q=0}}} = 0$, no Raman shift can be observed. Substituting $\bar{\bar{\alpha}}$ from Equation 6.4 and $\vec{E}$ from Equation 6.1 into Equation 6.3 leads to

$$\vec{P} = \bar{\bar{\alpha}}_0 \cdot \vec{E}_0 \cos\left(2\pi\nu_0 t\right) + \frac{1}{2} Q_0 \bar{\bar{\alpha}}' \cdot \vec{E}_0 [\cos(2\pi(\nu_0 + \nu)t) + \cos(2\pi(\nu_0 - \nu)t)] \tag{6.5}$$

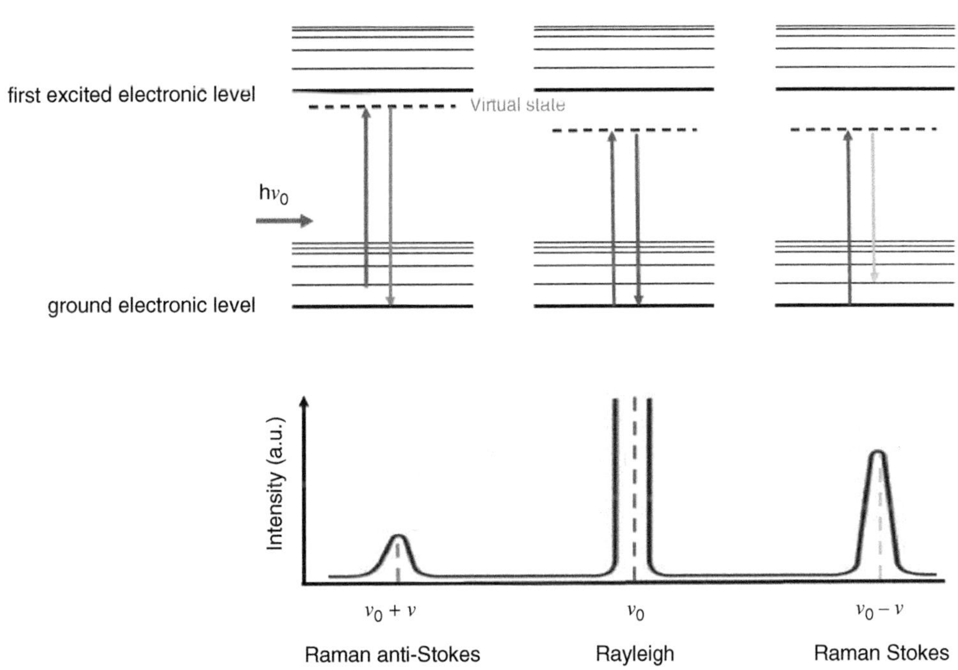

Figure 6.1 Raman band positions related to Raman anti-Stokes, Rayleigh, and Raman Stokes transitions as well as their corresponding relative energy levels.

From this equation, one clearly recognizes the three oscillating contributions of the induced dipole moment involved in the scattering phenomenon:

- The Rayleigh scattering (of frequency ν_0),
- The Stokes Raman scattering (of frequency $\nu_0 - \nu$),
- The anti-Stokes Raman scattering (of frequency $\nu_0 + \nu$).

For a vibrating diatomic molecule treated in the harmonic oscillator approximation, the vibrational frequency ν of a stretching mode is given by

$$\nu = \bar{\nu} c = \frac{1}{2\pi}\sqrt{\frac{k}{m}} \qquad (6.6)$$

where $\bar{\nu}$ is the vibrational wavenumber (cm^{-1}), c is the light velocity (m/s), k is the force constant of the harmonic oscillator (N/m), and m is its reduced mass (kg). In the frame of the harmonic approximation (Equation 6.6), two basic properties regarding the vibrational modes must be kept in mind. The vibrational frequency depends on the mass of the vibrator: that is, heavy vibrators have Raman signatures at lower frequencies than light vibrators. Such a property is of particular interest when considering various isotopic species. For instance, molecular hydrogen H_2 has its vibron at about 4070 cm^{-1} at 150 GPa, and it is observed at about 2990 cm^{-1} for molecular deuterium D_2 at the same pressure (Ross et al., 2013). The force constant could be assimilated to the strength of the involved chemical bonding: the simple $C-C$ bond possesses stretching wavenumbers ranging between 1000 cm^{-1} and 1300 cm^{-1}, and the triple $C\equiv C$ bond is observed at about 2140 cm^{-1}. In all discussed expressions, we have only considered an isolated harmonic oscillator, not interacting with its surroundings. For an isolated molecule constituted of N atoms, one has to consider the 3 N-6 normal modes (or 3 N-5 for a linear molecule). Such a molecule thus gives rise to several active Raman bands corresponding to internal vibrations (intramolecular modes) with wavenumbers typically ranging between 300 and 4000 cm^{-1}. Bending modes are typically observed below 1500 cm^{-1}, and stretching modes involving light atoms (H, C, N, O) often appear at wavenumbers higher than 1000 cm^{-1}. These internal vibrations thus constitute spectroscopic signatures of the chemical composition of a sample. Table 6.1 lists the typical frequency range expected for chemical groups relevant in atmospheric chemistry (i.e., discerned in Raman spectra of atmospheric particles).

In the case of a molecular crystal (unit cell constituted of M molecules), 3 M phonon modes (intermolecular modes typically observed below 500 cm^{-1}) must be considered in addition to the intramolecular modes. These crystalline modes are highly damped or even not observed in the case of an amorphous molecular organization due to the absence of long-range ordering. In atmospheric chemistry, this difference in the phonon signature could be relevant to identify the long-range molecular arrangement (crystalline or not) of a given compound, for instance. Moreover, a frequency shift of a given vibrational mode can be observed due to specific interactions (van der Waals, ionic interactions, hydrogen bonding, etc.) of the considered molecule with its surroundings (Petuya et al., 2018). In addition, it is possible to get information regarding the type of water cage occupied by nitrogen molecules, provided that Raman measurements are performed with very high spectral resolution (details on the spectral resolution are given

Table 6.1 Characteristic Raman Frequencies for the Most-Encountered Chemical Groups Found in Atmospheric Chemistry

Major peak Raman shift frequency ranges (cm^{-1})	Chemical groups of interest	Typical species in atmospheric particles
Broad band centered ~3,400	ν (O—H)	Water-containing or OH-containing particles
3,000–3,200	ν (N—H)	$(NH_4)NO_3$, $(NH_4)_2SO_4$, organics
~2,800–3,000	ν (CH_2) and ν (CH_3)	Alkyl groups in organic-containing particles
~1,660–1,750	ν (C=O)	Carboxylic acids and carboxylates in oxidized organic particles
~1,410–1,420	ν(C—O)	Carboxylic acids and carboxylates in oxidized organic particles
1,300 and 1,600 broad bands	D and G bands	Carbonaceous particles (i.e., soots, graphite-like particles)
~1,080–1,090	ν (CO_3^{2-})	Carbonates: $CaCO_3$, (Mg, Ca)CO_3, Na_2CO_3, $FeCO_3$
1,050 (free)–1,070	ν (NO_3^-)	$NaNO_3$, HNO_3, $Ca(NO_3)_2$, $(NH_4)NO_3$
~1,000	Carbon ring breathing	Carbonaceous particles
970–1,050	ν (SO_4^{2-}) and ν (HSO_4^-)	Na_2SO_4, $CaSO_4$, $(NH_4)_2SO_4$, H_2SO_4, HSO_4^-
464	ν (Si—O)	Quartz – SiO_2
<500	Metal oxides	Iron oxides (Fe_2O_3, FeOOH, Fe_3O_4), lead oxides (PbO), zinc oxide (ZnO), etc.

in Section 6.2). Raman scattering thus constitutes a nondestructive spectroscopic technique with very high chemical sensitivity (i.e., the ability to detect a chemical species under a specific molecular environment).

In addition to the frequencies, it is interesting to focus on the Raman scattering intensity to illustrate the opportunities for quantitative analysis. The Raman intensity depends on the transition polarizability between the initial state and the final state of the system, which involves the square of the differential polarizability tensor. Omitting detailed calculations of the scattered intensity, the Raman intensity of a given vibrational band can be summarized as

$$I_{\text{Raman}} = F\,C\,\sigma_{\text{Raman}} \text{ with } \sigma_{\text{Raman}} = |\alpha'|^2 \qquad (6.7)$$

where σ_{Raman} ($\text{cm}^2\ \text{sr}^{-1}$) corresponds to the Raman scattering cross-section, α' corresponds to the element of the differential polarizability tensor $\overline{\overline{\alpha'}}$ involved in the considered Raman band, C is the concentration of the probed molecular "vibrators" involved in the considered Raman band, and F is an experimental factor depending on the incident intensity, the scattered wavelength, and geometrical factors (e.g., probed volume) associated with the measurement setup (a detailed expression of the Raman intensity is provided for the case of a microspectrometer in Section 6.3.1). This "simplified" expression of the Raman band intensity leads to three important remarks regarding Raman scattering:

- The activity of a Raman band depends on the probed elements of the differential polarizability tensor. Raman activity is observed only if the probed element differs from zero (selection rule). The tensor element values depend on the symmetry of the considered molecule and are treated with the help of the group theory, which is not detailed in this chapter (the reader is referred to many existing textbooks for details). The Raman activity of a molecule (as well as its IR activity) can be deduced using the well-known character tables.
- Knowledge of the molecular symmetry leads to accessing the symmetry of the differential polarizability tensor. By considering Equation 6.3, one can observe that each induced dipole moment component (P_X, P_Y, or P_Z) depends on a given combination of the differential polarizability components (α_{ij}) and of the electric field (E_X, E_Y, or E_Z). It follows that the use of a linearly polarized incident beam leads to "turning off" two incident electric field components. Moreover, it is possible to analyze the scattered light with a linear polarizer (i.e., to probe only a selected component of the induced dipole moment, namely P_X, P_Y, or P_Z). Such polarization analysis yields to probe only selected differential polarizability tensor components of the considered molecules. In the case of an oriented sample (e.g., a single crystal or layer deposit on a substrate), it turns out that polarization analysis can provide information about the molecular orientation (this point will be detailed in Section 6.3.2).
- Finally, Equation (6.7) shows that the Raman intensity depends on the concentration of the probed chemical species. In principle, any measured Raman intensity thus contains information about the chemical species concentration. Any quantitative analysis requires the knowledge of specific parameters that are "summarized" in the experimental factor F and the Raman scattering cross-section σ_{Raman}. It follows that any quantitative analysis has as a prerequisite a reference sample for which the concentration is known, to measure the product $F.\ \sigma_{\text{Raman}}$ for given experimental conditions and to extract any concentration variation induced by external parameters in the sample (e.g., variation of a chemical species in a gas or a liquid). However, the quantitative analysis of microheterogeneities in a sample remains quite easy with the help of confocal micro-Raman imaging, for instance. In such a case, the product $F.\ \sigma_{\text{Raman}}$ can be considered a constant on the probed sample area at a constant sample depth so that any intensity variation of a selected Raman band intensity will be a direct signature of the heterogeneities of the probed species at a micrometer scale. The details regarding the spatial resolution and the refraction effect on Raman imaging are provided in the following section.

6.3. MICROANALYSIS AND IMAGING

6.3.1. Raman Microspectrometers: Description, Resolutions, and Confocality

The chemical characterization of microscale particles requires the use of a Raman microspectrometer: that is, an instrument coupling a Raman spectrometer to an upright or an inverted optical microscope. Figure 6.2 shows a sketch of a typical confocal RMS in the widely used backscattering geometry, but other scattering configurations are possible. Confocal Raman microspectrometers contain six crucial elements (Figure 6.2): (a) an excitation laser with conveniently filtered UV, visible, or IR emission; (b) an air, water-immersion, or oil immersion microscope objective (with magnifications from 5× to 100× and numerical apertures from 0.1 to 1.45) to focus the laser beam on the sample and collect scattered photons, in the case of the backscattering configuration; (c) a notch or edge filter to remove Rayleigh scattering; (d) a confocal hole allowing for in-depth sample sectioning (Sourisseau, 2004); (e) a diffraction grating, typically with 600, 1,200, 1,800, or 2,400 grooves/mm; and (f) a CCD camera as a detector. A half-wave plate can also be added behind the laser source to shift the (usually) linear polarization of the laser light.

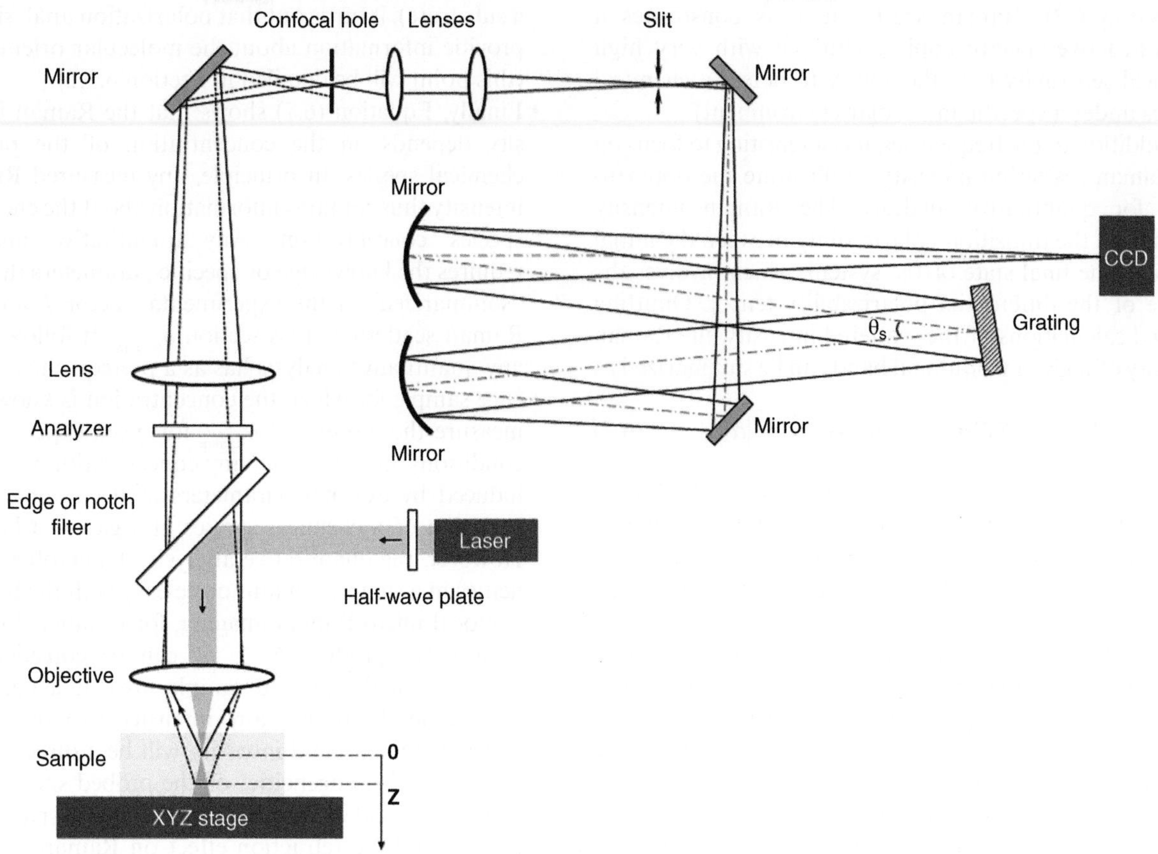

Figure 6.2 Confocal RMS instrument in backscattering configuration. θ_S, spectrometer angle; Z distance of focalization.

The spectral resolution of a Raman microspectrometer, defined as the ability to separate the light signal originating from two neighbor grooves of the diffraction grating, can be estimated from the $\Delta\bar{\nu}$ spectral separation using the expression

$$\Delta\bar{\nu}\left(\mathrm{cm}^{-1}\right)=\frac{\cos\left(\sin^{-1}\left(\frac{N_G\left(\mathrm{mm}^{-1}\right)\times\lambda_0\left(\mathrm{nm}\right)\times10^{-6}}{2\cos\theta_S}\right)+\theta_S\right)}{\lambda_0^2\left(\mathrm{nm}\right)\times N_G\left(\mathrm{mm}^{-1}\right)\times f_S\left(\mathrm{mm}\right)\times10^{-10}}\times W_i\left(\mu\mathrm{m}\right) \tag{6.8}$$

where λ_0 is the excitation laser wavelength, N_G is the number of grooves per millimeter of the diffraction grating, θ_S is the spectrometer angle, f_S is the exit focal length of the spectrometer, and W_i is the dimension of the image of the spectrometer entrance slit on the CCD detector that is given by

$$W_i\left(\mu\mathrm{m}\right)=W_S\left(\mu\mathrm{m}\right)\frac{\cos\left(\sin^{-1}\left(\frac{N_G\left(\mathrm{mm}^{-1}\right)\times\lambda_0\left(\mathrm{nm}\right)\times10^{-6}}{2\cos\theta_S}\right)-\theta_S\right)}{\cos\left(\sin^{-1}\left(\frac{N_G\left(\mathrm{mm}^{-1}\right)\times\lambda_0\left(\mathrm{nm}\right)\times10^{-6}}{2\cos\theta_S}\right)+\theta_S\right)} \tag{6.9}$$

where W_S is the entrance slit width. Equation 6.8 shows that the spectral resolution increases (i.e., $\Delta\bar{\nu}$ decreases) when the groove density of the diffraction grating, the excitation wavelength, and the exit focal length increase.

The spatial resolution of a Raman microspectrometer is composed of a radial (lateral) and an axial (in-depth) contribution. The lateral resolution is determined in the *XY* sample plane. It is often defined by the Rayleigh criterion providing the minimal separation d_{min} allowing two point objects to be distinguished from each other and can be expressed as (Inoué, 2006; Webb, 1996)

$$d_{\min} = \frac{0.61\lambda_0}{NA} = \frac{0.61\lambda_0}{n\sin\theta_o} \quad (6.10)$$

where NA is the numerical aperture of the objective, n is the refractive index of the object medium, and θ_o is the maximal half-angle of the cone light that can enter or exit the objective lens. For large numerical apertures ($NA > 1$), the lateral resolution is about half the excitation wavelength and reaches the diffraction limit that cannot be overcome in conventional optical spectroscopy.

The axial resolution is determined in the Z direction orthogonal to the XY sample plane. In bright-field microscopy, it is approximately equal to the depth of field: that is, the distance from the nearest object plane in focus to the farthest object plane simultaneously in focus along the Z direction. The axial resolution $z_{\min}$ verifies the relation (Inoué, 2006; Webb, 1996):

$$z_{\min} = \frac{2n\lambda_0}{NA^2} \quad (6.11)$$

The axial resolution typically equals a few hundred nanometers to 1–2 μm. For confocal micro-Raman spectrometers, however, slightly reduced lateral and axial resolutions can be deduced from (Webb, 1996)

$$d_{\min} = \frac{0.44\lambda_0}{NA} \text{ and } z_{\min} = \frac{1.5n\lambda_0}{NA^2} \quad (6.12)$$

Considering a Gaussian-ellipsoid laser illumination profile, the confocal volume V_{conf} is then given by (Mertz, 1998; Rüttinger et al., 2007)

$$V_{\text{conf}} = \left(\frac{\pi}{2}\right)^{3/2} w_r^2 w_z \quad (6.13)$$

where w_r and w_z are the radial and axial waists of the profile. For large numerical apertures (NA > 0.8), we can write

$$w_r \approx \frac{0.52\lambda_0}{n\sin\theta_o} \text{ and } w_z \approx \frac{0.76\lambda_0}{n(1-\cos\theta_o)} \quad (6.14)$$

Depending on the experimental conditions and optical properties of probed materials (discussed shortly), this leads to confocal volumes ranging from 10 to 10^{-3} μm^3, which is convenient for microanalysis of model and atmospheric particles.

Additionally, in transparent solids, the refractive index of the probed material is a crucial parameter. Indeed, when the laser beam penetrates the sample, the refractive index modifies the optical path according to Snell's law ($n_1 \sin\theta = n_2 \sin\theta'$, where n_1 and n_2 are refractive indices of two media and θ and θ' are the angles of incidence and transmission, respectively). As a result, the focal point is downshifted to a position (h) different from position (p) expected in the absence of refractive effects, as illustrated in Figure 6.3.

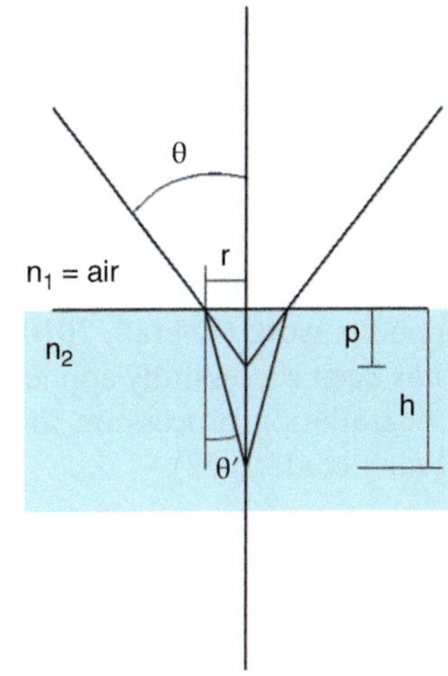

Figure 6.3 Optical pathways in a transparent sample with n refractive index. p would be the expected penetration depth of the laser if the sample (of n_2 refractive index) did not affect the focalization and h is the genuine penetration depth in the sample.

In Raman microscopy, the genuine laser penetration depth in the material h can be calculated based on Snell's law by

$$h = p\frac{\tan(\arcsin NA)}{\tan(\arcsin NA/n_2)} \quad (6.15)$$

For example, when using a 100× (NA = 0.9) objective, the penetration depth strongly depends on the p position and can be significantly lengthened in atmospheric particles containing common transparent (solid or liquid) compounds (see Table 6.2 for examples) because of the particle topography. This experimental issue can complicate and even impede any reliable surface analysis. It

Table 6.2 Laser Focal Distance (h in μm) into Transparent Solid Compounds with Atmospheric Relevance (Objective 100× – NA = 0.9)

	Calcite ($CaCO_3$) $n_2 = 1.65$	Anhydrite ($CaSO_4$) $n_2 = 1.57$	Quartz (SiO_2) $n_2 = 1.544$	Nitrocalcite ($Ca(NO_3)_2$) $n_2 = 1.465$
$p = 0.1$ μm	0.31	0.30	0.29	0.26
$p = 1$ μm	3.17	2.95	2.87	2.65
$p = 2$ μm	6.34	5.90	5.75	5.30

explains why the substrate contribution cannot be neglected when <5 μm particles composed of transparent materials are analyzed. Obviously, in the case of a heterogeneous mixture of compounds within the particle, the effect could be even worse. Common errors and artifacts when using confocal RMS, including spherical aberrations, being out of focus, and so on, have been reviewed in detail in a previous work (Everall, 2010). This confocality property has been successfully applied to reveal the photochemical degradation process on the surface of a single droplet (Wang et al., 2022).

6.3.2. Polarization Phenomenon

When incident light interacts with a molecule, there is an induced dipole moment $\vec{P} = \bar{\bar{\alpha}}\,\vec{E}$, where $\vec{E}$ is the electric field of the incident light and $\bar{\bar{\alpha}}$ is the second-rank molecular polarizability tensor that is created. This tensor can be decomposed into two contributions, as shown by Equation 6.4.

More precisely than in Equation 6.7, in the laboratory (XYZ) coordinate system, the Raman scattering intensity in a microprobe instrument can be written as (Sobanska et al., 2006; Sourisseau, 2004; Tanaka & Young, 2006)

$$I_{\text{Raman}} = 64\pi^2 I_{\text{inc}} \nu^4_{\text{Raman}} \iint \left| \vec{E}^T_{\text{inc}} \cdot \bar{\bar{\alpha}}'_{XYZ} \cdot \vec{E}_s \right|^2 d\Omega dV \quad (6.16)$$

where I_{inc} is the intensity of the incident light, ν_{Raman} is the Raman radiation wavenumber, $\vec{E}^T_{\text{inc}}$ is the transpose of the electric field vector $\vec{E}_{\text{inc}}$ of the incident light, $\vec{E}_s$ is the electric field vector of the scattered light, and $\bar{\bar{\alpha}}'_{XYZ}$ is the second-rank differential polarizability tensor leading to the Raman scattering phenomenon. V is the total scattering volume ($V \approx V_{\text{conf}}$), and Ω is the solid angle defined by the angular half-aperture of the objective. The differential polarizability tensor can be also expressed in the internal (xyz) coordinate system where the z-axis is often chosen as the principal axis of symmetry and becomes

$$\bar{\bar{\alpha}}'_{XYZ} = [R]^{\text{T}} \cdot \bar{\bar{\alpha}}'_{xyz} \cdot [R] \quad (6.17)$$

where $[R]$ (respectively, $[R]^{\text{T}}$) is the rotation matrix (and its transpose) that can be defined with respect to Euler angles (Sourisseau, 2004; Tanaka & Young, 2006). The resulting internal molecular/crystal differential polarizability tensor is expressed as a matrix:

$$\bar{\bar{\alpha}}'_{xyz} = \begin{pmatrix} \alpha'_{xx} & \alpha'_{xy} & \alpha'_{xz} \\ \alpha'_{yx} & \alpha'_{yy} & \alpha'_{yz} \\ \alpha'_{zx} & \alpha'_{zy} & \alpha'_{zz} \end{pmatrix} \quad (6.18)$$

Each α'_{ij} (i, j = x, y, or z) value is typical of the molecular/crystal vibration. Depending on the orientation ([R]) relative to the electric field vector of the incident and scattered light (i.e., light polarization), only certain α'_{ij} elements can contribute to the Raman scattering intensity. Experimentally, the insertion of a polarizer (an analyzer) on the optical path of a Raman spectrometer allows the incident (scattered) light polarization to be controlled (Figure 6.2), to determine the different α'_{ij} elements as well as the depolarization ratio reflecting the polarization direction dependence of the Raman scattering (Sourisseau, 2004; Tanaka & Young, 2006). This technique constitutes the well-named polarized micro-Raman spectroscopy, which is a powerful tool to characterize molecular and crystal orientations at the microscale. In particular, Raman spectra of crystallized particles with crystal sizes larger than the lateral spatial resolution depend on the orientation ([R]), and Raman spectra of amorphous compounds and randomly oriented nanocrystals are similar to Raman signatures observed for powder samples (Sobanska et al., 2006).

6.3.3. Species Identification and Quantification

RMS analysis of a single particle may be performed using a manual spot mode (i.e., each analytical point corresponds to a single particle and a single spectrum) or the imaging technique. The imaging methodology for particle analysis will be detailed in a dedicated paragraph. As previously mentioned, the Raman activity, and thus the cross-section of compounds, depends on Raman selection rules. For example, sodium chloride (NaCl), one of the most abundant compounds in sea salt aerosols, belongs to the O_h^5 (Fm3m) space group and is not active in Raman. In contrast, TiO_2 (anatase or rutile), which is commonly identified as a trace in clay minerals, is very active in Raman spectroscopy and can be detected even at trace levels in analyzed volumes. Finally, clay minerals (mainly swelling clay), which are constituents of soil-derived particles, exhibit an intense fluorescent signal due to impurities trapped in interlayer spaces (such as humic substances, for instance). The identification of species consists in assigning each vibrational mode, or combination of vibrational modes, observed in the Raman spectrum, to functional groups that are related to a specific compound, as described in detail in Section 6.2. This spectroscopic signature (i.e., the band position and relative intensities between the Raman bands in the spectrum) is typical of a single compound, making identification robust even with a single Raman band (called a marker band). As an example, solid carbonate compounds are identified through the four dominant lines in the spectral region between 200 and 2,100 cm^{-1} that are due to internal vibrational modes of the carbonate

CO_3^{2-} ions (>600 cm^{-1}) and to lattice vibrational modes (<350 cm^{-1}) that are influenced by the corresponding cation. Indeed, the symmetric stretching mode of the CO_3^{2-} group around 1080 cm^{-1} (which is the most intense band in the spectrum) is found at 1087 cm^{-1} for calcite ($R\bar{3}c$) but shifts to 1,099 or 1,053 cm^{-1} for dolomite ($CaMg(CO_3)_2$) and cerussite ($PbCO_3$), respectively (Herman et al., 1987). The Raman band positions of a single compound also depend on the physical state of the compound itself (i.e., liquid or solid). Typically, nitrate-like compounds found in collected aerosol can be in liquid or solid form with a stretching mode of NO_3^- at 1050 cm^{-1} or at 1068 cm^{-1} for free NO_3^- and $NaNO_3$, respectively. The sensitivity of Raman band positions to the physical state of compounds has been advantageously used to characterize the ability of an aerosol to absorb water (i.e., hygroscopic properties of aerosols) (Yeung et al., 2009, 2010; Treuel et al., 2011; Seng et al., 2018; Wu et al., 2020, to cite a few). RMS is sensitive to unique species or molecular ions, so in solution, it has the advantage of directly measuring the acidity and identifying acid conjugates. For instance, SO_4^{2-} and HSO_4^- anions have been well characterized through Raman vibrational shifts for stretching modes at ~985 and ~1050 cm^{-1}, respectively. Therefore, the pH of individual particles could be determined from the concentration of each anion within laboratory-generated aerosol particles based on integrated peak area (Rindelaub et al., 2016; Craig et al., 2017; Chang et al., 2020; and reviewed in Ault, 2020).

Most of the time, the identification of a compound in a sample is achieved by comparing its Raman spectrum with either a reference spectrum from a spectral library or a Raman spectrum acquired on pure compounds. Typically, species identification is performed manually by comparing the position and relative intensities of the bands in the experimental spectrum with those in a reference spectrum. In the case of a mixture of species—that is, when a Raman spectrum is composed of several species—the identification can be performed similarly by comparing the positions and relative intensities of the bands in the spectrum of the pure reference compound with the bands observed in the experimental spectrum.

Dedicated commercial software packages may provide an automatic spectral search by using Raman band wavenumbers, intensities, areas, or bandwidths that are compared with integrated reference databases. The automated search is achieved using various algorithms (Kwiatkowski et al., 2010, as an example). The spectra are compared by summing up, according to different norms (e.g., absolute or squared values), the differences between the measured spectra (or derivatives) and the reference spectrum from the library. In addition, spectral search techniques can include computation of the dot products or the use of machine learning techniques (Carey et al., 2015) for database sorting.

However, Raman spectra associated with analyzed collected particles contain not only valuable Raman fingerprints but also other contributions such as fluorescence backgrounds, cosmic radiation, Gaussian noise, and other effects dependent on experimental parameters. These interference contributions have to be removed prior to the identification process to ensure that compound identification is based on Raman measurement and not other parasitic effects. Many preprocessing methodologies are available in spectrometer software, including background correction, filtering, smoothing, and scaling (see Section 4). The signal-to-noise ratio (S/N ratio or SNR) is one of the most important criteria needed for a suitable identification of Raman bands. Stray light, environmental light, and the inherent interior noise of electronic or optical devices from the spectrometer result in noise adding up over a long integration time. These factors influence SNR, thus further affecting the feature extraction of the valid Raman signal. In practice, a Raman band can be considered real when its intensity is about three times higher than the noise signal and it is centered at the same wavenumber irrespective of the duration of the integration time to collect the Raman signature, under nondestructive irradiation conditions (unable to induce photochemical effects).

The increasing importance of the combination of Raman spectroscopy and chemometrics gave rise to the development of various preprocessing procedures adapted for atmospheric particle analysis (Bocklitz et al., 2011 and references therein). In the case of heterogeneous particles, the Raman spectrum can be considered the sum of several pure compound spectra that can be known as described previously.

The main compounds identified in particles collected during field campaigns are dependent on particle sources and are presented in Table 6.3. Inorganic compounds are easily identified in collected samples, and organic ones are mostly identified thanks to the intense band assigned to CH_2 and CH_3 stretching vibrations around 3000 cm^{-1}, and sometimes thanks to the C=O and C—O stretching vibrations around 1700 and 1400 cm^{-1} indicating oxygenated compounds, or a typical band of carbon ring breathing mode at 1000 cm^{-1}. Carbonaceous particles—that is, carbon-containing particles such as soots and related carbonaceous materials—are well distinguished through the typical D and G Raman bands of graphite-like compounds at ~1300 and ~1600 cm^{-1} (see as an example Zheng et al., 2023 and references therein). Note that particles are often composed of a mixture of inorganic and organic species, the detection of organic ones being, most of the time, difficult (Bondy et al., 2018; Sobanska et al., 2014; Wang et al., 2021). Finally, in most

Table 6.3 Typical Compounds Identified in Samples Collected During Field Campaigns Using RMS, as Function of Their Sources

Sources	Main compounds identified using RMS
Marine	$NaNO_3$, Na_2SO_4, $Ca(NO_3)_2$, $CaSO_4$ (anhydrite), $CaSO_4.2H_2O$ (gypsum), organics (oxygenated compounds)
Terrestrial	Carbonates such as $CaCO_3$ (calcite), (Ca, Mg)CO_3 (dolomite), SiO_2 (quartz), $CaSO_4$ (anhydrite), $CaSO_4.2H_2O$ (gypsum), $Ca(NO_3)_2$, Fe_2O_3 (hematite), TiO_2 (anatase associated with clay minerals), feldspar
Urban	$(NH_4)_2SO_4$, carbonaceous (graphitic C, soots), metal oxides (mainly iron), nitrate-rich compounds, organics (oxygenated compounds mainly)
Ferrous and nonferrous smelting industries	Iron oxides: Fe_2O_3 (hematite), FeOOH (goethite), Fe_3O_4 (magnetite), FeO M-oxides, sulfides, and sulfates (M: Pb, Cd, Zn)
Mining	M-oxides

cases, the compounds identified by RMS in different real samples from the same source are mainly the same. Only the proportions of species within these samples vary from one sampling campaign to another.

Theoretically, RMS helps identify the molecular species within each particle because Raman intensity can be related to the concentration of chemical species. Because of the confocal RMS theory detailed in previous paragraphs, any quantitative analysis requires specific cautions:

- The analyzed material is required to be homogeneous and isotropic. The relative concentration of species within a liquid droplet can be quantified if a calibration curve is determined (Rindelaub et al., 2016).
- The penetration depth of the laser beam inside the particles needs to be less than the particle size. The influence of the signal due to the collection substrate can be a strong limitation for the quantification (Falgayrac et al., 2018; Godoi et al., 2006; Zhou et al., 2014). Constraints on the analyzed volume and the penetration depth prevent the analysis of transparent particles with diameters less than 5 μm (i.e., composed of nitrates, sulfates, carbonates, water, and so on).
- The size of the laser spot (typically 1 μm^2) needs to be larger than the size of the solids analyzed with RMS to discard any polarization phenomenon (Batonneau et al., 2003). This point limits the quantification of species within solid particles.
- In the case of heterogeneous particles, the Raman cross-section of each compound needs to be comparable. Otherwise, a semiquantification procedure after some normalization process has to be employed (Bocklitz et al., 2011).

6.3.4. 2D and 3D Imaging

The imaging technique allows the description of the chemical heterogeneity within a scanned area of the sample. Global analyses of particle samples (i.e., collected on filters) provide a useful global chemical composition but do not give any realistic features of the chemical mixing of the particles (Ault & Axson, 2017, as a review). Indeed, in collected samples with particles of diameter less than 10 μm, a Raman spectrum of a single particle often reflects a mixture of several compounds. The characterization of such a chemical heterogeneity within a single particle is crucial to better understanding the impacts of particles on health and climate change. Manually selecting and finding particles for analysis (called *manual spot mode*) is commonly used for characterizing the molecular composition of individual particles collected in ambient air but gives partial information on the particle composition. Only a few studies have described the chemical heterogeneity using the more adequate (but time-consuming) Raman imaging technique (Sobanska et al., 2006, 2014; Lee et al., 2020 as examples), but a restricted number of particles has been analyzed since then. In laboratory-based studies, Raman imaging is, however, useful to unravel heterogeneous processes for large particles (> 5 μm).

Two-dimensional (2D) Raman imaging consists of mapping a defined area in the XY sample plane. The scanning of the analyzed area of dimensions $X \times Y$ (i.e., $m \times n$ pixels) allows the generation of a three-dimensional (3D) cube of data of type ($m \times n \times \bar{\nu}$: that is, $m \times n$ spectra of $\bar{\nu}$ =2,040 points). Fast imaging can be achieved using a combination of scanning mirrors that scan the laser beam along a line for line (i.e., one-dimensional [1D]) profiles or over an area for 2D mapping. This imaging mode, which reduces acquisition time for a large sample area, is offered by most RMS manufacturers. This acquisition mode has the advantage of allowing hyperspectral imaging (i.e., the collection of a spectrum per image pixel). However, this methodology is not widely used for the analysis of atmospheric particles (Doughty & Hill, 2020a, 2020b).

Two conventional methods of 2D Raman image reconstruction are routinely used: (a) intensity integration within a spectral interval and (b) decomposition of model spectra. With the development of chemometrics for spectroscopy, the multivariate data processing of images also constitutes a suitable alternative because the image data cube may represent a huge number of spectra that have to be processed simultaneously.

Intensity Integration

Intensity integration within a spectral interval consists of calculating the area under the peak of a characteristic Raman band recorded on a judiciously chosen spectral domain. Considering that the line of the compound lies between wavenumbers $\bar{\nu}_1$ and $\bar{\nu}_2$, the sub-peak area $A(m,n)$ for each spectrum (SP) recorded and located at the point of coordinates (m,n) is given by

$$A(m,n)=\int_{\bar{\nu}_1}^{\bar{\nu}_2} SP(m,n,\bar{\nu})\,\mathrm{d}\bar{\nu} \tag{6.19}$$

The image can then be built up from the values of A for each (m,n) pixel. Other characteristics of the spectral lines, such as their full width at half maximum (fwhm), can also be utilized. The advantage of this method lies in its applicability during the acquisition of the spectra. Nevertheless, it can only be applied to spectrally well-resolved Raman lines that are typical of a compound, which constitutes a major drawback. Finally, regarding image reconstruction, only three compounds can be mapped because only three colors are available in the RGB (red, green, blue) imaging representation. The color of each pixel is then determined by the combination of the red, green, and blue intensities.

Decomposition of Model Spectra

This process assumes a priori knowledge of the compounds in the mixture. Considering that n compounds are present, this method decomposes each experimental SP spectrum as the sum of model spectra, weighted by the k_n coefficients associated with each compound. From the n model spectra (reference) $M_1, M_2, \ldots, M_n$, each spectrum of the SP pixels is expressed as follows:

$$SP(m,n,\bar{\nu})=\sum_{i=1}^{n} k_i(m,n)M_i(\bar{\nu}) \tag{6.20}$$

The determination of the coefficients $k_1, k_2, \ldots, k_n$ is obtained by minimizing the error by the method of least squares. The spatial distribution of the compound n will thus be established from the coefficients k_n calculated for each pixel of coordinates (x, y).

Both of these methods have significant drawbacks. The second one requires a priori knowledge of the compounds contained in the sample, and the first one requires nonoverlapping of the characteristic Raman bands.

Multivariate Data Processing for Imaging

The computer-microcontrolled XY scanning of Raman images provides many pixel Raman spectra that reflect the complex mixture at the single particle level. This large heterogeneity generates a strong spatial overlap of spectral signatures associated with pure components because of the diffraction-limited spatial resolution of RMS. To solve this problem, self-modeling mixture analysis tools such as multivariate curve resolution (MCR) methods (Hamilton & Gemperline, 1990; Windig, 1992) have been popularized for spectral imaging applications (Windig, 1997, Windig et al., 2002; Dupuy & Batonneau, 2003; Batonneau et al., 2004, 2006, to cite initiators). The data processing of Raman images using multivariate methods simultaneously takes into account a huge number of spectra and allows single spectral components to be distinguished from each other in heterogeneous samples, which cannot be achieved by conventional Raman imaging methods. Among MCR methods, the Simplisma (simple-to-use interactive self-modeling mixture analysis) approach is particularly adapted for the data processing of Raman images (Windig, 1992). When highly overlapping spectral features and/or baselines are present in the spectra, second-derivative spectra can be used to resolve the data properly (Guilment et al., 1994; Windig, 1994; Windig et al., 2002; Windig & Stephenson, 1992). This approach uses pure spectra as a first estimate and derives pure variables from the resolved contribution profiles. The 3D cube of imaging data of type $m \times n \times \bar{\nu}$ is expressed as

$$D = C \times P^{\mathrm{T}} + E \tag{6.21}$$

where D is the matrix of experimental spectra, C is the matrix of spectral concentrations, and P^{T} is the transposed matrix of pure spectra. E represents the error matrix.

The resolution of Equation 6.21 allows pure spectra (i.e., a single spectrum per species) to be extracted from the analyzed area. The algorithm for pure variables in Simplisma is described in detail elsewhere (Windig, 1992, 1997). Raman images of molecular species are reconstructed pixel by pixel from the spectral concentration matrix C. The validation of the calculation is performed by calculating the differences between the experimental matrix D and the reconstructed one, $D^{\text{reconstructed}}$, by means of the relative square root of the sum of square differences (RRSSQ):

$$RRSSQ=\sqrt{\frac{\sum_{i=1}^{nspec}\sum_{j=1}^{nvar}\left(d_{ij}-d_{ij}^{\text{recons}}\right)^2}{\sum_{i=1}^{nspec}\sum_{j=1}^{nvar} d_{ij}^2}} \tag{6.22}$$

where $d_{i,j}$ is the ith row and jth column element of D, d_{ij}^{recons} is the ith row and jth column element of $D^{\text{reconstructed}}$, *nspec* is the number of mixture spectra, and *nvar* is the number of recorded intensities.

The RRSSQ value thus expresses the difference between reconstructed and experimental intensities. For a perfect match, the value is 0.

The effectiveness of a combined use of computer-controlled Raman microspectroscopy mapping and MCR methods to determine from Raman images chemical and heterogeneous characteristics of individual collected and model aerosol particles is well recognized (Batonneau et al., 2004, 2006; Falgayrac et al., 2006, 2013, 2014; Sobanska et al., 2006; Sobanska et al., 2014).

The potential of confocal RMS for 3D imaging has not yet been systematically exploited for the study of particles. However, this method, by considering the limitation described above, allows 3D maps of the distribution of species in a particle to be obtained and thus gives a description of the mixing states in particles for different thicknesses (z) of the order of the micron. The data cube D is then composed of $(m \times n \times \bar{\nu} \times z)$ spectra and can be processed using MCR methods as described previously. A significant example is given in Figure 6.4 for a particle taken during one of the campaigns carried out in a coastal area in France. In this case, the composition of the particle deduced by Raman spectroscopy depends on its thickness. An image of $19 \times 20\,\mu m^2$ was acquired with a confocal hole opening equal to $150\,\mu m$ (to reduce the volume of analysis) with a $1\,\mu m$ pitch in the X and Y directions and a Z-position step of $1\,\mu m$. Four images were recorded, each containing 380 spectra. The fusion of the spectral data relative to these images led to 1,520 spectra. The data processing of the whole dataset with MCR shows six typical spectra (Figure 6.4c) corresponding to five Raman spectra assigned to $CaCO_3$, deliquescent $NaNO_3$, solid $NaNO_3$, $CaSO_4.2H_2O$, and TiO_2 (anatase) and a luminescence (or fluorescence) spectrum that can be attributed to the presence of clay minerals (Batonneau et al., 2004). It should be noted that the MCR procedure extracted the background signal from the entire data matrix in a preprocessing step, allowing the fluorescent

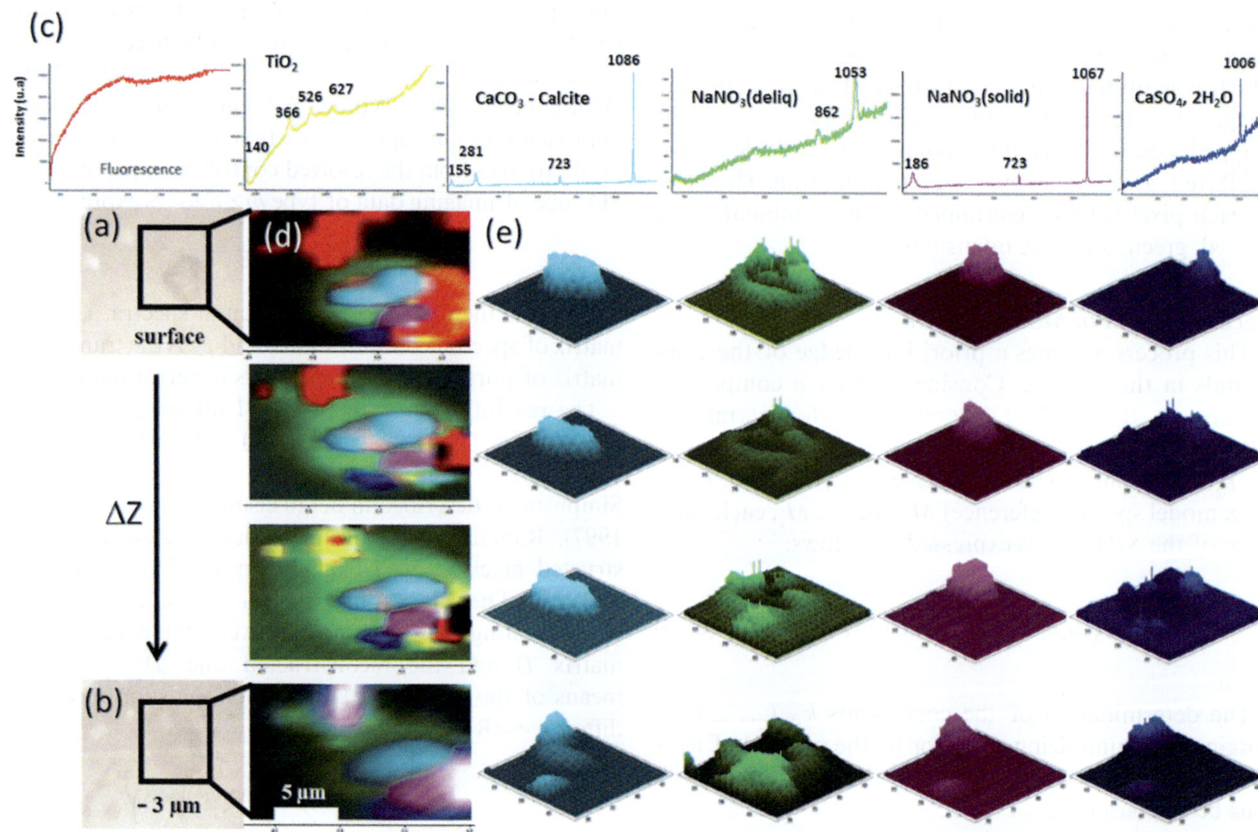

Figure 6.4 3D reconstruction of the composition of an aerosol particle by confocal Raman imaging ($\lambda = 633$ nm). (a) Optical image (×100) of the surface of the particle and definition of the analyzed area, (b) optical image (×100) of the particle with a Z focus of $1\,\mu m$ in the particle, (c) spectra extracted after MCR processing of four images, (d) 2D reconstruction of four images for Z varying from the particle surface to $3\,\mu m$, (e) 3D reconstruction corresponding to Raman spectra assigned to $CaCO_3$, deliquescent $NaNO_3$, solid $NaNO_3$, and $CaSO_4.2H_2O$.

signal to be distinguished (Windig et al., 2002). The reconstruction of the 2D images (Figure 6.4d), including information on intensities (Figure 6.4e) for each *Z*-position, shows that the distribution of species varies with the depth of analysis. In particular, we can highlight the presence of TiO_2, which is often associated with clay minerals, inside the particle but not at its surface. However, we must keep in mind that Raman images are distorted compared with the real species distribution due to the refractive-index dependence of the laser penetration depth (see Section 6.1).

6.4. AUTOMATION AND DATA PROCESSING USING STATISTICAL TOOLS

Determining the composition of particles collected in ambient air using SPA, as opposed to bulk analysis, presents essential advantages that have been demonstrated in numerous works. An automation of measurements is, however, needed to study a sample size above the previous level of statistical significance. Additionally, the development of robust data processing methods is helpful because the large data matrix entails the classification of objects in a reduced number of groups. In the case of atmospheric particles, data processing leads to classifying particles as groups based on their chemical composition. It can provide reliable spatial and temporal chemical evolution of the particle composition because each particle group can be interpreted with respect to its source and/or a specific atmospheric process. This methodology was largely developed for the SPA using electron microscopy. The classification of particles based on their Raman spectra is performed from the classification of Raman intensities for each given wavenumber, which is defined as a variable in data processing algorithms. Combining RMS with chemometric methods requires (1) the automation of measurements (i.e., construction of the data cube), (2) the preprocessing of the Raman spectra (i.e., the determination of reliable variables), and (3) data processing of the dataset (i.e., classification).

6.4.1. Automation of the Measurement

Automating the measurement consists of either a semi-automatic line scanning imaging measurement as recently performed (Doughty & Hill, 2017, 2020a, 2020b) or an offline analysis based on the automated imaging mode provided by the instrument. The recent development of appropriate software by instrument companies allows computer-controlled Raman (CC-Raman) acquisition, as described by Doughty and Hill, or automated image-based acquisition. These two acquisition methodologies apply automated mapping to increase analysis speed, allowing for up to 100 particles to be analyzed in an hour, as was demonstrated (Craig et al., 2017). Acquisition parameters (wavelength of the laser, laser power, laser focalization, integration time, number of acquisitions, confocal hole opening) have to be carefully chosen to provide a Raman spectrum with an acceptable SNR ratio in a reasonable time. Previous works based on scanning electron microscopy (SEM) have demonstrated that SPA of ~300 particles is enough to properly describe the main species contained in a collected sample (Jambers et al., 1995).

6.4.2. Preprocessing of Spectroscopic Data

Preprocessing is required to eliminate the effects of unwanted signals such as fluorescence, Mie scattering, detector noise, calibration errors, cosmic rays, laser power fluctuations, substrate effects, and so on. It mainly concerns baseline correction, filtering, scaling, and normalization. The preprocessing procedure has to be carefully considered because the results may have an influence on the MCR processing (Bocklitz et al., 2011). Indeed, baseline correction and normalization are crucial steps because they influence the intensity of spectra and thus variables used for MCR. A large variety of preprocessing methods for Raman spectra have been developed and are included within the software of commercial instruments. For instance, asymmetric least squares smoothing (AsLS) is attractive for the estimation of the Raman baseline (Eilers & Boelens, 2005; He et al., 2014). AsLS background correction was used to estimate the fluorescence background from 32,718 Raman spectra in automatic mode, which significantly improved the process of semicontinuous automated detection of airborne single particles especially (Doughty and Hill, 2017).

6.4.3. Combination of RMS and Chemometrics for the Processing of Datasets

This has become an important issue in the field of analytical aerosol chemistry (Gautam et al., 2015; Ofner et al., 2015). Multivariate analysis methods are applied to obtain information about the sample composition, the intercorrelation between particles, and specifications of the main particle groups. The description of particles based on just a few representative components facilitates possible monitoring of their chemical composition in relation to particular variables, such as meteorological conditions or the aerodynamic diameter (Yotova et al., 2016, as an example). Several important statistical methods widely used in data analysis (mainly linear transformation methods such as principal component analysis [PCA] and clustering [hierarchical or *k*-means]) were applied to SPA.

One of the fundamental aspects to be considered in linear transformations concerns the correlations between variables, by which a transformation of an original set into another space, in which new variables are uncorrelated or statistically independent, can be made. PCA is a statistical method that defines a linear transformation, which is able to support the description of a stationary stochastic process given in the form of a set of N-dimensional vectors with reduced dimensions to a lower K-dimensional set ($N > K$). This transformation takes place through the W matrix of dimensions $K \times N$, where the output space y of the reduced dimension retains the most important information about the original process. In other words, PCA replaces a high amount of information contained in a mutually correlated input into a set of statistically independent components, classified according to their relative importance. PCA can be exploited to process large datasets. For example, it has been used to realize single particle (SEM-EDX) and bulk (X-ray fluorescence) analysis of a combined dataset related to approximately 25,000 individual particles collected in Lake Balaton in Hungary (Osan et al., 2001). Such data processing was applied to determine potential sources of the collected aerosol particles.

Clustering data into meaningful groups is an important task of chemometrics (Ziegel, 2004). Clustering is considered to be an unsupervised classification of data. A number of clustering algorithms have been described in the literature (Everitt & Hothorn, 2011). Two popular clustering algorithms are generally applied to SPA datasets: the agglomerative hierarchical clustering algorithm (HCA) and the partitional k-means algorithm. HCA seeks to build a hierarchy of clusters without a fixed number of clusters, and k-means analysis requires a prespecified number of clusters. The first method is useful when the target is to arrange the clusters into a natural hierarchy (e.g., classifying a group of spectra), and the second is well suited to very large datasets and uses a median or mean to represent each cluster (e.g., one spectrum representing a typical composition). The k-means cluster analysis was performed for ~5,000 individual aerosol particles analyzed by computer-controlled SEM and RMS techniques (Moffet et al., 2012). By application of k-means clustering, Fe^{2+} was found to be an insignificant enrichment in ambient anthropogenic particles, except in standard Asian mineral dust. Chemometric analysis based on clustering of the spectroscopic data of model salts demonstrated the suitability of RMS to classify and identify the inorganic composition of atmospheric aerosol samples (Jentzsch et al., 2013). Interestingly, automated online RMS was used recently to measure approximately 100,000 particles during a 24-hour sampling period. In this case, the Raman spectral data cube was processed using a clustering method for classification of particle types (Doughty & Hill, 2020a, 2020b).

A combination of HCA, k-means, and PCA methods was employed to scrutinize the multisensory data cube (i.e., fused hyperspectral data cubes) that allowed a detailed assignment of chemical species and their relationship in atmospheric particles (Ofner et al., 2015).

More recently, an effective analytical algorithm based on MCR, PCA, and HCA has been employed for processing Raman spectra and describing the chemical mixing and quantitative composition of aerosol particle samples (Siepka et al., 2018). Through the application of the MCR approach on the Raman spectra dataset, the identification of pure compounds was possible. The methodology was constructed to distinguish particles composed of only one species (pure molecular compounds) from mixed composite particles and to define particle types as a function of their chemical similarity. Finally, the quantification of each group of particles according to their composition and chemical mixing was presented to describe their chemical composition and heterogeneity.

Because RMS is an offline technique, particles must be collected on suitable substrates for analysis. Thus, a careful choice of the analytical substrate for SPA is necessary. The substrate should be characterized to allow for an optimal optical contrast (if the realization of optical imaging is planned) and check its chemical inertness (to avoid any modification of the chemical composition and morphology of the particles). The substrate contribution to the overall Raman signal may be strong with respect to the relevant information on the sample and thus impair the final results extracted from SPA. This is particularly crucial when the particle size is less than the laser spot size, a situation that is typically encountered for atmospheric micron-sized particles. The substrate commonly used for RMS analysis of individual particles is a glass (or quartz) slide onto which particles are impacted or transferred. A glass slide often has a background signal that must be subtracted from the Raman signal for proper particle characterization. In addition, Al- or Ag-foil, Si-wafer, and transmission electron microscopy (TEM) grids with an ultrathin C layer are popular substrates as they are used for complementary SPA techniques. The evaluation of these substrates was carried out in a combined RMS and electron microscopy analysis, concluding that the TEM grid presented the best compromise (Godoi et al., 2006). Recently, the evaluation of usual particle-collecting substrates (i.e., Ag, In, Si, SiO_2, microscope slide, and TEM grid) in the context of an MCR approach applied to the processing of Raman spectra has been presented (Falgayrac et al., 2018). The evaluation of the spectral particle contribution exported by MCR was based on the benchmark specification of the optimal analytical substrate for RMS imaging of species within aerosol particles.

6.5. BEYOND THE DIFFRACTION LIMIT

6.5.1. Oversampling versus Super Resolution

In the case of optical microspectroscopy, the limit size for individual particle analysis has typically been set at 1 μm due to the Abbe diffraction limit. Many studies were focused on the analysis of larger aerosol particles (1–10 μm diameter), providing valuable information on internal particle processes as described previously. Thanks to technological advances, the lower limits of optical microscopy are approaching the diffraction limit, allowing spectral analysis of fine aerosol particles. However, resolving submicron particles remains challenging (Ault & Axson, 2017; Laskin et al., 2019). In the field of molecular imaging, chemometric methods are an integral part of an analytical protocol for pushing back the analytical limits. The oversampling methodology consists of mapping the area of interest beyond the lateral resolution: that is, typically less than 1 μm. The combination of this oversampling with MCR data processing of the image data cube provides the map of chemical species with a better resolution. Alternatively, using both the super-resolution concept and MCR algorithms in confocal Raman imaging allows instrumental limits to be surpassed and chemical components of atmospheric aerosols to be characterized at the single particle level (Offroy et al., 2015). The possibility to go beyond the diffraction limit with this algorithmic approach was demonstrated. Indeed, the spatial resolution was improved by 65% to achieve 200 nm (size of the pixel on the Raman image) for the considered far-field spectrophotometer.

6.5.2. Enhanced Raman Spectroscopies: Principle and Application in Organic Particle Analysis

It is now well recognized that submicron aerosol particles (mainly organic enriched) have a major impact on human health and climate change. There is a need to understand the surface chemistry of such particles, as this determines the physical and chemical properties of the aerosol in the atmosphere. Although abundant inorganic compounds are often easily detected, organic species in lower concentrations mostly remain undetectable. Despite many advantages, Raman spectroscopy presents two severe drawbacks to the study of atmospheric aerosols. The insufficient detection limits of this technique for both particle size and analyte concentrations may indeed impede the characterization of submicronic atmospherically relevant aerosol particles and the detailed description of the composition of larger particles. Furthermore, the lateral resolution is limited by diffraction to a few hundreds of nanometers, which hinders any precise nanoscale compositional characterization of aerosol particles as a function of the position of the laser spot inside particles. Note that it is not impossible to detect the Raman signature of aerosol nanoparticles. However, all their molecular compounds contribute to the Raman signal if the nanoparticles are smaller than the laser spot size.

Surface-enhanced Raman spectroscopy (SERS) has been developed for 40 years to improve the molecular sensitivity of Raman techniques. The SERS technique is based on the laser light excitation of the localized surface plasmon resonance (LSPR) of metal nanoparticles at a nanoscale distance (less than 10 nm typically) from the analyte. Localized surface plasmons are quantum quasi-particles associated with elementary excitations of transverse charge density in the plasma at the interface between the metal nanoparticle and the surrounding dielectric medium. Their resonant light excitation leads to the strong electromagnetic enhancement (up to 10^{10}) and the nanoscale confinement of the Raman signal. The LSPR strongly depends on the nanoparticle size, shape, and composition (gold or silver in general) and on the dielectric properties of the surrounding medium (Schlücker, 2014; Sharma et al., 2012). SERS is thus highly sensitive to experimental conditions. SERS and spontaneous Raman selection rules are not the same, which may lead to different intensities for similar vibrational modes. Nevertheless, SERS and spontaneous Raman wavenumbers are often comparable unless charge transfer between the nanoparticle metal and analyte molecular electronic states occurs (Morton et al., 2011). In this latter case, the SERS signal is not only electromagnetically but also chemically enhanced, and band shifts can be observed. Thanks to these electromagnetic and chemical enhancements, SERS allows the analyte detection limit to be significantly improved with respect to conventional RMS and information associated with nanoscale sample areas to be collected.

SERS was applied by several authors to determine the composition of the model and atmospheric aerosol particles. Their strategies mainly differed from each other by the type of SERS-active platforms used to enhance the Raman scattered signal: silver colloids deposited on filter paper (Ayora et al., 1997), silver nanoparticle-coated quartz substrates (Craig et al., 2015; Tirella et al., 2018) or Si wafer (Yoo et al., 2022), Klarite commercial nanostructured gold substrates (Fu et al., 2017), electrospray-deposited silver nanoparticles (Gen et al., 2019; Gen & Chan, 2017), commercial silver-coated silicon nanopillars (Kunihisa et al., 2020), Cu/Ag sphere segment void arrays (Dong et al., 2019), Ag nanocube-based 3D metal-organic frameworks (Phan-Quang et al., 2019), silver foils (Chen et al., 2021; Sun et al., 2019a; Sun et al., 2019b) and aerosolized plasmonic silver colloidosomes (Phan-Quang et al., 2018). These investigations allowed the identification of inorganic compounds, such as ammonium and sodium sulfates/nitrates, black carbon, silicates and aluminosilicates, and organic contaminants, such as 2,4-dinitrophenol, 4-nitrophenol, 9-aminoacridine, acridine, BrC coatings, naphthalene, quinacrine, and toluene (Table 6.4). In addition, many SERS bands in the 1,000–1,800 cm^{-1} spectral range could be assigned to C—H, C—C, C—O, O—H, C=C, and C=O groups (among others), and some of them revealed the presence of organonitrates and carboxylic acids (Craig et al., 2015; Gen & Chan, 2017; Tirella et al., 2018; Wei et al., 2018). Recent advances in SERS allowed surface-adsorbed water to be probed on ammonium sulfate particles (Gen & Chan, 2017) and stable pH gradients to be demonstrated in aerosol microdroplets (Wei et al., 2018). A novel approach combining dark-field microscopy and SERS was developed for the detection of individual submicron-size particles (Yoo et al., 2022) and validated on model particles. However, no real particle of size less than 150 nm was examined using SERS (Tirella et al., 2018). The random positioning of SERS-active metal nanoparticles on (or below) aerosol nanoparticles renders their full chemical characterization with sufficient lateral resolution difficult.

To improve not only the molecular sensitivity but also the lateral optical resolution of far-field RMS, tip-enhanced Raman spectroscopy (TERS) has been proposed (Bonhommeau et al., 2022; Schmid et al., 2013; Verma, 2017). TERS is a near-field optical microscopy technique that benefits from both the high (nanoscale) spatial resolution of scanning probe microscopy (SPM) and the high chemical specificity and sensitivity of SERS. The high lateral resolution of SPM techniques, such as atomic force microscopy (AFM) and scanning tunneling microscopy (STM) (Hamers, 1996; Poggi et al., 2002), is then directly related to the morphology of the apex of nanostructured or tapered SPM probes, the so-called TERS tips (Bonhommeau et al., 2022; Verma, 2017). As in SERS, the enhancement in TERS originates from two contributions: one electromagnetic and the other chemical (Bonhommeau et al., 2022; Shi et al., 2017). The electromagnetic enhancement can be decomposed into three main effects. The plasmonic effect that requires the light excitation of the LSPR of metal TERS tips is the prevalent one. The lightning-rod effect is a nonresonant phenomenon originating from the accumulation of surface charges at the end of TERS tips. Finally, the resonant antenna effect adds another dependency on the geometry and dielectric properties of metal tips. As in SERS, the chemical enhancement originates from charge transfer between the metal and molecular electronic states.

TERS experiments can be performed using TERS instruments that differ from each other by the irradiation conditions of the TERS tip (Bonhommeau et al., 2022; Verma, 2017). In a bottom-illumination configuration, an SPM is coupled to an inverted optical microscope. The TERS tip is then perfectly aligned with the linearly or radially polarized laser beam that is tightly focused

Table 6.4 SERS Wavenumbers of Inorganic and Organic Compounds of Atmospheric Aerosols

Inorganic compounds	SERS wavenumbers (cm^{-1})	References
Na_2SO_4	454, 621, 985, 1061	Dong et al. (2019)
$(NH_4)_2SO_4$	440–450, 520, 605–633, 963–988, 1,100, 1,461, 1,700, 3,130–3,153	Fu et al. (2017), Tirella et al. (2018), Gen & Chan (2017), Sun et al. (2019b), Dong et al. (2019), Kunihisa et al. (2020)
$(NH_3CH_3)_2SO_4$	970, 1,000, 1,086, 1,109	Gen et al. (2019)
$(NH_3CH_3)Cl$	1,000, 1,020, 1,086, 1,109	Gen et al. (2019)
$NaNO_3$	717–725, 1,051–1,054, 1,067, 1,386	Tirella et al. (2018), Dong et al. (2019)
NH_4NO_3	614, 979, 1,045, 1,051	Sun et al. (2019a), (2019b)
$(NH_3CH_3)NO_3$	1,000, 1,052	Gen et al. (2019)
Black carbon	1,341–1,350, 1,589–1,598	Gen & Chan (2017), Sun et al. (2019a, 2019b)
Silicates/aluminosilicates	500–900	Tirella et al. (2018), Craig et al. (2015)
Organic compounds		
2,4-dinitrophenol	1,119, 1,131, 1,271, 1,335	Ayora et al. (1997)
4-nitrophenol	1,135, 1,165, 1,259, 1,326, 1,371	Ayora et al. (1997)
9-aminoacridine	1,263, 1,375, 1,401, 1,456	Ayora et al. (1997)
Acridine	1,135, 1,173, 1,266, 1,382, 1,408, 1,475	Ayora et al. (1997)
BrC coatings	1,400, 1,440, 3,150	Gen et al. (2019)
Levoglucosan	1,600, 2,950	Kunihisa et al. (2020)
Naphtalene	510, 759–763, 815, 1020, 1,054, 1,302, 1,378–1,380, 1,576, 1,684	Fu et al. (2017), Phan-Quang et al. (2018, 2019)
Quinacrine	1,256, 1,368, 1,427, 1,482	Ayora et al. (1997)
Toluene	786–787, 1,004–1,005, 1,084, 1,212, 1,603	Phan-Quang et al. (2018), (2019)

through a transparent substrate using a high numerical aperture objective ($NA \geq 1.4$). Recently, a promising bottom-illumination configuration in total internal reflection has also been reported (Talaga et al., 2020). This geometry would be particularly interesting to locate aerosol nanoparticles using dark-field Rayleigh scattering microscopy prior to their TERS characterization (Talaga et al., 2015). To study optically opaque samples, a side- or top-illumination configuration must be preferred. In side illumination, a long-working-distance objective ($NA < 0.6$) focuses the linearly polarized laser light on the sample. The collection efficiency is thus lower, and the asymmetric spot size is larger (2–3 µm) than their counterparts in bottom-illumination configuration, but larger laser powers can be employed. Top illumination enables symmetric focusing and can be achieved by means of a parabolic mirror ($NA = 1$) or a microscope objective ($NA = 0.7$) to minimize the "shadowing effect" of the TERS tip that reduces the excitation and collection efficiency. More experimental details on optical geometries in TERS instruments can be found in the literature (Stadler et al., 2012).

In TERS, the enhancement of the electromagnetic field is directly linked to the characteristics of the TERS tip. Several preparation methods for such tips have been proposed, depending on the type of TERS instrument (Verma, 2017). For AFM-based TERS systems that are particularly relevant for the study of aerosol particles (Ofner et al., 2016; Lipiec et al., 2014; Lipiec et al., 2018; Lipiec et al., 2021; Cooney et al., 2023; Talaga et al., 2020), the traditional preparation method is the noble metal (mostly gold or silver) deposition on insulating AFM probes by means of thermal evaporation or sputtering. To tune the LSPR of TERS tips, to protect them from potential oxidation/sulfuration in air, and to improve the metal adhesion on the AFM probe, ultrathin metal oxide capping layers (alumina or silicon oxides) and insulating/metal buffer layers can be added. This cost-effective preparation method provides, however, very variable performance regarding the fabrication yield, reproducibility, and durability of TERS tips. Therefore, other manufacturing procedures based on electro-/electroless and colloidal depositions, the tip decoration with metal nanoparticles, and microfabrication methods (e.g., focused ion beam milling and photolithography) have also been developed to obtain high enhancement factors (up to 10^9) and spatial resolutions (down to 7 nm) and improve the aforementioned performance.

The TERS enhancement factor (EF) can be calculated from experimental values using Equation 6.23 (Schmid et al., 2013):

$$EF = \frac{I_{NF} \cdot V_{FF}}{I_{FF} \cdot V_{NF}} = \gamma \frac{V_{FF}}{V_{NF}} \text{ with } \gamma = \frac{I_{NF}}{I_{FF}} = \frac{I_{tip-in} - I_{tip-out}}{I_{tip-out}} \quad (6.23)$$

where I_{NF} (I_{FF}) is the near-field (far-field) intensity of a selected TERS band, V_{NF} (V_{FF}) is the near-field (far-field) analyzed volume ($V_{FF} \approx V_{conf}$), γ is the optical contrast, and $I_{tip\text{-}in}$ ($I_{tip\text{-}out}$) is the Raman intensity collected when the TERS tip is in (out of) contact with the sample. The optical contrast is a purely experimental value independent of the estimation of the probed volume. As explained in Section 6.2, V_{FF} can be estimated from the values of radial and axial waists of the laser illumination profile. In contrast, V_{NF} identifies with the excited volume. Therefore, it can be simply assessed from the lateral resolution deduced by measuring the TERS intensity variation of a sample marker band across a sample edge. A conservative value of 10 nm is generally determined, but sub-nanometer resolutions have been reported (Zhang et al., 2013; Zhang et al., 2017). V_{NF} thus typically ranges between 1 and 10^3 nm^3.

As in SERS, the TERS EF can be different for distinct vibrational modes. Therefore, its experimental assessment from a specific TERS band provides an order of magnitude rather than a highly accurate value. Furthermore, TERS intensities are subject to fluctuations owing to the high sensitivity of the TERS signal relative to the confined excited/probed volume, to sample compositional heterogeneity, and to molecular orientation (Bonhommeau et al., 2022; Cooney et al., 2023; Shi et al., 2017; Verma, 2017), as reported for aerosol nanoparticles (Ofner et al., 2016). This leads to additional difficulties for TERS imaging purposes even though multivariate analysis (PCA; Lipiec et al., 2014), hierarchical cluster analysis (Lipiec et al., 2018, etc.) methods can sometimes be used to process TERS data in an efficient and reliable way.

To date, a single TERS study has focused on the characterization of atmospherically relevant aerosol nanoparticles (Ofner et al., 2016). Nanoparticles of a few tens of nanometers were probed. TERS results revealed the presence of inorganic seed particles, variability in the chemical composition of equally sized particles in the nucleation mode, and the occurrence of chemical transformations during photochemical aging. In particular, spectral markers assigned to many organic constituents (organic salts, sulfates, nitrates, nitro compounds, carboxylic acid salts, sulfones, esters, and carboxylates) were identified, showing that TERS can be employed to describe the organic content of aerosol particles that is largely masked by the contribution of inorganic species in conventional Raman spectroscopy. Owing to the surface sensitivity of TERS, this technique is thus promising to analyze the compositional heterogeneity of atmospheric particles in the three space dimensions. Finally, considering recent advances in SERS (Gen et al., 2019; Gen & Chan, 2017; Wei et al., 2018), TERS might also help precise surface chemical reactions in the presence of liquid water inside aerosol particles, as well as the eventual role of relative humidity on the particle composition.

6.6. COUPLING WITH OTHER SINGLE-PARTICLE ANALYSIS TECHNIQUES

Besides the identification of species, the RMS analytical method suffers from several limitations, such as difficulty in detecting compounds with low Raman cross-sections and/or present in insufficient amounts in the sample. For a complete characterization of aerosol samples, the combination of various SPA techniques has been employed (Ault & Axson, 2017; Laskin et al., 2019; Tomlin et al., 2020). Coupled analyses using electron microscopy and RMS techniques carried out successively on the same sample are the most common methodology. This provides chemical analysis (i.e., elemental and molecular) and morphological characterization of particles from the same samples. The combination of colocalized imaging techniques provides more details on single particle heterogeneity than uncoupled techniques (Sobanska et al., 2012; Sobanska et al., 2014). In the same way, complementary to Raman spectroscopy, IR spectroscopy is suitable for probing functional groups in individual particles (Eom et al., 2013; Eom et al., 2016; Jung et al., 2014; Malek et al., 2011, 2019; Song et al., 2010, 2013; Wu et al., 2019). However, IR radiation limits the analysis to particles larger than 10 μm. New developments in nano-IR-based spectroscopy allowed for high lateral resolution better than the ~10 μm diffraction-limited value typically observed with far-field IR microspectroscopy (Bondy et al., 2017; Or et al., 2018). Olson et al. (2020) demonstrated that simultaneous nano-IR (Optical photothermal IR [O-PTIR]) and Raman spectroscopy analysis is a promising coupled SPA technique for the characterization and imaging of submicrometer particles. A recent application combining RMS, O-PTIR, and SEM-EDX for the characterization of sea-salt-like aerosols enabled detailed identification of organic substances in the particles that could not be detected using RMS alone (Mirrielees et al., 2022).

Albeit very useful for the full description of individual particles, the use of combined SPA techniques is a time-consuming and laborious procedure that cannot be applied to the analysis of a large number of samples.

GLOSSARY

AFM	Atomic force microscopy
$\overline{\overline{\alpha}}$	Molecular polarizability tensor ($C\,m^2\,V^{-1}$)
$\overline{\overline{\alpha}}'$	Differential polarizability tensor
c	Light velocity (m/s)
CCD	Charge-coupled device
Chemometrics	Scientific discipline using statistical tools for spectral data processing
D	Data cube composed of ($m \times n \times \lambda$). Raman scanning acquisition generates a three-dimensional data set ($m \times n \times \lambda$): that is, $m \times n$ pixel spectra and 2,040 spectral points.
$\vec{E}$	Oscillating electric field (V/m)
FTIR	Fourier Transformed Infrared Spectroscopy
Fwhm	Full width at half maximum for a given Raman band
H	Planck constant 6.626×10^{-34} J.s
HCA	Hierarchical component analysis
Hygroscopicity	Ability of a particle to absorb or release water
k	Force constant of the harmonic oscillator (N/m)
λ_0	Excitation laser wavelength (commonly given in nm)
LSPR	Localized surface plasmon resonance
m	Reduced mass (kg)
MCR	Multivariate curve resolution
NA	Numeral aperture of the objective lens
O-PTIR	Optical photothermal infrared
$\bar{v}$	Vibrational wavenumber (cm^{-1})
ν	Vibrational frequency (Hz)
$\vec{P}$	Induced dipole moment (C m)
PCA	Principal component analysis
Polarizability	Ability of the electronic cloud of a molecule to be deformed
Rayleigh scattering	Elastic light scattering
RH	Relative humidity; expressed as a percentage. Indicates a ratio between the partial pressure of water (P_{vap}) in ambient air and the value of saturated pressure (P_{sat}) for a given temperature (T).
RMS	Raman microspectroscopy
RRSSQ	Relative square root of the sum of square differences given by

$$RRSSQ = \sqrt{\frac{\sum_{i=1}^{nspec}\sum_{j=1}^{nvar}\left(d_{ij} - d_{ij}^{\text{recons}}\right)^2}{\sum_{i=1}^{nspec}\sum_{j=1}^{nvar} d_{ij}^2}} \tag{6.22}$$

where $d_{i,j}$ is the ith row and jth column element of D, the ith row and jth column element of $D^{\text{reconstructed}}$, $nspec$ is the number of mixture spectra, and $nvar$ is the number of recorded intensities

σ_{Raman}	Raman scattering cross-section (cm^2/sr)
SEM	Scanning electron microscopy
SEM-EDX	Scanning electron microscopy equipped with an energy-dispersive X-ray spectrometer
SERS	Surface-enhanced Raman spectroscopy
Simplisma	Simple-to-use interactive self-modeling mixture analysis method
SPA	single particle analysis, including all techniques based on analysis at the micrometer scale of unique particles. This includes mainly electron microscopy-based techniques, AFM, RMS, and FTIR (see review from Ault & Axson, 2017).
SPM	Scanning probe microscopy
STM	Scanning tunneling microscopy
TEM	Transmission electron microscopy
TERS	Tip-enhanced Raman spectroscopy

REFERENCES

An, T., Liang, Z., Chen, Z., & Li, G. (2023). Recent progress in online detection methods of bioaerosols. *Fundamental Research*. in press.

Ault, A. P. (2020). Aerosol acidity: Novel measurements and implications for atmospheric chemistry. *Accounts of Chemical Research*, *53*(9), 1703–1714.

Ault, A. P., & Axson, J. L. (2017). Atmospheric aerosol chemistry: Spectroscopic and microscopic advances. *Analytical Chemistry*, *89*, 430–452.

Ault, A. P., Guasco, T. L., Ryder, O. S., Baltrusaitis, J., Cuadra-Rodriguez, L. A., Collins, D. B., et al. (2013). Inside versus outside: Ion redistribution in nitric acid Reacted Sea spray aerosol particles as determined by single particle analysis. *Journal of the American Chemical Society*, *135*, 14528–14531.

Ault, A. P., Guasco, T. L., Baltrusaitis, J., Ryder, O. S., Trueblood, J. V., Collins, D. B., et al. (2014). Heterogeneous reactivity of nitric acid with Nascent Sea spray aerosol: Large differences observed between and within individual particles. *The Journal of Physical Chemistry Letters*, *5*, 2493–2500.

Ayora, M. J., Ballesteros, L., Pérez, R., Rupérez, A., & Laserna, J. J. (1997). Detection of atmospheric contaminants in aerosols by surface-enhanced Raman spectrometry. *Analytica Chimica Acta*, *355*, 15–21.

Batonneau, Y., Bremard, C., Laureyns, J., Merlin, J. C., & Windig, W. (2003). Polarization effects of confocal Raman microspectrometry of crystal powders using interactive self-modeling analysis. *The Journal of Physical Chemistry B*, *107*, 1502–1513.

Batonneau, Y., Bremard, C., Gengembre, L., Laureyns, J., Le Maguer, A., Le Niaguer, D., et al. (2004). Speciation of PM10 sources of airborne nonferrous metals within the 3-km zone of lead/zinc smelters. *Environmental Science and Technology*, *38*, 5281–5289.

Batonneau, Y., Sobanska, S., Laureyns, J., & Bremard, C. (2006). Confocal microprobe Raman imaging of urban tropospheric aerosol particles. *Environmental Science and Technology*, *40*, 1300–1306.

Bocklitz, T., Walter, A., Hartmann, K., Rösch, P., & Popp, J. (2011). How to pre-process Raman spectra for reliable and stable models? *Analytica Chimica Acta*, *704*, 47–56.

Bondy, A. L., Kirpes, R. M., Merzel, R. L., Pratt, K. A., Banaszak Holl, M. M., & Ault, A. P. (2017). Atomic force microscopy-infrared spectroscopy of individual atmospheric aerosol particles: Subdiffraction limit vibrational spectroscopy and morphological analysis. *Analytical Chemistry*, *89*, 8594–8598.

Bondy, A. L., Craig, R. L., Zhang, Z., Gold, A., Surratt, J. D., & Ault, A. P. (2018). Isoprene-derived Organosulfates: Vibrational mode analysis by Raman spectroscopy, acidity-dependent spectral modes, and observation in individual atmospheric particles. *The Journal of Physical Chemistry A*, *122*, 303–315.

Bonhommeau, S., Cooney, G. S., & Huang, Y. (2022). Nanoscale chemical characterization of biomolecules using tip-enhanced Raman spectroscopy. *Chemical Society Reviews*, *51*, 2416–2430.

Carey, C., Boucher, T., Mahadevan, S., Bartholomew, P., & Dyar, M. D. (2015). Machine learning tools formineral recognition and classification from Raman spectroscopy. *Journal of Raman Spectroscopy*, *46*, 894–903.

Chang, P., Chen, Z., Zhang, Y., & Liu, Y. (2020). Direct measurement of aerosol pH in individual malonic acid and citric acid droplets under different relative humidity conditions via Raman spectroscopy. *Chemosphere*, *241*, 124960–124969.

Chen, H., Duan, F., Du, J., Yin, R., Zhu, L., Dong, J., et al. (2021). Surface-enhanced Raman scattering for mixing state characterization of individual fine particles during a haze episode in Beijing, China. *Journal of Environmental Sciences*, *104*, 216–224.

Cooney, G. S., Talaga, D., Ury-Thiery, V., Fichou, Y., Huang, Y., Lecomte, S., & Bonhommeau, S. (2023). Chemical imaging of RNA-tau amyloid fibrils at the nanoscale using tip-enhanced Raman spectroscopy. *Angewandte Chemie International Edition*, *62*, e202314369.

Craig, R. L., Bondy, A. L., & Ault, A. P. (2015). Surface enhanced Raman spectroscopy enables observations of previously undetectable secondary organic aerosol components at the individual particle level. *Analytical Chemistry*, *87*, 7510–7514.

Craig, R. L., Bondy, A. L., & Ault, A. P. (2017). Computer-controlled Raman microspectroscopy (CC-Raman): A method for the rapid characterization of individual atmospheric aerosol particles. *Aerosol Science and Technology*, *51*(9), 1099–1112.

Delhaye, M., & Dhamelincourt, P. (1975). Raman microprobe and microscope with laser excitation. *Journal of Raman Spectroscopy*, *3*, 33–43.

Delhaye, M., Dhamelincourt, P., & Wallart, F. (1979). Analysis of particulates by Raman microprobe. *Toxicological & Environmental Chemistry Reviews*, *3*, 73–87.

DeMott, P. J., Hill, T. C. J., Moore, K. A., Perkins, R. J., Mael, L. E., Busse, H. L., et al. (2023). Atmospheric oxidation impact on sea spray produced ice nucleating particles. *Environmental Science: Atmospheres*, *3*(10), 1513–1532.

Dong, X., Ohnoutek, L., Yang, Y., Feng, Y., Wang, T., Tahir, M. A., et al. (2019). Cu/Ag sphere segment void Array as efficient surface enhanced Raman spectroscopy substrate for detecting individual atmospheric aerosol. *Analytical Chemistry*, *91*, 13647–13657.

Doughty, D. C., & Hill, S. C. (2017). Automated aerosol Raman spectrometer for semi-continuous sampling of atmospheric aerosol. *Journal of Quantitative Spectroscopy and Radiative Transfer*, *188*, 103–117.

Doughty, D. C., & Hill, S. C. (2020a). Raman spectra of atmospheric aerosol particles: Clusters and time-series for a 22.5hr sampling period. *Journal of Quantitative Spectroscopy and Radiative Transfer*, *248*, 106907–106921.

Doughty, D. C., & Hill, S. C. (2020b). Raman spectra of atmospheric particles measured in Maryland, USA over 22.5 h using an automated aerosol Raman spectrometer. *Journal of Quantitative Spectroscopy and Radiative Transfer*, *244*, 106839.

Dupuy, N., & Batonneau, Y. (2003). Reliability of the contribution profiles obtained through the SIMPLISMA approach and used as reference in a calibration process—application to Raman micro-analysis of dust particles. *Analytica Chimica Acta*, *495*, 205–215.

Eilers, P. H. C. & Boelens, H. F. M. (2005). Baseline correction with asymmetric least squares smoothing (report). Leiden University Medical Centre.

Eom, H.-J., Jung, H.-J., Sobanska, S., Chung, S.-G., Son, Y.-S., Kim, J.-C., et al. (2013). Iron speciation of airborne Subway particles by the combined use of energy dispersive Electron probe X-ray microanalysis and Raman microspectrometry. *Analytical Chemistry*, *85*, 10424–10431.

Eom, H. J., Gupta, D., Cho, H. R., Hwang, H. J., Hur, S. D., Gim, Y., & Ro, C. U. (2016). Single-particle investigation of summertime and wintertime Antarctic Sea spray aerosols using low-Z particle EPMA, Raman microspectrometry, and ATR-FTIR imaging techniques. *Atmospheric Chemistry and Physics*, *16*, 13823–13836.

Everall, N. J. (2010). Confocal Raman microscopy: Common errors and artefacts. *Analyst*, *135*, 2512–2522.

Everitt, B., & Hothorn, T. (2011). *An introduction to applied multivariate analysis with R*. Springer Science & Business Media.

Falgayrac, G., Sobanska, S., Laureyns, J., & Bremard, C. (2006). Heterogeneous chemistry between $PbSO_4$ and calcite microparticles using Raman microimaging. *Spectrochimica Acta Part a-Molecular and Biomolecular Spectroscopy*, *64*, 1095–1101.

Falgayrac, G., Sobanska, S., & Brémard, C. (2013). Heterogeneous microchemistry between $CdSO_4$ and $CaCO_3$ particles under humidity and liquid water. *Journal of Hazardeous Material*, *248–249*, 415–423.

Falgayrac, G., Sobanska, S., & Brémard, C. (2014). Raman diagnostic of the reactivity between $ZnSO_4$ and $CaCO_3$ particles in humid air relevant to heterogeneous zinc chemistry in atmosphere. *Atmospheric Environment*, *85*, 83–91.

Falgayrac, G., Siepka, D., Stefaniak, E. A., Penel, G., & Sobanska, S. (2018). Influence of collecting substrate on the Raman imaging of micron-sized particles. *Analytica Chimica Acta*, *1014*, 41–49.

Fu, Y., Kuppe, C., Valev, V. K., Fu, H., Zhang, L., & Chen, J. (2017). Surface-enhanced Raman spectroscopy: A facile and rapid method for the chemical component study of individual atmospheric aerosol. *Environmental Science & Technology*, *51*, 6260–6267.

Gautam, R., Vanga, S., Ariese, F., & Umapathy, S. (2015). Review of multidimensional data processing approaches for Raman and infrared spectroscopy. *EPJ Techniques and Instrumentation*, *2*, 8.

Gen, M., & Chan, C. K. (2017). Electrospray surface-enhanced Raman spectroscopy (ES-SERS) for probing surface chemical compositions of atmospherically relevant particles. *Atmospheric Chemistry and Physics*, *17*, 14025–14037.

Gen, M., Kunihisa, R., Matsuki, A., & Chan, C. K. (2019). Electrospray surface-enhanced Raman spectroscopy (ES-SERS) for studying organic coatings of atmospheric aerosol particles. *Aerosol Science and Technology*, *53*, 760–770.

Godoi, R. H. M., Potgieter-Vermaak, S., De Hoog, J., Kaegi, R., & Van Grieken, R. (2006). Substrate selection for optimum qualitative and quantitative single atmospheric particles analysis using nano-manipulation, sequential thin-window electron probe X-ray microanalysis and micro-Raman spectrometry. *Spectrochimica Acta Part B: Atomic Spectroscopy*, *61*, 375–388.

Guilment, J., Markel, S., & Windig, W. (1994). Infrared chemical Micro-imaging assisted by interactive self-modeling multivariate analysis. *Applied Spectroscopy*, *48*, 320–326.

Hamers, R. J. (1996). Scanned probes microscopics in chemistry. *The Journal of Physical Chemistry*, *100*, 13103–13120.

Hamilton, J. C., & Gemperline, P. J. (1990). Mixture analysis using factor analysis. II: Self-modeling curve resolution. *Journal of Chemometrics*, *4*, 1–13.

He, S., Zhang, W., Liu, L., Huang, Y., He, J., Xie, W., et al. (2014). Baseline correction for Raman spectra using an improved asymmetric least squares method. *Analytical Methods*, *6*, 4402–4407.

Herman, R. G., Bogdan, C. E., Sommer, A. J., & Simpson, D. R. (1987). Discrimination among carbonate minerals by Raman spectroscopy using the laser microprobe. *Applied Spectroscopy*, *41*, 437–440.

Huang, Z., Yu, X., Liu, Q., Maki, T., Alam, K., Wang, Y., et al. (2024). Bioaerosols in the atmosphere: A comprehensive review on detection methods, concentration and influencing factors. *Science of the Total Environment*, *912*, 168818.

Inoué, S. (2006). Foundations of confocal scanned imaging in light microscopy. In J. B. Pawley (Ed.), *Handbook of biological confocal microscopy* (3rd ed.). Springer.

Ishizaka, S., Guo, F., Tian, X., Seng, S., Tobon, Y. A., & Sobanska, S. (2020). In situ observation of efflorescence and deliquescence phase transitions of single NaCl and $NaNO_3$ mixture particles in air using a laser trapping technique. *Bulletin of the Chemical Society of Japan*, *93*, 86–91.

Ivleva, N. P., Huckele, S., Weinzierl, B., Niessner, R., Haisch, C., & Baumann, T. (2013). Identification and characterization of individual airborne volcanic ash particles by Raman microspectroscopy. *Analytical and Bioanalytical Chemistry*, *405*, 9071–9084.

Ivleva, N. P., McKeon, U., Niessner, R., & Pöschl, U. (2007). Raman microspectroscopic analysis of size-resolved atmospheric aerosol particle samples collected with an ELPI: Soot, humic-like substances, and inorganic compounds. *Aerosol Science and Technology*, *41*, 655–671.

Jambers, W., Bock, L. D., & Grieken, R. V. (1995). Recent advances in the analysis of individual environmental particles: A review. *Analyst*, *120*, 681–692.

Jentzsch, P. V., Ciobotă, V., Kampe, B., Rösch, P., & Popp, J. (2012). Origin of salt mixtures and mixed salts in atmospheric particulate matter. *Journal of Raman Spectroscopy*, *43*, 514–519.

Jentzsch, P. V., Kampe, B., Ciobota, V., Rosch, P., & Popp, J. (2013). Inorganic salts in atmospheric particulate matter: Raman spectroscopy as an analytical tool. *Spectrochimica Acta Part a-Molecular and Biomolecular Spectroscopy*, *115*, 697–708.

Jung, H.-J., Eom, H.-J., Kang, H.-W., Moreau, M., Sobanska, S., & Ro, C.-U. (2014). Combined use of quantitative ED-EPMA, Raman microspectrometry, and ATR-FTIR imaging techniques for the analysis of individual particles. *Analyst*, *139*, 3949–3960.

Kunihisa, R., Iwata, A., Gen, M., Chan, C. K., & Matsuki, A. (2020). Application of SERS on the chemical speciation of individual Aitken mode particles after condensational growth. *Aerosol Science and Technology*, *54*, 826–836.

Kwiatkowski, A., Gnyba, M., Smulko, J., & Wierzba, P. (2010). Algorithms of chemicals detection using Raman spectra. *Metrology and Measurements Systems*, *4*, 549–560.

Laskin, A., Moffet, R. C., & Gilles, M. K. (2019). Chemical imaging of atmospheric particles. *Accounts of Chemical Research*, *52*, 3419–3431.

Laskina, O., Morris, H. S., Grandquist, J. R., Qin, Z., Stone, E. A., Tivanski, A. V., & Grassian, V. H. (2015). Size matters in the water uptake and hygroscopic growth of atmospherically relevant multicomponent aerosol particles. *The Journal of Physical Chemistry A*, *119*, 4489–4449.

Lee, H. D., Wigley, S., Lee, C., Or, V. W., Hasenecz, E. S., Stone, E. A., et al. (2020). Physicochemical mixing state of sea spray aerosols: Morphologies exhibit size dependence. *ACS Earth and Space Chemistry*, *4*, 1604–1611.

Liang, Z., Chu, Y., Gen, M., & Chan, C. K. (2022). Single-particle Raman spectroscopy for studying physical and chemical processes of atmospheric particles. *Atmospheric Chemistry and Physics*, *22*, 3017–3044. doi: 10.5194/acp-22-3017-2022

Lipiec, E., Kaderli, J., Kobierski, J., Riek, R., Skirlińska—Nosek, K., Sofińska, K., et al. (2021). Nanoscale hyperspectral imaging of amyloid secondary structures in liquid. *Angewandte Chemie International Edition*, *60*, 4545–4550.

Lipiec, E., Perez-Guaita, D., Kaderli, J., Wood, B. R., & Zenobi, R. (2018). Direct nanospectroscopic verification of the amyloid aggregation pathway. *Angewandte Chemie*, *130*, 8655–8660.

Lipiec, E., Sekine, R., Bielecki, J., Kwiatek, W. M., & Wood, B. R. (2014). Molecular characterization of DNA double strand breaks with tip-enhanced Raman scattering. *Angewandte Chemie International Edition*, *53*, 169–172.

Mael, L. E., Busse, H., & Grassian, V. H. (2019). Measurements of immersion freezing and heterogeneous chemistry of atmospherically relevant single particles with Micro-Raman spectroscopy. *Analytical Chemistry*, *91*, 11138–11145.

Malek, M. A., Kim, B., Jung, H.-J., Song, Y.-C., & Ro, C.-U. (2011). Single-particle mineralogy of Chinese soil particles by the combined use of low-Z particle Electron probe X-ray microanalysis and attenuated Total reflectance-FT-IR imaging techniques. *Analytical Chemistry*, *83*, 7970–7977.

Malek, M. A., Nakazawa, T., Kang, H.-W., Tsuji, K., & Ro, C.-U. (2019). Multi-modal compositional analysis of layered paint chips of automobiles by the combined application of ATR-FTIR imaging, Raman microspectrometry, and SEM/EDX. *Molecules*, *2019*(24), 1381–1397.

Mertz, J. (1998). Molecular photodynamics involved in multi-photon excitation fluorescence microscopy. *The European Physical Journal D*, *3*, 53–66.

Mirrielees, J. A., Kirpes, R. M., Haas, S. M., Rauschenberg, C. D., Matrai, P. A., Remenapp, A., et al. (2022). Probing individual particles generated at the freshwater–seawater Interface through combined Raman, Photothermal infrared, and X-ray spectroscopic characterization. *ACS Measurement Science Au*, *2*(6), 605–619.

Moffet, R. C., Furutani, H., Rödel, T. C., Henn, T. R., Sprau, P. O., Laskin, A., et al. (2012). Iron speciation and mixing in single aerosol particles from the Asian continental outflow. *Journal of Geophysical Research*, *117*, D07204–D07215. doi: 10.1029/2011JD016746

Morton, S. M., Silverstein, D. W., & Jensen, L. (2011). Theoretical studies of plasmonics using electronic structure methods. *Chemical Reviews*, *111*, 3962–3994.

Offroy, M., Roggo, Y., Milanfar, P., & Duponchel, L. (2010). Infrared chemical imaging: Spatial resolution evaluation and super-resolution concept. *Analytica Chimica Acta*, *674*, 220–226.

Offroy, M., Moreau, M., Sobanska, S., Milanfar, P., & Duponchel, L. (2015). Pushing back the limits of Raman imaging by coupling super-resolution and chemometrics for aerosols characterization. *Scientific Reports*, *5*, 12303.

Ofner, J., Kamilli, K. A., Eitenberger, E., Friedbacher, G., Lendl, B., Held, A., & Lohninger, H. (2015). Chemometric analysis of multisensor Hyperspectral images of precipitated atmospheric particulate matter. *Analytical Chemistry*, *87*, 9413–9420.

Ofner, J., Deckert-Gaudig, T., Kamilli, K. A., Held, A., Lohninger, H., Deckert, V., & Lendl, B. (2016). Tip-enhanced Raman spectroscopy of atmospherically relevant aerosol nanoparticles. *Analytical Chemistry*, *88*, 9766–9772.

Olson, N. E., Xiao, Y., Lei, Z., & Ault, A. P. (2020). Simultaneous optical Photothermal infrared (O-PTIR) and Raman spectroscopy of submicrometer atmospheric particles. *Analytical Chemistry*, *92*, 9932–9939.

Or, V. W., Estillore, A. D., Tivanski, A. V., & Grassian, V. H. (2018). Lab on a tip: Atomic force microscopy—photothermal infrared spectroscopy of atmospherically relevant organic/inorganic aerosol particles in the nanometer to micrometer size range. *Analyst*, *143*, 2765–2774.

Osan, J., de Hoog, J., Worobiec, A., Ro, C. U., Oh, K. Y., Szaloki, I., & Van Grieken, R. (2001). Application of chemometric methods for classification of atmospheric particles based on thin-window electron probe microanalysis data. *Analytica Chimica Acta*, *2001*(446), 211–222.

Petuya, C., Damay, F., Chazallon, B., Bruneel, J. L., & Desmedt, A. (2018). Guest partitioning and Metastability of the nitrogen gas hydrate. *The Journal of Physical Chemistry C*, *122*(1), 566–573.

Phan-Quang, G. C., Lee, H. K., Teng, H. W., Koh, C. S. L., Yim, B. Q., Tan, E. K. M., et al. (2018). Plasmonic hotspots in air: An omnidirectional three-dimensional platform for stand-off in-air SERS sensing of airborne species. *Angewandte Chemie International Edition*, *57*, 5792–5796.

Phan-Quang, G. C., Yang, N., Lee, H. K., Sim, H. Y. F., Koh, C. S. L., Kao, Y.-C., et al. (2019). Tracking airborne molecules from Afar: Three-dimensional metal–organic framework-surface-enhanced Raman scattering platform for stand-off and real-time atmospheric monitoring. *ACS Nano*, *13*, 12090–12099.

Poggi, M. A., Bottomley, L. A., & Lillehei, P. T. (2002). Scanning probe microscopy. *Analytical Chemistry*, *74*, 2851–2862.

Rindelaub, J. D., Craig, R. L., Nandy, L., Bondy, A. L., Dutcher, C. S., Shepson, P. B., & Ault, A. P. (2016). Direct measurement of pH in individual particles via Raman microspectroscopy and variation in acidity with relative humidity. *The Journal of Physical Chemistry A*, *120*, 911–917.

Rosen, H., & Novakov, T. (1977). Raman scattering and the characterisation of atmospheric aerosol particles. *Nature*, *266*, 708–710.

Ross, T., Gregoryanz, H. E., & Goncharov, F. A. (2013). Hydrogen (deuterium) vibron frequency as a pressure comparison gauge at multi-Mbar pressures. *Journal of Applied Physics*, *114*, 073505–0730511.

Rüttinger, S., Buschmann, V., Krämer, B., Erdmann, R., Macdonald, R., & Koberling, F. (2007). Determination of the confocal volume for quantitative fluorescence correlation spectroscopy. *Proceedings of SPIE*, *6630*, 66300D.

Schill, G. P., & Tolbert, M. A. (2014). Heterogeneous ice nucleation on simulated sea-spray aerosol using Raman microscopy. *The Journal of Physical Chemistry C*, *118*, 29234–29241.

Schlücker, S. (2014). Surface-enhanced Raman spectroscopy: Concepts and chemical applications. *Angewandte Chemie International Edition*, *53*, 4756–4795.

Schmid, T., Opilik, L., Blum, C., & Zenobi, R. (2013). Nanoscale chemical imaging using tip-enhanced Raman spectroscopy: A critical review. *Angewandte Chemie International Edition*, *52*, 5940–5954.

Seng, S., Guo, F., Tobon, Y. A., Ishikawa, T., Moreau, M., Ishizaka, S., & Sobanska, S. (2018). Deliquescence behavior of photo-irradiated single $NaNO_3$ droplets. *Atmospheric Environment*, *183*, 33–39.

Sharma, B., Frontiera, R. R., Henry, A.-I., Ringe, E., & Van Duyne, R. P. (2012). SERS: Materials, applications, and the future. *Materials Today*, *15*, 16–25.

Shi, X., Coca-López, N., Janik, J., & Hartschuh, A. (2017). Advances in tip-enhanced near-field Raman microscopy using Nanoantennas. *Chemical Reviews*, *117*, 4945–4960.

Siepka, D., Uzu, G., Stefaniak, E. A., & Sobanska, S. (2018). Combining Raman microspectrometry and chemometrics for determining quantitative molecular composition and mixing state of atmospheric aerosol particles. *Microchemical Journal*, *137*, 119–130.

Sobanska, S., Falgayrac, G., Laureyns, J., & Brémard, C. (2006). Chemistry at level of individual aerosol particle using multivariate curve resolution of confocal Raman image. *Spectrochimica Acta Part A*, *64*, 1102–1109.

Sobanska, S., Hwang, H., Choel, M., Jung, H.-J., Eom, H.-J., Kim, H., et al. (2012). Investigation of the chemical mixing state of individual Asian dust particles by the combined use of Electron probe X-ray microanalysis and Raman microspectrometry. *Analytical Chemistry*, *84*, 3145–3154.

Sobanska, S., Falgayrac, G., Rimetz-Planchon, J., Perdrix, E., Brémard, C., & Barbillat, J. (2014). Resolving the internal structure of individual atmospheric aerosol particle by the combination of atomic force microscopy, ESEM–EDX, Raman and ToF–SIMS imaging. *Microchemical Journal*, *114*, 89–98.

Song, Y.-C., Ryu, J., Malek, M. A., Jung, H.-J., & Ro, C.-U. (2010). Chemical speciation of individual airborne particles by the combined use of quantitative energy-dispersive Electron probe X-ray microanalysis and attenuated Total reflection Fourier transform-infrared imaging techniques. *Analytical Chemistry*, *82*, 7987–7998.

Song, Y. C., Eom, H. J., Jung, H. J., Malek, M. A., Kim, H. K., Geng, H., & Ro, C. U. (2013). Investigation of aged Asian dust particles by the combined use of quantitative ED-EPMA and ATR-FTIR imaging. *Atmospheric Chemistry and Physics*, *13*, 3463–3480.

Sourisseau, C. (2004). Polarization measurements in macro- and micro-Raman spectroscopies: Molecular orientations in thin films and azo-dye containing polymer systems. *Chemical Reviews*, *104*, 3851–3891.

Stadler, J., Schmid, T., & Zenobi, R. (2012). Developments in and practical guidelines for tip-enhanced Raman spectroscopy. *Nanoscale*, *4*, 1856–1870.

Sun, Z., Duan, F., He, K., Du, J., & Zhu, L. (2019a). Sulfate-nitrate-ammonium as double salts in $PM_{2.5}$: Direct observations and implications for haze events. *Science of the Total Environment*, *647*, 204–209.

Sun, Z., Duan, F., He, K., Du, J., Yang, L., Li, H., et al. (2019b). Physicochemical analysis of individual atmospheric fine particles based on effective surface-enhanced Raman spectroscopy. *Journal of Environmental Sciences*, *75*, 388–395.

Talaga, D., Comesaña-Hermo, M., Ravaine, S., Vallée, R. A. L., & Bonhommeau, S. (2015). Colocalized dark-field scattering, atomic force and surface-enhanced Raman scattering microscopic imaging of single gold nanoparticles. *Journal of Optics*, *17*, 114006.

Talaga, D., Bremner, A., Buffeteau, T., Vallée, R. A. L., Lecomte, S., & Bonhommeau, S. (2020). Total internal reflection tip-enhanced Raman spectroscopy of cytochrome c. *The Journal of Physical Chemistry Letters*, *11*, 3835–3840.

Tanaka, M., & Young, R. J. (2006). Polarised Raman spectroscopy for the study of molecular orientation distributions in polymers. *Journal of Materials Science*, *41*, 963–991.

Tirella, P. N., Craig, R. L., Tubbs, D. B., Olson, N. E., Lei, Z., & Ault, A. P. (2018). Extending surface enhanced Raman spectroscopy (SERS) of atmospheric aerosol particles to the accumulation mode (150-800 nm). *Environmental Science: Processes & Impacts*, *20*, 1570–1580.

Tomlin, J. M., Jankowski, K. A., Rivera-Adorno, F. A., Fraund, M., China, S., Stirm, B. H., et al. (2020). Chemical imaging of fine mode atmospheric particles collected from a research aircraft over agricultural fields. *ACS Earth and Space Chemistry*, *4*, 2171–2184.

Treuel, L., Sandmann, A., & Zellner, R. (2011). Spatial separation of individual substances in effloresced crystals of ternary ammonium Sulphate/Dicarboxylic acid/water aerosols. *ChemPhysChem*, *12*, 1109–1117.

Yotova, G. I., Tsitouridou, R., Tsakovski, S. L., & Simeonov, V. D. (2016). Urban air quality assessment using monitoring data of fractionized aerosol samples, chemometrics and meteorological conditions. *Journal of Environmental Science and Health – Part A Toxic Hazardous Substances and Environmental Engineering*, *51*, 544–552.

Verma, P. (2017). Tip-enhanced Raman spectroscopy: Technique and recent advances. *Chemical Reviews*, *117*, 6447–6466.

Wang, M., Zheng, N., Zhao, D., Shang, J., & Zhu, T. (2021). Using Micro-Raman spectroscopy to investigate chemical composition, mixing states, and heterogeneous reactions of individual atmospheric particles. *Environmental Science & Technology*, *55*(15), 10243–10254.

Wang, Y., Brigante, M., Mailhot, G., Talaga, D., Wu, Y., Dong, W., & Sobanska, S. (2022). Toward a better understanding of ferric-oxalate complex photolysis: The role of the aqueous/air interface of droplet. *Chemosphere*, *289*, 133127.

Webb, R. H. (1996). Confocal optical microscopy. *Reports on Progress in Physics*, *59*, 427–471.

Wei, H., Vejerano, E. P., Leng, W., Huang, Q., Willner, M. R., Marr, L. C., & Vikesland, P. J. (2018). Aerosol microdroplets exhibit a stable pH gradient. *Proceedings of the National Academy of Sciences of the United States of America*, *115*, 7272–7277.

Windig, W. (1992). Self-modeling mixture analysis of spectral data with continuous concentration profiles. *Chemometrics and Intelligent Laboratory Systems*, *16*, 1–16.

Windig, W. (1994). The use of second-derivative spectra for pure-variable based self-modeling mixture analysis techniques. *Chemometrics and Intelligent Laboratory Systems*, *23*, 71–86.

Windig, W. (1997). Spectral data files for self-modeling curve resolution with examples using the Simplisma approach. *Chemometrics and Intelligent Laboratory Systems*, *36*, 36–16.

Windig, W., & Stephenson, D. A. (1992). Self-modeling mixture analysis of second-derivative near-infrared spectral data using the SIMPLISMA approach. *Analytical Chemistry*, *64*, 2735–2742.

Windig, W., Antalek, B., Lippert, J. L., Batonneau, Y., & Brémard, C. (2002). Combined use of conventional and second-derivative data in the SIMPLISMA self-modeling mixture analysis approach. *Analytical Chemistry*, *74*, 1371–1379.

Wu, L., & Ro, C. U. (2020). Aerosol Hygroscopicity on a single particle level using microscopic and spectroscopic techniques: A review. *Asian Journal of Atmospheric Environment*, *14*(3), 177–209.

Wu, L., Li, X., Kim, H., Geng, H., Godoi, R. H. M., Barbosa, C. G. G., et al. (2019). Single-particle characterization of aerosols collected at a remote site in the Amazonian rainforest and an urban site in Manaus, Brazil. *Atmospheric Chemistry and Physics*, *19*(2), 1221–1240.

Wu, L., Becote, C., Sobanska, S., Flaud, P. M., Perraudin, E., Villenave, E., et al. (2020). Hygroscopic behavior of aerosols generated from solutions of 3-methyl-1,2,3-butanetricarboxylic acid, its sodium salts, and its mixtures with NaCl. *Atmospheric Chemistry and Physics*, *20*, 14103–14122.

Yeung, M. C., Lee, A. K. Y., & Chan, C. K. (2009). Phase transition and hygroscopic properties of internally mixed ammonium sulfate and Adipic acid (AS-AA) particles by optical microscopic imaging and Raman spectroscopy. *Aerosol Science and Technology*, *43*, 387–399.

Yeung, M. C., Chan, C. K., & Water Content and Phase (2010). Transitions in particles of inorganic and organic species and their mixtures using Micro-Raman spectroscopy. *Aerosol Science and Technology*, *44*, 269–280.

Yoo, H., Lee, H., Park, C., Shin, D., & Ro, C. U. (2022). Novel single-particle analytical technique for submicron atmospheric aerosols: Combined use of dark-field scattering and surface-enhanced Raman spectroscopy. *Analytical Chemistry*, *94*(38), 13028–13035.

Zhang, R., Zhang, Y., Dong, Z. C., Jiang, S., Zhang, C., Chen, L. G., et al. (2013). Chemical mapping of a single molecule by plasmon-enhanced Raman scattering. *Nature*, *498*, 82–86.

Zhang, R., Zhang, X., Wang, H., Zhang, Y., Jiang, S., Hu, C., et al. (2017). Distinguishing individual DNA bases in a network by non-resonant tip-enhanced Raman scattering. *Angewandte Chemie International Edition*, *56*, 5561–5564.

Zheng, L., Birch, M. E., Johnson, B., Breitenstein, M., Snawder, J., & Kulkarni, P. (2023). Correlation between graphitic carbon and elemental carbon in diesel particulate matter in workplace atmospheres. *Analytical Chemistry*, *95*(6), 3283–3290.

Zhou, Q., Pang, S. F., Wang, Y., Ma, J. B., & Zhang, Y. H. (2014). Confocal Raman studies of the evolution of the physical state of mixed Phthalic acid/ammonium sulfate aerosol droplets and the effect of substrates. *The Journal of Physical Chemistry B*, *118*, 6198–6205.

Ziegel, E. (2004). Statistics and Chemometrics for analytical chemistry: Statistics and Chemometrics for analytical chemistry. *Technometrics*, *46*, 498–499.

7

Atomic Force Microscopy of Individual Particles

Hansol D. Lee and Alexei V. Tivanski

ABSTRACT

Challenges remain in predicting the overall atmospheric aerosol effect on the climate and environment. This is particularly true for submicrometer-sized aerosols, whose effect depends on properties such as the mixing state, hygroscopicity, phase state, and surface tension. The aerosol composition, relative humidity, and temperature of the surrounding environment all influence these properties. Due to the physical and chemical complexity of the aerosol, they can exhibit significant particle-to-particle variability. Therefore, experimental tools that can directly study individual aerosols are needed. This chapter focuses on the emerging role of atomic force microscopy (AFM) in directly measuring the mixing state, hygroscopicity, phase state, and surface tension of submicrometer-sized aerosols. Recent applications of AFM on both model and real sea-spray aerosols are discussed.

7.1. INTRODUCTION

The atmosphere contains countless aerosols. Although they are too small to see with the naked eye, submicrometer-sized atmospheric aerosols can have significant influences on the climate and the environment. For example, primary aerosols (e.g., sea spray, dust, volcanic ash, and soot) can directly scatter solar radiation or seed clouds that reflect the sunlight, leading to global cooling (Fuentes et al., 2010; Jacobson, 2001; Pöschl, 2005). Atmospheric aerosols of varying sizes can originate from a variety of different sources that yield chemically complex aerosols. Moreover, depending on the relative humidity (RH) and temperature (T) of their immediate environment, chemical complexity can also give rise to physical complexity. For example, the complex mixing of different chemical entities can yield different mixing states for the aerosol population (Riemer et al., 2019). Furthermore, the phase state and surface tension of the aerosol are strongly influenced by the chemical composition, RH, and T (Lee et al., 2020a). These physicochemical properties are dynamic with respect to the origin of the aerosol, the composition of the aerosol, and the environment in which the aerosols reside (Lee et al., 2018; Reid et al., 2018).

The atmospheric community has focused a significant amount of effort on studying a variety of aerosols and their physicochemical properties to elucidate the aerosols' role in the climate. This has been typically done using an ensemble-average approach, where millions of aerosols are sampled at once to report the average of a particular measured property. In addition, the use of single-particle techniques is becoming more common in the atmospheric community. These techniques are helpful when studying a population of aerosols that exhibit high selectivity for specific atmospheric processes. An example of this can be seen in sea-spray aerosols (SSA), where only one out of a million particles can nucleate ice at −25 °C (DeMott et al., 2016; DeMott et al., 2010).

One example of a single-particle technique that has been applied to study aerosols is atomic force microscopy

Department of Chemistry, University of Iowa, Iowa City, IA, USA

Microanalysis of Atmospheric Particles: Techniques and Applications, Geophysical Monograph 288, First Edition.
Edited by Joseph M. Conny and Peter R. Buseck.

DOI: 10.1002/9781119554318.ch7

(AFM). Unlike other high-resolution microscopy techniques, a vacuum is not required, allowing for single-particle-level measurements under ambient pressure and at varying environmental conditions of RH and T (Giessibl, 2003). As an offline microscopy technique, the aerosols are collected onto a solid surface (substrate) for a variety of measurements. Sharp AFM probes (as sharp as 2–10 nm tip radius of curvature) are used to raster scan the substrate containing substrate-deposited aerosol particles using contact, intermittent contact, or noncontact imaging methods (Barkay et al., 2005). Intermittent contact mode is often used to obtain 3D height, amplitude, and phase images (Poggi et al., 2004). Afterward, the technique allows for measurements of force between the AFM probe and the particles of interest (Morris et al., 2015). Measurements of force are compiled as force plots that contain measured force as a function of the vertical position of the AFM probe as it moves toward and away from the particle surface. The force plots are analyzed to quantify atmospherically relevant physicochemical properties of the aerosol particles.

In this chapter, we summarize the latest works that have used AFM to study SSA. AFM has been used to perform a variety of measurements to quantify atmospherically relevant physical-chemical properties of both model and nascent submicrometer-sized SSA on a single-particle level. The following sections will outline each property that has been measured using AFM. Each property of the aerosol will be discussed, followed by the AFM methodology used to quantify that property. Several works that highlight the applicability of AFM will be mentioned.

7.2. MIXING STATE OF AEROSOLS

Aerosol mixing state refers to the distribution of various properties within the population of aerosols (Riemer et al., 2019). In this chapter, we focus on the physicochemical mixing state, which refers to the distribution of morphologies within a particle population. The physicochemical mixing state is the result of complex mixing of various chemical entities within each aerosol—originating from seawater for SSA—under particular environmental conditions (Lee et al., 2020b). Observing the mixing state of the aerosol is important to accurately predict aerosols' effect on the climate because the mixing state can change the optical properties, cloud nucleating ability, and ice nucleation of the aerosol population (Li et al., 2016; Riemer et al., 2019). For SSA, which consists of a complex mixture of organic, inorganic, and biological entities from seawater, the physicochemical mixing state has been shown to vary depending on size (Cochran et al., 2017; Lee et al., 2020b). The morphology of individual particles can be studied with AFM 3D height and phase imaging to understand the physicochemical mixing state (Martinez & Garcia, 2006; Nagao & Dvorak, 1999; Scott & Bhushan, 2003). By supplying a driving voltage to the AFM cantilever, the AFM probe raster scans the substrate with the deposited particles. While scanning, the closed-feedback-loop system observes how the drive voltage phase shifts due to the probe–particle interactions. When the AFM probe interacts with viscoelastic particles, a phase shift is observed in comparison to the original driving voltage. This phase shift is quantified for individual pixels to construct a phase micrograph. Due to its inherent sensitivity, phase imaging can be used to identify phase separations within the particle (e.g., between a solid inorganic core and semisolid or liquid organic shell of a core-shell particle).

Figure 7.1a shows AFM 3D height and phase micrographs of 1 : 8 (M) glucose:NaCl mixture particle. Both glucose and NaCl have been found in nascent SSA previously; thus, this mixture can be considered a simplified model system of SSA (Jayarathne et al., 2016). Phase images exhibit a core-shell morphology, where the inorganic core is encased in an organic shell. Core-shell is one of six different types of main morphologies available to SSA. In addition, smaller SSA were shown to exhibit a greater proportion of rounded morphology, whereas larger SSA exhibited a greater fraction of prism-like morphology (Lee et al., 2020b). Core-shell SSA fall somewhat in between the two morphologies in relative abundance. From these images, a phase image mask can be applied to identify the boundaries of the core and shell regions of the particle. Subsequently, the mask is applied to the 3D height image of the particle to quantify the core and total particle volumes. Subtracting the core volume from the total particle volume produces the organic shell volume, which, if divided by the corresponding total particle volume, quantifies the organic volume fraction (OVF) of the particle. The OVF is reflective of the relative amount of organic material present within the particle (Lee et al., 2019). A value close to one indicates that all or nearly all of the particle is comprised of organic matter, whereas a value close to zero is indicative of predominantly inorganic particles. However, AFM alone does not provide chemical composition data for these particles. Recent innovations such as the photothermal AFM-IR aim to solve this problem by allowing for both AFM and IR spectroscopy measurements (Ault & Axson, 2017; Bondy et al., 2017; Or et al., 2018; Slade et al., 2019).

Figure 7.1b shows the OVF measured against the volume-equivalent diameter of 1 : 8 (M) glucose:NaCl particles. The dashed black line is the expected OVF value, calculated using the mass of the glucose and NaCl within the bulk solution prior to aerosolization. The bulk OVF was measured using high-performance ion chromatography on filters collected during the same time that AFM substrates were collected. The data

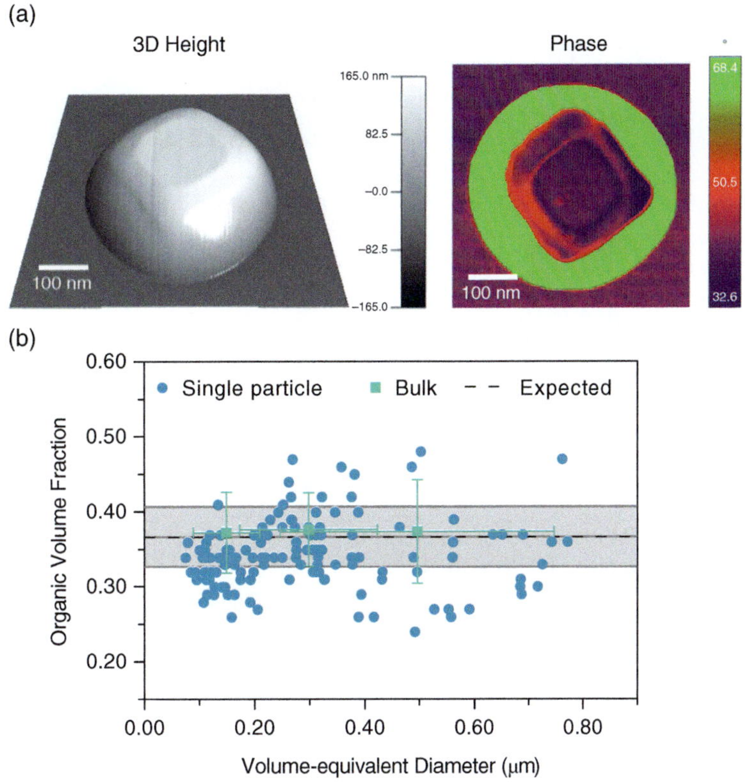

Figure 7.1 (a) AFM 3D height and phase micrographs of 1 : 8 (M) glucose:NaCl particle. (b) Quantified organic volume fraction versus volume-equivalent diameter from single-particle AFM (circles), bulk high-performance ion chromatography (squares), and expected values from the bulk solution (dashed line) with +/− 10% error shown as the grey area. Lee et al. (2019)/Copernicus Publications/CC BY 4.0.

shows a good overlap between the single-particle AFM and the bulk data, as well as the expected OVF across the entire size range. This illustrates the applicability of AFM in quantifying the single-particle OVF and thereby can be used to identify the particle-to-particle variability and the overall mixing state of the aerosols within a population. The same methodology was used to quantify the OVF of nascent SSA (Lee et al., 2020a). When measured, AFM data showed a good correlation between the increased OVF of SSA and increased chlorophyll-a concentrations in the seawater. This showed that in certain circumstances, phytoplankton blooms in the seawater can enrich the SSA with increased relative amounts of organic matter, which can influence the hygroscopicity, phase state, and surface tension of SSA.

7.3. HYGROSCOPICITY OF AEROSOLS

Hygroscopicity refers to the water uptake of the aerosol (Kreidenweis et al., 2005). SSA that exhibit a complex physicochemical mixing state can have both soluble and insoluble chemical components. The mixture influences how much water is absorbed into the aerosol at a particular RH (Cochran et al., 2017; Estillore et al., 2017). Soluble chemical components uptake more water than insoluble components (Hodas et al., 2015). The hygroscopicity is important because it can alter various concentration-dependent aerosol properties such as the phase state and surface tension, as increasing extent of water uptake leads to decreasing solute concentration (Lee et al., 2017a; Lee et al., 2017b). Furthermore, the hygroscopicity determines the size of the aerosol at a certain RH and T, which affects the aerosols' ability to scatter light and nucleate into clouds. The latter is exemplified in the κ-Köhler theory (Petters & Kreidenweis, 2007)

$$\frac{RH}{100} = \frac{D_{\mathrm{w}}^3 - D_{\mathrm{d}}^3}{D_{\mathrm{w}}^3 - D_{\mathrm{d}}^3(1-\kappa)} \exp\left(\frac{4\sigma M_w}{RT\rho_w D_w}\right) \tag{7.1}$$

where κ is the hygroscopicity parameter, D_{w} is the wet diameter, D_{d} is the dry diameter, M_{w} is the molecular weight of water, ρ_{w} is the density of water, R is the gas constant, T is the temperature, and σ is the surface tension. A large κ is indicative of a hygroscopic aerosol,

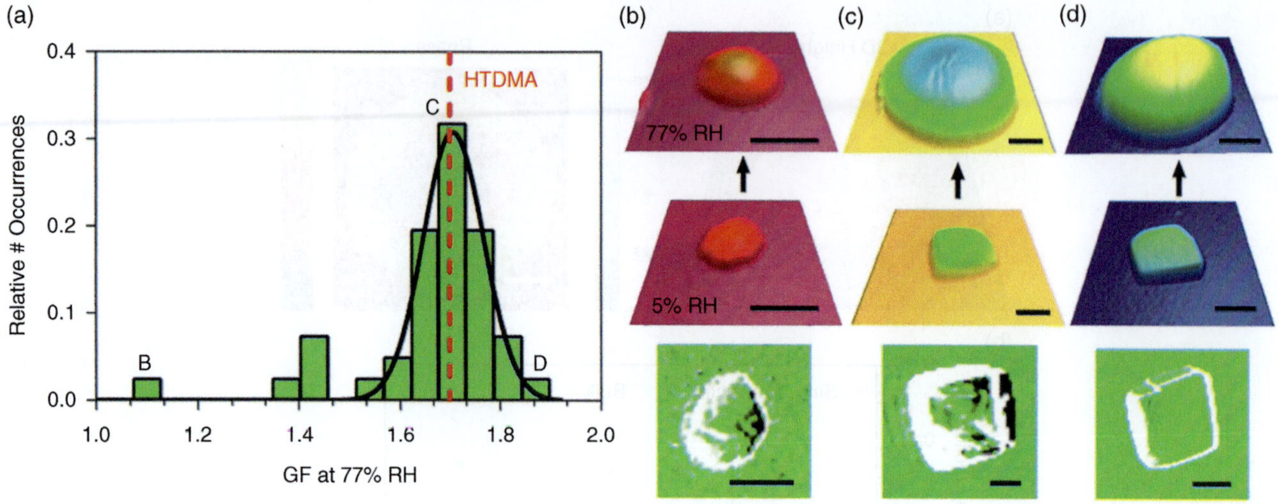

Figure 7.2 (a) Histogram of AFM measured GF values of 6 : 1 (M) NaCl:nonanoic acid individual particles compared against the ensemble-average HTDMA. (b–d) AFM 3D micrographs of three selected particles from dry (middle) to wet (top) states at varying RH. Corresponding phase micrographs are shown in the dry state (bottom). Morris et al. (2016)/with permission of American Chemical Society.

and it directly influences D_w. For example, the D_w of NaCl ($\kappa = 1.12$–1.28) at 85% RH would be approximately two times greater than D_d (Estillore et al., 2017).

The hygroscopicity of the aerosol can reveal particle-to-particle variability within the aerosol population (Morris et al., 2016). In Figure 7.2, a model system of 6 : 1 (M) NaCl:nonanoic acid was studied. Single-particle AFM imaging and hygroscopic tandem differential mobility analyzer (HTDMA) were used to quantify the growth factor (GF), which is defined as the ratio of D_w to D_d. Particle diameters are measured directly using AFM height imaging at a particular RH. A high GF indicates a significant increase in the particle size due to water uptake. Single-particle AFM data produces an average GF similar to that of the ensemble-average result from HTDMA, and AFM also reveals uniquely different hygroscopic nature of individual aerosols in the study. This is shown despite the relatively simple binary mixture of NaCl and nonanoic acid. This indicates that single-particle AFM is an ideal experimental tool to study SSA under atmospherically relevant conditions to accurately capture particle-to-particle variability. It is noteworthy that the method does not require prior knowledge of the aerosol chemical composition. Furthermore, it was also used to quantify the hygroscopic growth of individual sub- and supermicrometer nascent SSA produced during two phytoplankton blooms (Cochran et al., 2017). The GF data were consistent with the known composition of the SSA as a complex mixture containing a variety of organic, inorganic, and biological entities (Lee, et al., 2020a).

7.4. VISCOSITY AND PHASE STATE OF AEROSOLS

Depending on a variable RH, T, and hygroscopicity, aerosols can absorb varying amounts of water, which in turn modulates their concentration-dependent viscosity and phase state (Forestieri et al., 2018; Lee et al., 2017b; Marshall et al., 2016; Ruehl et al., 2016; Song et al., 2016). Viscosity values are commonly used to assign the phase state of atmospheric aerosols, which can be solid (viscosity $>10^{12}$ Pa s), semisolid (viscosity range 10^2–10^{12} Pa s), or liquid (viscosity $<10^2$ Pa s). The viscosity and phase state of atmospheric aerosols are central to understanding and predicting atmospheric impacts of aerosols because these properties can suppress water uptake (Mikhailov et al., 2009; Murray, 2008; Zobrist et al., 2008), modulate molecular diffusion rates (Kuwata & Martin, 2012; Shiraiwa et al., 2011), and alter ice nucleating efficiency (Baustian et al., 2013; Berkemeier et al., 2014; Murray et al., 2010). As the aerosols often display size-dependent chemical composition and mixing states, measurements that enable quantification of the viscosity and identification of the phase state on a single-particle basis are important to assess the extent of particle-to-particle variability of these properties (Lee et al., 2020b). Furthermore, because atmospheric aerosols exist in both sub- and supermicrometer sizes, such measurements must be accessible for both size ranges.

In this regard, AFM can be used to directly identify the phase state of individual substrate-deposited sub- and supermicrometer particles as a function of RH. Specifically, the AFM tip can be used to first image a

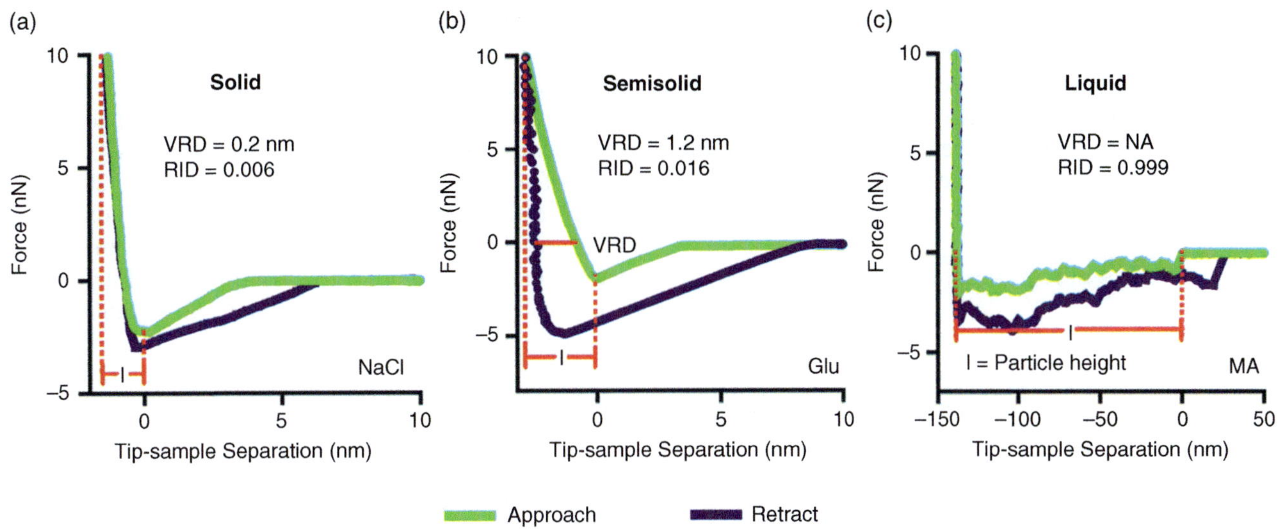

Figure 7.3 Examples of AFM force plots obtained on individual particles of (a) solid NaCl, (b) semisolid glucose, and (c) liquid malonic acid to quantify the VRD and RID at a certain RH. Adapted from Ray et al. (2019).

particle of interest, and then by utilizing AFM force spectroscopy, forces acting on the same imaging tip can be measured as a function of tip–sample separation during vertical tip movement toward and away from the particle surface (Martinez et al., 2006). The resultant force plots can be obtained with a controlled scan rate, maximum applied force, and overall distance at a selected particle position and a specific RH (Figure 7.3). Here, a decrease in the tip–sample separation corresponds to the AFM tip vertically approaching the particle surface with zero separation indicative of an approximate contact point with the particle. After the contact, the tip starts to indent into the particle (negative tip–sample separation) until the measured force increases to a predefined value. Afterward, the probe retracts until the tip and particle are no longer in contact, with the tip returning to the initial position above the particle. The force plots over the same particle position can be repeatedly collected at varying RH during the hydration or dehydration modes (Lee et al., 2020a; Lee et al., 2017b; Ray et al., 2019). To identify the phase state at a particular RH, the corresponding force plot is utilized to quantify two parameters: viscoelastic response distance (VRD) and relative indentation depth (RID). The VRD accounts for the hysteresis in the tip–sample separation between the approach and retract curves at zero force and originates from the viscoelasticity of the particle, where larger VRD values generally correspond to lower particle viscosity. The RID is defined as a ratio of maximum indentation (I) determined from the force plot at a specified force divided by the height of the particle (H) that can be directly obtained from AFM height imaging.

If the RID value is significantly less than 1, then the indentation is insignificant compared to the height of the particle (Figure 7.3a), which is indicative of a solid or semisolid particle phase state. If the RID value approaches 1, then the AFM probe indented completely through the particle and reaches the underlying solid substrate, implying a liquid particle phase state (Figure 7.3c). To differentiate a solid from a semisolid phase state (Figure 7.3b), the VRD measurement is utilized. Here, a VRD value greater than 0.5 nm is indicative of a semisolid, and less than 0.5 nm is indicative of a solid. Based on this approach, a step-by-step framework was established for identifying the solid, semisolid, and liquid phase states of individual particles (Figure 7.4). Although this approach requires direct contact between AFM tip and particle, it is applicable for both sub- and supermicrometer individual particles and enables identification of the phase states from 10^{-3} to 10^{16} Pa s in viscosity. Furthermore, the single-particle phase-state measurement can be performed without prior knowledge of the aerosol chemical composition, making it a suitable approach in identifying the phase state of real atmospheric aerosols. In that regard, the method was recently used to determine the phase state of nascent core-shell SSA produced from wave breaking of natural seawater (Lee et al., 2020a). Specifically, the organic shell surrounding the core of the SSA was determined to be predominantly semisolid at both 20% and 60% RH.

It is important to emphasize that AFM force spectroscopy-based method is time-consuming. Thus, a limited number of individual particles are typically studied. Instead, an alternative semiquantitative approach that can assess the phase state for many substrate-deposited (via an impactor) individual particles is based on AFM height imaging (Ray et al., 2019). Although it is less accurate than force spectroscopy, the imaging-based method is much quicker. Specifically, the aspect ratio (AR)—height divided by the area equivalent diameter—of substrate-deposited SSA-relevant particles can be directly measured from the AFM height image. A higher AR is indicative of a solid-like phase state, as the particle resisted the particle flattening effect of the impaction on a solid substrate. In a similar manner, a lower AR is indicative of a liquid-like phase state (Ray et al., 2019). In this case, the particle succumbed to the impaction forces. This was demonstrated on several SSA-relevant model system particles from 0.5 to 1.0 μm in size (Figure 7.5). Specifically, two inorganic salts (NaCl and $MgSO_4$), four organic acids (malonic, glutaric, azelaic, and palmitic acids), three saccharides (glucose, sucrose, and raffinose), and lipopolysaccharide from *Escherichia coli* were studied. Additionally, three inorganic–organic binary chemical mixtures (NaCl–malonic acid, NaCl–glucose, and $MgSO_4$–glucose) at 1:3 and 3:1 mass ratios were also studied. Particles with ARs less than 0.165 were liquid, those between 0.165 and 0.365 were semisolid, and those greater than 0.365 were solid. This was used to suggest specific AR cut-offs to semi-quantitatively assess the expected phase state. The AR cut-offs should not be considered absolute values, as additional factors such as the impaction method, the substrate type used for deposition, and significantly smaller particles (e.g., nanoparticles) need to be studied to further validate the correlation between the phase state and AR. However,

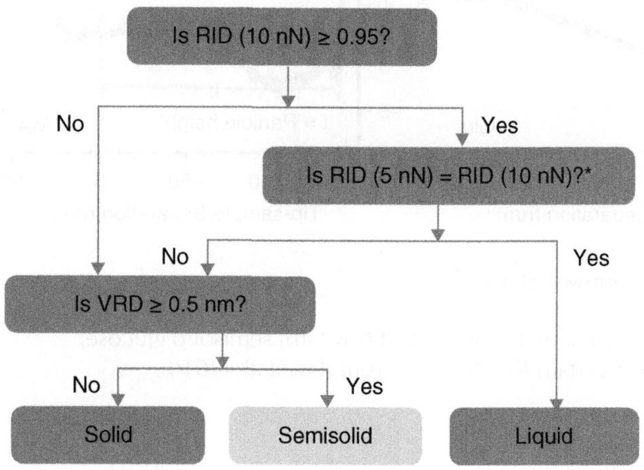

Figure 7.4 Phase state framework is used to interpret the measured RID and VRD values from force plots to identify the solid, semisolid, or liquid phase state of the particle. Ray et al. (2019)/with permission of American Chemical Society.

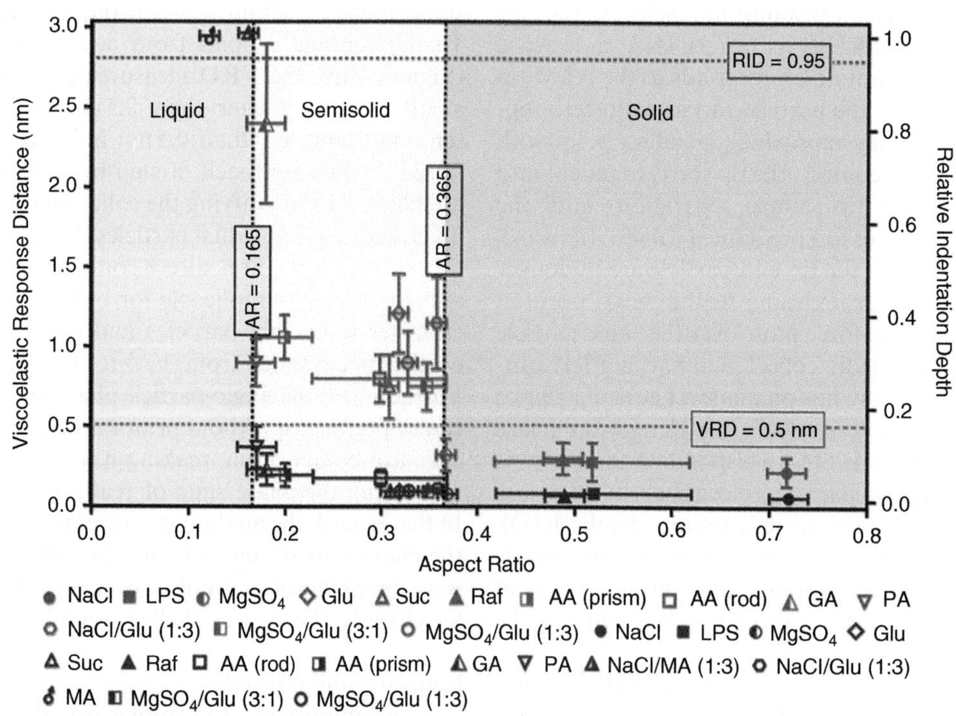

Figure 7.5 Comparison of measured AR versus VRD and RID for multiple model SSA relevant systems with abbreviations: AA, azelaic acid; GA, glutaric acid; Glu, glucose; LPS, lipopolysaccharide; MA, malonic acid; PA, palmitic acid; Raf, raffinose; Suc, sucrose. Ray et al. (2019)/with permission of American Chemical Society.

this methodology provides a quick, semiquantitative assessment of the phase state and can also be applicable to other microscopy techniques that can measure the AR of substrate-deposited particles.

7.5. SURFACE TENSION OF AEROSOLS

The κ-Köhler theory (Equation 7.1) is utilized to predict cloud formation at high RH by atmospheric aerosols. A key parameter in the theory is the humidity-dependent surface tension of liquid droplets, which is commonly assumed to be a fixed value of pure water (72.0 mN/m at 25 °C). However, atmospheric aerosols often contain surface-active compounds that can significantly decrease the surface tension of aerosol droplets, potentially leading to inaccurate predictions of their cloud-forming properties (Boyer & Dutcher, 2016; Bzdek et al., 2016; Forestieri et al., 2018; Lee et al., 2017a; Lee et al., 2020a; Morris et al., 2015; Ovadnevaite et al., 2017). For example, a lower aerosol surface tension can lead to lowering of the critical supersaturation, increasing the probability of activation as the aerosol grows into a cloud droplet. Thus, experimental methods that enable direct quantification of the aerosol surface tension as a function of RH are important. Furthermore, such measurements must be accessible for both sub- and supermicrometer size ranges and ideally should also be performed on a single-particle basis to assess the particle-to-particle variability of atmospheric aerosols.

In that regard, AFM has been used to directly quantify the surface tension of individual submicrometer aerosols in a liquid phase state as a function of RH (Lee et al., 2017a; Lee et al., 2020a; Morris et al., 2015). AFM methodology is based on force spectroscopy technique and utilizes a constant-diameter nanoneedle (ca. 50–200 nm) to both image an individual submicrometer liquid droplet and quantify the surface tension by measuring the maximum absolute force (retention force, F_{ret}) required to break the meniscus between the droplet and nanoneedle. Figure 7.6 shows an illustrative force plot collected over an approximate center of a liquid droplet as a function of vertical piezo position as the nanoneedle moves toward and away from the droplet surface. Symbols A–F correspond to the nanoneedle position relative to the droplet. At the beginning of the force plot measurement, the nanoneedle is initially positioned approximately 2 μm away from the droplet surface (position A). The nanoneedle is lowered at constant speed until it contacts the droplet surface (position B) and then continues to push all the way through the droplet until it reaches the underlying solid surface (position C). Afterward, the nanoneedle starts to push on the solid substrate below the droplet until a predefined maximum force is reached (position D). Upon

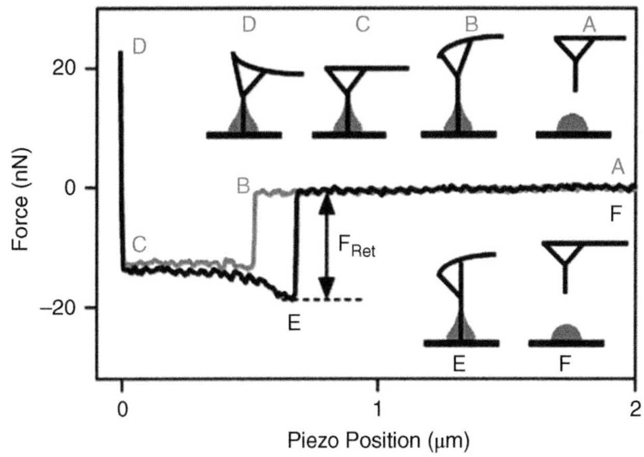

Figure 7.6 Illustrative example force plot obtained on a submicrometer liquid droplet to quantify the surface tension using a custom-made constant diameter AFM nanoneedle. Adapted from Morris et al. (2015).

reaching this force, the nanoneedle is retracted until the meniscus between it and the liquid particle breaks (position E). The nanoneedle is then returned to its original position above the droplet (position F). The retention force can thus be directly measured and corresponds to the contact angle approaching zero. By quantifying the retention force measured by indenting the nanoneedle into the droplet at a specific RH, the surface tension can be quantified using Equation 7.2 (Yazdanpanah et al., 2008)

$$F_{ret} = 2\pi r\sigma \quad (7.2)$$

where r is the radius of the nanoneedle and σ is the surface tension of the droplet. Although the force plot shown in Figure 7.6 corresponds to the nanoneedle indenting all the way through the droplet and pushing into the underlying solid surface, the approach is similarly applicable if the indentation distance is less than the height of the droplet.

This methodology was previously developed on SSA-relevant systems (Morris et al., 2015) and further validated on a variety of model systems (Lee et al., 2017a) that included surface-inactive and surface-active chemical systems such as inorganic salts, carboxylic acids, and saccharides (Figure 7.7). In this case, the single-particle AFM surface tension data on submicrometer-sized particles overlapped well with the bulk tensiometer studies. The deviation in the single-particle surface tension relative to the bulk observed when the solute concentration is beyond the solubility limit (e.g., Figure 7.7d) is attributed to increasing droplet viscosity and packing of surfactants on the nanoneedle-droplet interface. This methodology was also successfully applied to quantify

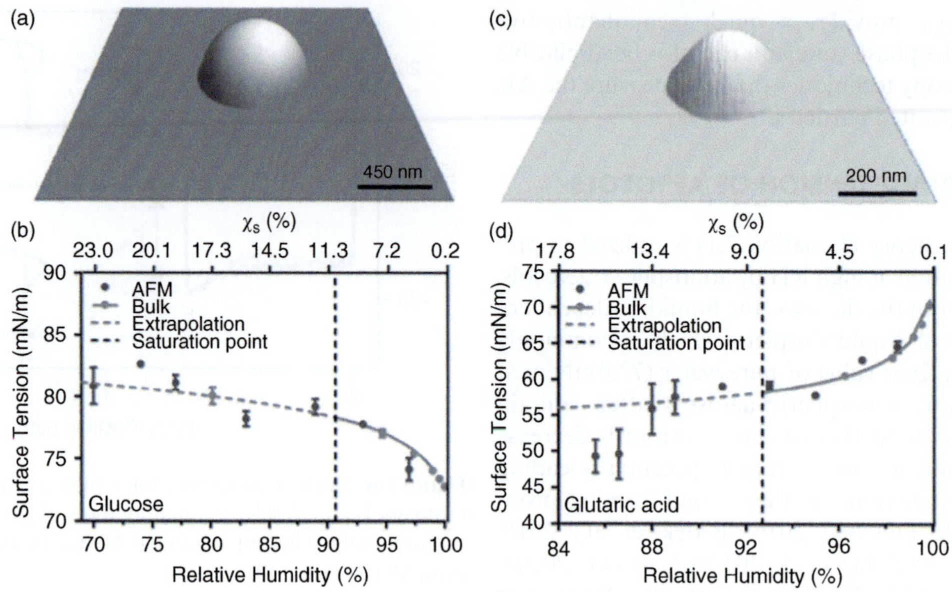

Figure 7.7 (a, c) 3D AFM micrographs of glucose and glutaric acid individual particles. AFM measured the surface tension of individual glucose (b) and glutaric acid (d) particles versus RH and solute mole percentage. Error bars correspond to two standard deviations. Single-particle data is compared against the bulk from a force tensiometer (red dots and solid lines, dashed red lines have been extrapolated beyond the saturation point). Black dashed vertical indicates the saturation point for each system. Lee et al. (2017a)/with permission of American Chemical Society.

the surface tension of nascent deliquesced SSA particles at 80% RH (Figure 7.8). SSA that were produced by the breaking of waves of isolated seawater were collected for AFM studies (Lee et al., 2020a). The surface tension data revealed a significant particle-to-particle variability that was greater than the analytical uncertainty of the methodology. Furthermore, a significant depression of the average SSA surface tension was observed that corresponded to the greatest biological activity in the seawater.

The advantage of the AFM approach is that the surface tension measurement methodology can be used to quantify the surface tension of individual submicrometer-sized particles at viscosity values of less than 10^2 Pa s or when the RID is close to or equal to one. Additionally, the method is based on direct measurement and does not require additional assumptions. However, such direct measurement can also be disadvantageous. Because the nanoneedle must come into direct contact with the particle, the nanoneedle may become coated with various organic or inorganic species over multiple uses or may even change shape/size due to repeated force measurements. Thus, the methodology requires careful nanoneedle calibration before and after each experiment using a reference liquid with known surface tension, such as ultrapure water (Lee et al., 2017a).

7.6. CONCLUSION

This chapter explored the use of single-particle AFM in studying the physicochemical properties of individual SSA particles. Although controlling for the RH, AFM was shown to enable accurate measurements of the mixing state, hygroscopicity, phase state, and surface tension of individual substrate-deposited aerosol particles without prior knowledge of their chemical identity. Several selected examples that demonstrate such measurements were presented for both the model and nascent SSA. Although the AFM method requires a substrate and single-particle studies can be time-consuming compared to ensemble-average measurements, it provides the ability to study the particle-to-particle variability within an aerosol population directly. The methodologies discussed in this work were shown to be applicable for both sub- and supermicrometer-sized aerosols over a wide range of subsaturated RH (~3% to ~97% RH) or viscosity range between 10^{-3}–10^{16} Pa s. With further developments, such as performing AFM experiments while modulating both the RH and T or introducing atmospheric gasses into the AFM chamber for in situ measurements, it is expected that the technique will continue to be a suitable tool to further elucidate the role of aerosols in the climate and the atmosphere.

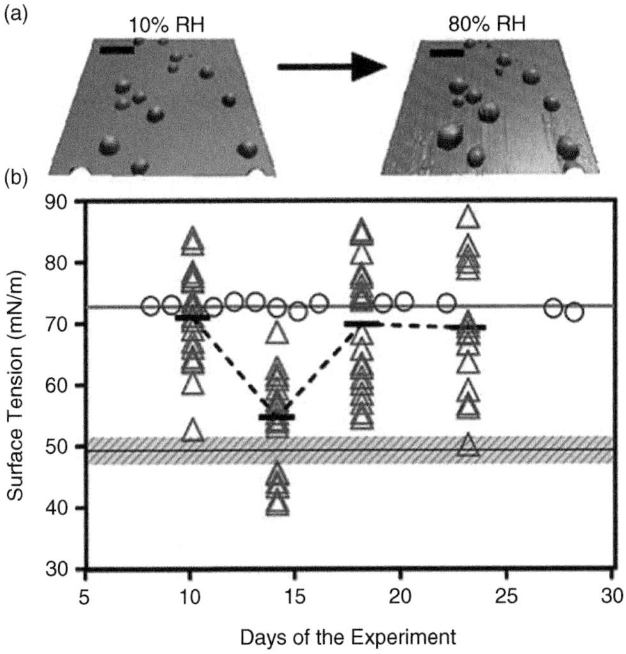

Figure 7.8 (a) AFM 3D micrographs of nascent sea-spray aerosol at low (left) and high (right) RH. (b) Measured surface tension of individual nascent SSA at 80% RH. Open triangles refer to the single-particle data. Thick black horizontal lines are used to denote the average OVF values. Open circles refer to measurements of the seawater using a force tensiometer. The solid blue line refers to the surface tension of pure water at room temperature. The thin solid black line is data taken from glutaric acid at 86% RH, with the striped green area showing the standard deviation of the measurement technique. Lee et al. (2020a)/with permission of the American Chemical Society.

REFERENCES

Ault, A. P., & Axson, J. L. (2017). Atmospheric aerosol chemistry: Spectroscopic and microscopic advances. *Analytical Chemistry*, *89*(1), 430–452. WOS:000391346600024

Barkay, Z., Teller, A., Ganor, E., Levin, Z., & Shapira, Y. (2005). Atomic force and scanning electron microscopy of atmospheric particles. *Microscopy Research and Technique*, *68*(2), 107–114. WOS:000233026000008

Baustian, K. J., Wise, M. E., Jensen, E. J., Schill, G. P., Freedman, M. A., & Tolbert, M. A. (2013). State transformations and ice nucleation in amorphous (semi-)solid organic aerosol. *Atmospheric Chemistry and Physics*, *13*(11), 5615–5628. WOS:000320380200013

Berkemeier, T., Shiraiwa, M., Pöschl, U., & Koop, T. (2014). Competition between water uptake and ice nucleation by glassy organic aerosol particles. *Atmospheric Chemistry and Physics*, *14*(22), 12513–12531.

Bondy, A. L., Kirpes, R. M., Merzel, R. L., Pratt, K. A., Banaszak Holl, M. M., & Ault, A. P. (2017). Atomic force microscopy-infrared spectroscopy of individual atmospheric aerosol particles: Subdiffraction limit vibrational spectroscopy and morphological analysis. *Analytical Chemistry*, *89*(17), 8594–8598. doi: 10.1021/acs.analchem.7b02381

Boyer, H. C., & Dutcher, C. S. (2016). Statistical thermodynamic model for surface tension of aqueous organic acids with consideration of partial dissociation. *Journal of Physical Chemistry A*, *120*(25), 4368–4375. WOS:000379456300013

Bzdek, B. R., Power, R. M., Simpson, S. H., Reid, J. P., & Royall, C. P. (2016). Precise, contactless measurements of the surface tension of picolitre aerosol droplets. *Chemical Science*, *7*(1), 274–285. WOS:000366826900031

Cochran, R. E., Laskina, O., Trueblood, J. V., Estillore, A. D., Morris, H. S., Jayarathne, T., et al. (2017). Molecular diversity of sea spray aerosol particles: Impact of ocean biology on particle composition and Hygroscopicity. *Chem*, *2*(5), 655–667. WOS:000408621100012

DeMott, P. J., Prenni, A. J., Liu, X., Kreidenweis, S. M., Petters, M. D., Twohy, C. H., et al. (2010). Predicting global atmospheric ice nuclei distributions and their impacts on climate. *Proceedings of the National Academy of Sciences of the United States of America*, *107*(25), 11217–11222. WOS:000279058000014

DeMott, P. J., Hill, T. C. J., McCluskey, C. S., Prather, K. A., Collins, D. B., Sullivan, R. C., et al. (2016). Sea spray aerosol as a unique source of ice nucleating particles. *Proceedings of the National Academy of Sciences of the United States of America*, *113*(21), 5797–5803. WOS:000376779900035

Estillore, A. D., Morris, H. S., Or, V. W., Lee, H. D., Alves, M. R., Marciano, M. A., et al. (2017). Linking hygroscopicity and the surface microstructure of model inorganic salts, simple and complex carbohydrates, and authentic sea spray aerosol particles. *Physical Chemistry Chemical Physics*, *19*(31), 21101–21111. WOS:000407763700080

Forestieri, S. D., Staudt, S. M., Kuborn, T. M., Faber, K., Ruehl, C. R., Bertram, T. H., & Cappa, C. D. (2018). Establishing the impact of model surfactants on cloud condensation nuclei activity of sea spray aerosol mimics. *Atmospheric Chemistry and Physics*, *18*(15), 10985–11005. WOS:000440940600001

Fuentes, E., Coe, H., Green, D., de Leeuw, G., & McFiggans, G. (2010). Laboratory-generated primary marine aerosol via bubble-bursting and atomization. *Atmospheric Measurement Techniques*, *3*(1), 141–162. WOS:000278118500011

Giessibl, F. J. (2003). Advances in atomic force microscopy. *Reviews of Modern Physics*, *75*(3), 949–983. doi: 10.1103/RevModPhys.75.949

Hodas, N., Zuend, A., Mui, W., Flagan, R. C., & Seinfeld, J. H. (2015). Influence of particle-phase state on the hygroscopic behavior of mixed organic-inorganic aerosols. *Atmospheric Chemistry and Physics*, *15*(9), 5027–5045. WOS:000355289100004

Jacobson, M. Z. (2001). Global direct radiative forcing due to multicomponent anthropogenic and natural aerosols. *Journal of Geophysical Research-Atmospheres*, *106*(D2), 1551–1568. WOS:000166596400008

Jayarathne, T., Sultana, C. M., Lee, C., Malfatti, F., Cox, J. L., Pendergraft, M. A., et al. (2016). Enrichment of saccharides and divalent Cations in sea spray aerosol during two phytoplankton blooms. *Environmental Science & Technology*, *50*(21), 11511–11520. WOS:000386991100010

Kreidenweis, S. M., Koehler, K., DeMott, P. J., Prenni, A. J., Carrico, C., & Ervens, B. (2005). Water activity and activation diameters from hygroscopicity data—part I: Theory and application to inorganic salts. *Atmospheric Chemistry and Physics*, *5*, 1357–1370. WOS:000229615500001

Kuwata, M., & Martin, S. T. (2012). Phase of atmospheric secondary organic material affects its reactivity. *Proceedings of the National Academy of Sciences of the United States of America*, *109*(43), 17354–17359. WOS:000311147800018

Lee, H. D., Estillore, A. D., Morris, H. S., Ray, K. K., Alejandro, A., Grassian, V. H., & Tivanski, A. V. (2017a). Direct surface tension measurements of individual sub-micrometer particles using atomic force microscopy. *The Journal of Physical Chemistry. A*, *121*(43), 8296–8305. https://www.ncbi.nlm.nih.gov/pubmed/28981283

Lee, H. D., Ray, K. K., & Tivanski, A. V. (2017b). Solid, semisolid, and liquid phase states of individual submicrometer particles directly probed using atomic force microscopy. *Analytical Chemistry*, *89*(23), 12720–12726. WOS:000417549600018

Lee, H. D., Ray, K. K., & Tivanski, A. V. (2018). Directly probing the phase states and surface tension of individual submicrometer particles using atomic force microscopy. *Multiphase Environmental Chemistry in the Atmosphere*, *1299*, 245–259.

Lee, H. D., Kaluarachchi, C. P., Hasenecz, E. S., Zhu, J. Z., Popa, E., Stone, E. A., & Tivanski, A. V. (2019). Effect of dry or wet substrate deposition on the organic volume fraction of core–shell aerosol particles. *Atmospheric Measurement Techniques*, *12*(3), 2033–2042. https://www.atmos-meas-tech.net/12/2033/2019

Lee, H. D., Morris, H. S., Laskina, O., Sultana, C. M., Lee, C., Jayarathne, T., et al. (2020a). Organic enrichment, physical phase state, and surface tension depression of nascent Core–Shell Sea spray aerosols during two phytoplankton blooms. *ACS Earth and Space Chemistry*. doi: 10.1021/acsearthspacechem.0c00032

Lee, H. D., Wigley, S., Lee, C., Or, V. W., Hasenecz, E. S., Stone, E. A., et al. (2020b). Physicochemical mixing state of sea spray aerosols: Morphologies exhibit size dependence. *ACS Earth and Space Chemistry*, *4*(9), 1604–1611. doi: 10.1021/acsearthspacechem.0c00153

Li, W. J., Shao, L. Y., Zhang, D. Z., Ro, C. U., Hu, M., Bi, X. H., et al. (2016). A review of single aerosol particle studies in the atmosphere of East Asia: Morphology, mixing state, source, and heterogeneous reactions. *Journal of Cleaner Production*, *112*, 1330–1349. WOS:000368206800008

Marshall, F. H., Miles, R. E. H., Song, Y. C., Ohm, P. B., Power, R. M., Reid, J. P., & Dutcher, C. S. (2016). Diffusion and reactivity in ultraviscous aerosol and the correlation with particle viscosity. *Chemical Science*, *7*(2), 1298–1308. WOS:000368835300058

Martinez, N. F., & Garcia, R. (2006). Measuring phase shifts and energy dissipation with amplitude modulation atomic force microscopy. *Nanotechnology*, *17*(7), S167–S172. WOS:000236979000012

Martinez, N. F., Patil, S., Lozano, J. R., & Garcia, R. (2006). Enhanced compositional sensitivity in atomic force microscopy by the excitation of the first two flexural modes. *Applied Physics Letters*, *89*(15). WOS:000241247900107

Mikhailov, E., Vlasenko, S., Martin, S. T., Koop, T., & Poschl, U. (2009). Amorphous and crystalline aerosol particles interacting with water vapor: Conceptual framework and experimental evidence for restructuring, phase transitions and kinetic limitations. *Atmospheric Chemistry and Physics*, *9*(24), 9491–9522. WOS:000273060200010

Morris, H. S., Grassian, V. H., & Tivanski, A. V. (2015). Humidity-dependent surface tension measurements of individual inorganic and organic submicrometre liquid particles. *Chemical Science*, *6*(5), 3242–3247. WOS:000353223100071

Morris, H. S., Estillore, A. D., Laskina, O., Grassian, V. H., & Tivanski, A. V. (2016). Quantifying the hygroscopic growth of individual submicrometer particles with atomic force microscopy. *Analytical Chemistry*, *88*(7), 3647–3654. WOS: 000373656300033

Murray, B. J. (2008). Inhibition of ice crystallisation in highly viscous aqueous organic acid droplets. *Atmospheric Chemistry and Physics*, *8*(17), 5423–5433. WOS:000259221400023

Murray, B. J., Wilson, T. W., Dobbie, S., Cui, Z. Q., Al-Jumur, S. M. R. K., Mohler, O., et al. (2010). Heterogeneous nucleation of ice particles on glassy aerosols under cirrus conditions. *Nature Geoscience*, *3*(4), 233–237. WOS:000276159600011

Nagao, E., & Dvorak, J. A. (1999). Phase imaging by atomic force microscopy: Analysis of living homoiothermic vertebrate cells. *Biophysical Journal*, *76*(6), 3289–3297. WOS:000080556700040

Or, V. W., Estillore, A. D., Tivanski, A. V., & Grassian, V. H. (2018). Lab on a tip: Atomic force microscopy—photothermal infrared spectroscopy of atmospherically relevant organic/inorganic aerosol particles in the nanometer to micrometer size range. *Analyst*, *143*(12), 2765–2774. WOS:000435113300006

Ovadnevaite, J., Zuend, A., Laaksonen, A., Sanchez, K. J., Roberts, G., Ceburnis, D., et al. (2017). Surface tension prevails over solute effect in organic-influenced cloud droplet activation. *Nature*, *546*(7660), 637–641. WOS:000404332000044

Petters, M. D., & Kreidenweis, S. M. (2007). A single parameter representation of hygroscopic growth and cloud condensation nucleus activity. *Atmospheric Chemistry and Physics*, *7*(8), 1961–1971. WOS:000246455500005

Poggi, M. A., Gadsby, E. D., Bottomley, L. A., King, W. P., Oroudjev, E., & Hansma, H. (2004). Scanning probe microscopy. *Analytical Chemistry*, *76*(12), 3429–3444. doi: 10.1021/ac0400818

Pöschl, U. (2005). Atmospheric aerosols: Composition, transformation, climate and health effects. *Angewandte Chemie International Edition*, *44*(46), 7520–7540. doi: 10.1002/anie.200501122

Ray, K. K., Lee, H. D., Gutierrez, M., Chang, F. J., & Tivanski, A. V. (2019). Correlating 3D morphology, phase state and viscoelastic properties of individual substrate-deposited particles. *Analytical Chemistry*, *91*(12). doi: 10.1021/acs.analchem.9b00333

Reid, J. P., Bertram, A. K., Topping, D. O., Laskin, A., Martin, S. T., Petters, M. D., et al. (2018). The viscosity of atmospherically relevant organic particles. *Nature Communications*, *9*. WOS:000426659300002

Riemer, N., Ault, A. P., West, M., Craig, R. L., & Curtis, J. H. (2019). Aerosol mixing state: Measurements, modeling, and impacts. *Reviews of Geophysics*, *57*(2), 187–249. doi: 10.1029/2018RG000615

Ruehl, C. R., Davies, J. F., & Wilson, K. R. (2016). An interfacial mechanism for cloud droplet formation on organic aerosols. *Science*, *351*(6280), 1447–1450. WOS:000372756200048

Scott, W. W., & Bhushan, B. (2003). Use of phase imaging in atomic force microscopy for measurement of viscoelastic contrast in polymer nanocomposites and molecularly thick lubricant films. *Ultramicroscopy*, *97*(1–4), 151–169. WOS:000183783200019

Shiraiwa, M., Ammann, M., Koop, T., & Poschl, U. (2011). Gas uptake and chemical aging of semisolid organic aerosol particles. *Proceedings of the National Academy of Sciences of the United States of America*, *108*(27), 11003–11008. WOS: 000292376700022

Slade, J. H., Ault, A. P., Bui, A. T., Ditto, J. C., Lei, Z., Bondy, A. L., et al. (2019). Bouncier particles at night: Biogenic secondary organic aerosol chemistry and sulfate drive Diel variations in the aerosol phase in a mixed Forest. *Environmental Science & Technology*, *53*(9), 4977–4987. doi: 10.1021/acs.est.8b07319

Song, Y. C., Haddrell, A. E., Bzdek, B. R., Reid, J. P., Barman, T., Topping, D. O., et al. (2016). Measurements and predictions of binary component aerosol particle viscosity. *Journal of Physical Chemistry A*, *120*(41), 8123–8137. WOS:000386107400015

Yazdanpanah, M. M., Hosseini, M., Pabba, S., Berry, S. M., Dobrokhotov, V. V., Safir, A., et al. (2008). Micro-Wilhelmy and related liquid property measurements using constant-diameter Nanoneedle-tipped atomic force microscope probes. *Langmuir*, *24*(23), 13753–13764. WOS:000261216500076

Zobrist, B., Marcolli, C., Pedernera, D. A., & Koop, T. (2008). Do atmospheric aerosols form glasses? *Atmospheric Chemistry and Physics*, *8*(17), 5221–5244. WOS:000259221400011

Part III
Particle Types and Transformations

8

Transmission Electron Microscopy for Studying Atmospheric Soot Particles

Kouji Adachi

ABSTRACT

Soot particles are emitted from the combustion of fossil fuels and biofuels, such as diesel engine emissions and biomass burning. Smoke that appears black contains a large fraction of soot particles. The darkness of the soot particles causes dimming on the surface and warming in the atmosphere. Soot particles consist of aggregates of carbon nanospherules and have fractal shapes with sizes ranging from hundreds of nanometers to micrometers. Because soot particles have high surface-to-volume ratios, they become a venue for gas condensation. In addition, soot particles coagulate with other aerosol particles, forming internally mixed particles. Shapes, mixing states, and compositions are of interest to understand their influences on climate and human health. Because of their small sizes and complicated shapes and compositions, transmission electron microscopy (TEM) is commonly used to investigate the chemical and physical properties of individual soot particles. TEM has a high spatial resolution down to nanometers or angstroms. In addition, TEM can be used to measure various chemical and physical properties of materials with nanometer sizes with various detectors, such as an energy-dispersive X-ray spectrometer and an electron energy loss spectrometer. This chapter aims to review TEM analyses and their applications to atmospheric soot particles.

8.1. INTRODUCTION

Soot particles are ubiquitous in the global atmosphere, including over the ocean, at high altitudes, and in Arctic regions (Schwarz et al., 2010, 2017; Schill et al., 2020). These particles are emitted through the incomplete combustion of fossil fuels and biomass burning (Bond et al., 2004, 2013). Soot particles absorb visible light and, in turn, emit heat (Bond and Bergstrom, 2007). The light-absorbing property of soot particles makes them an important climate factor (Ramanathan and Carmichael, 2008). Soot particles are the second- or third-largest contributor to positive radiative forcing in the atmosphere (Ramanathan & Carmichael, 2008; Jacobson, 2001; Bond et al., 2013; IPCC, 2013). Additionally, soot particles can be transported over thousands of kilometers and are removed from the atmosphere by becoming cloud condensation and ice nuclei. They are washed out by rain droplets or snowflakes and then settle down to the ground surface (Liu et al., 2020). When soot particles deposit on snow, they contaminate the snow, reduce its albedo, and accelerate melting (Hansen & Nazarenko, 2004).

In general, *soot* does not have a precise definition, and its definition is based on instruments (Buseck et al., 2014; Petzold et al., 2013). When measuring soot particles based on their light absorption properties, the light-absorbing materials are commonly called black carbon (BC). When measuring soot particles using thermal–optical analysis, the refractory materials are called elemental carbon (EC). When viewing "soot" particles

Department of Atmosphere, Ocean, and Earth System Modeling Research, Meteorological Research Institute, Tsukuba, Japan

Microanalysis of Atmospheric Particles: Techniques and Applications, Geophysical Monograph 288, First Edition.
Edited by Joseph M. Conny and Peter R. Buseck.

DOI: 10.1002/9781119554318.ch8

using electron microscopy techniques, their light-absorbing properties or thermal–optical properties are not directly known, so it is inappropriate to refer to "soot" as BC and EC in the imaging analysis based on their strict definitions. Buseck et al. (2014) proposed the term *ns-soot*, where "ns" refers to carbon nanospheres and is defined as "particles that consist of nanospheres, typically with diameters < 100 nm, that possess distinct structures of concentrically wrapped, graphene-like layers of carbon and with grape-like morphologies" for TEM analysis. Here, we apply the definition of ns-soot to *soot*.

The shapes and compositions of soot particles vary depending on the emission sources, condensation of gases, collision with other particles, and hygroscopic history (i.e., water absorption and drying out) in the atmosphere (Tumolva et al., 2010; Moffet et al., 2016). Such particle shapes and mixtures with other materials significantly change the optical and hygroscopic properties of soot (Zhang et al., 2008; Wang et al., 2014; Liu et al., 2017; China et al., 2015). The optical properties of soot influence its radiative forcing, and the hygroscopicity determines its cloud condensation nuclei (CCN) activity and atmospheric lifetime (Oshima et al., 2009). For example, when soot particles are coated with light-scattering materials (e.g., sulfate and organic materials), they amplify the light absorption of soot by focusing the incident light on the soot core (Bond et al., 2006). When soot particles have large, compact shapes, the surface area exposed to light relative to their masses becomes smaller than those having open, fractal shapes, resulting in less light mass absorption efficiency (Cappa et al., 2012; Adachi et al., 2010). The presence of mesopore (pore diameters from 2 to 50 nm) on soot particles also influences their ability to act as ice nucleating particles (INPs) through a pore condensation and freezing mechanism (Marcolli et al., 2021; Mahrt et al., 2018) and can change their ice cloud-forming properties.

Changes in the chemical and physical properties of soot occur on its surface at a scale of one hundred nanometers or smaller. Transmission electron microscopy (TEM) is a powerful technique to analyze such small-scale phenomena (Buseck & Pósfai, 1999; Li et al., 2016; Zhu et al., 2013). Scanning electron microscopy (SEM) has also been applied to measure soot particles. The SEM technique measures particle surface shapes and compositions using a scanning electron beam. The advantage of SEM over TEM is that SEM measures particles that are collected on various substrates (e.g., quartz fiber filters and membrane filters), whereas TEM measures only thin particles (usually thinner than several hundred nanometers) collected only on TEM grid substrates. SEM generally uses lower accelerating voltages (<~30 keV) than TEM (>~80 keV), resulting in less beam damage caused by the accelerating voltage. On the other hand, TEM has a better spatial resolution than SEM and measures internal structures and various physical and chemical properties of the target materials.

TEM is equipped with various detectors to measure shapes, compositions, crystallinity, and other chemical and physical properties that arise through the interactions between materials and the electron beam. Although TEM is useful for aerosol particle measurements, the knowledge of basic X-ray physics and crystallography is needed to interpret the results. This chapter provides a brief introduction to such knowledge for applications related to atmospheric soot particle measurements and shows TEM imaging for shapes and mixing states (Section 8.3.1), diffraction for crystallography measurements (Section 8.3.2), energy-dispersive X-ray spectroscopy (EDS or EDX) for compositional measurements (Section 8.3.3), electron energy loss spectroscopy (EELS) for compositions, chemical bonding, and electronic state measurements (Section 8.3.4), electron tomography (ET) for three-dimensional configuration analyses (Section 8.3.5), and environmental TEM techniques for in situ experiments (Section 8.3.6).

8.2. PRINCIPLES AND LIMITATIONS OF TEM

8.2.1. TEM and STEM

TEM is a technique that measures thin specimens by using a controlled parallel electron beam and shows an image with the transmitted electrons (Figure 8.1). When

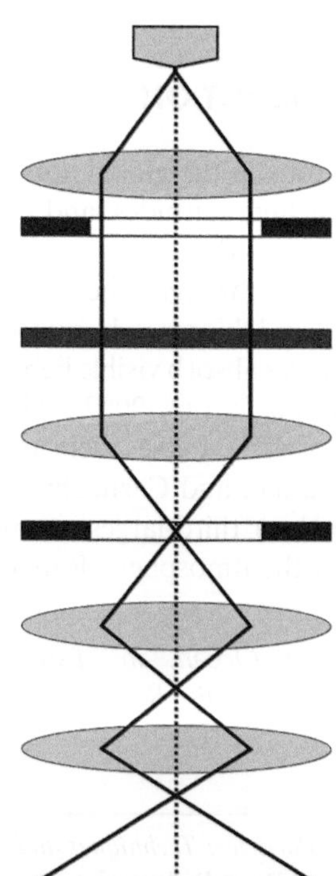

Figure 8.1 Schematic ray path in the TEM mode. Various lenses and apertures are used to control the electron beam.

the electron beam is transmitted through the specimens, the beam is scattered and absorbed by the specimens, which changes its energy from the incident beam (Figure 8.2). Furthermore, characteristic X-rays that have various energies are also emitted, depending on the elements that comprise the specimen. By measuring the electrons and characteristic X-rays using specified detectors, one can analyze the shapes, compositions, crystal structures, chemical bonding, and optical properties of nanometer-sized materials.

TEM is used in two different modes: conventional TEM (Figure 8.3a) and scanning transmission electron microscopy (STEM; Figures 8.3c,d). To obtain TEM and STEM images, the former uses a film or charge-coupled device (CCD) camera with a spread electron beam, and the latter uses a STEM detector with a focused electron beam scanned over the specimen and forms the image

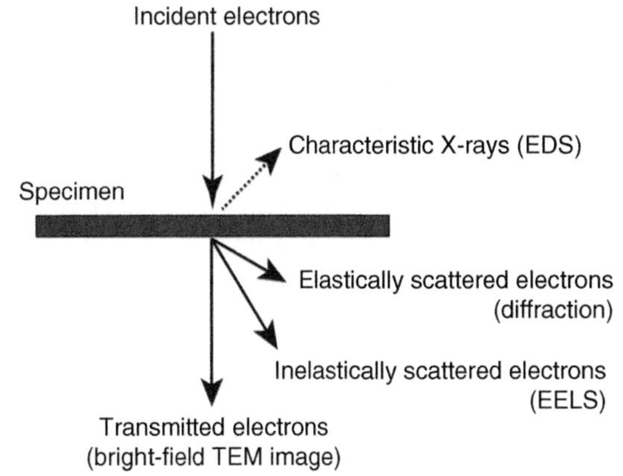

Figure 8.2 Signals generated by the interactions of electrons with a specimen used for bright-field imaging, EDS, diffraction pattern, and EELS in TEM and STEM.

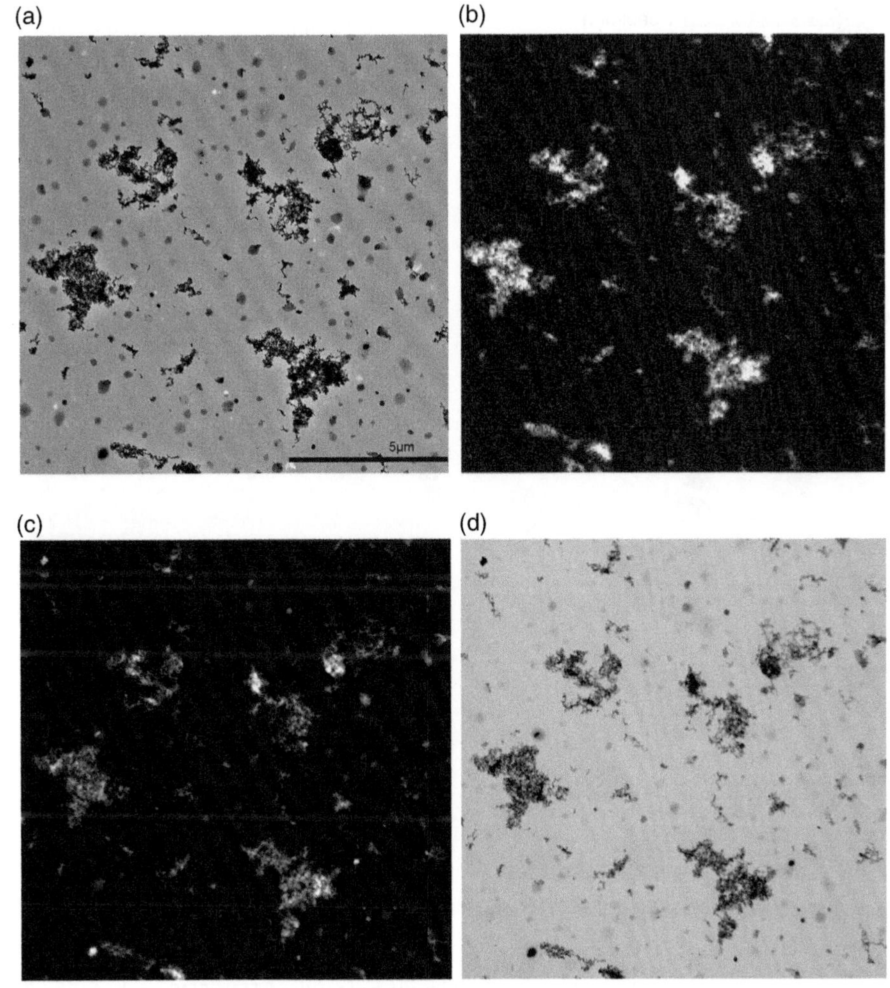

Figure 8.3 (a) Bright-field TEM image, (b) elemental mapping image (carbon) based on energy-dispersive X-ray spectra collected in STEM mode, (c) dark-field STEM image, and (d) bright-field STEM image of soot particles. All images were taken from the same area using a JEM-1400 (JEOL, Tokyo, Japan). The sample was collected from a traffic site at Kawasaki, Japan, on 20 August 2016 (JST).

pixel by pixel using the signal collected at each beam position (Figure 8.4). STEM images are shown in either dark-field (Figure 8.3c) or bright-field (Figure 8.3d) mode. A bright-field STEM image uses a transmitted electron beam so that the more transmitting portions (e.g., thin, light elements) become brighter (Figures 8.2 and 8.4). On the other hand, a dark-field STEM image is constructed from the intensities of scattered electrons, as measured in each beam position (Figure 8.4), and has a Z-contrast, which indicates that high atomic number elements appear bright. A high-angle annular dark-field (HAADF) detector, which detects thermal diffuse scattering (or inelastic scattering) at high angles, is also used to view dark-field STEM images with high Z-contrast, which strongly correlated to the atomic mass (i.e., heavier elements contribute to brighter contrast) (Utsunomiya & Ewing, 2003) (Figure 8.4). When soot particles are shown in a TEM image, it is essential to understand what TEM mode is used to take their photos, as the contrasts have different meanings (Figure 8.3).

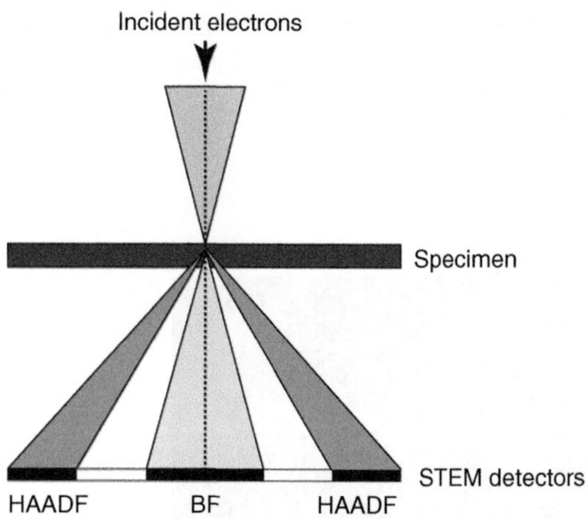

Figure 8.4 Electron detection systems in the STEM mode. BF, bright-field STEM detector; HAADF, high-angle annular dark-field STEM detector.

8.2.2. TEM Imaging

The TEM image quality depends on the TEM instruments, settings, and adjustments (Figure 8.1). The electron gun (e.g., LaB_6, tungsten, cold field emission, and thermally assisted field emission), CCD camera, aberration correction, accelerating voltage, apertures, beam current and density, alignments (e.g., gun, aperture, beam, and lens), focusing, and sample preparation all influence TEM imaging. In addition, when obtaining a bright and high-magnification TEM image, the specimen will be damaged by the intense electron beam, and the shapes and structures of beam-sensitive materials such as sulfate or organic aerosol particles will change. Thus, an adequate beam intensity and quick operation are necessary to obtain high-quality TEM images.

8.2.3. Electron Beam Damage

Many aerosol particles are sensitive to electron beam exposure (Figure 8.5). Electron beam irradiation produces heating, electrostatic charging, ionization damage

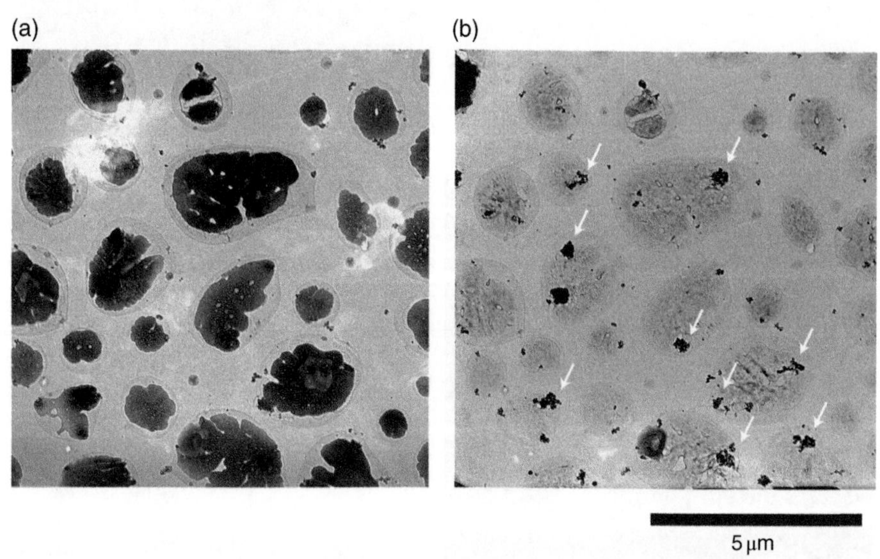

Figure 8.5 Beam damage and removal of sulfate. (a) TEM image before the STEM-EDS analysis and (b) TEM image after the STEM-EDS analysis. Sulfate was removed by the electron beam, and internally mixed soot particles, indicated by white arrows, became apparent (b). This sample was collected from 23:00 to 23:30 on 22 March 2012 at a mountain site in Japan. Adachi et al. (2014)/with permission of the American Geophysical Union.

(radiolysis), displacement damage, and sputtering (Egerton et al., 2004), resulting in the decomposition of sulfate, organic matter, sea salts, and other aerosol particles. The increase in sample temperature is estimated to range from 0.5 to 1.4 °C for a 5 nA electron beam with beam diameters from 1 μm to 1 nm on a carbon film (Egerton et al., 2004). Thus, evaporation by heating is not a major source of beam damage, but other irradiation damage causes decomposition. To reduce beam damage, for example, a low-dose imaging technique (Veghte et al., 2018) and cooling a TEM specimen (Egerton et al., 2004) are available.

Soot particles are relatively stable when exposed to an electron beam. However, their coatings can decompose after intensive beam exposure (Figure 8.5). Thus, soot particles embedded within other particles become apparent after removing the coating material by exposing it (Adachi et al., 2014). In general, a broad electron beam at lower magnifications (<~× 10 K) with an adequate exposure time (~10 sec) does not significantly damage soot particles and other aerosol particles.

8.2.4. TEM Chamber Vacuum

TEM samples need to be measured in a vacuum to eliminate the collision of the electron beam with air molecules. The vacuum level varies depending on the type of electron gun source and is commonly 10^{-5} Pa or lower. In a vacuum, volatile and semivolatile organic and inorganic matter will be lost, and only nonvolatile materials can be measured by TEM. If the aerosol particle samples contain water, the water is lost in the TEM chamber. Although soot particles do not have water, their coating can have water in the atmosphere, which will be lost in the TEM chamber.

8.2.5. TEM Sample Collection

An impactor sampler and an electrostatic sampler are commonly used for aerosol particle collections on TEM grids. An impactor sampler collects aerosol particle samples by impacting particles on TEM grids with airflow speeds of, generally, 1 to 3 l/min with sampling times of several to 30 minutes. The airflow speed, nozzle diameter, and distance between the jet nozzle and TEM grids determine the 50% cutoff aerodynamic diameter sizes. An electrostatic sampler collects charged particles with a wide size range from nano- to micron-sized particles. Generally, it takes several hours or longer to collect enough particles on TEM grids. For both samplers, most aerosol particles are small enough to transmit an electron beam and can be analyzed using TEM without any sample treatments after sampling.

The TEM samplings have disadvantages compared to measurements by online instruments. First, TEM analysis requires particles that are collected on TEM grid substrates. This means TEM cannot measure aerosol particles that are collected on, for example, quartz fiber filters and membrane filters, both of which are commonly used in aerosol sampling. Second, TEM grid sampling has lower sampling time resolutions (from several minutes to 30 minutes) than online instruments. Last, TEM substrates (e.g., formvar, collodion, and lacey) are small (~3 mm in diameter), thin, and fragile and must be carefully handled so that they do not break during the samplings.

Various types of TEM grids and substrates are available for aerosol particle samplings. Continuous carbon substrates (e.g., formvar and collodion substrates; Figure 8.3), flat carbon substrates with holes (holey carbon substrates), and lacey carbon substrates are commonly used for TEM sampling (Chen et al., 2017). The measurement of aerosol particles with continuous carbon substrates using EDS is generally easier than that with lacey carbon substrates because lacey substrates have a similar contrast with the aerosol particles. On the other hand, soot particles collected on the edges of holey or lacey carbon substrates have minimal interaction with the substrate and a higher TEM resolution than continuous substrates.

Cu TEM grids are widely applied for aerosol particle sampling. In addition, Mo, Ni, Al, Ti, and stainless grids are available. The elements used in grids are always detected in the EDS analysis as background signals and cannot be measured in the samples. Mo TEM grids can be used for TEM heating measurements because of their higher melting temperature than Cu (Adachi et al., 2018).

8.3. TEM TECHNIQUES

8.3.1. Imaging

The imaging of small particles is the most common and advanced technique in TEM. In bright-field TEM imaging, electrons are transmitted through the specimen and bombard the fluorescent screen, the CCD camera, or the film camera to create a TEM image. The contrast in bright-field TEM imaging depends primarily on the particle thickness and is also influenced by the particle composition and crystallinity. Thin particles scatter and absorb fewer electrons and thus appear bright in the TEM image. On the other hand, thick particles look dark (Figure 8.3a). When no electron beam is transmitted through a particle, it appears black in the bright-field TEM images.

Soot particles are generally thin enough to allow us to observe their structures without any pre-analysis treatments. When soot particles are thinly coated with other materials, internally mixed soot particles can still be

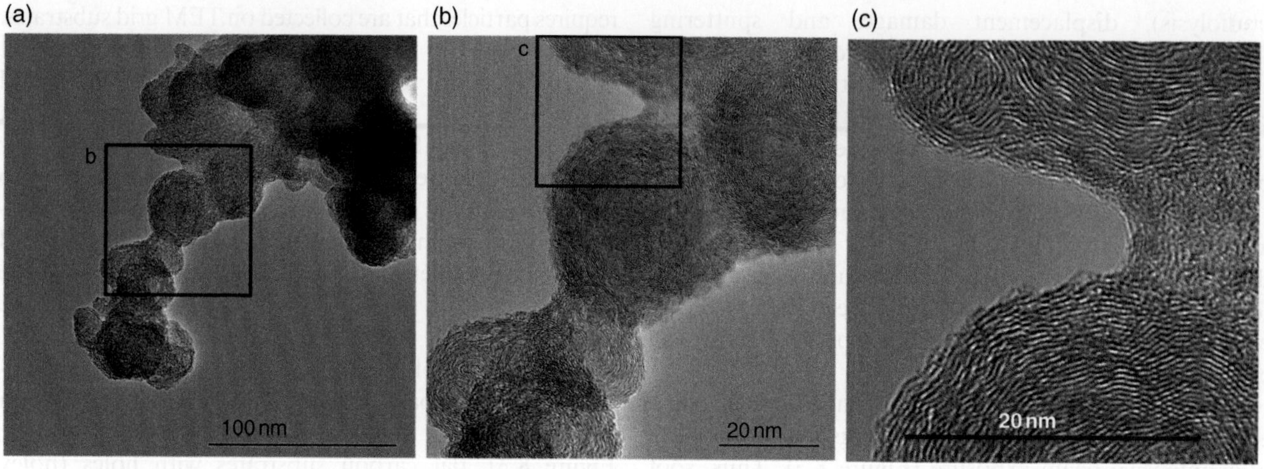

Figure 8.6 High-resolution TEM images of a soot particle. Graphitic structures of the soot particle can be seen in (c). The sample was collected between 15:50 and 15:54 on 30 July 2014 (UTC) during the BBOP campaign. TEM images were taken using an ARM-200 (JEOL, Tokyo, Japan). Adachi et al. (2019) with permission of National Academy of Sciences.

observed (Ishimoto et al., 2019; Buseck et al., 2014). When soot particles are thickly coated or embedded within other materials, the soot particles are difficult to see unless the coating materials are removed by beam irradiation or the water dialysis method (Okada, 1983).

High-resolution imaging of soot particles reveals that they consist of carbon nanospherules with diameters from ~10 to 100 nm (Figures 8.6a, b) (Buseck et al., 2014). The nanospherules have several graphitic layers with lattice spacing from ~0.36 to 0.38 nm (Figure 8.6c), which is larger than that of graphite (0.335 nm; Pósfai et al., 1999). The numbers of nanospherules are typically several to hundreds. These nanospherules form aggregates with open or closed structures. The degrees of particle compactness can be evaluated using particle shape factors such as aspect ratio (large axis/minor axis) and circularity ($4\pi A/(P^2)$, where P is the perimeter and A is the area) (Zhang et al., 2020). The degree of particle compactness with a fractal shape can also be indicated using the fractal dimension (D_f): that is, a fractal particle follows the statistical scaling law, $N = k_a(R_g/a)^{Df}$, where N is the number of monomers that consist of the fractal particle, k_a is the fractal prefactor, R_g is the radius of gyration, and a is the monomer radius (Adachi et al., 2007). D_f can be determined from soot TEM images and is an important parameter in estimating their optical properties (Adachi et al., 2010) and atmospheric aging processes (Wang et al., 2017).

8.3.2. A Selected Area Diffraction Pattern

A selected area diffraction (SAD) pattern in TEM contains information on the spatial frequencies (such as periodic lattice spacings in crystals) within the specimen (Figure 8.7a). A SAD pattern within the area limited by the selected area aperture forms in the back focal plane of the objective lens and can be viewed by using lens settings to project this back focal plane onto the screen. The SAD patterns are unique to the crystal structures of the target materials; they are a series of spots for crystal materials (e.g., metal particles) or ring patterns for defective crystalline and amorphous materials such as soot particles (Figure 8.7a). The distances of the lattice spaces can be obtained from the spots and ring patterns by calibrating the camera length used for SAD measurements with a standard material (e.g., gold).

8.3.3. Energy Dispersive X-Ray Spectroscopy

EDS is a technique to measure the compositions of materials from characteristic X-rays using an EDS detector (Figures 8.2, 8.3, and 8.7b). Both EDS and EDX are commonly used for the acronym; *EDS* is used here. Boron (B) or heavier elements, depending on the EDS detectors and materials used for the X-ray detector window, can be measured from the area where the electron beam is exposed. When a specimen is exposed to an electron beam, the atoms in the specimen are excited and emit specific wavelengths of X-rays that are characteristic of the atomic structure of the elements. The X-ray intensity depends on its relative amount, absorption by the specimen, and its atomic number (Figure 8.7b). A focused beam in TEM or STEM mode is used for the EDS measurement. Acquisition times for the EDS measurements of aerosol particles depend on the purpose of the measurements, and the average composition of an individual particle can be obtained using an acquisition time from 10 to 100 seconds (e.g., 20 seconds [Niemi

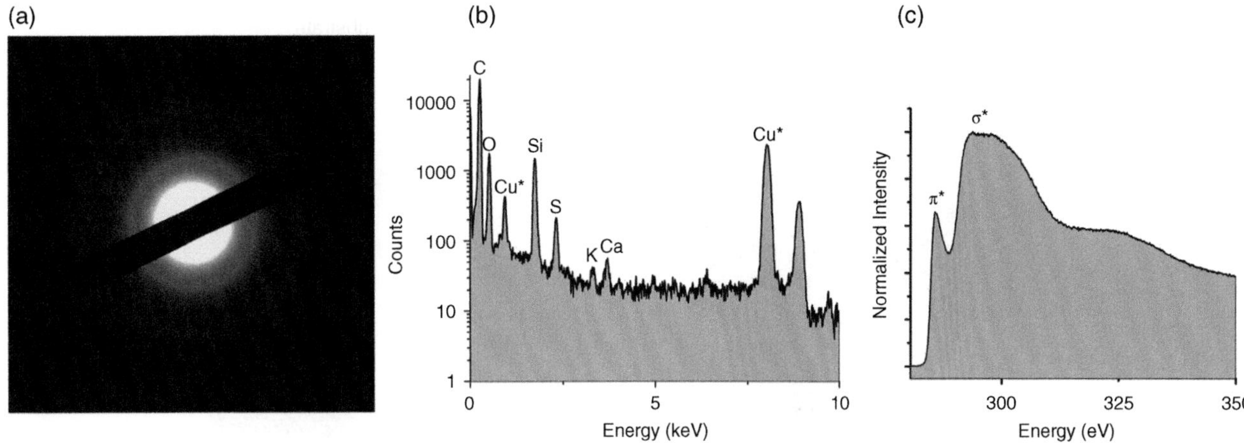

Figure 8.7 (a) A selected area diffraction pattern, (b) an EDS spectrum, and (c) an EELS spectrum of soot particles. The diffraction pattern in a and EDS spectrum in (b) were obtained from soot particles shown in Figure 8.3. The beam stop bar shown in the diffraction image (a) covers the bright central beam spot to protect the CCD camera from the focused electron beam and to give better contrast in the image. Asterisks in (b) indicate elements originating from the Cu grid. Note that the y-axis is indicated by a log scale, indicating a strong C signal in the soot particle with minor elements (O, Si, S, K, and Ca). The soot particle used in (c) was obtained from biomass burning. A JEM-1400 TEM (JEOL, Tokyo, Japan) was used for the diffraction pattern and EDS measurements. An ARM 200 (JEOL, Tokyo, Japan) was used for the EELS. Adapted from Adachi et al. (2019)/with permission of the National Academy of Sciences.

et al., 2006]). Under the STEM mode, one can obtain elemental mapping images (Figure 8.3b) or a line profile by acquiring characteristic X-ray spectra from each pixel forming the image or line.

Precise quantification in EDS is challenging, especially for aerosol samples. The electron beam intensity and particle thickness influence the quantification (Geng et al., 2010). Particle thicknesses vary particle by particle: that is, thick particles emit greater X-ray intensities and absorb more X-rays from light elements than thin particles. Thus, the relative mass fractions of selected elements within each particle are commonly measured as a semi-quantification method of aerosol particles in STEM-EDS (Ching et al., 2019).

For aerosol particle measurements, the classifications of aerosol species in EDS use characteristic elements for specific aerosol particle types (e.g., S for sulfate) (Ching et al., 2019) with prior knowledge of the information from other types of measurements (e.g., an aerosol mass spectrometer). In the case of soot particles, as their compositions are similar to most organic materials, both shape and composition are applied to distinguish them from other carbonaceous particles (Fu et al., 2012; Adachi et al., 2021). Because soot particles typically emit greater C and O signals (~7 times) than the substrate (Adachi et al., 2016), they can be identified using EDS. However, its accurate quantification is challenging because of interference from the carbon substrate.

8.3.4. Electron Energy Loss Spectroscopy

In EELS, the energy loss of transmitted electrons can be used for the analysis of compositions, chemical bonding, and electronic states (Figure 8.7c). EELS only measures compositions of selected elements from each spectrum, whereas EDS can obtain an average particle composition with a wide range of elements. On the other hand, as an advantage over EDS, EELS can distinguish different valences, e.g., Fe^{2+} and Fe^{3+}, in aerosol Fe-oxide particles (Maher et al., 2016; Moteki et al., 2017) and can measure light elements that cannot be detected using EDS (e.g., Li and Be). EELS also has a better spatial resolution (<1 nm) than EDS.

EELS has been used to identify soot particles by detecting their structure (Deboudt et al., 2010; Chen et al., 2006; Katrinak et al., 1992). Soot particles mainly consist of graphitic structures, which have a strong π peak originating from sp2 bonding (Figure 8.7c). The ratio of π/σ peaks is higher in soot than in organic materials. This graphitic soot structure has also been measured by near-edge X-ray absorption fine structure spectroscopy (NEXAFS) via scanning transmission X-ray microscopy (STXM). In general, EELS has better spatial resolution than NEXAFS, and NEXAFS has better energy resolution and less beam damage than EELS when measuring soot particles (Braun et al., 2005, 2009). EELS with a monochromated electron beam has been applied to evaluate the refractive index of soot particles, yielding 1.8–0.5*i* at a wavelength of 550 nm (Zhu et al., 2014).

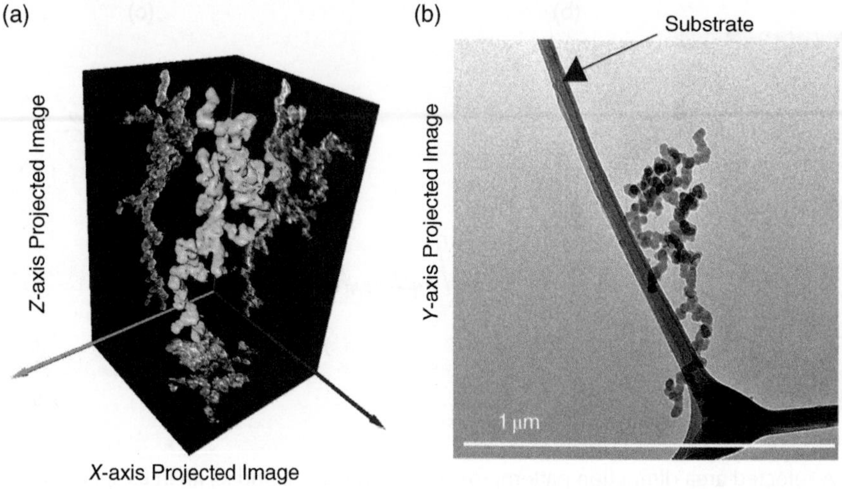

Figure 8.8 (a) Three-dimensional soot particle image with projected images and (b) the corresponding soot TEM image. Adapted from Adachi et al. (2007)/with permission of the American Geophysical Union.

8.3.5. Electron Tomography

ET is a technique to reconstruct the three-dimensional (3D) shapes of submicron particles using a tilt series of TEM images (Figure 8.8) (van Poppel et al., 2005; Adachi et al., 2007). TEM images of an object are acquired from ~+70° to −70° using a high-tilt TEM holder. The series of images are used for reconstructing 3D data using, for example, the sequential iterative reconstruction technique (SIRT) algorithm. Using 3D objects, one can view the 3D shape and determine the surface area, volume, and 3D parameters (e.g., D_f, R_g, and mass center).

Using ET, van Poppel et al. (2005) showed that the surface areas of soot particles were high because of their aggregate shapes and thus influenced their atmospheric lifetime. In addition, the optical properties of individual soot particles have been calculated based on their 3D configurations and coating materials in the sensitivity study of their radiative forcing with various assumptions of the soot mixing states (Adachi et al., 2010).

8.3.6. Environmental TEM Techniques

In environmental TEM (ETEM), the temperatures and relative humidity (RH) are controlled during the TEM observation to measure the changes in shape, size, and composition of the sample in response to the monitored temperature and RH. For aerosol particle samples, the volatility and refractoriness at specific temperatures can be measured using a furnace heating holder (Adachi et al., 2018) or a micro-electro mechanical system (MEMS) (Veghte et al., 2018). Soot particles are relatively stable for a temperature change and did not show changes up to 800 °C (Adachi et al., 2018), indicating their refractoriness.

By introducing water vapor in the ETEM chamber with a differential pumping system or in a holder with electron-transparent windows, shape and volume changes of hygroscopic aerosol particles can be directly observed through water uptake or loss, deliquescence, and efflorescence (Wise et al., 2007). Soot particles do not absorb water or change their shape by increasing RH (Semeniuk et al., 2006). On the other hand, the coating materials (e.g., sulfate) can increase their volume by absorbing water, resulting in changing soot light mass absorption efficiency.

8.4. CONCLUSION

In this chapter, various TEM techniques were introduced by focusing on measurements of atmospheric soot particles. Because the abundances, shapes, mixing states, and graphitic structures of soot particles are important to understand their climate influences, various TEM techniques have been used for soot particle measurements. As TEM measurements have both advantages and limitations, proper knowledge is necessary. Because soot particles are ubiquitous, have complicated shapes, and are stable to electron beam irradiation, TEM is a powerful technique for their analysis.

REFERENCES

Adachi, K., Chung, S. H., Friedrich, H., & Buseck, P. R. (2007). Fractal parameters of individual soot particles determined using electron tomography: Implications for optical properties. *Journal of Geophysical Research*, *112*(D14202). doi: 10.1029/2006jd008296

Adachi, K., Chung, S. H., & Buseck, P. R. (2010). Shapes of soot aerosol particles and implications for their effects on climate. *Journal of Geophysical Research*, *115*(D15206). doi: 10.1029/2009jd012868

Adachi, K., Zaizen, Y., Kajino, M., & Igarashi, Y. (2014). Mixing state of regionally transported soot particles and the coating effect on their size and shape at a mountain site in Japan. *Journal of Geophysical Research: Atmospheres*, *119*(9), 5386–5396. doi: 10.1002/2013JD020880

Adachi, K., Moteki, N., Kondo, Y., & Igarashi, Y. (2016). Mixing states of light-absorbing particles measured using a transmission electron microscope and a single-particle soot photometer in Tokyo, Japan. *Journal of Geophysical Research: Atmospheres*, *121*(15), 9153–9164. doi: 10.1002/2016JD025153

Adachi, K., Sedlacek, A. J., Kleinman, L., Chand, D., Hubbe, J. M., & Buseck, P. R. (2018). Volume changes upon heating of aerosol particles from biomass burning using transmission electron microscopy. *Aerosol Science and Technology*, *52*(1), 46–56. doi: 10.1080/02786826.2017.1373181

Adachi, K., et al. (2019). Spherical tarball particles form through rapid chemical and physical changes of organic matter in biomass-burning smoke. *Proceedings of the National Academy of Sciences of the United States of America*, *116*(39), 19336–19341. doi: 10.1073/pnas.1900129116

Adachi, K., Oshima, N., Ohata, S., Yoshida, A., Moteki, N., & Koike, M. (2021). Compositions and mixing states of aerosol particles by aircraft observations in the Arctic springtime, 2018. *Atmospheric Chemistry and Physics*, *21*(5), 3607–3626. doi: 10.5194/acp-21-3607-2021

Bond, T. C., & Bergstrom, R. W. (2007). Light absorption by carbonaceous particles: An investigative review. *Aerosol Science and Technology*, *40*(1), 27–67. doi: 10.1080/02786820500421521

Bond, T. C., Streets, D. G., Yarber, K. F., Nelson, S. M., Woo, J.-H., & Klimont, Z. (2004). A technology-based global inventory of black and organic carbon emissions from combustion. *Journal of Geophysical Research*, *109*, D14203. doi: 10.1029/2003JD003697

Bond, T. C., Habib, G., & Bergstrom, R. W. (2006). Limitations in the enhancement of visible light absorption due to mixing state. *Journal of Geophysical Research*, *111*(D20211). doi: 10.1029/2006jd007315

Bond, T. C., et al. (2013). Bounding the role of black carbon in the climate system: A scientific assessment. *Journal of Geophysical Research: Atmospheres*, *118*(11), 5380–5552. doi: 10.1002/jgrd.50171

Braun, A., Huggins, F. E., Shah, N., Chen, Y., Wirick, S., Mun, S. B., et al. (2005). Advantages of soft X-ray absorption over TEM-EELS for solid carbon studies—A comparative study on diesel soot with EELS and NEXAFS. *Carbon*, *43*(1), 117–124. doi: 10.1016/j.carbon.2004.08.029

Braun, A., Kubatova, A., Wirick, S., & Mun, S. B. (2009). Radiation damage from EELS and NEXAFS in diesel soot and diesel soot extracts. *Journal of Electron Spectroscopy and Related Phenomena*, *170*(1–3), 42–48. doi: 10.1016/j.elspec.2007.08.002

Buseck, P. R., & Pósfai, M. (1999). Airborne minerals and related aerosol particles: Effects on climate and the environment. *Proceedings of the National Academy of Sciences of the United States of America*, *96*(7), 3372–3379. doi: 10.1073/pnas.96.7.3372

Buseck, P. R., Adachi, K., Gelencsér, A., Tompa, É., & Pósfai, M. (2014). Ns-soot: A material-based term for strongly light-absorbing carbonaceous particles. *Aerosol Science and Technology*, *48*(7), 777–788. doi: 10.1080/02786826.2014.919374

Cappa, C. D., et al. (2012). Radiative absorption enhancements due to the mixing state of atmospheric black carbon. *Science*, *337*(6098), 1078–1081. doi: 10.1126/science.1223447

Chen, Y., Shah, N., Huggins, F. E., & Huffman, G. P. (2006). Microanalysis of ambient particles from Lexington, KY, by electron microscopy. *Atmospheric Environment*, *40*(4), 651–663. doi: 10.1016/j.atmosenv.2005.09.036

Chen, C., Enekwizu, O., Ma, Y., Zakharov, D., & Khalizov, A. (2017). The impact of sampling medium and environment on particle morphology. *Atmosphere*, *8*(12), 162. doi: 10.3390/atmos8090162

China, S., et al. (2015). Morphology and mixing state of aged soot particles at a remote marine free troposphere site: Implications for optical properties. *Geophysical Research Letter*, *42*, 1243–1250. doi: 10.1002/2014GL062404

Ching, J., Adachi, K., Zaizen, Y., Igarashi, Y., & Kajino, M. (2019). Aerosol mixing state revealed by transmission electron microscopy pertaining to cloud formation and human airway deposition. *npj Climate and Atmospheric Science*, *2*(1). doi: 10.1038/s41612-019-0081-9

Deboudt, K., Flament, P., Choël, M., Gloter, A., Sobanska, S., & Colliex, C. (2010). Mixing state of aerosols and direct observation of carbonaceous and marine coatings on African dust by individual particle analysis. *Journal of Geophysical Research: Atmospheres*, *115*(D24207). doi: 10.1029/2010jd013921

Egerton, R. F., Li, P., & Malac, M. (2004). Radiation damage in the TEM and SEM. *Micron*, *35*(6), 399–409. doi: 10.1016/j.micron.2004.02.003

Fu, H., Zhang, M., Li, W., Chen, J., Wang, L., Quan, X., & Wang, W. (2012). Morphology, composition and mixing state of individual carbonaceous aerosol in urban Shanghai. *Atmospheric Chemistry and Physics*, *12*(2), 693–707. doi: 10.5194/acp-12-693-2012

Geng, H., Kang, S., Jung, H.-J., Choël, M., Kim, H., & Ro, C.-U. (2010). Characterization of individual submicrometer aerosol particles collected in Incheon, Korea, by quantitative transmission electron microscopy energy-dispersive X-ray spectrometry. *Journal of Geophysical Research*, *115*(D15306). doi: 10.1029/2009jd013486

Hansen, J., & Nazarenko, L. (2004). Soot climate forcing via snow and ice albedos. *Proceedings of the National Academy of Sciences of the United States of America*, *101*(2), 423–428. doi: 10.1073/pnas.2237157100

IPCC (2013). Climate change 2013: The physical science basis. In T. F. Stocker, et al. (Eds.), 1535 p*Contribution of Working Group I to the Fifth Assessment Report of the Intergovernmental Panel on Climate Change*. Cambridge, UK, and New York: Cambridge Univ. Press.

Ishimoto, H., Kudo, R., & Adachi, K. (2019). A shape model of internally mixed soot particles derived from artificial surface tension. *Atmospheric Measurement Techniques*, *12*(1), 107–118. doi: 10.5194/amt-12-107-2019

Jacobson, M. Z. (2001). Strong radiative heating due to the mixing state of black carbon in atmospheric aerosols. *Nature*, *409*(6821), 695–697. doi: 10.1038/35055518

Katrinak, K. A., Rez, P., & Buseck, P. R. (1992). Structural variations in individual carbonaceous particles from an urban aerosol. *Environmental Science & Technology*, *26*(10), 1967–1976. doi: 10.1021/es00034a014

Li, W., Shao, L., Zhang, D., Ro, C.-U., Hu, M., Bi, X., et al. (2016). A review of single aerosol particle studies in the atmosphere of East Asia: Morphology, mixing state, source, and heterogeneous reactions. *Journal of Cleaner Production*, *112*, 1330–1349. doi: 10.1016/j.jclepro.2015.04.050

Liu, D., Whitehead, J., Alfarra, M. R., Reyes-Villegas, E., Spracklen, D. V., Reddington, C. L., et al. (2017). Black-carbon absorption enhancement in the atmosphere determined by particle mixing state. *Nature Geoscience*, *10*(3), 184–188. doi: 10.1038/ngeo2901

Liu, D., He, C., Schwarz, J. P., & Wang, X. (2020). Lifecycle of light-absorbing carbonaceous aerosols in the atmosphere. *npj Climate and Atmospheric Science*, *3*(1). doi: 10.1038/s41612-020-00145-8

Maher, B. A., Ahmed, I. A. M., Karloukovski, V., MacLaren, D. A., Foulds, P. G., Allsop, D., et al. (2016). Magnetite pollution nanoparticles in the human brain. *Proceedings of the National Academy of Sciences*, *113*(39), 10797–10801. doi: 10.1073/pnas.1605941113

Mahrt, F., Marcolli, C., David, R. O., Grönquist, P., Barthazy Meier, E. J., Lohmann, U., & Kanji, Z. A. (2018). Ice nucleation abilities of soot particles determined with the horizontal ice nucleation chamber. *Atmospheric Chemistry and Physics*, *18*, 13363–13392. doi: 10.5194/acp-18-13363-2018

Marcolli, C., Mahrt, F., & Kärcher, B. (2021). Soot PCF: Pore condensation and freezing framework for soot aggregates. *Atmospheric Chemistry and Physics*, *21*, 7791–7843. doi: 10.5194/acp-21-7791-2021

Moffet, R. C., O'Brien, R. E., Alpert, P. A., Kelly, S. T., Pham, D. Q., Gilles, M. K., et al. (2016). Morphology and mixing of black carbon particles collected in Central California during the CARES field study. *Atmospheric Chemistry and Physics*, *16*(22), 14515–14525. doi: 10.5194/acp-16-14515-2016

Moteki, N., Adachi, K., Ohata, S., Yoshida, A., Harigaya, T., Koike, M., & Kondo, Y. (2017). Anthropogenic iron oxide aerosols enhance atmospheric heating. *Nature Communications*, *8*, 15329. doi: 10.1038/ncomms15329

Niemi, J. V., Saarikoski, S., Tervahattu, H., Mäkelä, T., Hillamo, R., Vehkamäki, H., et al. (2006). Changes in background aerosol composition in Finland during polluted and clean periods studied by TEM/EDX individual particle analysis. *Atmospheric Chemistry and Physics*, *6*(12), 5049–5066. doi: 10.5194/acp-6-5049-2006

Okada, K. (1983). Nature of individual hygroscopic particles in the urban atmosphere. *Journal of the Meteorological Society of Japan. Ser. II*, *61*(5), 727–736. doi: 10.2151/jmsj1965.61.5_727

Oshima, N., Koike, M., Zhang, Y., & Kondo, Y. (2009). Aging of black carbon in outflow from anthropogenic sources using a mixing state resolved model: 2. Aerosol optical properties and cloud condensation nuclei activities. *Journal of Geophysical Research*, *114*(D18202). doi: 10.1029/2008jd011681

Petzold, A., et al. (2013). Recommendations for reporting "black carbon" measurements. *Atmospheric Chemistry and Physics*, *13*(16), 8365–8379. doi: 10.5194/acp-13-8365-2013

Pósfai, M., Anderson, J. R., Buseck, P. R., & Sievering, H. (1999). Soot and sulfate aerosol particles in the remote marine troposphere. *Journal of Geophysical Research: Atmospheres*, *104*(D17), 21685–21693. doi: 10.1029/1999JD900208

Ramanathan, V., & Carmichael, G. (2008). Global and regional climate changes due to black carbon. *Nature Geoscience*, *1*(4), 221–227. doi: 10.1038/ngeo156

Schill, G. P., et al. (2020). Widespread biomass burning smoke throughout the remote troposphere. *Nature Geoscience*, *13*(6), 422–427. doi: 10.1038/s41561-020-0586-1

Schwarz, J. P., Spackman, J. R., Gao, R. S., Watts, L. A., Stier, P., Schulz, M., et al. (2010). Global-scale black carbon profiles observed in the remote atmosphere and compared to models. *Geophysical Research Letters*, *37*(18), L18812. doi: 10.1029/2010gl044372

Schwarz, J. P., Weinzierl, B., Samset, B. H., Dollner, M., Heimerl, K., Markovic, M. Z., et al. (2017). Aircraft measurements of black carbon vertical profiles show upper tropospheric variability and stability. *Geophysical Research Letters*, *44*, 1132–1140. doi: 10.1002/2016GL071241

Semeniuk, T. A., Wise, M. E., Martin, S. T., Russell, L. M., & Buseck, P. R. (2006). Hygroscopic behavior of aerosol particles from biomass fires using environmental transmission electron microscopy. *Journal of Atmospheric Chemistry*, *56*(3), 259–273. doi: 10.1007/s10874-006-9055-5

Tumolva, L., Park, J.-Y., Kim, J.-s., Miller, A. L., Chow, J. C., Watson, J. G., & Park, K. (2010). Morphological and elemental classification of freshly emitted soot particles and atmospheric ultrafine particles using the TEM/EDS. *Aerosol Science and Technology*, *44*(3), 202–215. doi: 10.1080/02786820903518907

Utsunomiya, S., & Ewing, R. C. (2003). Application of high-angle annular dark field scanning transmission electron microscopy, scanning transmission electron microscopy-energy dispersive X-ray spectrometry, and energy-filtered transmission electron microscopy to the characterization of nanoparticles in the environment. *Environmental Science & Technology*, *37*(4), 786–791. doi: 10.1021/es026053t

van Poppel, L. H., Friedrich, H., Spinsby, J., Chung, S. H., Seinfeld, J. H., & Buseck, P. R. (2005). Electron tomography of nanoparticle clusters: Implications for atmospheric lifetimes and radiative forcing of soot. *Geophysical Research Letters*, *32*(24). doi: 10.1029/2005gl024461

Veghte, D. P., China, S., Weis, J., Lin, P., Hinks, M. L., Kovarik, L., et al. (2018). Heating-induced transformations of atmospheric particles: Environmental transmission Electron microscopy study. *Analytical Chemistry*, *90*(16), 9761–9768. doi: 10.1021/acs.analchem.8b01410

Wang, Q. Y., Huang, R. J., Cao, J. J., Han, Y. M., Wang, G. H., Li, G. H., et al. (2014). Mixing state of black carbon aerosol in a heavily polluted urban area of China: Implications for light absorption enhancement. *Aerosol Science and Technology*, *48*(7), 689–697. doi: 10.1080/02786826.2014.917758

Wang, Y., Liu, F., He, C., Bi, L., Cheng, T., Wang, Z., et al. (2017). Fractal dimensions and mixing structures of soot particles during atmospheric processing. *Environmental Science & Technology Letters*, *4*(11), 487–493. doi: 10.1021/acs.estlett.7b00418

Wise, M. E., Biskos, G., Martin, S. T., Russell, L. M., & Buseck, P. R. (2007). Phase transitions of single salt particles studied using a transmission Electron microscope with an environmental cell. *Aerosol Science and Technology*, *39*(9), 849–856. doi: 10.1080/02786820500295263

Zhang, C., Zhang, Y., Wolf, M. J., Nichman, L., Shen, C., Onasch, T. B., et al. (2020). The effects of morphology, mobility size, and secondary organic aerosol (SOA) material coating on the ice nucleation activity of black carbon in the cirrus regime. *Atmospheric Chemistry and Physics*, *20*(22), 13957–13984. doi: 10.5194/acp-20-13957-2020

Zhang, R., Khalizov, A. F., Pagels, J., Zhang, D., Xue, H., & McMurry, P. H. (2008). Variability in morphology, hygroscopicity, and optical properties of soot aerosols during atmospheric processing. *Proceedings of the National Academy of Sciences of the United States of America*, *105*(30), 10291–10296. doi: 10.1073/pnas.0804860105

Zhu, J., Crozier, P. A., & Anderson, J. R. (2013). Characterization of light-absorbing carbon particles at three altitudes in east Asian outflow by transmission electron microscopy. *Atmospheric Chemistry and Physics*, *13*(13), 6359–6371. doi: 10.5194/acp-13-6359-2013

Zhu, J., Crozier, P. A., Ercius, P., & Anderson, J. R. (2014). Derivation of optical properties of carbonaceous aerosols by monochromated Electron energy-loss spectroscopy. *Microscopy and Microanalysis*, *20*(3), 748–759. doi: 10.1017/S143192761400049X

9

Microscopy Methods for Organic Composition of Atmospheric Aerosol Particles

Veronica Berta and Lynn M. Russell

ABSTRACT

Atmospheric particles have various origins and processes that lead to diverse organic chemical components, which are critical to quantify the indirect and direct effects of aerosols that impact our climate as well as to better understand air quality. Offline microscopic techniques such as scanning transmission X-ray microscopy with near-edge X-ray absorption fine structure (STXM-NEXAFS), scanning or transmission electron microscopy with energy-dispersive X-ray spectroscopy (TEM-EDX or SEM-EDX or STEM-EDX), and Raman microspectroscopy (RMS) can provide a variety of complementary chemical information about individual particles. STXM-NEXAFS and RMS are able to identify functional groups based on electronic transitions and vibrational transitions, respectively, and electron microscopy methods allow for carbon detection without chemically distinguishing between elemental and organic forms. Additional techniques are now also available for application to atmospheric particles. Because each technique is unique in both instrumentation and functionality, single-particle analyses that use multiple techniques provide the most complete information.

9.1. INTRODUCTION

Atmospheric aerosols exert an unprecedented influence on the environment through radiative forcing and cloud formation (Poschl, 2005). The extent of such influence, however, varies greatly as the composition, size distributions, and morphologies of aerosols are diverse and complex. For example, light-colored aerosols, such as sulfate and sea salt, scatter radiation and contribute to the negative radiative forcing, whereas dark-colored aerosols, such as black carbon (BC), may both scatter and absorb radiation and contribute to positive radiative forcing. These inherent properties can change when aerosols are internally mixed with organic components (Schwarz et al., 2008). Aerosol particles that contain organic aerosol (OA) components can arise from both biogenic and anthropogenic sources. Primary organic aerosols (POA) originate from direct emission from combustion sources (e.g., incomplete fuel combustion, biomass burning) as well as wind-driven sources. Secondary organic aerosols (SOA) form from the condensation of products of volatile organic compound (VOC) oxidation onto pre-existing particles and from nucleation of VOCs. The vast array of OA sources has resulted in a great variety of OA compositions, necessarily linking composition to sources. Photochemical processes in the atmosphere also affect OA composition. In addition, organic compounds can be oxidized via heterogeneous uptake of oxidants such as OH, O_3, NO_2, and NO_3 (Rudich et al., 2007). Ultimately, the many sources and processes influencing the organic composition of particles can alter optical properties and hygroscopicity, which determine the radiative forcing and cloud condensation

Scripps Institution of Oceanography, University of California at San Diego, La Jolla, CA, USA

Microanalysis of Atmospheric Particles: Techniques and Applications, Geophysical Monograph 288, First Edition.
Edited by Joseph M. Conny and Peter R. Buseck.

DOI: 10.1002/9781119554318.ch9

nuclei (CCN) potential (Farmer et al., 2015; Shrivastava et al., 2017). The extent to which OA particles may impact the environment depends on particle morphology and composition, making these properties crucial to understanding aerosol impacts in the atmosphere.

OA composition can be measured in ambient conditions in real time and after collection in laboratories. A commonly employed method for online analysis is mass spectrometry (MS). This technique ionizes aerosol particles, converting individual compounds into charged fragment ions and separating fragments by mass-to-charge ratios using a mass analyzer. The resulting mass spectrum contains organic peaks that have fragments of the original organic components, and it can reveal real-time chemical changes with limited collection artifacts. MS cannot measure particle structure or morphology because the particles are destroyed in volatilizing the components.

Microscopy and other offline techniques can preserve the particles collected, allowing the potential for repeated and complementary analyses. Complementary techniques are particularly relevant for organic particles, given their complexity and the current absence of a single technique for speciating and quantifying the complete array of organic compounds found in atmospheric particles.

Several current microscopy techniques provide images that contain sufficient spatial resolution for the detection of single particles with enough sensitivity to identify areas within particles with different chemically distinct features (Baustian et al., 2012; Posfai et al., 1999; Russell et al., 2002). Such higher-quality microscopic images reveal both the structure and composition of multicomponent particles. In addition, microscopy is commonly coupled with spectroscopy. Spectroscopy can reveal various compositional characteristics such as elemental, molecular, and functional features of chemical compounds present. Thus, spectroscopy effectively complements microscopy by coupling a detailed chemical composition with insight into particle structure. This information can be used to find particle origins as well as to improve our understanding of chemical processes in the atmosphere.

For microscopic analysis, OA particles can be collected together onto a substrate in "bulk" samples. Size-selective inlets may use cyclones or impactors to collect aerosol particles below a certain diameter, such as in the sampling of PM 2.5 or PM 10 (Keywood et al., 1999). Size-segregated sampling can also use cascade impactors that rely on diameter-dependent inertia during curvilinear motion to collect particles within a specific diameter range. These inlets work on the principle that particles of larger diameters (and inertia) suspended in air flows do not follow air in tight turns at high speeds, but smaller particles will follow the air and can be collected as air is pulled through a filter. After collection, changes in composition and morphology during handling, storage, and analysis are minimized by sealing and freezing the samples.

This review summarizes the microspectroscopic techniques that can analyze and "map" submicron organic composition within individual aerosol particles, which include scanning transmission X-ray microscopy with near-edge X-ray absorption fine structure (STXM-NEXAFS), scanning or transmission electron microscopy with energy-dispersive X-ray spectroscopy (TEM-EDX or SEM-EDX or STEM-EDX), and Raman microspectroscopy (RMS). STXM-NEXAFS utilizes a soft X-ray beamline that transmits a range of high-energy photons through a particle of interest to produce X-ray absorption spectra at a series of points on a two-dimensional grid that can be illustrated as maps of areas with different compositions. The absorption of specific energies reveals the abundance and speciation of light elements (C, N, O) and associated functional groups. TEM-EDX and SEM-EDX both use an electron beam to excite electrons within specific atoms, inducing an electron transition and releasing energy in the form of X-rays. Detected X-rays then generate maps of elements within individual particles. RMS uses UV and near IR as an excitation source to induce vibrational transitions rather than electronic transitions. These vibrational transitions occur when energy is transferred between the molecule and the incident photon, thus shifting the energy of the photon. The resulting spectrum shows energy peaks that correlate to specific bonding energies that can be unique to specific compounds. Further descriptions of these techniques and specific sample preparation, measurements, and limitations are provided in Sections 9.2, 9.3, and 9.4. There are additional microspectroscopic techniques that provide chemical information on small samples, such as micro-Fourier transform infrared spectroscopy (FTIR) (Cain et al., 2010; Popovicheva et al., 2019; Wang & Laskin, 2014), that are not included here because they lack the size resolution sufficient to resolve the composition of individual submicron particles. A few methods with limited atmospheric applications to date, such as tip-enhanced Raman spectroscopy (TERS), surface-enhanced Raman spectroscopy (SERS), atomic force microscopy with infrared (AFM-IR), and optical photothermal infrared (O-PTIR), are discussed in Section 9.5, with a summary in Section 9.6.

9.2. SCANNING TRANSMISSION X-RAY MICROSCOPY WITH NEAR-EDGE X-RAY MICROSCOPY (STXM-NEXAFS)

STXM-NEXAFS is an effective technique for analyzing the chemical form of lighter elements such as C, N, and O because it can use very precise energies to differentiate the peaks associated with different types of bonds. It is unique in its ability to identify organic functional groups by distinguishing energy peaks near the carbon edge that provide a relative and semi-quantitative characterization of organic composition.

9.2.1. Method

For STXM analysis, aerosol particles are collected by impaction onto substrates that are largely transparent in the soft X-ray range. Common substrates include 100-nm-thick silicon nitride (Si_3N_4) windows or carbon-coated TEM grids. In order to probe the composition of OAs, STXM specifically utilizes highly monochromatic soft X-rays that are generated from a synchrotron light source and selected by a high-spectral resolution monochromator to limit diffraction. The photon flux that is transmitted through the sample is collected by an X-ray-sensitive detector, which is often a thin layer of phosphor coating on a Lucite light pipe coupled to a photodiode. Samples are placed on stepping motor and piezoelectric stages to be raster scanned at the focal point and to obtain STXM images one pixel at a time at each selected photon energy. The spatial resolution of these images is determined by both the coherence of the X-rays and the width of the outermost region of the zone plate. For most aerosol particle analyses, the spatial resolution that is typically achieved is 30–35 nm (Takahama et al., 2011; Wang et al., 2016). A stack of STXM images is produced as the sample is scanned across a range of energies. At every point in a designated grid, the absorption associated with a specified energy is recorded. Repeating this procedure for a series of energies provides a NEXAFS absorption spectrum. The measured change in light intensity can be converted to optical depth (OD) using the Beer–Lambert Law, $OD = \ln\left(\frac{I}{I_0}\right) = \mu\rho t$, where I is the intensity of photon flux, I_0 is the intensity of the photon flux through the substrate without a sample, μ is the mass absorption coefficient, ρ is the sample mass density, and t is the sample thickness (for $OD < 1.5$).

STXM-NEXAFS quantifies differences in organic functional groups by distinguishing energy peaks near the carbon edge that are characteristic of the electron structure. This is accomplished by selecting specific energies that excite the photon-absorbing inner-shell electrons into unoccupied valence orbitals. This excitation-triggering absorption is called the *X-ray absorption edge* and occurs at an energy below which ionization occurs. Absorption edges are referred to as the *K-edge*, *L-edge*, and *M-edge*, depending on the electron orbital corresponding to the transition. For example, carbon contains a transition in the s electron orbital, which corresponds to the carbon K-edge frequently used in STXM-NEXAFS analysis. Other "edges" used for NEXAFS include the nitrogen and oxygen K-edges for their s orbitals as well as the sulfur, calcium, potassium, and chlorine L-edges for their p electron orbitals.

For X-ray absorption spectra of the carbon K-edge, samples are often scanned for energy levels between ~278 and 320 eV. This range includes a pre-edge region (278–283 eV), which quantifies contributions from inorganic elements, and a post-edge region (320 eV), which contains both organic and inorganic contributions (Moffet et al., 2010a). The near-edge region (~284–290 eV) contains contributions from 1s transitions. Russell et al. (2002) showed that STXM-NEXAFS could be used to detect and map organic functional groups in individual aerosol particles based on these transitions and their corresponding absorptions. The identified functional groups included alkyl ($R(CH_n)R'$), ketonic carbonyl ($R(C{=}O)R$), carboxylic carbonyl $R(C{=}O)OH$, and alkene ($R(C{=}C)R'$) groups, which are characterized by absorptions at 287.7 ± 0.7 eV, 286.7 ± 0.2 eV, 288.7 ± 0.3 eV, and 285.0 ± 0.2 eV, respectively (Table 9.1). Additional identification of an amide-like functional group (CHN) at 289.5 ± 0.1 eV, carbonate (CO_3^{2-}) at 290.4 ± 0.2 eV, potassium (K^+) at 294.6 and 297.2 ± 0.2 eV, and calcium (Ca^{2+}) at 347.9 and 351.4 ± 0.4 eV provided spatial maps that highlighted the contributions of these organic groups and inorganic ions for individual particles (Figure 9.1). These maps of individual particles revealed features that could only be seen by spectromicroscopy, including surfaces enhanced in more oxygenated groups (such as carboxylic carbonyl) that indicated contributions from oxidation by heterogeneous processes (Russell et al., 2002). Subsequent STXM-NEXAFS analyses used this approach to show that different atmospheric mechanisms affect particle size distributions differently, with smaller particles being more affected by surface-limited oxidation, resulting in enhanced carbonyl surfaces, but larger particles reflecting volume-limited oxidation and resulting in a uniform increase in carbonyl groups (Maria et al., 2004).

9.2.2. Application to Aerosol Particles

Using STXM-NEXAFS analyses of hundreds of individual particles, Takahama et al. (2007) proposed a classification scheme for particles based on STXM images and NEXAFS spectra according to particle size, shape, and organic functional group composition. Their analyses included ambient particles collected from the PELTI, ACE-ASIA, INTEX-B, MILAGRO, and DYCOMS II campaigns. Nearly 1,000 particles were divided into 14 spectral types (Figure 9.2) and then further categorized into four metaclasses that included combustion-derived, biogenic, carboxylic-acid dominated and unidentified particles. These

Table 9.1 Organic Functional Groups and Corresponding Energies from STXM-NEXAFS

	Alkene/ aromatic (R(C=C)R′)	Ketonic carbonyl (R(C=O)R′)	Alkyl (R(CHn)R′)	Carboxylic carbonyl (R(C=O)OH)	Amine (RNH-ROH) or alcohol (R-COH)	Benzoquinone R(C=C)R′	Amide carbonyl (R-NH(C=O)R)
Moffet et al. (2010a)	285.4 eV			288.5 eV			
Russell et al. (2002)	285.0 ± 0.2 eV	286.7 ± 0.2 eV	287.7 ± 0.7 eV	288.7 ± 0.3 eV	289.5 ± 0.1 eV		
Maria et al. (2004)	285 eV	286.6 eV	287.7 eV	288.8 eV	289.4 eV		
Takahama et al. (2007)	285 eV	286.7 eV	287.7 eV	288.7 eV			
Bahadur et al. (2010)	285 eV	286.7 eV	287.7 eV	288.7 eV	289.5 eV	284 eV	
Shakya et al. (2013)	285 eV	286.7 eV	287.7 eV	288.7 eV	289.5 eV		
Liu et al. (2011)	285 eV		287.7 eV	288.7 eV			
Takahama et al. (2010)	285.1 eV	286.7 eV	287.7 eV	288.7 eV	289.5 eV	284.0 eV	
Tivanski et al. (2007)	285.1 eV	286.7 eV		288.5 eV	289.5 eV		
Moffet et al. (2013)	285.1 eV			288.7 eV			
Hopkins et al. (2007)	285.3 eV						
O'Brien et al. (2015)	285.4 eV			288 eV			
Hawkins et al. (2010)	284.4–286.4 eV	286.2–290.9 eV	287.4–288.5 eV (C1s-σ* C–H transition), 290.8–293 eV (C 1s-σ* C–C)	288.2–288.9 ± 0.3 eV	289.5 ± 0.3		288.3 ± 0.2 eV
Ault et al. (2013a)	285.4 eV	286.5 eV		288.5 eV			
Laskin et al. (2012)				288.5 eV			

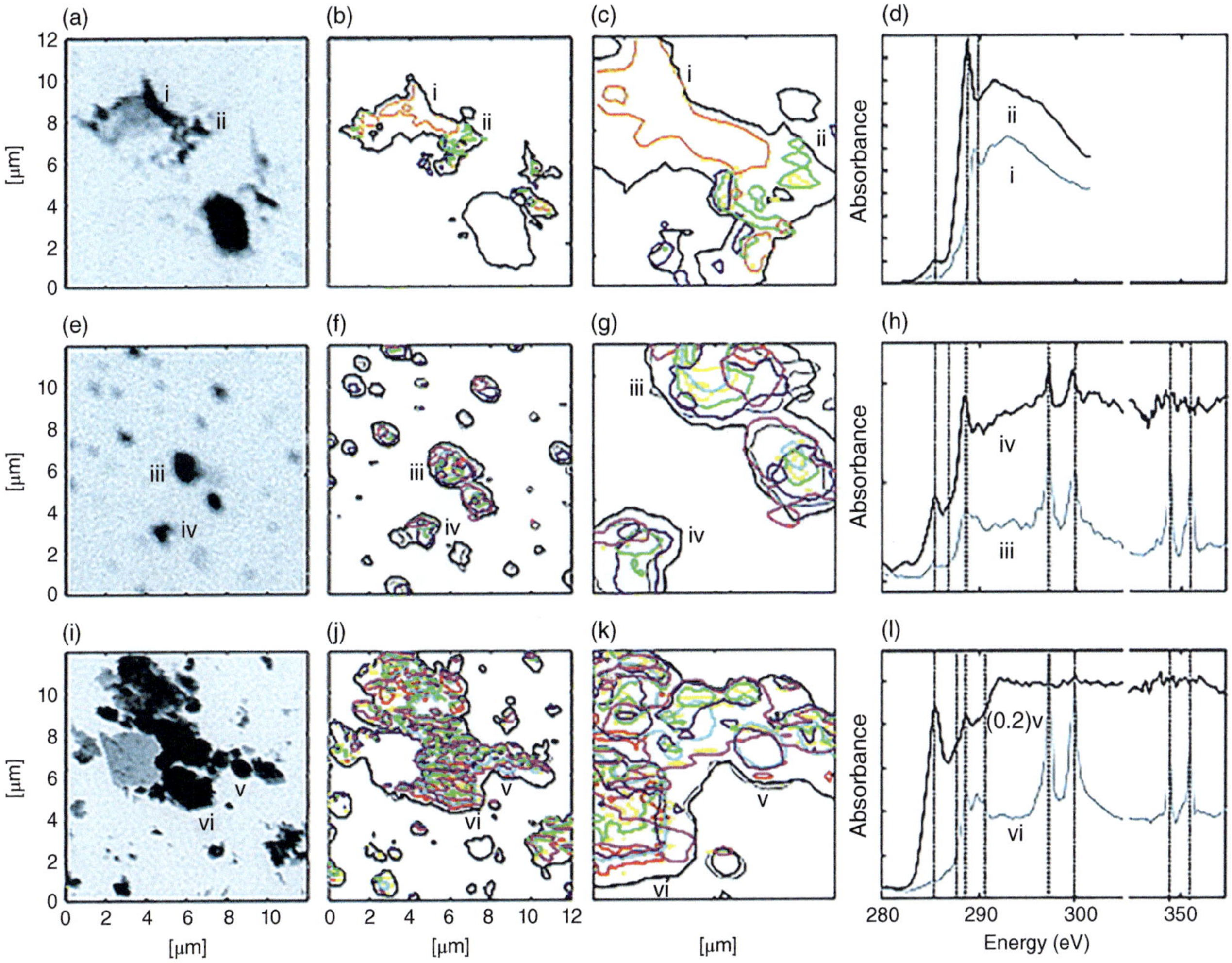

Figure 9.1 Images, speciated maps of detectable regions, and representative spectra of measured organic and inorganic species for marine boundary layer particles. High-resolution soft X-ray images (a, e, i) show (a) one absorbance image at 300.0 eV (in which Ca^{2+} was not measured) for particles collected in the marine boundary layer near St. Croix at 310 m from 1601 GMT on 19 July 2000 and (e, i) two images at 289.9 eV for particles at 30 m from 1613 GMT on 16 July 2000. Threshold contours (b, f, j) at the detection limits are drawn for the same three samples R(C=C)R′-purple (285.0 ± 0.2 eV), R(C=O)R-cyan (286.7 ± 0.2 eV), R(CH_n)R′-blue (287.7 ± 0.7 eV), R(C=O)OH-green (288.7 ± 0.3 eV), * transition for CNH (in an amide-like group connected to a carbon)-orange (289.5 ± 0.1 eV), CO_3^{2-}-yellow (290.4 ± 0.2 eV), K-red (294.6, 297.2 ± 0.2 eV), and Ca^{2+}-magenta (347.9, 351.4 ± 0.4 eV). Enlarged maps (c, g, k) and spectra (d, h, l) for representative particles labeled (i–vi) illustrate detailed composition characteristics for each sample. Reproduced from Russell et al. (2002)/with permission of John Wiley & Sons.

categories provide insight into possible sources by comparing the individual particle spectra to reference spectra from standards and source-specific tracers. Combustion-derived particles contain a significant alkene region absorption indicative of sp^2 hybridized graphitic carbon, which was a unique feature of BC or soot identified previously (Braun, 2005; Braun et al., 2006; Braun et al., 2007; Braun et al., 2004; Braun et al., 2005). Biogenic aerosols that may originate from biomass combustion are identified by multiple spectra absorbances, similar to humic or fulvic acid and soil substances. Carboxylic-acid particles had high carbonyl absorption and are likely to have formed by oxidation of VOCs or through extensive aging of POA. Carboxylic-acid particles with additional carbonate and potassium absorption peaks were indicative of mineral contributions that were confirmed with reference spectra of collected soil and dust samples (Takahama et al., 2007).

The Takahama et al. (2007) classification relied on a *k*-means clustering algorithm to provide efficient grouping of similar particle spectra. Subsequent studies (Bahadur et al., 2010; Day et al., 2009; Hawkins et al., 2010; Leaitch

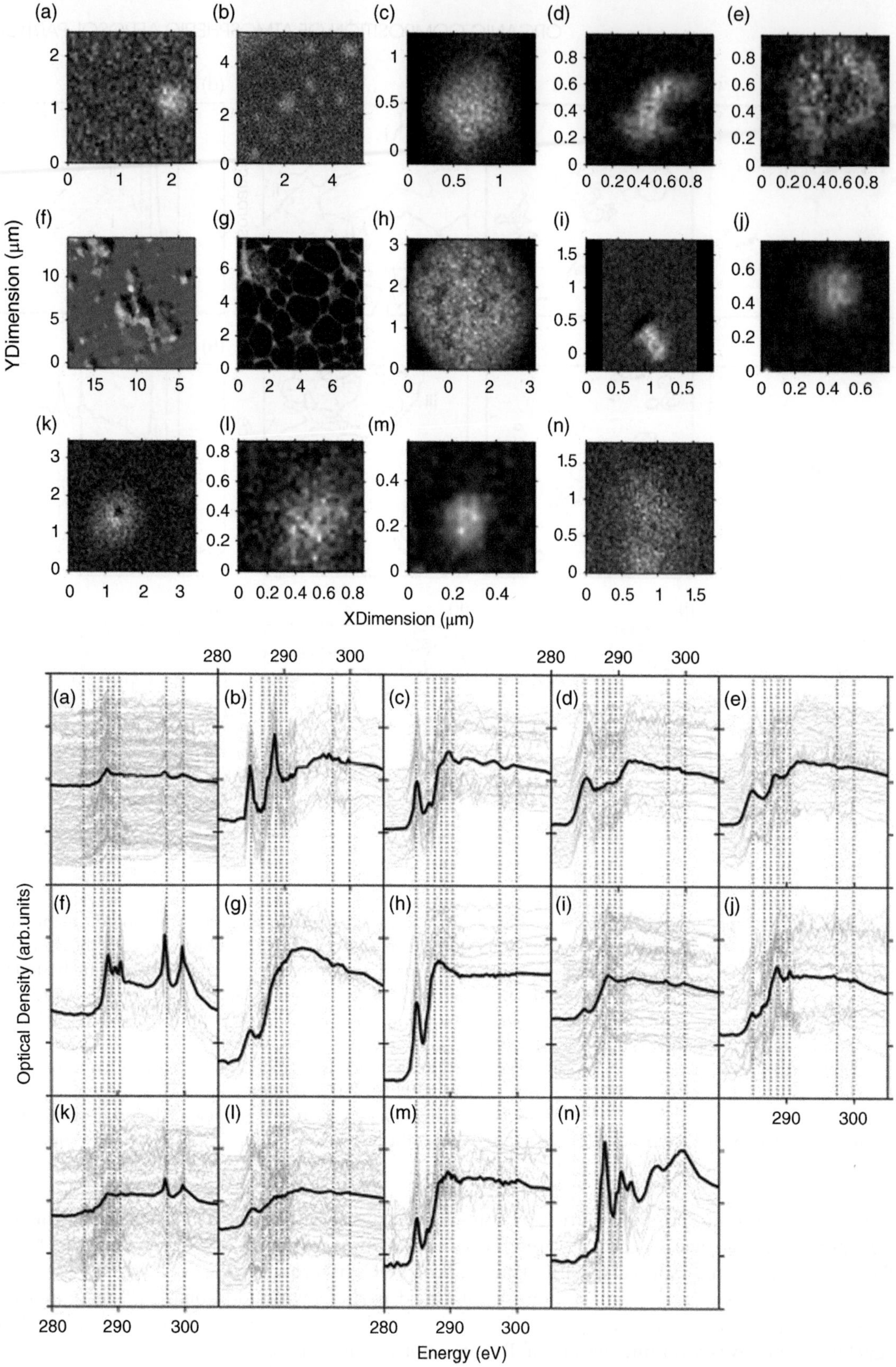

Figure 9.2 Top: Example images for each category; each panel corresponds to the respective panels in Figure 9.1. Particles are from (a) ACE-ASIA, (b) MILAGRO (A), (c) ACE-ASIA, (d) MILAGRO (A), (e) MILAGRO (U), (f) PELTI, (g) Princeton, (h) MILAGRO (U), (i) ACE-ASIA, (j) MILAGRO (A), (k) ACE-ASIA, (l) MILAGRO (A), (m) MILAGRO (A), (n) MILAGRO (U). Bottom: Fourteen classifications of spectra. Gray lines are scaled individual spectra; dark lines are averages of all spectra. Individual spectra include arbitrary shifts on the vertical axis to display them separately. Vertical lines at 285, 286.7, 287.7, 288.7, and 290.4 eV represent R(C=C)R′, R(C=O)R, $R(CH_n)R'$, R(C=O)OH, and CO_3^{2-} transitions, respectively. Reproduced from Takahama et al. (2007)/with permission of Elsevier.

et al., 2009; Liu et al., 2012; Liu et al., 2011; Liu et al., 2009; Moffet et al., 2010a, b; Saliba et al., 2020; Schwartz et al., 2010; Shakya et al., 2013; Takahama et al., 2010) have successfully used the same approach, providing a growing data set that links composition and morphology from a variety of campaigns in different parts of the world. There were also some unique findings from different campaigns. For example, OA particles collected from the INTEX-B campaign contained strong carboxyl, carbonate, and potassium peaks in particles greater than 0.5 μm diameter. The large (>1 μm) diameter of many of these particles suggests that the organic carboxyl group contribution was added by the scavenging of VOCs rather than the suspension of dust preconditioned with organic material (Leaitch et al., 2009). Airborne studies also showed that these soil dust particles were similar to particles collected in the ACE-ASIA campaign, suggesting long-range transport (Day et al., 2009). This classification system (Takahama et al., 2007) showed that particles from the MILAGRO campaign were from biomass-burning sources based on their similarity to fulvic acid (Bahadur et al., 2010; Liu et al., 2009; Tivanski et al., 2007). Similarly, biogenic OA identified as amorphous tar balls with ketone absorptions are thought to come from the oxidation of ketone-containing molecules during biomass burning events and possibly from biogenic volatile organic compound (BVOC) emissions (Schwartz et al., 2010; Takahama et al., 2011).

To take more complete advantage of the spatial information provided by STXM-NEXAFS, algorithms were developed to quantify and categorize different features within particles (Takahama et al., 2010). OAs were categorized by chemical heterogeneities that included carbon surrounded by an aqueous phase, organics on dust, particles surface-enriched in carboxylic acid, and inorganic cores encapsulated by an organic outer layer (Figure 9.3). The morphology of these particles was examined by measurements of their sphericity that distinguished agglomerations from those with inclusions or liquid films of nonspherical particles. Particles were categorized as carbon surrounded by an aqueous phase that contained strong aromatic absorbances as well as carboxylic carbonyl and potassium absorbances. These particles were identified to be BC that were insoluble but had been covered in aqueous carboxylic acid and potassium, which suggested that these particles were products of biomass burning. Dust particles were identified by the presence of carbonate absorbance. An additional carboxylic carbonyl absorbance, which appeared as clumps in spatial mappings, suggested that these particles had formed by condensation or coagulation of organic components onto pre-existing dust particles. On the other hand, particles that had surfaces enriched with carboxylic acid were suggested to have been formed

Description	Illustration	Number of particles
a Black carbon with aqueous phase		90 particles
b Organics on dust		106 particles
c Particles surface-enriched in carboxylic acid		23 nonspherical, 31 spherical
d Inorganic core with organic outer layer		10 particles, all spherical

Figure 9.3 Illustration of categorized particle types. Reproduced from Takahamal et al. (2010)/with permission of John Wiley & Sons.

through oxidative processing. Köhler theory calculations showed that the organic coatings found on these particles could potentially alter the CCN activity of aerosols (Takahama et al., 2010).

Because some organic particles are formed by photochemical reactions in the atmosphere, Shakya et al. (2013) compared STXM-NEXAFS spectra from lab-chamber SOA from five precursors (glyoxal, α-pinene, isoprene, 1,2,4-trimethylbenzene (1,2,4-TMB), and D-limonene) to spectra from particles collected from six field campaigns (MILAGRO, VOCALS, Whistler-2008, Scripps Pier-2009, CalNex (Bakersfield-2010), and Whistler-2010). Functional groups that were probed included an alcohol group with energy of 289.05 ± 0.3 eV, in addition to those identified by Russell et al. (2002). SOA particles formed by glyoxal SOA condensing onto ammonium bisulfate, magnesium sulfate, sodium nitrate, and sodium sulfate seed particles had well-defined alcohol, amide carbonyl, and alkyl group peaks, in addition to carbon-nitrogen groups. SOA formed from the photooxidation of α-pinene in the presence of BC and ammonium bisulfate seeds had alkyl group contributions that resembled soot-like particles. SOA formed from coagulation of isoprene and α-pinene contained fewer ketonic carbonyl and more carboxylic carbonyl, alkyl, and organic hydroxyl groups. SOA formed from 1,2,4-TMB photooxidation contained a larger contribution from carboxylic carbonyl, alkyl, and alcohol groups compared to ketonic carbonyl and aromatic groups, whereas SOA formed from limonene photooxidation had distinct alkyl and carboxylic carbonyl group contributions. Ultimately, ambient particle spectra were most frequently correlated to spectra from α-pinene mixed with isoprene (ammonium bisulfate and BC seeds) or glyoxal (sodium nitrate seeds) and from 1,2,4-TMB. Particles containing similar spectra to 1,2,4-TMB were found in more urban, anthropogenic areas and had an additional absorption indicative of BC inclusions, suggesting combustion as a major source of 1,2,4-TMB. Similarly, particles collected in more rural locations, such as from Whistler campaigns, had a greater resemblance to SOA from biogenic precursors like α-pinene and isoprene (Shakya et al., 2013).

STXM-NEXAFS can also be used to create spatial maps that specify elemental carbon, organic carbon (OC), and inorganic components by using spectra from the carbon K-edge to identify the carboxyl group as OC, the alkene group as elemental carbon, as well as spectra from the pre-edge region to identify inorganic components. Moffet et al. (2010a) developed an automated analytical technique to observe the mixing state of particles by applying detection limits when using these identifiers (Figure 9.4). Individual STXM pixels were examined to distinguish OC by an $OD_{320} - OD_{278}$ (post-edge minus pre-edge) value greater than zero, inorganic components by an OD_{278}/OD_{320} (pre-edge over post-edge) ratio greater than 0.5, and elemental carbon by a $\%sp^2$ hybridization value greater than 35% based on Hopkins et al. (2007). This identification of elemental carbon could also indicate the presence of components that have been identified by a variety of other names, including BC, soot, and graphite (Petzold et al., 2013). By probing the potassium L_2 and L_3 edges, mineral contributions were also identified (Moffet et al., 2010a). Spatial mapping was specifically utilized to probe the mixing state as a function of time of aerosols collected from the MILAGRO campaign.

As aerosols moved from the city center to suburban and rural areas, contributions from inorganic components and elemental carbon decreased while OC contributions increased, providing evidence of photochemical contributions of SOA to so-called *particle aging*. The formation of SOA was also indicated by the increase in homogeneously mixed organic particles in the total population as aerosols traveled from the city (Moffet et al., 2010b). These results revealed that the compositional effects of particle aging were similar to those seen during other campaigns (Fraund et al., 2017; Moffet et al., 2013). O'Brien et al. (2015) also used these imaging parameters to analyze particles collected during one of the same field campaigns, finding that the observed particle aging was intensified by soot aggregation.

Recent studies have analyzed marine aerosols that contained organic components produced from biological organisms within the ocean. Hawkins and Russell (2010) categorized primary marine aerosols based on functional group contributions, dividing them into carboxylic acid-containing polysaccharides, low-solubility polysaccharides, calcareous phytoplankton fragments, and proteinaceous material. Nearly all the particles labeled as carboxylic acid-containing polysaccharides were collected in the Arctic except for a few from California that were in marine air masses. The spectra of these particles contained strong carboxylic carbonyl peaks and weak alcohol, carbonate, and potassium peaks, similar to spectra of hygroscopic alginic acid found in kelp and algae in the region. Low-solubility polysaccharide particles collected in the Pacific contained alcohol, aromatic, ketonic, and carboxylic carbonyl carbon peaks similar to marine particulate organic matter (POM) found in ocean sediment. The spectra of other particles collected from over the Pacific contained notable inorganic carbonate and alkyl (both pi and sigma transitions) absorptions and were in the submicron diameter range. The small pre-edge absorption and strong carbonate peak indicate that these particles are calcareous and are likely composed of calcium carbonate (Hawkins & Russell, 2010). Certain phytoplankton are known to produce carbonate as well as alkenones, making these particles likely to be calcareous phytoplankton fragments. Finally, the spectra of proteinaceous particles that were collected off the coast of California and downwind of northeastern Asia consisted of aromatic/alkene, alkyl, and, most importantly, amide carbon absorptions, which could indicate peptide bonds within proteins (Hawkins & Russell, 2010). Building on these four kinds of marine particles, measurements of aerosol particles over the Southern Ocean showed five different kinds of organic mixtures that were associated with sea-spray particle sources (McFarquhar et al., 2020; Saliba et al., 2020; Twohy et al., 2021).

The classification of carbon into organic, elemental, and inorganic components by Moffet et al. (2010a) was used to determine the mixing state of marine aerosols at a coastal location. The particles that contained inorganic cores with organic coatings were likely due to organic compounds condensing onto pre-existing sea salt particles. By using their classification method in combination with mapping, marine particle aging was tracked to highlight the change in composition of particles that included NaCl and SOA (Laskin et al., 2012). Similar to previous studies on particle aging, particles were found to be enhanced with organic components. Yet this method also revealed a simultaneous depletion of chlorine. This chlorine depletion was found to be a consequence of particle drying when HCl is released from sodium chloride (Laskin et al., 2012). Ault et al. (2013a) reported composition changes from particle aging by using STXM to probe the carbon K-edge and sulfur and chlorine L-edges

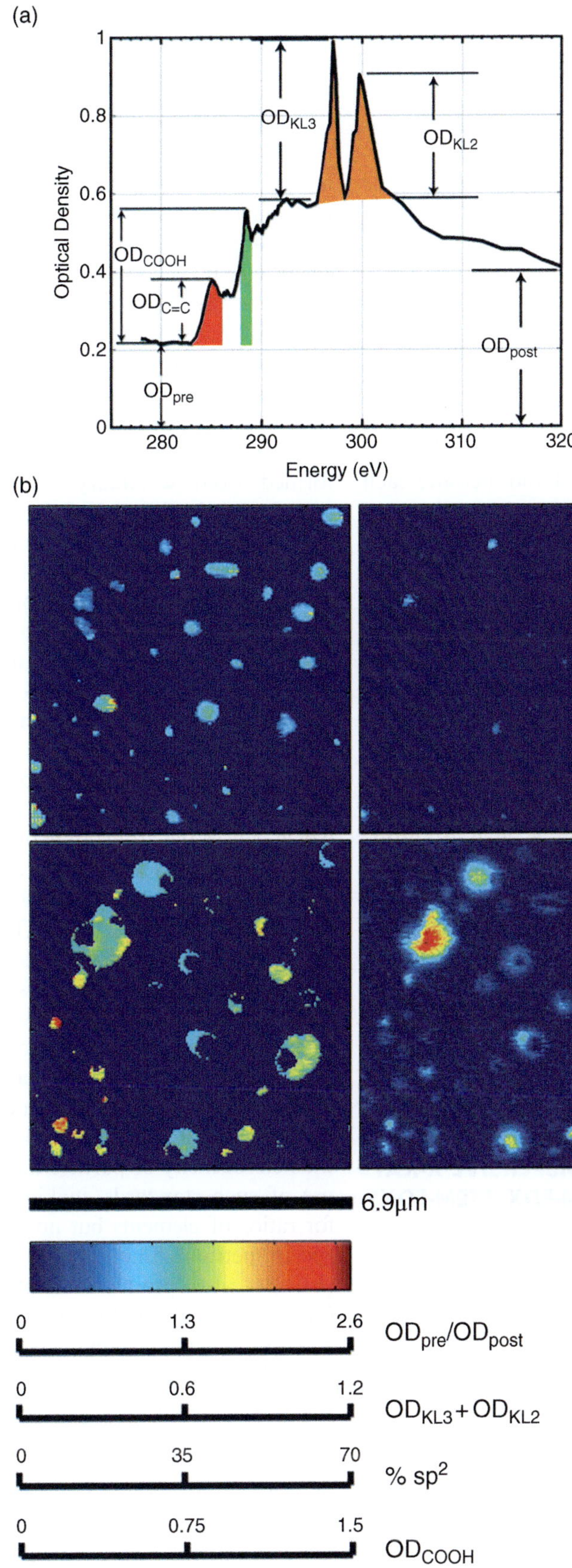

Figure 9.4 (a) Characteristic single-particle carbon K-edge near-edge X-ray fine structure (NEXAFS) and spectral feature assignments: pre-edge (OD_{pre}) occurs between 278 and 283 eV, carbon-carbon double bonds ($OD_{C=C}$) shown in red, carboxylic acids (OD_{COOH}) shown in green, potassium L_2 and L_3 edges (OD_{KL3}, OD_{KL2}) shown in orange. OD_{post} is the post-absorption edge at 320 eV. (b) STXM particle maps indicating different spectral components identified in (a): OD_{pre}/OD_{post} represents total carbon absorption, $OD_{KL3}+OD_{KL2}$ is a potassium map (orange peaks in (a)), % sp^2 shows carbon-carbon double bonds (red peak in (a)), and OD_{COOH} is a map of carboxylic acid groups (green peak in (a)). Reproduced from Moffet et al. (2010a)/with permission of the American Chemical Society.

of lab-generated sea-spray aerosols (SSA). The spectra of SSA particles below 1 μm contained a significant carboxylate absorbance, and many, but not all, contained alkene and carbonate absorbances (Ault et al., 2013a).

9.2.3. Limitations

STXM-NEXAFS limitations include the limited availability of STXM facilities: there are only a few beamlines around the world that provide the required X-rays, and only a couple have the appropriate end station for this type of measurement. This limited availability is compounded by the time required to find and measure each particle, precluding statistically significant sample numbers. This limitation additionally restricts OA analyses to be offline and sometimes at vacuum, meaning that the aerosol composition and structure may change from when it was in the atmosphere. The X-rays required by STXM may also inflict radiation damage that could potentially change chemical bonding so that the analyzed particles have a different composition than the originally collected particles. Nonetheless, STXM-NEXAFS has been shown to provide effective resolution without evidence of damaging particles (Russell et al., 2002). Although STXM energy resolution is able to probe functional groups, the energy at which an electron transition occurs can vary (e.g., alkyl groups are observed to vary in an energy range of ±0.7 eV), which adds uncertainty to the identification of specific groups. STXM is also not suited to provide a large range of energies at a single beam station, and therefore studies are usually limited to focusing on one or only a few elements rather than measuring all elements present.

9.3. SCANNING AND TRANSMISSION ELECTRON MICROSCOPY WITH ENERGY-DISPERSIVE X-RAY SPECTROSCOPY (SEM-EDX, TEM-EDX, STEM-EDX)

Electron microscopy uses a detector with high resolution that detects carbon as an element. It is well suited for analyzing carbonaceous particles of sizes smaller than 200 nm and for identifying mixtures of carbon with other elements within particles.

9.3.1. Method

Transmission electron microscopy (TEM) typically requires aerosols to be collected on small grids (~3 mm) that consist of copper, nickel, or molybdenum coated with a carbon-film or lace substrate. Scanning electron microscopy (SEM) often uses collection onto semi-conductive silicon wafers. For both TEM and SEM, a cathode supplies the source of energy needed to probe aerosol composition. This source, equipped with an electron gun, produces a high-voltage beam of electrons (60–300 keV for TEM, up to 50 keV for SEM) that is accelerated by an anode under a high-vacuum system and focused onto the sample using electromagnetic lenses. For TEM, the incident electrons are scattered by the sample, modifying the transmitted electron beam. This modification forms an image revealing the properties of the volume through which the beam was transmitted. An objective lens is used to magnify this image onto an imaging device, such as a fluorescent phosphor screen or a charge-coupled device (CCD) camera for detection. In contrast, SEM requires a much finer electron beam that is raster scanned across the sample. Images are then formed from secondary electrons scattered from the sample exterior instead of the transmitted electron beam used by TEM. SEM can produce atomic-resolution images that correspond to the sample surface. Often, TEM is equipped with a scanning mode to perform scanning transmission electron microscopy (STEM), a technique that provides both internal and external information pertaining to the sample composition and structure.

Both TEM and SEM can be paired with energy-dispersive X-ray spectrometers (EDS or EDX) to provide an elemental analysis of aerosol particles. EDS excites atoms within the sample with an incident electron beam, which causes the atoms to emit inner-shell electrons. This prompts electrons from higher energy orbitals to fill in the low-energy vacancy, consequently releasing additional energy in the form of X-rays. These X-rays have energies that correspond to individual elements that can be detected with EDS silicon drift detectors. The X-ray signals include peaks in the spectrum that can be identified by the elements that correspond to the sample energy. The contribution of a particular element within a particle is indicated by peaks at specific energies, providing a chemical analysis that can be quantified if a suitable standard is available. Typically, the heterogeneity of particle shapes and sizes prevents the use of such standards, making the method quantitative for ratios of elements but not for masses. An alternative spectrometric technique, electron energy loss spectroscopy (EELS), also produces a similar emission spectrum (Hitchcock et al., 2008). This technique is used with TEM and STEM because it extracts compositional information by the loss of energy when the electron beam is transmitted through the sample. Both EDS and EELS, however, are limited by their inability to distinguish between elemental and organic carbon, as they can only identify the presence of carbon atoms, not the type of carbon bonds.

9.3.2. Application to Aerosol Particles

Studies that first employed electron microscopy were often used to examine the physical features of lab-generated particles, particularly those emitted from diesel and gasoline engines as exhaust. McKee and McMahon

(1960) analyzed particles emitted from an automobile engine during idling and cruise conditions. The particles collected consisted of agglomerates that were more pronounced for an idling engine and were more spherical in shape and larger in size for cruising conditions. The presence of OC in these particles was confirmed by extracting in a benzene solution to identify carbonyl and olefin (McKee & McMahon, 1960). Frey and Corn (1967) first produced electron micrographs of diesel exhaust particles and additionally quantified their inorganic portion. Similar to particles emitted from gasoline engines, these particles were aggregates that ranged from 0.1 to 0.3 μm (Frey & Corn, 1967). Subsequent diesel-based studies have also found a size dependence on exhaust temperature (Soderholm, 1981). Vuk et al. (1976) suggested that the volatile nature of organic matter was responsible for the decreased agglomerate size found at higher temperatures (Vuk et al., 1976).

Early electron microscopy studies did not detect elements lighter than sodium, including carbon, and instead inferred organic contributions by the lack of X-ray emission (Parungo et al., 1990). To better analyze organic particles, Sheridan (1989a, 1989b) coupled an electron microscope with an X-ray spectrometer. However, this development did not resolve the ambiguity about OC because methods for determining the organic nature of carbon varied (Sheridan, 1989a, 1989b). Since then, OC has been identified by its instability in electron microscopes (in which the electron beam causes deterioration of the material). Other studies have used the hygroscopic tendencies of particles to identify the presence of OC, identifying those particles that take up less water than sulfate or salts as organic components (Pitchford & McMurry, 1994; Rausch et al., 1999; Zhang et al., 1993). Posfai et al. (1999) determined the organic nature of a carbon peak given by the EDS spectra at an energy of around 284 keV by using TEM images. The images of compact or chain-like associations of spherules were identified as carbonaceous soot aggregates that consisted of elemental carbon. The presence of an OC film was also identified on some particles (Posfai et al., 1999). Adachi and Buseck (2008) suggested that the organic coatings on soot particle surfaces were SOA.

EDS peaks of carbon and oxygen are present in organic components and tar balls. Various studies have differentiated these particle types by the detection of additional inorganic inclusions (including sulfur, potassium, magnesium, calcium, chlorine, and phosphorus). These inclusions were most common for OA and have even been found to change the hygroscopicity of OA, enhancing their ability to act as CCN (Li & Shao, 2010; Posfai et al., 2004; Semeniuk et al., 2007). Using computer-controlled SEM (CCSEM) with EDS, Moffet et al. (2010a, 2010b) analyzed carbonaceous particles containing inorganic inclusions, finding possible formation mechanisms based on particle morphology. For example, particles containing sulfur and potassium-enhanced cores (Figure 9.5) suggested a possible biomass-burning origin, whereas increased carbon contributions at the particle surface were interpreted as SOA formation (Moffet et al., 2010b).

Twohy et al. (2005) also inspected internally mixed organic particles collected from clouds during polluted conditions using both SEM and TEM. The carbon detected by EELS was determined to be organic based on the homogeneous structure of the particles. EELS also identified additional oxygen, sulfur, and nitrogen contributions. The presence of OC and nitrogen in evaporated cloud drop samples suggested the possibility that organonitrogen compounds could nucleate cloud droplets but also that OA particles may act as CCN without inorganic inclusions (Twohy et al., 2005). OA particles can be further categorized as secondary particles by the inclusion of beam-sensitive inorganic cores shown in microscopic images and X-ray spectra (W. H. Wang et al., 2018).

OA may be primary particles, which are often identified by having one of the following two shapes: (1) irregular or spherical morphologies, with chain-like spherules coated in OC; and (2) tar balls, particles containing OC that are spherical in shape (Xu et al., 2017).

TEM-EDS is useful for categorizing OA based on combustion sources, as OA particles have been found to dominate ignition and flaming stages in experimental burning chambers (W. H. Wang et al., 2019). Liu et al. (2017) burned 16 kinds of materials varying from crop residues and wood to solid wastes to reveal the morphology, mixing state, and composition of the particles emitted in controlled laboratory conditions (Figure 9.6). TEM-EDS analyses directly linked the composition of particles with the composition of their sources, particularly for the combustion of materials containing brominated flame retardant (BFR) that produced OA with Br-rich inclusions. Notable changes in morphology and composition arose from the burning of crop residues. During the smoldering phase, crop residue particles were spherical OA, either by itself or with potassium inclusions that were interpreted from darker and lighter areas in TEM images. During the flaming stages, these particles became irregular in shape and mixed with soot. Similarly, the flaming stages of wood and solid waste burning produced both soot coated in organic matter as well as pure soot particles that were irregularly shaped (L. Liu et al., 2017). The physical coagulation of organic particles with soot and inorganic particles may be responsible for producing these irregularly shaped particles, whereas gas condensation can produce internally mixed, spherical SOA particles (Freney et al., 2018; Li et al., 2015).

Using *k*-means clustering and single-particle EDS spectra, Moffet et al. (2013) were able to reveal a size

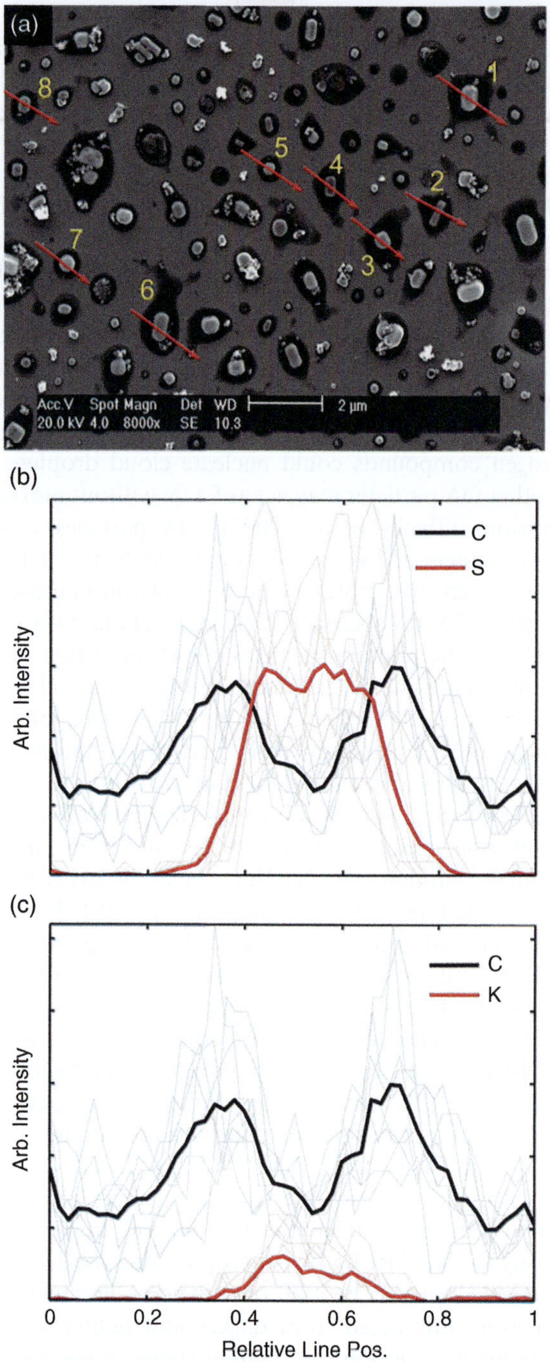

Figure 9.5 SEM image together with carbon, sulfur, and potassium line scans for particles from the T0 site. The locations and directions of line scans are shown by the arrows in panel (a). The intensities of characteristic X-rays along the line scans are shown in panels (b) and (c). The line scans confirm that particle cores contain S and K, coated by carbonaceous material. X axis was normalized by the total scan distance. The bold lines represent the X-ray intensities averaged for a number of individual line scans represented by faint lines. Moffet et al. (2010b)/European Geosciences Union/ CC BY 3.0.

dependence of the composition of carbonaceous aerosols collected during a field campaign in northern California. Fine-mode particles (<1 µm) included nitrogen-containing carbonaceous with contributions of nitrogen, oxygen, and sulfur above 1%, secondary carbonaceous with a contribution from carbon followed by oxygen, nitrogen, and sulfur (but lacking an inorganic core), and refractory carbonaceous with enhanced contributions from sodium, magnesium, iron, potassium, and chloride (indicating condensation of secondary organic precursors onto sea salt or biomass burning emissions). Coarse-mode carbonaceous particles (>1 µm) were classified as coarse carbonaceous/nitrogenous particles, which resembled the refractory carbonaceous group but had an enhanced nitrogen contribution. These particles were collected at three separate sites ranging from urban to rural, and the carbon content of individual particles increased with plume age, providing an indication of increasing SOA content (Moffet et al., 2013).

Laskin and Cowin (2001) improved the capabilities of electron microscopy by developing an automated SEM/EDX analytical technique that more efficiently presented quantitative compositional information of particles down to the submicron level. This technique was the first to require a carbon-film substrate, which allowed for elemental mappings of particles to include carbon due to the low background of X-rays emitted from the film (Laskin & Cowin, 2001). As a result, this technique has been employed in various studies to produce spatial mappings of organic particles. For example, Ault et al. (2013b) used TEM and SEM-EDS to produce mappings of elements, including carbon, oxygen, chlorine, sulfur, calcium, magnesium, and sodium, of single lab-generated SSA particles. These elemental mappings highlighted an inorganic sodium chloride core with an organic coating and carbon, oxygen, and magnesium inclusions (Figure 9.7). This study revealed that the crystallization of the inorganic salt core becomes depleted as the seawater is augmented with additional biological and organic material, which could, in turn, influence CCN activity (Ault et al., 2013a; Collins et al., 2013). Depletion of chloride has also been observed in aging SSA, indicated by enhanced carbon and oxygen coatings with diminished and even nonexistent inorganic cores, perhaps the result of heterogeneous chemical reactions triggered by inorganic and organic acids (Bondy et al., 2018; Chi et al., 2015; Kirpes et al., 2019; Kirpes et al., 2020; Laskin et al., 2012).

Cryogenic TEM (cryo-TEM) has also been used to investigate the organic composition of atmospheric particles. This method entails flash-freezing particles to preserve their components, which can help in probing soft biological structures. Patterson et al. (2016) have recently employed this technique, identifying fully intact structures that include hydrated bacteria, diatoms, virus particles, marine vesicles, and gel networks in SSA.

Material	Combustion phase	Particle type	Particle number	Percentage
Maize straw	smoldering	OM-K	141	100
	flaming	OM-K	57	53
		soot-OM-K	50	47
Wheat straw	smoldering + flaming[a]	OM	35	16
		OM-K	124	57
		soot-OM-K	60	27
Cotton straw	smoldering	OM	71	100
	flaming	OM-K	42	70
		soot-OM-K	18	30
Wild grass	smoldering	OM	102	100
	flaming	OM	26	45
		soot-OM	24	41
		Cl-rich-OM/soot	8	14
Common reeds	flaming	OM-K[b]	/	/
Apple wood	smoldering	OM	115	100
	flaming	OM-K	70	68
		soot-OM-K	33	32
	burnout[c]	soot-OM	29	100
Cypress wood	flaming	soot-OM	177	100
Pine wood	flaming	OM-K	84	100
Pear wood	flaming	soot	66	100
Bamboo	flaming	soot	64	46
		soot-OM-K	75	54
Cardboard	flaming	OM	153	91
		S-rich-OM	15	9
Foam boxes	flaming	soot	106	100
Waste rubber tires	flaming	soot-OM	55	100
Plastic bottles/bags	flaming	soot-OM	105	100
Printed circuit boards	flaming	OM-Br	104	71
		OM-Br-Pb/P/Zn	42	29
Copper-core cables	flaming	OM-Br-soot	83	100

[a] The combustion phases were mixed between smoldering and flaming phases, so smoldering + flaming here.
[b] The particle number of common reeds cannot be counted because the gel-like OM were connected together.
[c] Burnout refers to the end of the flaming phase.

Figure 9.6 Particle classification, analyzed particle number and percentage from different materials in the smoldering/flaming phase. Liu, L. et al. (2017)/Springer Nature/CC BY 4.0.

9.3.3. Limitations

Electron microscopy is unique in its ability to measure the presence of almost all elements in particles by EDS/EDX and to combine that composition with high-resolution images that reveal morphology. TEM can show particles and their features down to 10 nm, providing images with a spatial resolution of <0.1 nm and magnification of up to 50 million times. SEM shows particles down to 50 nm in diameter and provides images with a spatial resolution of <10 nm and magnification of up to 1–2 million times. SEM and STEM, however, have the advantage

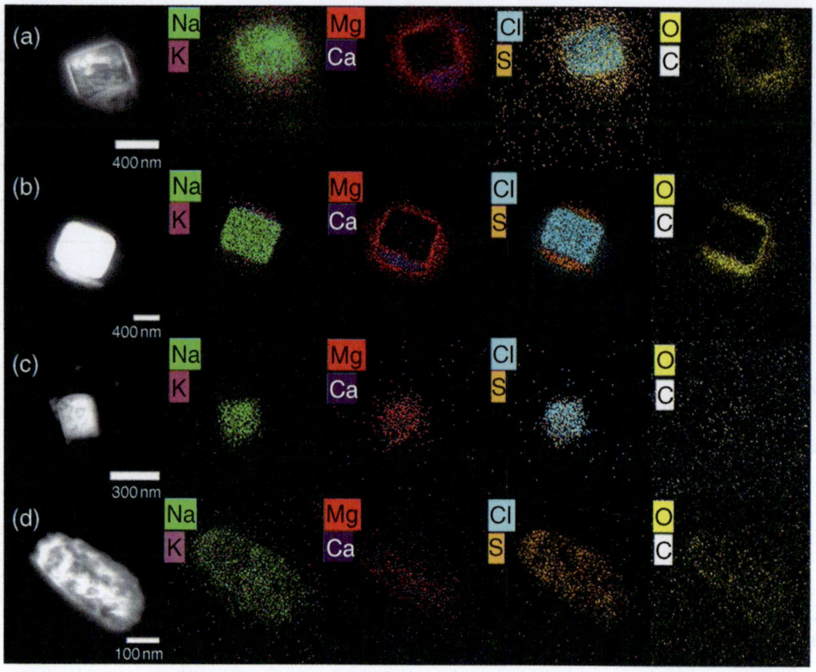

Figure 9.7 TEM images (dark field) and elemental maps of common particle types observed below 1 μm: (a) SS-OC with Ca and S not isolated to any specific region within the particle, (b) SS-OC with Ca–S rod structure ($CaSO_4$), (c) SS-OC with S and other cations distributed homogeneously throughout the particle, and (d) an OC particle which also contains inorganic elements including Na, Mg, Ca, and S but not Cl. Reproduced with permission from Ault et al. (2013a)/American Chemical Society.

that they have been computer-controlled, which allows for time-efficient analysis of more particles. Both TEM and SEM are limited by their high voltage, which can damage atmospheric particles, high-vacuum environments that can lose volatile components of OA composition, and interference from carbon filters that make detection of carbon less sensitive compared to other elements. Most importantly, electron microscopy with EDS/EDX lacks the ability to determine the chemical bonding and oxidative state of elements, which ultimately means it cannot chemically distinguish between elemental and organic carbon components.

9.4. RAMAN MICROSPECTROSCOPY (RMS)

RMS provides a powerful tool for characterizing chemical bonds using molecular vibrations. Raman bands produced from Raman scattering correlate to the vibrational states of molecules and can be used to identify and quantify organic functional groups. The specificity and range of this method mean in many cases, specific organic molecules can be identified.

9.4.1. Method

Particles that are to be analyzed using RMS (also called *micro-Raman*) are collected by impaction onto various substrates, including titanium and aluminum foils, silicon wafers, and quartz fiber filters. This technique shines a monochromatic light through the sample to measure composition. Laser rejection filters are used to produce a single wavelength from light sourced from an excitation laser in the visible and near-infrared range. RMS additionally uses a confocal optical microscope rather than a traditional fluorescence microscope to eliminate out-of-focus light with a pinhole aperture. This aperture focuses light onto a small section of the sample instead of the entire sample. The monochromatic light that is directed onto the sample is absorbed by fluorophores within the sample and re-emitted as a weaker fluorescent light shifted in frequency. This resulting fluorescence is then distinguished from the excitation source by emission filters and focused by a second pinhole aperture onto a detector, allowing for enhanced optical resolution. Common detectors include charge-coupled or photomultiplier devices that convert the incident light into an electrical signal. 2D and 3D images can be collected by placing samples onto a motorized stage and raster-scanning over a designated area. The spatial resolution of these images depends on the laser wavelength and the numerical aperture of the microscope focusing pinhole but can reach down to ~200 nm laterally and ~500 nm vertically (Dieing et al., 2010).

RMS requires a confocal optical microscope to be paired with a Raman spectrometer that quantifies the emission of energy-shifted light. This scattering, known as *Raman scattering*, arises from the interaction between the incident photons of the laser and the electron cloud of functional group bonds. More specifically, electrons within the cloud are initially excited into a virtual energy state by the incident photons but are subsequently relaxed into excited vibrational states of lower energy, changing the polarizability and inducing dipole moments within the molecule. The signal produced from this Raman shift is then measured by the spectrometer and plotted against the wavenumber (cm^{-1}) to produce a Raman spectrum. The peaks within a Raman spectrum, referred to as *Raman bands*, correspond to the energy levels of various functional group vibrations, and the peak intensity is proportional to the concentration of each group within the sample. RMS studies of aerosol particles typically produce Raman spectra spanning from 400 to 4,000 cm^{-1} with spectral resolution of 1.5–6.0 cm^{-1}.

9.4.2. Application to Aerosol Particles

Since the 1960s, standards of organic compounds have been analyzed by RMS, and the resulting spectra serve as references to allow identification of new spectra. Specific vibrational frequencies have been assigned to the Raman bands for a range of reference standards. These peak assignments, including C—H stretching peaks in the 2,700–3,000 cm^{-1} range, have been employed to identify organic components in atmospheric aerosols. Rosen and Novakov (1977, 1978) were the first to observe a Raman band at,2700 cm^{-1} due to the C—H stretching for lab-generated graphitic soot particles. Sadezky et al. (2005) expanded on the characterization of graphitic soot particles by producing a Raman spectrum of polycyclic aromatic hydrocarbon (PAH) hexabenzocoronene (HBC), an organic compound often found within soot particles. HBC was found to contain Raman shifts at ~1,600 cm^{-1}, ~1,320 cm^{-1}, and ~1,250 cm^{-1}. These peaks occurred at similar positions in the soot spectra, demonstrating the inclusion of PAH in soot (Gaffney et al., 2015; Ivleva et al., 2007).

Baustian et al. (2012) also used C—H stretching peaks to identify organic components in aerosol particles as well as to produce spatial mappings that illustrate the variability of the organic particle mixtures. These spatial maps show the location of the particle chemical components, in some cases exhibiting distinct morphologies (Figure 9.8). These morphologies include an organic particle in which an inorganic core is completely enclosed by an organic coating and an organic particle with both amorphous inorganic and organic inclusions. The particle containing an organic coating may have been formed from condensation from both anthropogenic and biogenic sources during transport. In contrast, the amorphous particle may have been a carbonate-mineral particle that has scavenged organic and nitrate compounds (Baustian et al., 2012). Ault et al. (2013b) observed a similar contribution from the C—H stretching region within the Raman spectra of lab-generated SSA.

This region, spanning from 2,800 to 3,000 cm^{-1}, included contributions from various organic groups. Peaks at 2,850 and 2,930 cm^{-1} were identified as the symmetric and asymmetric stretches of C—H in methylene groups, respectively. The organic group variability was further recognized in the spectra of particles generated from seawater spiked with bacteria and phytoplankton. These particles contained increased C—H stretching modes in the range from 2,920 to 2,940 cm^{-1} and peaks in the range of 2,985 to 3,000 cm^{-1}, together providing possible tracers for marine life in atmospheric particles. These particles also showed a correlation between organic abundance and particle size, with the intensity of peaks within the C—H stretching region increasing with decreasing particle size (Ault et al., 2013b). Computer-controlled RMS has also been shown to reduce the user bias that arises from manually choosing particles of interest and to allow analysis of up to 100 particles within an hour (Craig et al., 2017).

Kirpes et al. (2019) were able to attribute spectra peaks to specific vibrational modes, identifying several organic compound classes within individual particles, including saccharides, short-chain fatty acids, surfactant long-chain fatty acids, and amino acids. Although some particle spectra did not contain any peaks that matched standard spectra, the ambient SSA particle spectrum representative of saccharides (1,014, 1,050, 1,095, 1,127, 1,436, 2,906, 2,942 cm^{-1}), short-chain fatty acids (1,062, 1,121, 2,847, 2,883, 2,910, 3,389 cm^{-1}), long-chain fatty acids (1,065, 1,130, 1,296, 1,444, 2,844, 2,880, 2,930 cm^{-1}), and amino acids with saccharides (858, 996, 1,045, 1,455, 1,606, 2,890, 2,938, 2,976, 3,391 cm^{-1}) can be seen in several spectra (Figure 9.9). Nearly all individual particles contained one or more of these biogenic compounds within an organic coating, with the most abundant compounds being saccharides and short-chain fatty acids. A possible source of these compounds is the production of exopolymeric substances (EPSs) by sea ice algae and bacteria during the winter. This mechanism was found to be consistent with the observation of a diminishing organic coating of particles collected during summertime, highlighting the impact of biological processes on the composition of OA (Kirpes et al., 2019). Marine sources may also contribute long-chain organic compounds that are not associated with continental anthropogenic origins (Deng et al., 2014).

Laskina et al. (2013) employed RMS to investigate the reactions between organic acids and minerals within dust

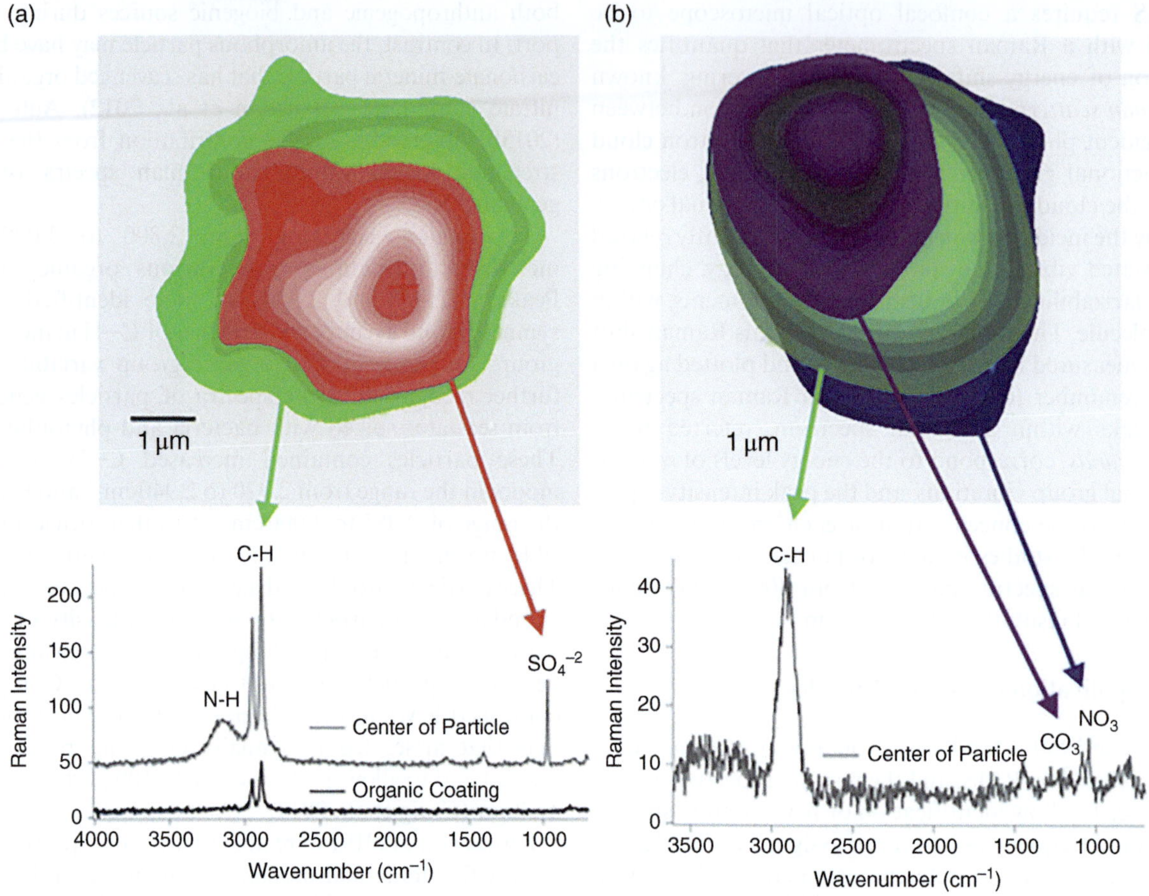

Figure 9.8 (a) Raman spectral map of an ammonium sulfate particle coated with organic material. The sulfate-containing region is highlighted in red, and the organic material is shown in green. (b) A spectral map of a complex uncoated particle with regions containing nitrate (blue) and organics (green) and a smaller region containing a carbonate signature (purple). The spectra shown, with the exception of the organic coating, were obtained near the center of the particle and therefore contain signatures from multiple constituents. Reproduced from Baustian et al. (2012)/with permission of John Wiley & Sons.

particles. The Raman spectra of these particles gave evidence of reactions between calcite and oxalic acid in Raman bands at 1,491, 1,464, and 897 cm^{-1} that can be assigned to the C═O stretching band of calcium oxalate monohydrate as well as C—O and C—C stretching modes associated with calcium oxalate formation. This finding was substantiated by the inclusion of a strong CO_3^{2-} stretching vibration at 1,087 cm^{-1} and a weaker CO_3^{2-} bending mode at 713 cm^{-1} due to calcite. 2D Raman mappings of these dust particles further revealed an uneven distribution of calcium oxalate on the calcium carbonate core, indicating the possible separation of carbonate and oxalate phases within the particle. An additional reaction between calcite and acetic acid, forming calcium acetate, was evident by a strong band at 2,933 cm^{-1}. This Raman shift is representative of symmetric, vibrational C—H stretches of acetic acid methyl groups that are found in calcium acetate. The Raman mappings produced for calcium acetate particles revealed a similar distribution of segregated product inclusions and uneven coatings to that of calcium oxalate particles. These segregations and non-uniform coatings suggest a possible influence from particle drying (Laskina et al., 2013).

9.4.3. Limitations

RMS is a mostly nondestructive technique that can be used to identify organic compounds based on observed Raman shifts. The Raman effect, however, is weak in comparison to other scattering effects, such as Rayleigh scattering. A high-resolution spectrometer is required for detection of the Raman signal but can be insufficient if the sample fluorescence interferes with the spectrum. Although the second pinhole aperture within confocal microscopes may block out sample fluorescence, signal intensity may be decreased. Longer exposures to the

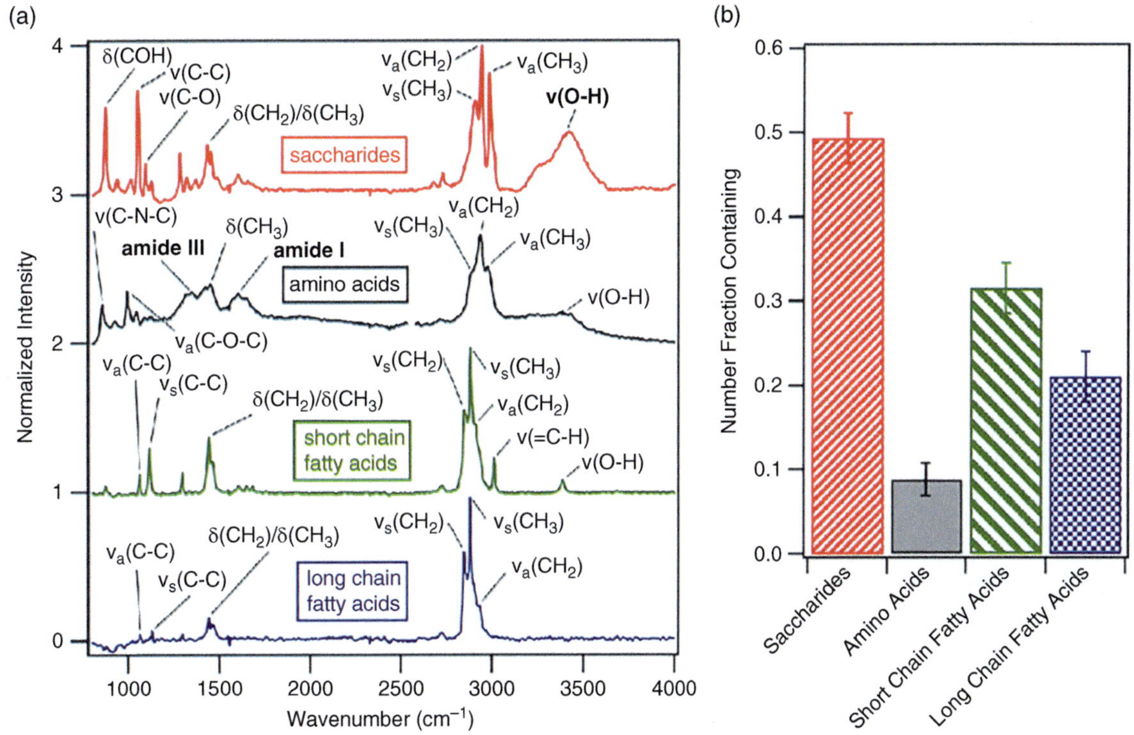

Figure 9.9 Marine-derived organic compound types observed in individual wintertime SSA particles. (a) Raman spectra of four representative individual SSA particles matching marine-derived saccharides ($\chi^2 = 0.013$), amino acids ($\chi^2 = 0.011$), short-chain fatty acids ($\chi^2 = 0.003$), and long-chain fatty acids ($\chi^2 = 0.0005$), respectively. (b) Number fractions (and associated standard errors) of 300 individual SSA particles containing saccharides, amino acids, short-chain fatty acids, and/or long-chain fatty acids. Each SSA particle Raman spectrum was allowed to be fitted to up to two organic compound types. Kirpes et al. (2019)/with permission of the American Chemical Society.

incident laser can be used to increase the signal but may risk damaging the sample.

Molecular characterization by RMS is accomplished by comparing Raman spectra to known standards, which allow the bands associated with specific organic functional groups to be identified and associated with specific molecules. However, information is limited for spectra that do not match available reference spectra.

The spatial resolution of RMS is worse than that of TEM, SEM, and STXM-NEXAFS techniques because confocal microscopes are a diffraction-limited system. RMS is unable to provide detailed molecular-level characterization of particles within the submicron range. RMS can also be used with cryo-electron microscopy techniques in order to preserve the composition of collected particles.

9.5. OTHER TECHNIQUES

Although this review has largely focused on techniques that have been applied to understanding atmospheric processes of aerosol particles, there are a few emerging techniques that are of note for their potential application to providing organic chemical composition to elucidate the processes and properties of submicron atmospheric particles. Here, we briefly review four such methods.

9.5.1. Tip-Enhanced Raman Spectroscopy (TERS)

TERS combines scanning probe microscopy with Raman spectroscopy and can perform nanoscale spectroscopic imaging. TERS differs from standard RMS by using a scanning probe microscope rather than the typical confocal optical microscope to maximize lateral resolution (Ofner et al., 2016). TERS further differs from RMS in that spectra contain both Raman bands and IR bands for vibrational modes that may overlap. Due to this overlap, the relative abundance of observed vibrational bands can provide semi-quantitative compositions. TERS has been applied to individual atmospheric particles below 20 nm diameter, revealing photochemical aging and chemical diversity among particles of similar size (Ofner et al., 2016). Because this technique is surface-sensitive, it does not detect inorganic cores within larger particles that are coated with organic components, meaning that it would not provide a representative composition for an entire particle.

9.5.2. Surface-Enhanced Raman Spectroscopy (SERS)

SERS uses substrates with a surface that contains molecules surrounded by metallic nanostructures to enhance Raman signals up to 10^{10} (Craig et al., 2015). These localized surface plasmon resonances or "hot spots" activate between the metal nanostructures and are smaller than 10 nm. SERS spectra contain peaks with preliminary assignments that indicate the presence of carbonyls, aromatics, alcohols, methyl rocking and scissoring modes, and organonitrates, ν(N—O) at ~1,650 cm^{-1}. Raman mapping showed inter- and intraparticle variability of organic components of SOA that had been assumed to be homogeneous in composition (Craig et al., 2015). Kunihisa et al. (2020) used a differential mobility analyzer and a condensation growth tube to separate and collect atmospheric particles as small as 20 nm in diameter. This method of collection deposits small particles more directly on the substrate "hot spots". The C—H stretching peaks at 2,800–3,000 cm^{-1} from this method showed organic signals in particles that were enhanced compared to the unassisted deposits.

9.5.3. Atomic Force Microscopy with Infrared Spectroscopy (AFM-IR)

AFM-IR measures physical properties including phase, morphology, hygroscopicity, and surface tension by AFM and chemical composition with IR spectrometry. This method overcomes the diffraction limit of visible and IR light that is associated with optical microscopes and vibrational spectroscopy to achieve chemical measurements at a resolution of 50–100 nm. Bondy et al. (2017) generated SOA from standards of succinic acid and sucrose to obtain AFM-IR spectra with resolved peaks at 1,201, 1,308, 3,049, 1,691, and 1,404 cm^{-1}, where these peaks were consistent with ν(C—C), ν(C—O), $\nu(CH_2)$, ν(C=O), and $\delta(CH_2)$ peaks (respectively) that were observed in succinic acid particles. Peaks at 1,057, 1,439, 2,913, and 3,345 cm^{-1} consistent with ν(C—O), δ(C—O—H), $\nu(CH_2)$, and ν(O—H) peaks (respectively) were observed in sucrose particles. Imaging capabilities of AFM-IR were exemplified by probing 550 nm particles with inorganic ammonium sulfate cores coated with polyethylene glycol (PEG). The inorganic and organic components of these particles were distinguished by AFM-IR height and phase images that detect phase separation of particles, which was then confirmed by optical and spectral maps produced by RMS. AFM-IR also showed different organic compositions within individual submicron atmospheric particles (Bondy et al., 2017), suggesting promising possibilities for future studies of atmospheric processes.

9.5.4. Optical Photothermal Infrared Spectroscopy (O-PTIR)

O-PTIR uses a continuous-wave visible laser and a tunable IR laser to obtain IR spectra with spatial resolution below 100 nm from the photothermal expansion of the sample (Olson et al., 2020). This method is able to analyze different phase states including liquids because it does not use direct contact with the sample (as with AFM-IR). O-PTIR produces spectral maps of particles with complicated mixing states. Olson et al. (2020) used O-PTIR concurrently with RMS to provide more complete information about organic and inorganic composition. Their work showed images of atmospheric particles with diameters >1 μm after collection on substrates that provided spectra with organic and sulfate components.

9.6. SUMMARY

Ambient particles have various origins that lead to diverse compositions. These compositions are critical to quantify the indirect and direct effects of aerosols that impact our climate, as well as to better understand air quality. Offline microscopic techniques such as STXM-NEXAFS, TEM-EDX, SEM-EDX, and RMS can provide a variety of complementary chemical information about individual particles. STXM-NEXAFS and RMS are able to identify functional groups based on electronic transitions and vibrational transitions, respectively, whereas electron microscopy methods allow for carbon detection without chemically distinguishing between elemental and organic forms. Because each technique is unique in both instrumentation and functionality (Figure 9.10), single-particle analyses that use multiple techniques provide the most complete information.

STXM-NEXAFS uses a monochromatic X-ray beam of relatively low energy to nondestructively probe OA. A carbon K-edge spectrum is obtained by scanning across a range of energies between 278 and 320 eV. Peaks within the near-edge region of the spectra are indicative of the contributions of various carbon functional groups. Functional groups that have been identified with this method include alkene or aromatic (R(C=C)R′), ketonic carbonyl (R(C=O)R′), alkyl ($R(CH_n)R'$), carboxylic carbonyl (R(C=O)OH), amine (RNH-ROH) or alcohol (R-COH), benzoquinone R(C=C)R′, and amide carbonyl (R-NH(C=O)R). STXM produces images with spatial resolutions between 30 and 35 nm

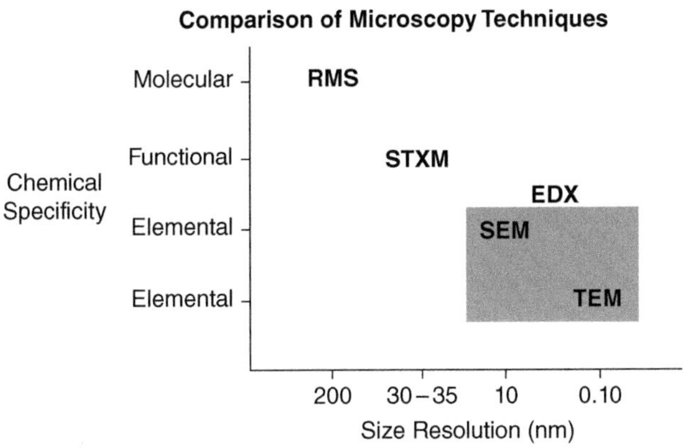

Figure 9.10 Comparison of the chemical specificity and size resolution of RMS, STXM-NEXAFS, and SEM/TEM-EDX.

that can be exploited to render spatial mappings of these functional groups. These nanoscale mappings are often used to reveal the distribution of various chemical components within OA. Other mappings can also reveal the inorganic and organic variability of particles that are down to submicron size.

Electron microscopy techniques, SEM-EDX and TEM-EDX, use an electron beam to obtain detailed images of aerosol particles. Although this beam can damage some particles, this method produces images with spatial resolutions of 0.1 nm for TEM-EDX and 10 nm for SEM-EDX. Studies that use this method scan across a much broader range of energies compared to STXM, detecting carbon in conjunction with many other elements. Carbon detection, however, is limited by the inability to distinguish between elemental and OC. OC may be identified by the particle morphology revealed in high-resolution images. By identifying various elements as well as the particle morphology, formation mechanisms and particle sources can be identified. This technique is ideal for analyzing the elemental composition of particles at submicron sizes, particularly those smaller than 200 nm that other methods cannot resolve.

RMS typically pairs a confocal microscope with a Raman spectrometer to focus monochromatic light onto the sample and thereafter record the resulting Raman scattering. The observed shifts in energy are seen as peaks in the Raman spectra and can be classified as specific organic bonds within the sample. Studies have used this method to identify possible organic functional groups and macromolecules within the sample by referencing documented spectra of known samples. Although this method provides the most comprehensive speciation of organic components in aerosol particles, the spatial resolution currently achieved using RMS is only ~200 nm laterally and ~500 nm vertically, which is insufficient for the majority of atmospheric particles. Enhanced micro-Raman techniques (SERS and TERS) are promising for identifying a more expansive scope of organic components in atmospheric aerosol if the number of particles can be improved.

Several studies have employed more than one of these imaging techniques to gain insight into both elemental and functional composition (Kirpes et al., 2018; Liati et al., 2019; Morillas et al., 2018; Pham et al., 2017; Saliba et al., 2020; Wagner et al., 2019). For example, RMS can identify various vibrational modes of organic groups detected by STXM-NEXAFS, and images of SEM-EDX and TEM-EDX can reveal the morphology as well as the variability of elements within particles. STXM-NEXAFS is effective for particles that are too small for RMS when chemical identification of organic properties is needed.

ACKNOWLEDGMENTS

The authors are grateful for the support of NSF grant AGS1660509.

REFERENCES

Adachi, K., & Buseck, P. R. (2008). Internally mixed soot, sulfates, and organic matter in aerosol particles from Mexico City. *Atmospheric Chemistry and Physics*, *8*(21), 6469–6481. doi: 10.5194/acp-8-6469-2008

Ault, A. P., Moffet, R. C., Baltrusaitis, J., Collins, D. B., Ruppel, M. J., Cuadra-Rodriguez, L. A., et al. (2013a). Size-dependent changes in sea spray aerosol composition and properties with different seawater conditions. *Environmental Science & Technology*, *47*(11), 5603–5612. doi: 10.1021/es400416g

Ault, A. P., Zhao, D. F., Ebben, C. J., Tauber, M. J., Geiger, F. M., Prather, K. A., & Grassian, V. H. (2013b). Raman microspectroscopy and vibrational sum frequency generation spectroscopy as probes of the bulk and surface compositions of size-resolved sea spray aerosol particles. *Physical Chemistry Chemical Physics*, *15*(17), 6206–6214. doi: 10.1039/c3cp43899f

Bahadur, R., Russell, L. M., & Prather, K. (2010). Composition and morphology of individual combustion, biomass burning, and secondary organic particle types obtained using urban and coastal ATOFMS and STXM-NEXAFS measurements. *Aerosol Science and Technology*, *44*(7), 551–562. doi: 10.1080/02786821003786048

Baustian, K. J., Cziczo, D. J., Wise, M. E., Pratt, K. A., Kulkarni, G., Hallar, A. G., & Tolbert, M. A. (2012). Importance of aerosol composition, mixing state, and morphology for heterogeneous ice nucleation: A combined field and laboratory approach. *Journal of Geophysical Research-Atmospheres*, *117*, 13. doi: 10.1029/2011jd016784

Bondy, A. L., Bonanno, D., Moffet, R. C., Wang, B. B., Laskin, A., & Ault, A. P. (2018). The diverse chemical mixing state of aerosol particles in the southeastern United States. *Atmospheric Chemistry and Physics*, *18*(16), 12595–12612. doi: 10.5194/acp-18-12595-2018

Bondy, A. L., Kirpes, R. M., Merzel, R. L., Pratt, K. A., Holl, M. M. B., & Ault, A. P. (2017). Atomic force microscopy-infrared spectroscopy of individual atmospheric aerosol particles: Subdiffraction limit vibrational spectroscopy and morphological analysis. *Analytical Chemistry*, *89*(17), 8594–8598. doi: 10.1021/acs.analchem.7b02381

Braun, A. (2005). Carbon speciation in airborne particulate matter with C (1s) NEXAFS spectroscopy. *Journal of Environmental Monitoring*, *7*(11), 1059–1065. doi: 10.1039/b508910g

Braun, A., Huggins, F. E., Kelly, K. E., Mun, B. S., Ehrlich, S. N., & Huffman, G. P. (2006). Impact of ferrocene on the structure of diesel exhaust soot as probed with wide-angle X-ray scattering and C(1s) NEXAFS spectroscopy. *Carbon*, *44*(14), 2904–2911. doi: 10.1016/j.carbon.2006.05.051

Braun, A., Mun, B. S., Huggins, F. E., & Huffman, G. P. (2007). Carbon speciation of diesel exhaust and urban particulate matter NIST standard reference materials with C(1s) NEXAFS spectroscopy. *Environmental Science & Technology*, *41*(1), 173–178. doi: 10.1021/es061044w

Braun, A., Shah, N., Huggins, F. E., Huffman, G. P., Wirick, S., Jacobsen, C., et al. (2004). A study of diesel PM with X-ray microspectroscopy. *Fuel*, *83*(7–8), 997–1000. doi: 10.1016/j.fuel.2003.08.015

Braun, A., Shah, N., Huggins, F. E., Kelly, K. E., Sarofim, A., Jacobsen, C., et al. (2005). X-ray scattering and spectroscopy studies on diesel soot from oxygenated fuel under various engine load conditions. *Carbon*, *43*(12), 2588–2599. doi: 10.1016/j.carbon.2005.05.017

Cain, J. P., Gassman, P. L., Wang, H., & Laskin, A. (2010). Micro-FTIR study of soot chemical composition-evidence of aliphatic hydrocarbons on nascent soot surfaces. *Physical Chemistry Chemical Physics*, *12*(20), 5206–5218. doi: 10.1039/b924344e

Chi, J. W., Li, W. J., Zhang, D. Z., Zhang, J. C., Lin, Y. T., Shen, X. J., et al. (2015). Sea salt aerosols as a reactive surface for inorganic and organic acidic gases in the Arctic troposphere. *Atmospheric Chemistry and Physics*, *15*(19), 11341–11353. doi: 10.5194/acp-15-11341-2015

Collins, D. B., Ault, A. P., Moffet, R. C., Ruppel, M. J., Cuadra-Rodriguez, L. A., Guasco, T. L., et al. (2013). Impact of marine biogeochemistry on the chemical mixing state and cloud forming ability of nascent sea spray aerosol. *Journal of Geophysical Research-Atmospheres*, *118*(15), 8553–8565. doi: 10.1002/jgrd.50598

Craig, R. L., Bondy, A. L., & Ault, A. P. (2015). Surface enhanced Raman spectroscopy enables observations of previously undetectable secondary organic aerosol components at the individual particle level. *Analytical Chemistry*, *87*(15), 7510–7514. doi: 10.1021/acs.analchem.5b01507

Craig, R. L., Bondy, A. L., & Ault, A. P. (2017). Computer-controlled Raman microspectroscopy (CC-Raman): A method for the rapid characterization of individual atmospheric aerosol particles. *Aerosol Science and Technology*, *51*(9), 1099–1112. doi: 10.1080/02786826.2017.1337268

Day, D. A., Takahama, S., Gilardoni, S., & Russell, L. M. (2009). Organic composition of single and submicron particles in different regions of western North America and the eastern Pacific during INTEX-B 2006. *Atmospheric Chemistry and Physics*, *9*(15), 5433–5446.

Deng, C. H., Brooks, S. D., Vidaurre, G., & Thornton, D. C. O. (2014). Using Raman microspectroscopy to determine chemical composition and mixing state of airborne marine aerosols over the Pacific Ocean. *Aerosol Science and Technology*, *48*(2), 193–206. doi: 10.1080/02786826.2013.867297

Dieing, T., Hollricher, O., & Toporski, J. (2010). *Confocal Raman microscopy*. Springer Series in Optical Sciences (Vol. 158). Springer International. doi: 10.1007/978-3-319-75380-5

Farmer, D. K., Cappa, C. D., & Kreidenweis, S. M. (2015). Atmospheric processes and their controlling influence on cloud condensation nuclei activity. *Chemical Reviews*, *115*(10), 4199–4217. doi: 10.1021/cr5006292

Fraund, M., Pham, D. Q., Bonanno, D., Harder, T. H., Wang, B. B., Brito, J., et al. (2017). Elemental mixing state of aerosol particles collected in Central Amazonia during GoAmazon2014/15. *Atmosphere*, *8*(9), 28. doi: 10.3390/atmos8090173

Freney, E., Sellegri, K., Chrit, M., Adachi, K., Brito, J., Waked, A., et al. (2018). Aerosol composition and the contribution of SOA formation over Mediterranean forests. *Atmospheric Chemistry and Physics*, *18*(10), 7041–7056. doi: 10.5194/acp-18-7041-2018

Frey, J., & Corn, M. (1967). Diesel exhaust particulates. *Nature*, *216*, 615–616.

Gaffney, J. S., Marley, N. A., & Smith, K. J. (2015). Characterization of fine mode atmospheric aerosols by Raman microscopy and diffuse reflectance FTIR. *Journal of Physical Chemistry A*, *119*(19), 4524–4532. doi: 10.1021/jp510361s

Hawkins, L. N., & Russell, L. (2010). Polysaccharides, proteins, and phytoplankton fragments: Four chemically distinct types of marine primary organic aerosol classified by single particle Spectromicroscopy. *Advances in Meteorology*, *2010*, 14. doi: 10.1155/2010/612132

Hawkins, L. N., Russell, L. M., Covert, D. S., Quinn, P. K., & Bates, T. S. (2010). Carboxylic acids, sulfates, and organosulfates in processed continental organic aerosol over the southeast Pacific Ocean during VOCALS-REx 2008. *Journal of Geophysical Research-Atmospheres*, *115*, 16. doi: 10.1029/2009jd013276

Hitchcock, A. P., Dynes, J. J., Johansson, G., Wang, J., & Botton, G. (2008). Comparison of NEXAFS microscopy and TEM-EELS for studies of soft matter (vol 39, pg 311, 2008). *Micron*, *39*(6), 741–748. doi: 10.1016/j.micron.2007.09.010

Hopkins, R. J., Lewis, K., Desyaterik, Y., Wang, Z., Tivanski, A. V., Arnott, W. P., et al. (2007). Correlations between optical, chemical and physical properties of biomass burn aerosols. *Geophysical Research Letters*, *34*(18), 5. doi: 10.1029/2007gl030502

Ivleva, N. P., McKeon, U., Niessner, R., & Poschl, U. (2007). Raman microspectroscopic analysis of size-resolved atmospheric aerosol particle samples collected with an ELPI: Soot, humic-like substances, and inorganic compounds. *Aerosol Science and Technology*, *41*(7), 655–671. doi: 10.1080/02786820701376391

Keywood, M. D., Ayers, G. P., Gras, J. L., Gillett, R. W., & Cohen, D. (1999). An evaluation of PM10 and PM2.5 size selective inlet performance using ambient aerosol. *Aerosol Science and Technology*, *30*(4), 401–407. doi: 10.1080/027868299304589

Kirpes, R. M., Bonanno, D., May, N. W., Fraund, M., Barget, A. J., Moffet, R. C., et al. (2019). Wintertime Arctic Sea spray aerosol composition controlled by sea ice lead microbiology. *ACS Central Science*, *5*(11), 1760–1767. doi: 10.1021/acscentsci.9b00541

Kirpes, R. M., Bondy, A. L., Bonanno, D., Moffet, R. C., Wang, B. B., Laskin, A., et al. (2018). Secondary sulfate is internally mixed with sea spray aerosol and organic aerosol in the winter Arctic. *Atmospheric Chemistry and Physics*, *18*(6), 3937–3949. doi: 10.5194/acp-18-3937-2018

Kirpes, R. M., Rodriguez, B., Kim, S., China, S., Laskin, A., Park, K., et al. (2020). Emerging investigator series: Influence of marine emissions and atmospheric processing on individual particle composition of summertime Arctic aerosol over the Bering Strait and Chukchi Sea. *Environmental Science: Processes & Impacts*, *22*(5), 1201–1213. doi: 10.1039/c9em00495e

Kunihisa, R., Iwata, A., Gen, M., Chan, C. K., & Matsuki, A. (2020). Application of SERS on the chemical speciation of individual Aitken mode particles after condensational growth. *Aerosol Science and Technology*, *54*(7), 826–836. doi: 10.1080/02786826.2020.1730298

Laskin, A., & Cowin, J. P. (2001). Automated single particle SEM/EDX analysis of submicrometer particles down to 0.1 microm. *Analytical Chemistry*, *73*(5), 1023–1029. doi: 10.1021/ac0009604

Laskin, A., Moffet, R. C., Gilles, M. K., Fast, J. D., Zaveri, R. A., Wang, B. B., et al. (2012). Tropospheric chemistry of internally mixed sea salt and organic particles: Surprising reactivity of NaCl with weak organic acids. *Journal of Geophysical Research-Atmospheres*, *117*, 12. doi: 10.1029/2012jd017743

Laskina, O., Young, M. A., Kleiber, P. D., & Grassian, V. H. (2013). Infrared extinction spectroscopy and micro-Raman spectroscopy of select components of mineral dust mixed with organic compounds. *Journal of Geophysical Research-Atmospheres*, *118*(12), 6593–6606. doi: 10.1002/jgrd.50494

Leaitch, W. R., Macdonald, A. M., Anlauf, K. G., Liu, P. S. K., Toom-Sauntry, D., Li, S. M., et al. (2009). Evidence for Asian dust effects from aerosol plume measurements during INTEX-B 2006 near Whistler, BC. *Atmospheric Chemistry and Physics*, *9*(11), 3523–3546.

Li, W. J., Chen, S. R., Xu, Y. S., Guo, X. C., Sun, Y. L., Yang, X. Y., et al. (2015). Mixing state and sources of submicron regional background aerosols in the northern Qinghai-Tibet Plateau and the influence of biomass burning. *Atmospheric Chemistry and Physics*, *15*(23), 13365–13376. doi: 10.5194/acp-15-13365-2015

Li, W. J., & Shao, L. Y. (2010). Mixing and water-soluble characteristics of particulate organic compounds in individual urban aerosol particles. *Journal of Geophysical Research-Atmospheres*, *115*, 9. doi: 10.1029/2009jd012575

Liati, A., Schreiber, D., Alpert, P. A., Liao, Y., Brem, B. T., Arroyo, P. C., et al. (2019). Aircraft soot from conventional fuels and biofuels during ground idle and climb-out conditions: Electron microscopy and X-ray micro-spectroscopy. *Environmental Pollution*, *247*, 658–667. doi: 10.1016/j.envpol.2019.01.078

Liu, L., Kong, S. F., Zhang, Y. X., Wang, Y. Y., Xu, L., Yan, Q., et al. (2017). Morphology, composition, and mixing state of primary particles from combustion sources – crop residue, wood, and solid waste. *Scientific Reports*, *7*, 15. doi: 10.1038/s41598-017-05357-2

Liu, S., Ahlm, L., Day, D. A., Russell, L. M., Zhao, Y. L., Gentner, D. R., et al. (2012). Secondary organic aerosol formation from fossil fuel sources contribute majority of summertime organic mass at Bakersfield. *Journal of Geophysical Research-Atmospheres*, *117*, 21. doi: 10.1029/2012jd018170

Liu, S., Day, D. A., Shields, J. E., & Russell, L. M. (2011). Ozone-driven daytime formation of secondary organic aerosol containing carboxylic acid groups and alkane groups. *Atmospheric Chemistry and Physics*, *11*(16), 8321–8341. doi: 10.5194/acp-11-8321-2011

Liu, S., Takahama, S., Russell, L. M., Gilardoni, S., & Baumgardner, D. (2009). Oxygenated organic functional groups and their sources in single and submicron organic particles in MILAGRO 2006 campaign. *Atmospheric Chemistry and Physics*, *9*(18), 6849–6863.

Maria, S. F., Russell, L. M., Gilles, M. K., & Myneni, S. C. B. (2004). Organic aerosol growth mechanisms and their climate-forcing implications. *Science*, *306*(5703), 1921–1924. doi: 10.1126/science.1103491

McFarquhar, G. M., Bretherton, C., Marchand, R., Protat, A., DeMott, P. J., Alexander, S. P., et al. (2020). Observations of clouds, aerosols, precipitation, and surface radiation over the Southern Ocean: An overview of CAPRICORN, MARCUS, MICRE and SOCRATES. *Bulletin of the American Meteorological Society*. doi: 10.1175/BAMS-D-20-0132.1

McKee, H. C., & McMahon, W. A. (1960). Automobile exhaust particulates—Source and variation. *Journal of the Air Pollution Control Association*, *10*(6). doi: 10.1080/00022470.1960.10467959

Moffet, R. C., Henn, T., Laskin, A., & Gilles, M. K. (2010a). Automated chemical analysis of internally mixed aerosol particles using X-ray spectromicroscopy at the carbon K-edge. *Analytical Chemistry*, *82*(19), 7906–7914. doi: 10.1021/ac1012909

Moffet, R. C., Henn, T. R., Tivanski, A. V., Hopkins, R. J., Desyaterik, Y., Kilcoyne, A. L. D., et al. (2010b). Microscopic characterization of carbonaceous aerosol particle aging in the outflow from Mexico City. *Atmospheric Chemistry and Physics*, *10*(3), 961–976. doi: 10.5194/acp-10-961-2010

Moffet, R. C., Rodel, T. C., Kelly, S. T., Yu, X. Y., Carroll, G. T., Fast, J., et al. (2013). Spectro-microscopic measurements of carbonaceous aerosol aging in Central California. *Atmospheric Chemistry and Physics*, *13*(20), 10445–10459. doi: 10.5194/acp-13-10445-2013

Morillas, H., Marcaida, I., Garcia-Florentino, C., Maguregui, M., Arana, G., & Madariaga, J. M. (2018). Micro-Raman and SEM-EDS analyses to evaluate the nature of salt clusters present in secondary marine aerosol. *Science of the Total Environment*, *615*, 691–697. doi: 10.1016/j.scitotenv.2017.09.299

O'Brien, R. E., Wang, B. B., Laskin, A., Riemer, N., West, M., Zhang, Q., et al. (2015). Chemical imaging of ambient aerosol particles: Observational constraints on mixing state parameterization. *Journal of Geophysical Research-Atmospheres*, *120*(18), 9591–9605. doi: 10.1002/2015jd023480

Ofner, J., Deckert-Gaudig, T., Kamilli, K. A., Held, A., Lohninger, H., Deckert, V., & Lendl, B. (2016). Tip-enhanced Raman spectroscopy of atmospherically relevant aerosol nanoparticles. *Analytical Chemistry*, *88*(19), 9766–9772. doi: 10.1021/acs.analchem.6b02760

Olson, N. E., Xiao, Y., Lei, Z. Y., & Ault, A. P. (2020). Simultaneous optical photothermal infrared (O-PTIR) and Raman spectroscopy of submicrometer atmospheric particles. *Analytical Chemistry*, *92*(14), 9932–9939. doi: 10.1021/acs.analchem.0c01495

Parungo, F. P., Nagamoto, C. T., Sheridan, P. J., & Schnell, R. C. (1990). Aerosol characteristics of Arctic haze sampled during AGASP-II. *Atmospheric Environment. Part A. General Topics*, *24*(4), 937–949. doi: 10.1016/0960-1686(90)90296-y

Patterson, J. P., Collins, D. B., Michaud, J. M., Axson, J. L., Sultana, C. M., Moser, T., et al. (2016). Sea spray aerosol structure and composition using cryogenic transmission electron microscopy. *ACS Central Science*, *2*(1), 40–47. doi: 10.1021/acscentsci.5b00344

Petzold, A., Ogren, J. A., Fiebig, M., Laj, P., Li, S. M., Baltensperger, U., et al. (2013). Recommendations for reporting "black carbon" measurements. *Atmospheric Chemistry and Physics*, *13*(16), 8365–8379. doi: 10.5194/acp-13-8365-2013

Pham, D. Q., O'Brien, R., Fraund, M., Bonanno, D., Laskina, O., Beall, C., et al. (2017). Biological impacts on carbon speciation and morphology of sea spray aerosol. *ACS Earth and Space Chemistry*, *1*(9), 551–561. doi: 10.1021/acsearthspacechem.7b00069

Pitchford, M. L., & McMurry, P. H. (1994). Relationship between measured water-vapor growth and chemistry of atmospheric aerosol for Grand-Canyon, ARIZONA, in winter 1990. *Atmospheric Environment*, *28*(5), 827–839. doi: 10.1016/1352-2310(94)90242-9

Popovicheva, O., Timofeev, M., Persiantseva, N., Jefferson, M. A., Johnson, M., Rogak, S. N., & Baldelli, A. (2019). Microstructure and chemical composition of particles from small-scale gas flaring. *Aerosol and Air Quality Research*, *19*(10), 2205–2221. doi: 10.4209/aaqr.2019.04.0177

Poschl, U. (2005). Atmospheric aerosols: Composition, transformation, climate and health effects. *Angewandte Chemie-International Edition*, *44*(46), 7520–7540. doi: 10.1002/anie.200501122

Posfai, M., Anderson, J. R., Buseck, P. R., & Sievering, H. (1999). Soot and sulfate aerosol particles in the remote marine troposphere. *Journal of Geophysical Research-Atmospheres*, *104*(D17), 21685–21693. doi: 10.1029/1999jd900208

Posfai, M., Gelencser, A., Simonics, R., Arato, K., Li, J., Hobbs, P. V., & Buseck, P. R. (2004). Atmospheric tar balls: Particles from biomass and biofuel burning. *Journal of Geophysical Research-Atmospheres*, *109*(D6), 9. doi: 10.1029/2003jd004169

Rausch, A., Ebert, M., Gorzawski, G., Weinbruch, S., Hoffmann, P., & Helas, G. (1999). High-resolution scanning electron microscopy of individual aerosol particles obtained during LACE 98. *Aerosol Science*, *30*, 516–517.

Rosen, H., & Novakov, T. (1977). Raman-scattering and characterization of atmospheric aerosol-particles. *Nature*, *266*(5604), 708–710. doi: 10.1038/266708a0

Rosen, H., & Novakov, T. (1978). Identification of primary particulate carbon and sulfate species by Raman-spectroscopy. *Atmospheric Environment*, *12*(4), 923–927. doi: 10.1016/0004-6981(78)90031-8

Rudich, Y., Donahue, N. M., & Mentel, T. F. (2007). Aging of organic aerosol: Bridging the gap between laboratory and field studies. *Annual Review of Physical Chemistry*, *58*, 321–352. doi: 10.1146/annurev.physchem.58.032806.104432

Russell, L. M., Maria, S. F., & Myneni, S. C. B. (2002). Mapping organic coatings on atmospheric particles. *Geophysical Research Letters*, *29*(16). doi: 10.1029/2002gl014874

Sadezky, A., Muckenhuber, H., Grothe, H., Niessner, R., & Poschl, U. (2005). Raman micro spectroscopy of soot and related carbonaceous materials: Spectral analysis and structural information. *Carbon*, *43*(8), 1731–1742. doi: 10.1016/j.carbon.2005.02.018

Saliba, G., Sanchez, K. J., Russell, L. M., Twohy, C. H., Roberts, G. C., Lewis, S., et al. (2020). Organic composition of three different size ranges of aerosol particles over the Southern Ocean. *Aerosol Science and Technology*, 1–23. doi: 10.1080/02786826.2020.1845296

Schwartz, R. E., Russell, L. M., Sjostedt, S. J., Vlasenko, A., Slowik, J. G., Abbatt, J. P. D., et al. (2010). Biogenic oxidized organic functional groups in aerosol particles from a mountain forest site and their similarities to laboratory chamber products. *Atmospheric Chemistry and Physics*, *10*(11), 5075–5088. doi: 10.5194/acp-10-5075-2010

Schwarz, J. P., Spackman, J. R., Fahey, D. W., Gao, R. S., Lohmann, U., Stier, P., et al. (2008). Coatings and their enhancement of black carbon light absorption in the tropical atmosphere. *Journal of Geophysical Research-Atmospheres*, *113*(D3). doi: 10.1029/2007jd009042

Semeniuk, T. A., Wise, M. E., Martin, S. T., Russell, L. M., & Buseck, P. R. (2007). Hygroscopic behavior of aerosol particles from biomass fires using environmental transmission electron microscopy. *Journal of Atmospheric Chemistry*, *56*(3), 259–273. doi: 10.1007/s10874-006-9055-5

Shakya, K. M., Liu, S., Takahama, S., Russell, L. M., Keutsch, F. N., Galloway, M. M., et al. (2013). Similarities in STXM-NEXAFS spectra of atmospheric particles and secondary organic aerosol generated from glyoxal, alpha-pinene, isoprene, 1,2,4-trimethylbenzene, and d-limonene. *Aerosol Science and Technology*, *47*(5), 543–555. doi: 10.1080/02786826.2013.772950

Sheridan, P. J. (1989a). Analytical electron-microscope studies of size-segregated particles collected during AGASP-II, flights 201-203. *Journal of Atmospheric Chemistry*, *9*(1–3), 267–282. doi: 10.1007/bf00052837

Sheridan, P. J. (1989b). Characterization of size segregated particles collected over Alaska and the Canadian high Arctic, AGASP-II flights 204-206. *Atmospheric Environment*, *23*(11), 2371–2386. doi: 10.1016/0004-6981(89)90250-3

Shrivastava, M., Cappa, C. D., Fan, J. W., Goldstein, A. H., Guenther, A. B., Jimenez, J. L., et al. (2017). Recent advances in understanding secondary organic aerosol: Implications for global climate forcing. *Reviews of Geophysics*, *55*(2), 509–559. doi: 10.1002/2016rg000540

Soderholm, S. C. (1981). Physical characterization of diesel exhaust particles in exposure chambers. *Environmental*

International, *5*(4–6), 325–330. doi: 10.1016/0160-4120(81)90078-7

Takahama, S., Gilardoni, S., Russell, L. M., & Kilcoyne, A. L. D. (2007). Classification of multiple types of organic carbon composition in atmospheric particles by scanning transmission X-ray microscopy analysis. *Atmospheric Environment*, *41*(40), 9435–9451. doi: 10.1016/j.atmosenv.2007.08.051

Takahama, S., Liu, S., & Russell, L. M. (2010). Coatings and clusters of carboxylic acids in carbon-containing atmospheric particles from spectromicroscopy and their implications for cloud-nucleating and optical properties. *Journal of Geophysical Research-Atmospheres*, *115*(D01202). doi: 10.1029/2009jd012622

Takahama, S., Schwartz, R. E., Russell, L. M., Macdonald, A. M., Sharma, S., & Leaitch, W. R. (2011). Organic functional groups in aerosol particles from burning and non-burning forest emissions at a high-elevation mountain site. *Atmospheric Chemistry and Physics*, *11*(13), 6367–6386. doi: 10.5194/acp-11-6367-2011

Tivanski, A. V., Hopkins, R. J., Tyliszczak, T., & Gilles, M. K. (2007). Oxygenated interface on biomass burn tar balls determined by single particle scanning transmission X-ray microscopy. *Journal of Physical Chemistry A*, *111*(25), 5448–5458. doi: 10.1021/jp070155u

Twohy, C. H., Anderson, J. R., & Crozier, P. A. (2005). Nitrogenated organic aerosols as cloud condensation nuclei. *Geophysical Research Letters*, *32*(19), 4. doi: 10.1029/2005gl023605

Twohy, C. H., DeMott, P. J., Russell, L. M., Toohey, D. W., Rainwater, B., Geiss, R., et al. (2021). Cloud-nucleating particles over the Southern Ocean in a changing climate. *Earth's Future*. doi: 10.1029/2020EF001673

Vuk, C. T., Jones, M. A., & Johnson, J. H. (1976). The measurement and analysis of the physical character of diesel particulate emissions. *SAE Technical Paper*, *85*, 556–597.

Wagner, J., Wang, Z. M., Ghosal, S., & Wall, S. (2019). Source identification on high PM2.5 days using SEM/EDS, XRF, Raman, and windblown dust modeling. *Aerosol and Air Quality Research*, *19*(11), 2518–2530. doi: 10.4209/aaqr.2019.05.0276

Wang, B. B., Harder, T. H., Kelly, S. T., Piens, D. S., China, S., Kovarik, L., et al. (2016). Airborne soil organic particles generated by precipitation. *Nature Geoscience*, *9*(6), 433-+. doi: 10.1038/ngeo2705

Wang, B. B., & Laskin, A. (2014). Reactions between water-soluble organic acids and nitrates in atmospheric aerosols: Recycling of nitric acid and formation of organic salts. *Journal of Geophysical Research-Atmospheres*, *119*(6), 3335–3351. doi: 10.1002/2013jd021169

Wang, W. H., Shao, L. Y., Li, J., Chang, L. L., Zhang, D. Z., Zhang, C. C., & Jiang, J. K. (2019). Characteristics of individual particles emitted from an experimental burning chamber with coal from the lung cancer area of Xuanwei, China. *Aerosol and Air Quality Research*, *19*(2), 355–363. doi: 10.4209/aaqr.2018.05.0187

Wang, W. H., Shao, L. Y., Xing, J. P., Li, J., Chang, L. L., & Li, W. J. (2018). Physicochemical characteristics of individual aerosol particles during the 2015 China victory day parade in Beijing. *Atmosphere*, *9*(2), 10. doi: 10.3390/atmos9020040

Xu, L., Liu, L., Zhang, J., Zhang, Y. X., Ren, Y., Wang, X., & Li, W. J. (2017). Morphology, composition, and mixing state of individual aerosol particles in Northeast China during wintertime. *Atmosphere*, *8*(3), 10. doi: 10.3390/atmos8030047

Zhang, X. Q., McMurry, P. H., Hering, S. V., & Casuccio, G. S. (1993). Mixing characteristics and water-content of submicron aerosols measured in Los-Angeles and at the grand-canyon. *Atmospheric Environment. Part A. General Topics*, *27*(10), 1593–1607. doi: 10.1016/0960-1686(93)90159-v

10

Microanalytical Studies of Multiphase Reactions and Particle Aging

Peter A. Alpert[1], Pablo Corral Arroyo[2], and Markus Ammann[1]

ABSTRACT

Research in aerosol chemistry and physics has significantly advanced our understanding of atmospheric processes, including air quality, radiation balance, cloud formation, and chemical transformations. Aerosol particles are composed of diverse compounds and exhibit various chemical and physical properties, such as reactivity, solubility, and morphology. Microanalytical methods, particularly microscopy and spectroscopy, have been instrumental in studying the chemical and photochemical reactions of single aerosol particles at the micro- and nanoscale. Despite their small scale, these studies have revealed critical atmospheric processes with global implications. This chapter highlights in situ microanalytical studies that track the chemical composition and morphology of aerosol particles in real time during reactions. We emphasize X-ray spectromicroscopy for its ability to map the evolving chemical morphology within single particles. Additionally, we discuss the need for advanced data processing techniques, including artificial neural networks and modeling, as microanalytical technology progresses.

10.1. INTRODUCTION

A tremendous diversity of particles and gases are either directly emitted to the atmosphere or formed in the air through secondary chemistry, all contributing to the amount and composition of aerosol particles suspended in air (Laskin et al., 2019). Aerosol particles can impact climate (Myhre et al., 2013a) and can cause detrimental effects upon exposure to humans, plants, and animals (Fuzzi et al., 2015). These particles contain water, in addition to metals, soot, organic compounds, or other inorganic compounds. However, airborne particles and gases are chemically altered due to gas-phase chemistry, particle-phase chemistry, and multiphase chemistry. The latter involves chemical reaction and exchange between reactants in the gas and condensed (particle) phase. The air can host a variety of gases that are organic, nitrogenated, oxygenated, and sulfonated, among others, that can change the composition of particles due to uptake, chemical reaction, or condensation (Curtis et al., 2006; D'Amato et al., 2010; Ziemann and Atkinson, 2012). Overall, aerosol particles severely exacerbate adverse health effects and impact the climate, based on estimates that rely heavily on accurately predicting gas and aerosol particle abundance, their impact on cloud and other microphysical properties, and atmospheric composition (Boucher et al., 2013; Fuzzi et al., 2015; Monks et al., 2009; Myhre et al., 2013b; Pöschl, 2005).

Particles composed of organic material and aqueous solutions can become highly viscous depending on their degree of oxidation, temperature, *T*, and relative humidity, *RH* (Koop et al., 2011). For example, the lifetime of carcinogenic benzo(a)pyrene (BaP) and other polycyclic aromatic hydrocarbons becomes significantly longer when they are associated with semisolid particles, thus

[1] *PSI Center for Energy and Environmental Sciences, Paul Scherrer Institute, Villigen, Switzerland*

[2] *Laboratory for Physical Chemistry, Swiss Federal Institute of Technology Zurich, Zurich, Switzerland*

Microanalysis of Atmospheric Particles: Techniques and Applications, Geophysical Monograph 288, First Edition.
Edited by Joseph M. Conny and Peter R. Buseck.

DOI: 10.1002/9781119554318.ch10

potentially leading to greater exposure than expected for more liquid-like particles (Mu et al., 2018). Furthermore, sunlight-generated radicals and reactive oxygen species (ROS) can become trapped in highly viscous particles when *RH* and *T* are low, potentially being released into the water-saturated lung environment when inhaled (Alpert et al., 2021). Sunlight-driven photochemical reaction of secondary organic aerosol (SOA) particles may increase their viscosity and the overall molecular mass and oxidation of organic molecules in tropospheric conditions (Baboomian et al., 2022). Traditionally, reactions involving gas-phase and condensed-phase (liquid or solid) particles have been termed *heterogeneous* when referring only to a surface process and *multiphase* when the near surface (single-nanometer-scale depth beneath the surface) or aerosol bulk is chemically altered (Ravishankara, 1997). However, we prefer to use the term *multiphase chemistry* more broadly as it is not always possible to unambiguously separate such steps. An important example of multiphase chemistry is the positive feedback mechanism linking water content, reaction turnover, and mass when *RH* and aerosol water content are high (Su et al., 2020). In these conditions, severe pollution events can become even more exacerbated, in which high aerosol particle numbers and mass lead to greater multiphase reaction turnover rates and thus even greater production of aerosol mass (Su et al., 2020). Clearly, the link between multiphase chemistry and aerosol microphysics is important to accurately represent and predict the evolution of multiphase chemical reactants and their products over time.

Knowledge of bulk composition is invaluable to the scientific community as it gives an overall representation of aerosol composition, health hazards, and chemical processes (Arangio et al., 2016; Daellenbach et al., 2020; Paulson et al., 2019; Q. Zhang et al., 2007; Renjian Zhang et al., 2013). However, single-particle techniques that measure compositional change take an additional step toward understanding, for example, aerosol complexity and mixing and its relevance to ice and liquid cloud formation and aerosol light absorption properties (Creamean et al., 2013; Cziczo et al., 2013; Fierce et al., 2017; Fierce et al., 2016; Yao et al., 2021; Yao et al., 2022; L. Zhou et al., 2022). Very often, the bulk composition of particles is measured either online using aerosol mass spectrometry or ion chromatography or offline using high-volume sampling filters coupled with similar or other analytical techniques. As a result, they intrinsically represent particles as completely mixed without considering the possibility of the heterogeneity of phases or compositional gradients within particles that may be caused by slow molecular diffusion and reaction. However, chemical morphology (knowing what various chemical compounds are where in a single particle) may be important for climate and human health impacts.

Figure 10.1 shows an illustration of atmospheric multiphase chemistry in a scenario where aerosol particles emitted from natural and anthropogenic sources can

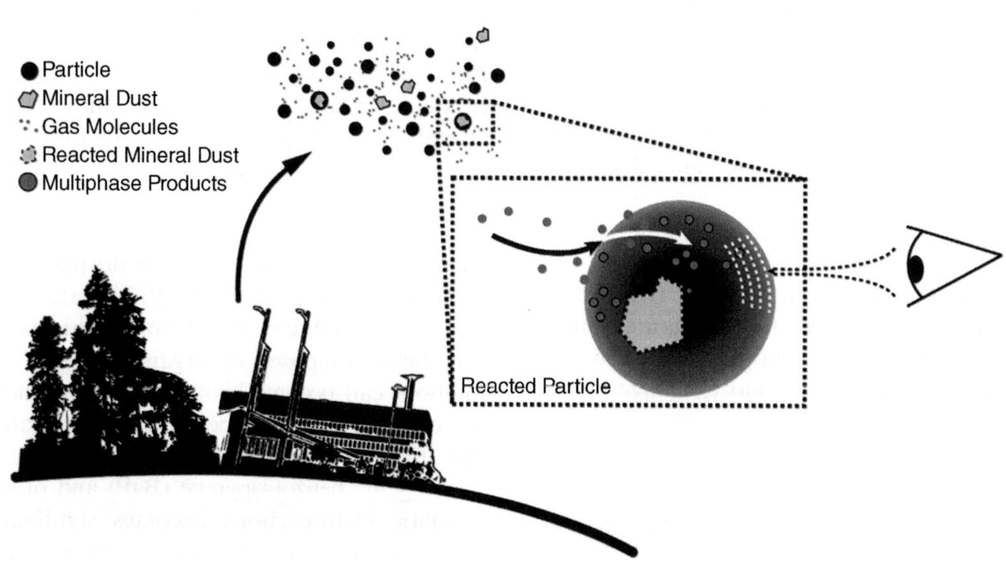

Figure 10.1 Illustration of multiphase chemistry between aerosol particles and reactive gases. Aerosol particles and mineral dust, for example, emitted from natural or anthropogenic sources, can be chemically altered due to gas uptake and chemical reactions. The white dotted curves are layers with a spatial thickness on the order of nanometers that conceptualize the scales microanalytical methods and models investigate. Condensed-phase products from multiphase chemistry can form that alter particle bulk or surface composition. Microanalytical techniques used to investigate multiphase chemistry observe the surface or internal structure of aerosol particles.

chemically change in the presence of gases that could condense onto particles or alter their composition through chemical reactions. This can include organic and inorganic aerosol particles, mineral dust, soot, and trace metals. Products of multiphase chemistry produce a variety of species that may be stable compounds, reactive compounds, or radical species, which can be carbon-centered radicals (CCRs) or ROS. In the condensed phase, these can undergo further reactions and chemical changes. Microanalytical techniques that investigate multiphase chemistry do so by observing and modeling the chemical change inside of particles. Here, careful attention is paid to surface and bulk reactions to elucidate fundamental chemical steps over time. This can be done by looking at the surface at molecular scales or using layers with a spatial thickness of nanometers, illustrated in Figure 10.1 as the white dashed lines. Surface or near-surface sensitivity having nanometer-scale resolution can give valuable insight into research questions about the initial steps in multiphase chemistry, such as gas adsorption and surface reactions. Instrumentation resolving composition on the scale of tens of nanometers can address other scientific questions of molecular reaction rates and transport inside of particles. Scanning transmission X-ray microscopy (STXM) coupled with near-edge X-ray absorption fine structure spectroscopy (NEXAFS) is such an instrument that we emphasize here for single particle in situ analysis. Multiphase chemistry may lead to gradual changes in composition occurring over the scale of the particle diameter, thus requiring resolution at least in the tens of nanometer range achievable with STXM/NEXAFS.

We focus particularly on microanalytical studies that detail multiphase chemistry on the sub-particle scale using experimental techniques or mathematical models. Microanalytical techniques in multiphase chemistry can refer to methods that aim to characterize compositional or morphological change on micrometer down to nanometer spatial scales. These methods can utilize high spatially resolving electron imaging and spectroscopy with scanning electron microscopy (SEM), transmission electron microscopy (TEM), energy dispersive X-ray analysis (EDX), computer-controlled SEM (CCSEM), and electron energy loss spectroscopy (EELS) (Laskin et al., 2012; Laskin et al., 2015). X-rays are used for STXM/NEXAFS (Moffet et al., 2010; Raabe et al., 2008). Photoelectron spectroscopy using X-ray (XPS) or UV (UPS) excitation has a probing depth on the single-nanometer scale, and the photoelectron angular distribution from an aerosol sample can be obtained with the use of a velocity map imaging (VMI) photoelectron spectrometer (Ahmed & Kostko, 2020; Orlando et al., 2019; Signorell & Winter, 2022). Other microanalytical methods include Raman spectromicroscopy for super-micrometer sizes (Liang et al., 2022) and atomic force microscopy (AFM) for nanometer-scale morphologies (Sobanska et al., 2014; Woodward et al., 2015). Other methods include diffuse reflectance infrared Fourier transform spectroscopy (DRIFTS), which has near-surface sensitivity (Yongchun Liu et al., 2010; Z. Wang et al., 2019); time-of-flight secondary ion mass spectrometry ToF-SIMS (Sobanska et al., 2014); direct analysis in real time mass spectrometry (DART-MS) (Davies and Wilson, 2015; Lam et al., 2019); and fluorescence lifetime imaging microscopy (FLIM) (Fitzgerald et al., 2016; Hosny et al., 2016). Methods not reviewed here are holography (Berg & Videen, 2011), tomography (Adler et al., 2013), a combination of phase-sensitive sum and difference frequency-generation spectroscopy (Balos et al., 2022), diffractive imaging; and other techniques at synchrotron light sources or free electron laser facilities (Bogan et al., 2010; Richard et al., 2010).

In this chapter, we review recent and relevant research that quantified the chemistry and physics occurring during multiphase chemical reactions. First, offline techniques are introduced when multiphase reactions are stopped at certain times and particles are investigated. Then, in situ investigations of aerosol particles during multiphase reactions are presented. We also explore the potential of applying artificial neural networks (ANNs) to microanalytical image analysis. Finally, model representation and investigations of multiphase chemistry are reviewed.

10.2. SELECTED OFFLINE TECHNIQUES

Multiphase chemistry investigations using offline microanalytical techniques offer unique snapshots of ongoing compositional changes to aerosol particles. The reaction of gas-phase HNO_3 with mineral dust particles was investigated with SEM to observe such snapshots of elemental and morphological changes (Krueger et al., 2003a; Krueger et al., 2003b; Krueger et al., 2004; Laskin et al., 2005). In particular, individual mineral dust particles from natural sources (e.g., China loess) were exposed to nitric acid vapors at different times, followed by offline analysis with computer-controlled scanning electron microscopy (CCSEM). The authors observed that individual calcite-type particles experienced a major morphological and compositional change after multiphase reactions. However, other particle types, such as dolomite, showed no signs of change (Krueger et al., 2004). This work emphasizes the importance of single-particle analysis and microanalytical techniques that are capable of nanometer-scale resolution because only some particles in a complex population may undergo reactions. In a later study, the uptake coefficient, γ, of gaseous HNO_3 reacting with $CaCO_3$ particles was quantified using CCSEM, and these values were in agreement with other work (Y. Liu et al., 2008). Sea salt particles reacting with HNO_3 were investigated using SEM/EDX, which was used to image elemental and morphological changes in particles (Y. Liu et al., 2007). The data was used to determine that gas-phase diffusion was the limiting case for the multiphase chemistry that occurred.

Complex multiphase processes of SOA formation on sea spray aerosol (SSA) particles and their ensuing multiphase chemistry were investigated using microanalytical methods after exposure and reaction under a range of atmospheric conditions (B. Wang et al., 2015). NaCl reacted with SOA coatings over hours, resulting in acid displacement and producing condensed-phase organic salts and HCl to the gas phase, as shown in Figure 10.2 (B. Wang et al., 2015). These results show that the dried particle residues featured salt cores dominated by cubic-like NaCl, which completely changed the composition to more "snowflake-like" crystalline patterns with a higher-contrast organic coating, akin to a salty halo seen in Figure 10.2b–d but absent in Figure 10.2a. Use of STXM/NEXAFS showed that organic material was pervasive throughout the particles and revealed alteration in Na—Cl bonding (B. Wang et al., 2015). This implies that multiphase SSA particle chemistry affects their optical properties, hygroscopicity, and cloud-forming potential and may impact atmospheric composition (Chi et al., 2015; Laskin et al., 2012; Ma et al., 2013; B. Wang et al., 2015). We note that links between sea salt and mineral dust in the marine boundary layer may also influence halogen chemistry and the ozone budget (Sullivan et al., 2007). Aged and fresh SSA particles with organic coatings are routinely identified in ambient air using single-particle analysis with SEM and STXM/NEXAFS (Kirpes et al., 2018; Kirpes et al., 2019; Laskin et al., 2012). Ambient-aged particle morphology and composition exhibit changes that are similar to laboratory-aged particles (Chi et al., 2015; Laskin et al., 2012; Quinn et al., 2015; B. Wang et al., 2015), indicating the applicability of microanalytical techniques for quantifying SSA multiphase chemistry.

Tar balls from biomass burning aerosol (BBA) emissions reacting with NO_x, O_3, and OH and undergoing photochemical reactions were investigated for their changing optical properties (Li et al., 2019, 2020); in particular, AFM, SEM, and EELS were used for single-particle characterization. BBA particles initially were spherical semiliquid particles with a relatively low viscosity (Laskin et al., 2015; Li et al., 2019, 2020). When aging with NO_3, condensed-phase products had greater light absorption and scattering properties, larger molecular size coinciding with decreased volatility, lower solubility, decreased susceptibility to photo-bleaching, and potentially higher viscosity. Photochemical reactions with OH or with NO_3 did lead to more particle bleaching than photolysis alone, which possibly resulted from volatile organic nitrogen-containing species being released into the gas phase (Li et al., 2019; Li et al., 2020). Ammonium sulfate particles coated with SOA were reacted with isoprene epoxy diols (IEPOX) and analyzed offline, which resulted in drastic morphological changes as observed with SEM and also with AFM at ambient pressure and $RH = 30–40\%$ (Olson et al., 2019). The authors found that there was initially little to no limitation of IEPOX diffusion through the SOAs to react with the aqueous ammonium sulfate core, potentially forming organosulfate compounds, which later resulted in an increase in viscosity to a more solid-like state (Olson et al., 2019). Morphological and compositional changes due to OH reaction with IEPOX-SOA have been investigated using AFM and photothermal infrared spectroscopy, with driving factors identified as variable thickness of coatings and the reactivity of oligomers (Yan et al., 2023). Multiple microanalytical techniques—AFM, SEM, Raman spectromicroscopic mapping, and ToF-SIMS—were used to illustrate the combined complexity of multiphase chemistry that ambient particles can simultaneously contain chemically distinct solids, liquids, or highly viscous states (Sobanska et al., 2014). Particles from air influenced by marine and urban sources were observed to host both emitted sea salts and their reaction products with NO_x, solid mineral iron inclusions, aqueous organic and inorganic solutions, and persistent organic coatings that remain no matter the *RH* history the particle experienced (Sobanska et al., 2014).

Soot particles are an important aerosol type extensively studied for their chemical and morphological changes as they reside in the atmosphere with links to their light absorption, cloud-forming properties, and climate impacts (Lohmann et al., 2020; Moffet & Prather, 2009; Renyi Zhang et al., 2008). Only a few studies are highlighted here, with particular emphasis on microanalytical investigation. Their reaction with O_3 was investigated with TEM offline for the morphological changes and online to observe in situ compositional changes using DRIFTS (Yongchun Liu et al., 2010). Increased hygroscopicity of reacted soot was observed in this study that coincided with a smaller diameter of primary soot spherules, more compacted particles, a decrease of amorphous and graphitic carbon, and an increase in the oxygenated functionalities ketone, lactone, and anhydride (Yongchun Liu et al., 2010). Soot particles from diesel car exhaust and a wood stove, before and after photochemical processing in a smog chamber, and humidity cycles were studied using STXM/NEXAFS to gain further understanding of how these particles reacted and became coated with secondary organic matter (Zelenay et al., 2011b). The authors found that individual aged soot particle

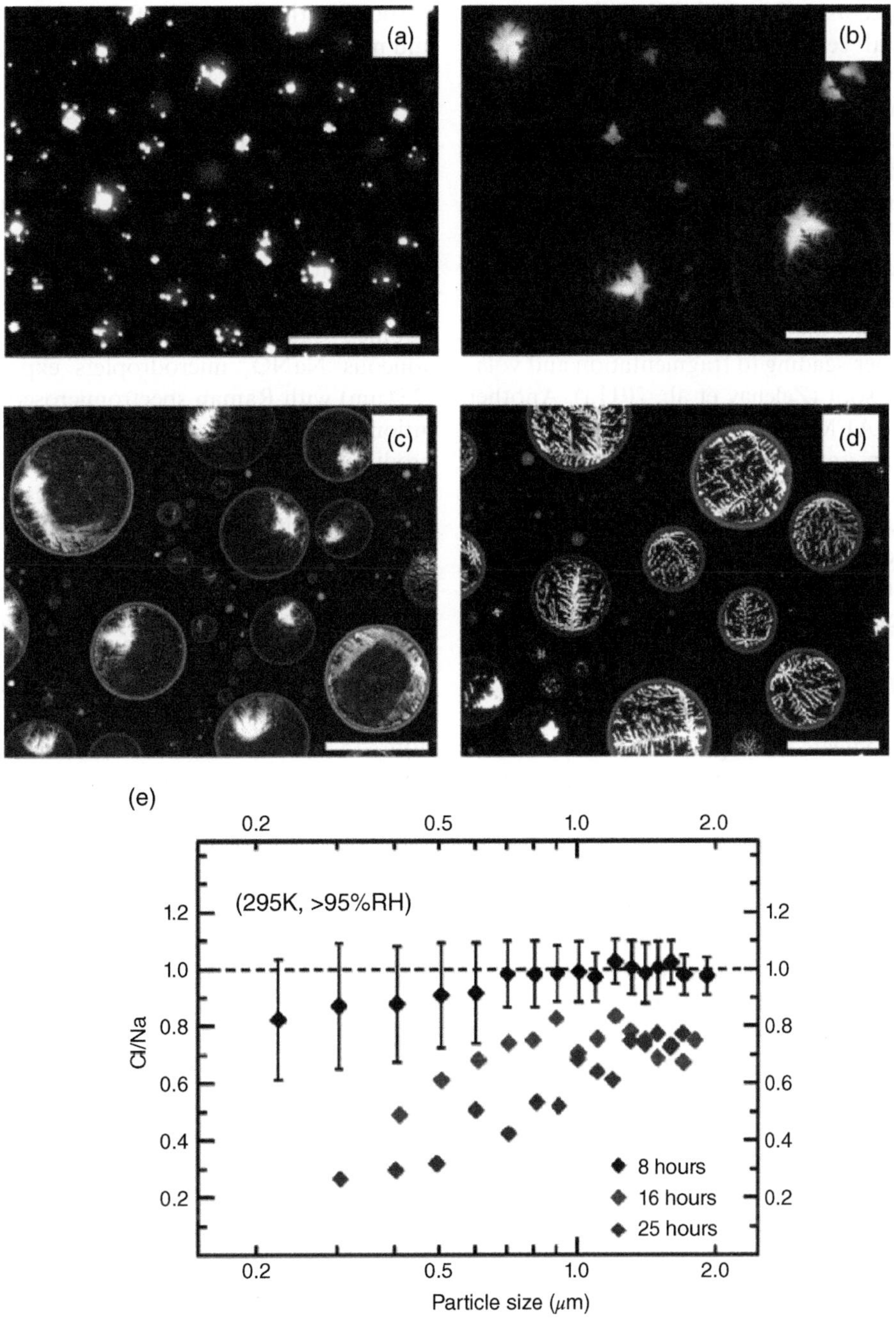

Figure 10.2 Time-dependent morphological and elemental changes of particles at a relative humidity >95% at 295 K. (a) STEM image of freshly prepared NaCl particles with coatings of α-pinene derived secondary organic matter. Images acquired after 8, 16, and 25 hours are shown as (b), (c), and (d), respectively. The scale bar for all images is 3 μm. Average Cl/Na atomic rations as a function of particle circle equivalent diameter (size) are shown in (e) for the various reaction times. The dashed line indicates Cl/Na = 1.0, and those values <1.0 indicate chlorine depletion. Representative standard deviation is shown as error bars. Wang et al. (2015)/with permission of the American Chemical Society.

composition inferred using STXM/NEXAFS in a humidified environmental cell was in close agreement with aerosol mass spectrometer (AMS) data and directly relatable to single-particle hygroscopicity. In another study with soot extracted from primary exhaust and then exposed to O_3 and UV light offline in a flow tube, STXM/NEXAFS, sensitive to carbon and oxygen functionalities, was used to determine a clear indication of reduced carboxyl and O-alkyl carbon functionalities in samples (Zelenay et al., 2011a). It was suspected that existing compounds that were slightly oxidized initially became oxidized even further, leading to fragmentation and volatilization from the soot (Zelenay et al., 2011a). Another offline study used AFM to observe a change in single-particle morphology and height to quantify the degradation of soot exposed to ozone and oxygen atoms (Lehmpuhl et al., 1999). The authors found that ozone reaction of ambient soot did not significantly result in mass loss or change in morphology; however, reaction with atomic oxygen did.

10.3. IN SITU MICROANALYTICAL METHODS

This section is dedicated to highlighting recent developments where microanalytical methods were used to investigate multiphase chemical reactions as they were occurring, referred to as in situ. Commonly, this would involve imaging a sample or gaining chemical detail on the micrometer or nanometer scale inside of a chemical reaction chamber. One focus is on Raman microscopy, which is a method that accomplishes this for micrometer-scale investigations. Multiphase chemistry has been resolved on the order of nanometers, typically using thin probing depths within a sample of particles or a film, but not from direct imaging. These are important indirect techniques for multiphase chemical investigations and will be discussed. Finally, STXM/NEXAFS is highlighted as it strikes a balance between spatial resolution and chemical selectivity, which is ideal for direct imaging and spectroscopic detection of chemical change in situ during multiphase reaction studies.

10.3.1. Mapping and Multiphase Chemistry in Droplets Using Raman Spectromicroscopy

Raman spectromicroscopy is a powerful and straightforward tool to access particle multiphase chemistry in situ on the micrometer scale. We note that in addition to the studies on in situ multiphase aerosol chemistry in this section, a complete review and application of single-particle Raman spectromicroscopy to atmospheric chemistry and physics can be found in this book as well as in previous literature (Liang et al., 2022). Sulfur oxidation reactions by NO_2 and O_2 on the surface of $CaCO_3$ particles with diameters of 3–7 μm were mapped in situ and revealed condensed-phase heterogeneity of products (Yu et al., 2018; Zhao et al., 2018). Raman spectroscopy and single-particle levitation have been used to study the multiphase reaction of O_3 with organic aerosol particles: for example, oleic acid (A. K. Y. Lee & Chan, 2007) and malic acid (Dennis-Smither et al., 2014). In another example, NO_3^- photochemistry in single levitated particles was investigated with a particular emphasis on in situ observation of compositional heterogeneity in particles (Gómez Castaño et al., 2019). The authors mapped aqueous $NaNO_3$ microdroplets exposed to UV light (254 nm) with Raman spectromicroscopy at a pixel resolution of $1 \times 1\ \mu m^2$ at $RH = 80\%$. After irradiation, nitrite produced from the reaction $NO_3^- \rightarrow NO_2^- + O^\bullet$ was observed to be heterogeneously distributed inside the drop, which the authors attributed to the heterogeneity of their illumination on the particle (Gómez Castaño et al., 2019). Although a more uniform illumination is certainly desirable in atmospheric multiphase chemistry studies, this work uniquely showed that their technique for stabilizing levitated particles resulted in low agitation (Gómez Castaño et al., 2019) and could be used in the future to track chemical reactions that may be limited by molecular transport resulting in compositional gradients of atmospheric relevance. Using Raman spectromicroscopy and synchrotron-based small angle X-ray scattering (SAXS), ozonolysis of fatty acid thin films and aerosol particles was investigated and shown to form an inhomogeneous, chemically distinct layer (Milsom et al., 2021a; Milsom et al., 2021b). These in situ investigations performed as a function of t, RH and film thickness, and other experimental parameters revealed that an unreactive crust could be produced, which, as the authors argued, was highly viscous and could inhibit further ozonolysis (Milsom et al., 2021a; Milsom et al., 2021b). In a series of papers, FLIM (Fitzgerald et al., 2016; Hosny et al., 2016) and Raman spectromicroscopy were used to identify in situ viscosity and compositional changes, respectively, in droplets of SOA material, oleic acid and squalene exposed to ozone (Athanasiadis et al., 2016). Figure 10.3 shows FLIM analysis for droplets 42 and 96 μm in diameter and for increasing O_3 exposure for up to 40 minutes (Athanasiadis et al., 2016). The fluorescence lifetime of Bodipy-C_{10} (a dye added to the droplets with a fluorescence lifetime proportional to viscosity) was mapped having values between about 2 and 5 ns, corresponding to the approximate range of viscosity of $\eta = 10^1$–10^3 mPa s shown on the scale bar (Athanasiadis et al., 2016). As O_3 exposure continued, a small increase in η was observed for the 42 μm particle and a much larger increase for the 96 μm particle, and a small gradient from the core toward the perimeter was present (Athanasiadis et al., 2016). This showed that products formed from

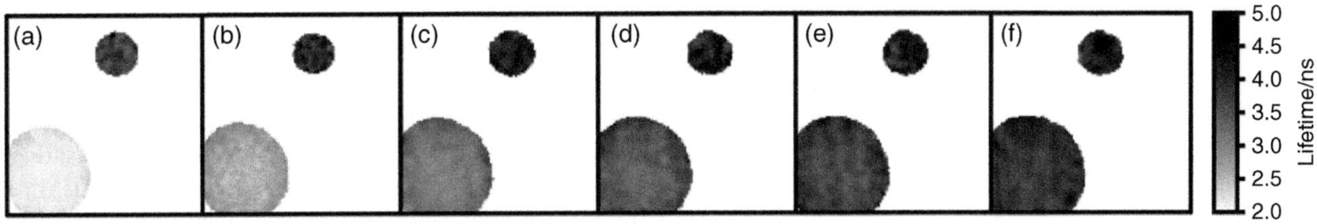

Figure 10.3 In situ fluorescence lifetime imaging microscopy (FLIM) analysis of squalene droplets with diameters of 42 and 96 μm during O_3 exposure over time *t*. Images of the same parties from left to right are exposures for *t* = 16, 20, 23, 27, 30, and 40 min. The gray scale corresponds to the fluorescence lifetime between 2 and 5 ns corresponding to a range of viscosity, η = 10^1–10^3 mPa s. Athanasiadis et al. (2016)/with permission of Royal Society of Chemistry.

ozonolysis have very different η, but it also revealed the limitations of molecular transport of products through the droplets. Therefore, the combined effect of reaction and diffusion interplaying with phase behavior may explain the trend of increasing η over *t*, but also the observed radial heterogeneity.

10.3.2. Indirect Microanalytical Methods of Multiphase Chemistry

Although in situ and offline imaging with compositional sensitivity can provide exceptional detail, other nonmicroscopic methods have resulted in a unique understanding of multiphase aerosol chemistry. The reaction between gas-phase OH and citric acid aerosol particles as a function of *RH* was investigated using DART-MS that probed entire particles or, alternatively, a layer from their surface to ~ 14 ± 5 nm toward their center, depending on the temperature in the ionization region (Davies & Wilson, 2015). The authors observed less citric acid degradation for entire particles compared to the near-surface layers, indicating evidence for nonuniform multiphase chemistry (Davies & Wilson, 2015). The formation of reaction products at 20% RH indicated multiple OH reaction steps, which demonstrated that multiple OH reactions and product generation were limited to nanometer-scale layers near the surface, but also that fresh citric acid could not diffusive toward the surface to react with OH. This was opposed to higher humidity experiments, where reaction products were the result of fewer oxidative reaction steps with OH but occurring over the whole volume of the particles, indicating a more well-mixed system (Davies & Wilson, 2015). DART-MS continues to be used as a successful tool for investigating multiphase chemistry, especially for OH as a reactive trace gas (Lam et al., 2021; Xu et al., 2020). Using VMI (Ahmed et al., 2021; Ahmed and Kostko, 2020; Kostko et al., 2017), the photoelectron velocity distribution from sub-micrometer-sized particles composed of either squalene or octacosene intersecting a beam of synchrotron radiation (UV or X-rays) was mapped, and thus ultraviolet or X-ray photoelectron spectra and electron yield X-ray absorption spectra of particles with near interfacial sensitivity could be derived (Jacobs et al., 2016; Jacobs et al., 2019). Photoelectrons observed with VMI emitted from C or O atoms only originate from nanometer-scale depths beneath a particle's surface, and the authors derived near-surface oxidation kinetics using their observed photoelectron spectra (Jacobs et al., 2016; Jacobs et al., 2019). The ability of oxidant gases to diffuse and react was teased apart in a multiphase chemistry study, which used particles of trans-aconitic acid exposed to O_3 and D-mannitol exposed to OH, both coated by either squalene or squalane (L. Lee & Wilson, 2016). Generating particles with a known coating thickness and knowledge of the bulk reaction kinetics from previous work, the authors could deconvolute reacto-diffusive length scales and diffusion coefficients of O_3 and OH through squalene and squalane (L. Lee & Wilson, 2016). Figure 10.4 shows the barrier coefficient, γ/γ_o, where γ and γ_o are the observed uptake coefficients for the particles with the core–shell morphology and the core only, respectively, consisting of trans-aconitic acid. γ_o was derived from the degradation of trans-aconitic acid, and γ is the corresponding uptake coefficient derived from the degradation of trans-aconitic acid when the core was coated with either squalane or squalene (L. Lee & Wilson, 2016). For the squalane coating, the only reactive sink for O_3 is the trans-aconitic acid in the core (circles in Figure 10.4). At unity, $\gamma/\gamma_o = 1$, diffusion of O_3 through the shell to the core is expected to be negligible due to an infinitesimally thin coating or a negligible limitation of molecular transport through the shell. As γ/γ_o decreases with increasing squalane thickness, the shell establishes a condensed-phase steady-state O_3 concentration gradient that limits mass transfer to the core. Thus, the decrease in γ reflects the increasing diffusive resistance, resulting in less degradation of trans-aconitic acid in the core (L. Lee & Wilson, 2016). In the presence of squalene as a coating, which is reactive toward O_3 by itself, γ/γ_o for

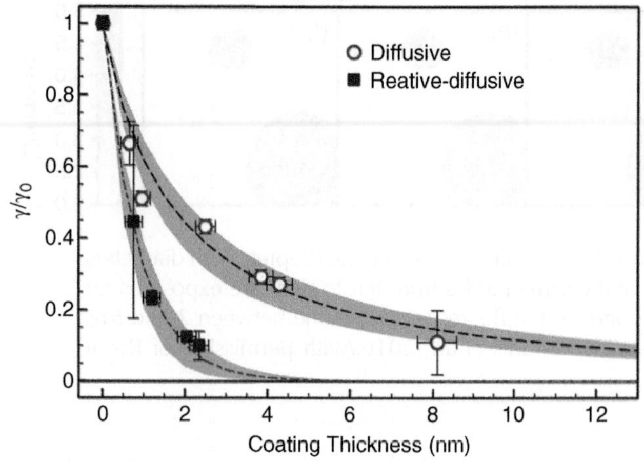

Figure 10.4 Ozone barrier coefficient (γ/γ0) through diffusive (squalane, round hollow) and reactive–diffusive (squalene, square filled) organic layers of increasing radial thickness. Best model fit is plotted in dashed lines with shaded areas representing a variation of diffusion constant by 30% (diffusive) and reactive–diffusive length by 50% (reactive–diffusive). L. Lee & Wilson, 2016/with permission of the American Chemical Society.

O_3 decreased even more, as shown as squares in Figure 10.4 and γ/γ_o approached zero for a coating thickness > 6 nm (L. Lee & Wilson, 2016). In this case, the loss of O_3 toward the core–shell interface is governed not only by diffusion but also by reaction with squalene. Thus, in this case, the degradation rate of trans-aconitic acid in the core maps the O_3 concentration remaining near the core–shell interface; or in other words, the coating thickness dependence of γ traces the reacto-diffusive length of O_3 in squalene.

10.3.3. Scanning Transmission X-Ray Microscopy

Microanalysis and multiphase chemistry of aerosol particles is exceptionally difficult due to small particle sizes. Therefore, techniques use either droplets at least several micrometers in size or filtration of submicrometer-sized particles to collect enough mass that their bulk composition can be obtained analytically. To obtain information on in-particle composition changes, on the other hand, it is necessary to use chemically selective instrumentation with spatial resolution finer than the size of particles. STXM/NEXAFS is one such technique that pairs spatial resolution and chemical selectivity and has been used for experiments in aerosol physics and chemistry to answer a broad scope of questions. A complete review of the STXM/NEXAFS technique can be found in this book and previous publications (Ade et al., 2003; Hitchcock et al., 2005; Kilcoyne et al., 2003; Laskin et al., 2019; Moffet et al., 2010; Raabe et al., 2008; Solomon et al., 2009), and a brief description relevant for multiphase atmospheric chemistry is presented here.

STXM/NEXAFS is particularly useful for chemical analysis of aerosol particles hundreds of nanometers in diameter (Figure 10.1), which outweighs the disadvantage in terms of spatial resolution compared to SEM and TEM combined with EELS. The operating principle of STXM/NEXAFS is the transmission of X-ray photons through a particle measured and converted to optical density, $OD = -\ln(I/I_o)$ following the Beer–Lambert law, where I and I_o are the transmitted and incident photon count, respectively. Single photons are counted, and their uncertainty is $\sqrt{I}$ and $\sqrt{I_o}$. The absorption of X-rays at a single energy is different for various elements, and a complete list of absorption coefficients and other data tables is available (Henke et al., 1993). A marked jump in the absorption coefficient for a given atom occurs in the vicinity of the binding energy of its core electronic levels. With exposure to X-rays at energies higher than the binding energy, ionization can occur. Particularly strong X-ray absorption over a narrow range in X-ray energy near the binding energy edge-step can be observed when electrons become resonantly excited into higher-energy, unoccupied molecular orbitals. These narrow peaks in an X-ray absorption spectrum as a function of energy make up the near-edge X-ray absorption fine structure (NEXAFS). The peak position and width of both resonant features and the ionization energy can vary based on the exact electronic level and the bonds that the atom makes with others, including the interaction of the molecular orbitals with surrounding molecules. For example, organic acids having double bonds or an aromatic ring structure (thus conjugated C=C), such as shikimic acid, a proxy for oxygenated organic material and present in BBA, strongly absorb at X-ray energies 284.4 eV and 288.2 eV (Steimer et al., 2014) corresponding to the 1 s–π* transitions that indicate aromatic or carbon double bonds and carboxyl functions (Moffet et al., 2010), respectively. Because relevant electronic levels span a large energy range between 160 eV and up to a few thousand eV, STXM/NEXAFS experiments require a tunable synchrotron X-ray source. In terms of atmospherically relevant species, soft X-ray beamlines (160–1600 eV) cover the K-edges of light elements (e.g., C, N, O, and Na) or the L-edges of heavier elements (e.g., S, Cl, K, Br, Ca, Fe, Cu, and Mn), and tender X-ray beamlines offer access to the K-edges of S, Cl, Fe, and P. Typical energy resolution during STXM/NEXAFS, ΔE, is about 0.2 eV but could be lower or higher depending on how a beamline is operated. Generally, ΔE is determined to be sufficient to observe a clear contrast between the sharp absorption peaks and satisfy the requirement for chemical selectivity. X-rays can be focused using a diffractive optic known as a Fresnel zone plate to a small spot size, typically

$35 \times 35\,nm^2$ for a standard operation, although this size can be much smaller, again, depending on the optimization of the X-ray beamline and zone plate fabrication. In fact, features as small as 7 nm have recently been resolved that exceed the fundamental limit of direct writing with electron beam lithography (Rösner et al., 2020). Indeed, zone plates for many applications and optical configurations are commercially available (www.xrnanotech.com). An *OD* image can be acquired when measuring photon transmission and scanning this small X-ray spot over a field of view containing particles. A chemical image can be obtained when rescanning the same particles over a range of X-ray energies and identifying peak absorption features in 2-D: that is, gaining quantitative chemical morphological information.

Figure 10.5 shows STXM/NEXAFS images of shikimic acid particles, where the gray scale is the measured *OD* at 284.4 eV normalized to the *OD* between 310 and 320 eV (Steimer et al., 2014). In the first experiment of its kind, particles were exposed to O_3 at different *RH* to image multiphase chemical reactions in situ inside single aerosol particles. This was possible due to the use of an environmental cell to ensure a controlled, humidified environment between X-ray-transparent windows to view the particles (Huthwelker et al., 2010). We note that environmental cells are reviewed elsewhere (Förster et al., 2020; Kelly et al., 2013; O'Brien et al., 2015; Ohigashi et al., 2016; Prabu et al., 2018; *J* Wang et al., 2011) and in this book. Discussion of the environmental cell here is restricted to multiphase chemical experiments in atmospheric aerosol science (Alpert et al., 2019; Alpert et al., 2021; Dou et al., 2021; Steimer et al., 2014). Using shikimic acid particles at $RH = 70\%$, Figure 10.5a shows the particle prior to O_3 exposure, and Figure 10.5b, c shows increasing O_3 exposure (Steimer et al., 2014). These images directly reveal the in situ decay of shikimic acid through a multiphase chemical reaction. This was determined from the continuous decrease in the 1 s–π* absorption peak over time. The authors calculated characteristic reacto-diffusive length scales for O_3 on the order of hundreds of nanometers at this *RH*. However, any gradient in the absorption peak extending from the perimeter to the center of the particle was not discernable (Steimer et al., 2014), which may have been the cause of the fast mixing of O_3 from near-surface layers to the center and that of unreacted shikimic acid from the center to the near-surface layers. Figure 10.5d–f shows the result of a similar experiment at $RH = 12\%$. Here, the decay of shikimic acid was not observed, likely due to the very short reacto-diffusive length calculated to be on the order of 1 nm, far smaller than the achievable resolution of STXM (Steimer et al., 2014). This was further corroborated later with a larger dataset and a macroscopic shikimic acid film combined with a depth-resolving model of transport and chemistry (Berkemeier et al., 2016; Steimer et al., 2014).

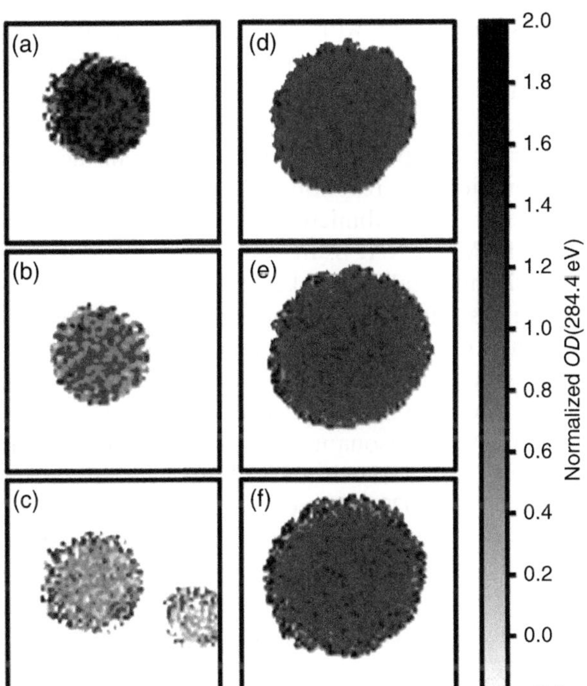

Figure 10.5 Depletion of shikimic acid over time. Normalized optical density, OD, images taken at 284.4 eV show particles before and during O_3 exposure at (a–c) RH = 71% RH and (d–f) RH = 12%. Exposures are (a) 0 atm s, (b) 5.0×10^{-2} atm s, (c) 7.7×10^{-2} atm s, (d) 0 atm s, (e) 1.4×10^{-2} atm s, and (f) 7.9×10^{-2} atm s (g). The scale bar denotes 1.0 µm. Steimer et al. (2014)/Copernicus Publications/CC BY 4.0.

In another study, the oxidation of iron(II) from $FeCl_2$ by O_3 through an organic matrix of xanthan gum was investigated using STXM/NEXAFS via monitoring the iron oxidation state at the iron L-edges (Alpert et al., 2019). In this way, the reaction of iron(II) to iron(III) was directly imaged as a marker for multiphase chemical reactions. There are a number of advantages to using iron as a marker, the greatest being the narrow width and strong absorption of the peaks that indicate oxidation state (Cressey et al., 1993). Additionally, the iron oxidation states of standard iron salts as reference minerals and of ambient aerosol particles have been extensively investigated and quantified (Garvie & Buseck, 1998; Majestic et al., 2006; Moffet et al., 2012; van Aken et al., 1998). A procedure referred to as *mapping* was used, where only a few images were acquired as opposed to obtaining multiple energy points to resolve a peak X-ray absorption shape (Alpert et al., 2019; Moffet et al., 2012). When imaging only at peak absorption energies 707.8 eV and 709.5 eV, indicating iron(II) and iron(III), respectively, the fraction of iron(II) remaining could be derived (Alpert et al., 2019; Moffet et al., 2012). Furthermore, a unique analytical method was created to derive 2D profiles of the iron(II) fraction, α, as a function of the distance from the particle perimeter (Alpert et al., 2019). The method involved first averaging α values that were identified only at the perimeter of the particles, corresponding to the average oxidation state of the particle layer nearest to the particle–air interface as defined by the resolution of the X-ray microscope. Then, concentric pixels toward the center of particles were averaged, indicating two or more pixels from the perimeter, yielding an average column-integrated profile. Applying this method to hundreds of particles at various *RH* resulted in the first direct evidence of in situ chemical gradients in micrometer-sized aerosol particles. In turn, these results were used to derive diffusion coefficients and test the application of reacto-diffusive kinetic regimes (Berkemeier et al., 2013) to model spatially resolved heterogeneous chemical degradation (Alpert et al., 2019).

Recent advances in multiphase chemistry and STXM/NEXAFS were made in a photochemical study utilizing particles containing iron(III)-citrate (Alpert et al., 2021; Dou et al., 2021). The photochemical reaction of this iron-carboxylate complex occurs in the condensed phase initiated first by light absorption, excitation, and ligand-to-metal charge transfer (LMCT) that leads to the reduction of iron and dissociation of the complex (Cieśla et al., 2004). Subsequently, decarboxylation of the dissociated citrate radical results in the release of CO_2 and the production of a CCR that can react with O_2 taken up by the particle from the gas phase. Therefore, O_2 uptake and reaction subsequent to light exposure were investigated. Particles before and after being exposed to UV light to initiate iron complex photolysis were imaged for their iron(III) fraction, $\beta = 1 - \alpha$, where the condensed-phase reactants were the photo-induced CCRs and the gas-phase reactant was O_2 (Alpert et al., 2021). Figure 10.6 shows the condensed-phase and multiphase reaction scheme, an image of particles immediately after light exposure ended in the presence of oxygen, and the reoxidation of iron after light was switched off at $RH = 40\%$, 50%, and 60%. Here, the same methods for determining chemical gradients in β were used as in the previous example.

Figure 10.6b clearly shows a trend toward surface oxidation after photochemical reduction of aerosol particles. This clear and in situ chemical contrast also reveals a lack of oxidation at the center of particles, indicating that gas-phase O_2 that was taken up by the particle was not present there. Profiles in β (Figures 10.6c, d) showed that at low humidity, $RH = 40\%$, the particle core was completely reduced by UV light and reoxidation only occurred slowly near the surface. At higher humidity, reoxidation was faster and chemical gradients were not as strong. The photochemical reaction and diffusion (PRAD) model was created to represent all chemical steps and derived 3D compositional gradients (Dou et al., 2021). The model is radially symmetric, and 3D profiles were easily derived, which could also be column integrated using a spherical geometry to determine 2D profiles that mimic the microscopy images of particles on a flat surface as in Figure 10.6b (Alpert et al., 2019; Alpert et al., 2021). Using the PRAD model, experimentally derived profiles in β (symbols in Figure 10.6c, d) could be directly compared with model-derived profiles (lines). We constrained the PRAD model using data from other techniques, such as coated wall flow tube experiments measuring the production of HO_2 and electrodynamic balance experiments that measured the reduction in mass due to the loss of CO_2 and any other volatile products (Alpert et al., 2021; Dou et al., 2019; Dou et al., 2021). STXM/NEXAFS constrained spatially resolved reoxidation kinetics, effectively giving quantitative information on the molecular transport and reaction of Fe and O_2. One main conclusion of these observations was that the lack of O_2 inside particles may have yielded a high concentration of CCRs (Alpert et al., 2021). This may have also explained ambient concentrations of environmentally persistent free radicals (Arangio et al., 2016) and bursts of ROS observed when ambient cloud water was first irradiated with UV (Paulson et al., 2019). Clearly, the entire volume of the particle is exposed to UV light to generate reactive species throughout the bulk; however, O_2 uptake as a driver for the peroxy radical chemistry is only possible from the particle surface.

Information on reaction and diffusion over small micrometer and submicrometer scales is obtainable using STXM/NEXAFS. In addition, the use of the in situ environmental cell makes it possible to use in situ imaging in studies of multiphase chemistry using atmospherically relevant aerosol and environments. We note that in situ condensed-phase chemical processing has also been recently investigated using STXM/NEXAFS to demonstrate resonant optical enhancement using the same iron(III)-citrate photochemical system (Arroyo et al., 2022). These particles in vacuum—that is, $RH = 0\%$—were highly viscous and irradiated with UV light directed perpendicular to the X-ray beam while β was mapped. Here, nanofocusing effects were investigated that were predicted to enhance light intensity inside of particles over nanoscopic regions of particles (Cremer et al., 2016; Nissenson et al., 2006; Signorell et al., 2016; Symes et al., 2004; Y. Zhang et al., 2018). Significant reduction of iron patterned across particles and featuring a hotspot due to nanofocusing was directly observed for the first time and confirmed by comparing with theoretical predictions (Arroyo et al., 2022). Future studies in multiphase photochemistry may need to consider these light

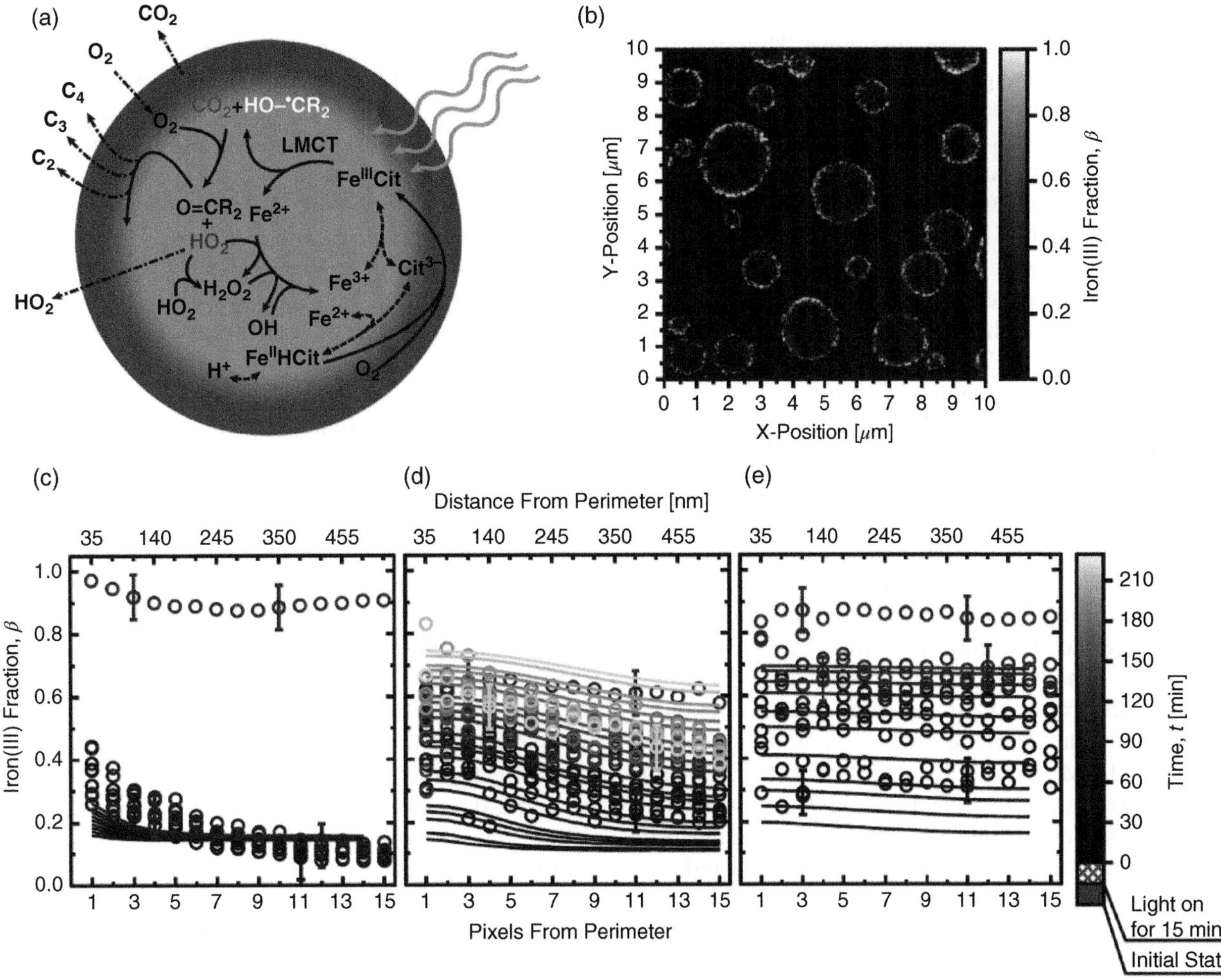

Figure 10.6 Imaged photochemical reactions of iron(III)-citrate ($Fe^{III}Cit$) inside single aerosol particles. (a) A sketch of the reaction scheme, depicting a particle oxidized near its surface (dark gray) and anoxic in its bulk (light gray). Solid lines are chemical reactions: for example, ligand-to-metal charge transfer (LMCT) reaction, dashed lines are equilibria conditions, and dash-dotted lines indicate gas evaporation or uptake. Citrate, the iron(II)-citrate complex, and CH_2COOH are abbreviated as Cit^{3-}, $Fe^{II}HCit$, and R, respectively. (b) STXM/NEXAFS image of the iron(III) fraction, *β*, directly demonstrating reoxidation only near the surface of particles. Gradients in *β* inside particles as a function of time, *t* (grayscale), and relative humidity, *RH*, during dark reoxidation at 20 °C after 15 minutes of UV light exposure are shown at *RH*= (c) 40%, (d) 50%, and (e) 60%. Initial values of *β* were acquired before UV light exposure. Solid lines are derived from the photochemical reaction and diffusion model (PRAD) model. Alpert et al., (2021) / Springer Nature / CC BY 4.0.

enhancement effects as they can also lead to increased reactant loss rates when diffusivity is increased (Arroyo et al., 2022).

10.4. IMAGING METHODS AND MODELING IN AEROSOL MULTIPHASE CHEMISTRY

10.4.1. Machine Learning and Image Processing

Microanalytical methods can rely heavily on image acquisition and processing to derive relevant parameters for atmospheric physics and chemistry. As previously mentioned, radial 3D chemical gradients from X-ray images have been derived from averaged 2D concentric pixels of particles (Alpert et al., 2019; Alpert et al., 2021). Here, we propose that future work on image processing could employ machine learning to identify chemical gradients in 2D or 3D more generally and when images have low signal-to-noise ratios. In STXM/NEXAFS studies of in situ environmental multiphase chemistry, particle diameters were limited to >0.2 μm, in part due to the noise in the STXM/NEXAFS images being too large when smaller particles were used (Alpert et al., 2019). Overcoming this limitation may become necessary in the future when using X-ray optics that result in an extremely small spot size down to 0.007 μm (Rösner et al., 2020) but also results in far less light on the sample. Thus, it is possible that strong image-processing methods, such as ANNs, will become crucial to improving image quality in future research. ANNs have found their way into many image-processing applications, such as image preprocessing and filtering, of which noise reduction (Egmont-Petersen et al., 2002) is the most applicable to X-ray absorption images as a function of X-ray energy. A type of ANN called a denoising autoencoder has been designed to recognize the dominant structural features of noisy input images, subtract the noise, and reconstruct only the underlying meaningful features in their output (Ronneberger et al., 2015). In recent literature, ANNs and STXM/NEXAFS were used to gain insight into the microscopic structural changes of developing hepatocellular carcinoma (a type of liver cancer) tumors (Wei et al., 2021) and hepatic stellate cells under exposure to arsenic (Xiu et al., 2020). Additionally, machine learning and ANNs have greatly assisted image segmentation: for example, identifying individual nanowires (Lin et al., 2022). Finally, the stability of the electron beam to generate synchrotron light has been greatly enhanced using a machine learning algorithm and applied to enhance STXM/NEXAFS image quality (Leemann et al., 2019). Image analysis using ANNs and machine learning has been utilized for diffractive imaging techniques such as tomography (Bührer et al., 2021) and ptychography (Cherukara et al., 2020). To our knowledge, the advanced application of ANNs to STXM/NEXAFS data has not been utilized for atmospheric aerosol multiphase chemistry until now.

Here, we present the performance of a denoising autoencoder for STXM/NEXAFS images. When suggesting a type of ANN, it is important to consider that very few training images are typically available due to the time constraints allotted for the use of STXM/NEXAFS at user-based synchrotron facilities. In addition, the ANN must be able to handle images of particle deformations. In other words, the 2D projection of single or coalesced particles may not be circular and could appear more akin to clusters of ovals. The ANN used here is based on the U-Net autoencoder structure and excels in noise reduction of biomedical imaging applications where the deformation of cells or separation of cells in contact with one another can be accounted for (Ronneberger et al., 2015). This architecture was originally introduced for image segmentation in microscopy but has been widely applied to several topics. The ANN chosen here is called a convolutional neural network (CNN) and consists of an encoder or contraction path with several contraction layers (CLs) that take a noisy image as an input. Each CL consists of convolution operations with a certain number of filters followed by a rectilinear function and a max-pooling operation. Then a decoder or expansion path with several expansion layers (ELs) takes the output of the encoder and converts it to a noise-free image. The ELs consist of transposed convolutions followed by a rectilinear function and a concatenation with the output of the corresponding layer in the contraction path. Further details of U-Net and examples of applying CNNs to image denoising and processing can be found in other sources (Egmont-Petersen et al., 2002; L. Fan et al., 2019; Ronneberger et al., 2015).

We wish to apply the CNN to demonstrate its capability to denoise images of β—for example, as shown in Figure 10.6b—and for this, we need an appropriate training set. As previously stated, experimental time using STXM/NEXAFS is quite limited, and it is not practical to obtain numerous images of various standard particles with known and distinctive chemical morphology. Therefore, synthetic training images were arbitrarily generated that were reminiscent of images of β by randomly sampling radial profiles. Synthetic datasets have been shown to be useful to overcome the challenges associated with limited STXM/NEXAFS training data (Lin et al., 2022). In fact, actual STXM/NEXAFS images (not synthetic trained images) are distinctive in terms of contrast generation and unknown to the trained model, providing an excellent test of this specific denoising application. First, training image profiles in β assumed a random sigmoidal functionality. Function parameters are randomly selected, and the resulting profile is calculated. Then, a hemispherical particle with a randomly selected radius between 0.15 and 0.7 μm was created, and a perfect 2D (column-integrated) β image was calculated. The resolution used here to generate images is a pixel size of $0.025 \times 0.025\,\mu m^2$. Note that each training image was a perfect circle with the center at a randomized position in the training images. Also, they were not allowed to touch the edge of the field of view. Noise was added to the perfect 2D image by imposing a random uncertainty derived from the empirically retrieved noise in β as a function of optical depth: that is, about ± 0.07 on the value of β (Alpert et al., 2019; Alpert et al., 2021). The CNN training input and output are these randomly generated noisy and perfect β images of single particles, respectively. The number of initial images generated was 1,440 and the number of images was increased by data augmentation to 5,760. Once the training images were generated, they were used in the CNN to learn parameters for all CLs and ELs that minimize the difference between the output of the network and the noise-free images. We constructed the network within the Keras framework in TensorFlow (Chollet, 2015) and trained the network with the Adam optimizer using mean squared error loss (Kingma & Ba, 2015). The training is performed for five epochs: that is, using the same set of images five times for training.

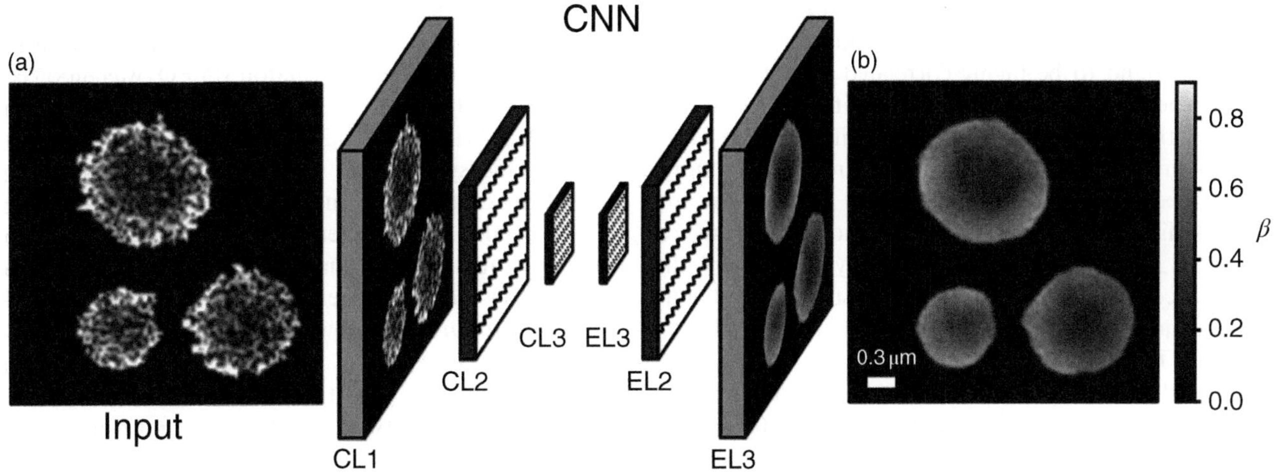

Figure 10.7 A schematic diagram that illustrates the architecture of the artificial neural network known as a convolutional neural network (CNN). (a) An experimental STXM image depicting the iron(III) fraction, β, of particles containing iron(III)-citrate after UV irradiation in an atmosphere at a relative humidity of 50%. The CNN consists of contraction layers (CL) and expansion layers (EL) to generate the corresponding denoised image after the processing of the experimental image shown in (b).

Two hundred images were excluded from the training data and used for validation.

Once trained, the CNN was used on actual STXM/NEXAFS images of β shown in Figure 10.7. The image in Figure 10.7a shows three particles with a relatively higher value of β at their perimeter and a lower value of β in the center. However, there is significant noise in the image, as seen as dark and light shades of gray appearing nearly adjacent to each other. Although this is likely the cause of the standard deviation of β of about ± 0.07, the image does have the appearance of a complex porous structure or highly localized hotspots for photochemical reactions in aerosol particles. It is especially important to reiterate that the CNN was trained using this uncertainty of ± 0.07 in β exactly due to the propagated error from X-ray photon counting statistics: that is, the noise used in training images represents the minimum allowable experimental uncertainty. This also does not consider other error sources, such as image misalignment, among others. Therefore, it is more likely that the structure seen in Figure 10.7b after denoising is more representative of reality than Figure 10.7a. Although this ANN example is highly simplistic and used only for demonstration, future work should consider ANNs in general to expand image analysis.

10.4.2. Representing Multiphase Chemistry

Experimental investigation and state-of-the-art micro-analytical tools reviewed here give unprecedented and direct evidence of the loss or production of compounds in aerosol particles. However, quantifying multiphase reactions can be difficult when considering fundamental physical and chemical gas-to-surface processes and how molecular diffusion can play a role in limiting them. These include mass transport, accommodation, solubility, and lifetime of reactants involved, as well as multigeneration surface and bulk reactions. Fortunately, a comprehensive kinetic model framework, termed the *PRA framework* (named after the scientists Pöschl, Rudich, and Ammann), has been developed to treat the complexity required for a consistent description of concentration and time dependence observed in experimental studies and for modeling of atmospheric multiphase aerosol processes (Pöschl et al., 2007; Ammann & Pöschl, 2007). The PRA framework models the time evolution of chemical reactions and molecular transport of the gas phase, near-surface gas phase, sorption layer, quasi-static surface, near-surface bulk, and bulk layers. In turn, these layers within particles and the concentration of products or reactants can be imaged using the micro-analytical techniques reviewed earlier.

Further development based on the PRA framework has led to the building of a kinetic multilayer model of aerosol surface and bulk chemistry (KM-SUB) (Shiraiwa et al., 2010) and applied to the multiphase reaction between O_3 and shikimic acid aerosol particles (Berkemeier et al., 2016). Experimental investigations resolved O_3 uptake coefficients over very short time scales immediately after exposure and, in addition, longer time scales over hours (Steimer et al., 2015). This allowed unique derivation of essential kinetic parameters and the identification of their independencies: for example, the O_3 desorption lifetime, $\tau_{d,O3}$, and the adsorption cross section, σ_{O3}, were found to be highly correlated with each other (Berkemeier et al., 2016). Furthermore, it could be shown uniquely that a gas-phase diffusion-limited surface reaction dominated initially, transitioning through a reaction-rate-limited surface reaction and finally evolving toward a bulk-diffusion-limited bulk reaction (Berkemeier et al., 2013; Berkemeier et al., 2016). Typically, the bulk diffusion coefficient, D, Henry's law constant, H, and bulk reaction-rate coefficient, k, are difficult to decouple for reactants. In the study with O_3, iron(II), and xanthan gum mentioned previously, STXM/NEXAFS imaging made it possible to image iron oxidation and constrain these bulk parameters for gas and condensed-phase reactants entirely due to the unprecedented monitoring of in-particle reactants (Alpert et al., 2019). Other applications of multiphase modeling using similar methods include multiphase reactions between OH and levoglucosan or abietic acid aerosol particles (Arangio et al., 2015), O_3 and protein bovine serum albumin (Shiraiwa et al., 2011), O_3 and benzo[a]pyrene or polycyclic aromatic hydrocarbons in aerosol composed of SOA and ammonium sulfate (Zhou et al., 2013; Zhou et al., 2019), O_3 and oleic acid (Berkemeier et al., 2021; Milsom et al., 2022; Pfrang et al., 2010) and O_3 and the neonicotinoid nitenpyram (Wang et al., 2020). Additionally, kinetic modeling has been successful in quantifying SOA formation in chambers using KM-GAP that treats the exchange of semivolatile species with the gas phase (Berkemeier et al., 2020; Shiraiwa et al., 2013), radical production in lungs from aerosol deposition (Lelieveld et al., 2021), O_3 reaction on a seawater surface to activate iodide (Schneider et al., 2020), and the reaction of O_3 with the human skin (Lakey et al., 2017; von Domaros et al., 2020; Won et al., 2020).

Another widely successful model of multiphase chemistry and microanalytics is Kinetiscope (Hinsberg & Houle, 2022), which is a stochastic model in contrast to KM-SUB (Shiraiwa et al., 2010) based on solving a system of differential equations. Kinetiscope uses an ensemble of various states, and there exists a probability that chemical reactions can change those states. This ensemble is then easily computed into a time trace of reactant and product concentrations to be directly compared with data. One major advantage of Kinetiscope is that it is easy to use and publicly available (Hinsberg & Houle, 2022). Another advantage is that it has been shown to represent scales of long and short reaction times, as well as dilute and concentrated species (Cheshire et al., 2020). Oxidative aging of squalane by OH was modeled with a focus on the reactive uptake onto the particle surface (Houle et al., 2015). Notable in this event-based model is that not only the concentration of species (e.g., peroxy radicals) could be derived but also the location of the specific reaction $R^\bullet + O_2$ was reported to occur mostly in the first 10 nm of particles (Houle et al., 2015; Wiegel et al., 2015). The OH uptake coefficient in this system was also derived and argued to depend, in part, on the molecular motion of squalene (Houle et al., 2015). When applied to triacontane, an isomer of squalane that is more viscous, Kinetiscope was used to show that first-generation products were confined to the aerosol surface, followed by unique and subsequent surface reactions (Houle et al., 2018a; Wiegel et al., 2017). The model was also used to scale laboratory and atmospheric conditions using a parameter known as the reaction-diffusion index, based on the Damköhler number, with the aim to simplify whether or not a multiphase process is limited more by molecular transport or chemical kinetics (Houle et al., 2018b), somewhat similar to the mixing parameters suggested by Berkemeier et al. (2013). The reaction between O_3 and squalene was modeled to quantify the chemical steps involving Criegee intermediates and the secondary ozonides, which can lead to adverse health effects upon exposure (Heine et al., 2017). Further work showing the versatility of Kinetiscope has also focused on other chemical systems (Prophet et al., 2024; Reynolds et al., 2023) and health-related impacts (Willis & Wilson, 2022), water evaporation (Houle et al., 2021), electrochemistry, and solar cells (Cheshire et al., 2020; Houle, 2019).

Other models of multiphase chemistry include development of a multilayer model for OH reaction with SOA formed from dark α-pinene ozonolysis (Li et al., 2018). It was shown that SOA diffusivity at low- and high RH were different, which could explain changes in composition. However, observed mass loss was found to be attributable more toward the RH-dependent OH uptake coefficient and the net probability of fragmentation (Li et al., 2018). In another model for the loss of methyl β-D-glucopyranoside particles with OH, it was shown that diffusion coefficients change drastically throughout the course of multiphase chemistry and can be different near the surface or in the bulk of particles, mimicking the water content over time (Fan et al., 2015). As the loss of methyl β-D-glucopyranoside continued over time, water activity changed inhomogeneously throughout the particle, driving changes in water content and, thus, water diffusion (Fan et al., 2015). OH oxidation of squalane and triacontane was revisited, and reaction rates were derived for a very wide temperature range down to 200 K (Li & Knopf, 2021). Use of derived reaction kinetics and phase state made it possible to predict OH penetration depths that decrease as a function of decreasing temperature.

10.5. OUTLOOK

Future research addressing multiphase chemistry and representing microscopic in-particle features, such as phase separations between crystalline, liquid, or glass-like phases, and chemical morphology, such as composition gradients within a single phase in models, should rely on a combination of experimental data taken from the gas phase, from the bulk of aerosol particles, and, most importantly, from the surface of particles or thin films. This may require using the microanalytical methods described here and, possibly, multiple methods simultaneously to better elucidate the physical and chemical parameters to further the fundamental understanding and appropriate parameterization of multiphase reaction processes. STXM/NEXAFS is a method balancing chemical selectivity and high (lateral) spatial resolution that is suited for submicrometer aerosol particles. However, surface information is out of reach unless operated with an electron detector generating a total electron yield (TEY) NEXAFS spectrum (Watts & Ade, 2012) or when paired with an electron energy analyzer to give a partial electron yield (PEY) NEXAFS spectrum (Roy et al., 2019). Briefly, when core electrons are excited to unoccupied states, filling the core hole results in the emission of an electron, termed an *Auger electron* (Briggs & Grant, 2003). TEY collects all Auger electrons and all other secondary electrons from nonresonant photoemission processes, secondary ionization, and inelastic scattering, regardless of their kinetic energy. In turn, measuring PEY requires an electron energy analyzer to select a particular kinetic energy range: for instance, that of the Auger electrons. The narrower electron kinetic energy range in PEY leads to a somewhat better-defined probing depth than the broad energy range in TEY measurements. In fact, STXM/NEXAFS, X-ray photoelectron spectroscopy (XPS), and Auger electron spectroscopy (AES) have been operated simultaneously and are thus capable of bulk and surface-sensitive measurements, with the potential of operating at ambient pressure for in situ studies (Roy et al., 2019). However, this is highly difficult and to our knowledge has not been attempted for aerosol multiphase chemical investigations. The difficulty arises due to the vacuum requirement for detecting secondary electrons, and therefore the use of an in situ environmental STXM/NEXAFS cell poses a challenge. However, these could be overcome by applying a potential difference between the cell windows and measuring current via lithographically patterned electrodes. Despite these difficulties, XPS and STXM methods have been coupled, advancing the ability for surface and bulk investigations, respectively (Roy et al., 2019).

Developments of near-ambient-pressure X-ray photoelectron spectroscopy (NAPP-XPS) are notable for the future microanalytical methods to access aerosol multiphase chemistry, where chemical changes within a few nanometers of samples can be observed (Ammann & Artiglia, 2022; Artiglia et al., 2017; Brown et al., 2013). Here, large samples exposed to relative humidity and trace gases are investigated in situ with XPS, those samples being aqueous, solid, or crystalline aerosol, ice particles, or a dilute aqueous solution jet. Direct evidence was found using NAPP-XPS for a bromide-ozone complex as an intermediate of the O_3 reaction with dissolved bromide salts and that this has a propensity for the interface (Artiglia et al., 2017). Similar experiments have been performed with organo-bromide compounds, where their reaction with O_3 was tracked over time (Chen et al., 2021). An investigation on solid ammonium sulfate salt exposed to *RH* was performed with NAPP-XPS. Although this was done without any trace gas oxidant, the authors discovered redox reactions that spontaneously occur when solvation began, and again, these reactions were enhanced at the surface (Kong et al., 2021). Other indirect methods have also discovered spontaneous interfacial chemistry due to strong intrinsic surface electric fields resulting in OH radical production without a prior known OH formation pathway (K. Li et al., 2023). The future of NAPP-XPS should work toward investigating surface and near-surface multiphase chemistry on aerosol particle samples with various probing depths. Doing so should not only confirm these findings but also lead atmospheric chemistry further toward understanding and quantifying new fundamental bulk and surface chemical mechanisms, especially when combining NAPP-XPS with the spatial resolution of STXM.

Knowledge of diffusion coefficients for various compounds or reaction-rate coefficients is invaluable to decrease the number of unknown variables in a multiphase kinetic model. New approaches for determining diffusion coefficients and glass transition temperatures (Evoy et al., 2020; Ingram et al., 2021; Katira et al., 2019) have been developed and may prove useful in future computational endeavors. For aerosol chemistry, in particular, previous work has been done to bridge the computationally expensive kinetic multilayer models to sub-modules in large-scale atmospheric models (Berkemeier et al., 2023). We have demonstrated that a multiphase reaction process limited by both molecular transport and chemical reaction can generate compositional gradients in particles. However, there remains a gap in imaging these gradients reliably on the nanometer scale. As X-ray optics come closer to this scale, ANNs may be required to denoise images to elucidate the internal aerosol particle structure and derive diffusion coefficients, reaction rates, or Henry's law constants. This is especially important when image acquisition time allotted to experimentalists is limited. Note that diffusion coefficients of multiple species in a wide distribution of products must be known to accurately predict their transport within particles at various relative humidity trajectories applicable to the atmosphere. We emphasized linking direct measurements of diffusion coefficients and viscosity with modeling and ANNs in future research using microanalytical methods to better predict the course of multiphase chemistry.

ACKNOWLEDGMENTS

This study was supported by the Swiss National Science Foundation (169176).

AUTHOR CONTRIBUTIONS

P.A.A. wrote the chapter, supervised by M.A. P.C.A. developed and verified the artificial neural network. P.A.A. generated training images. All authors commented on and proofread the manuscript.

AVAILABILITY STATEMENT

The data that support the findings of this study are openly available in the online Zenodo repository at http://doi.org/10.5281/zenodo.7113842.

REFERENCES

Ade, H., Kilcoyne, A. L. D., Tyliszczak, T., Hitchcock, P., Anderson, E., Harteneck, B., et al. (2003). Scanning transmission X-ray microscopy at a bending magnet beamline at the advanced light source. *Journal de Physique IV*, *104*, 3–8. doi: 10.1051/jp4:200300017

Adler, G., Koop, T., Haspel, C., Taraniuk, I., Moise, T., Koren, I., et al. (2013). Formation of highly porous aerosol particles by atmospheric freeze-drying in ice clouds. *Proceedings of the National Academy of Sciences of the United States of America*, *110*(51), 20414–20419. doi: 10.1073/pnas.1317209110

Ahmed, M., & Kostko, O. (2020). From atoms to aerosols: Probing clusters and nanoparticles with synchrotron based mass spectrometry and X-ray spectroscopy. *Physical Chemistry Chemical Physics*, *22*(5), 2713–2737.

Ahmed, M., Blum, M., Crumlin, E. J., Geissler, P. L., Head-Gordon, T., Limmer, D. T., et al. (2021). Molecular properties and chemical transformations near interfaces. *The Journal of Physical Chemistry. B*, *125*(32), 9037–9051. doi: 10.1021/acs.jpcb.1c03756

Alpert, P. A., et al. (2019). Visualizing reaction and diffusion in xanthan gum aerosol particles exposed to ozone. *Physical Chemistry Chemical Physics*, *21*(37), 20613–20627. doi: 10.1039/C9CP03731D

Alpert, P. A., et al. (2021). Photolytic radical persistence due to anoxia in viscous aerosol particles. *Nature Communications*, *12*(1), 1769. doi: 10.1038/s41467-021-21913-x

Ammann, M., & Artiglia, L. (2022). Solvation, surface propensity, and chemical reactions of solutes at atmospheric liquid–vapor interfaces. *Accounts of Chemical Research*, *55*(24), 3641–3651. doi: 10.1021/acs.accounts.2c00604

Ammann, M., & Pöschl, U. (2007). Kinetic model framework for aerosol and cloud surface chemistry and gas-particle interactions—Part 2: Exemplary practical applications and numerical simulations. *Atmospheric Chemistry and Physics*, *7*(23), 6025–6045. doi: 10.5194/acp-7-6025-2007

Arangio, A. M., Slade, J. H., Berkemeier, T., Pöschl, U., Knopf, D. A., & Shiraiwa, M. (2015). Multiphase chemical kinetics of OH radical uptake by molecular organic markers of biomass burning aerosols: Humidity and temperature dependence, surface reaction, and bulk diffusion. *The Journal of Physical Chemistry. A*, *119*(19), 4533–4544. doi: 10.1021/jp510489z

Arangio, A. M., Tong, H., Socorro, J., Pöschl, U., & Shiraiwa, M. (2016). Quantification of environmentally persistent free radicals and reactive oxygen species in atmospheric aerosol particles. *Atmospheric Chemistry and Physics*, *16*(20), 13105–13119. doi: 10.5194/acp-16-13105-2016

Arroyo, P. C., David, G., Alpert, P. A., Parmentier, E. A., Ammann, M., & Signorell, R. (2022). Amplification of light within aerosol particles accelerates in-particle photochemistry. *Science*, *376*(6590), 293–296. doi: 10.1126/science.abm7915

Artiglia, L., et al. (2017). A surface-stabilized ozonide triggers bromide oxidation at the aqueous solution-vapour interface. *Nature Communications*, *8*(1), 700. doi: 10.1038/s41467-017-00823-x

Athanasiadis, A., Fitzgerald, C., Davidson, N. M., Giorio, C., Botchway, S. W., Ward, A. D., et al. (2016). Dynamic viscosity mapping of the oxidation of squalene aerosol particles. *Physical Chemistry Chemical Physics*, *18*(44), 30385–30393. doi: 10.1039/C6CP05674A

Baboomian, V. J., Crescenzo, G. V., Huang, Y., Mahrt, F., Shiraiwa, M., Bertram, A. K., & Nizkorodov, S. A. (2022). Sunlight can convert atmospheric aerosols into a glassy solid state and modify their environmental impacts. *Proceedings of the National Academy of Sciences of the United States of America*, *119*(43), e2208121119. doi: 10.1073/pnas.2208121119

Balos, V., Garling, T., Duque, A. D., John, B., Wolf, M., & Thämer, M. (2022). Phase-sensitive vibrational sum and difference frequency-generation spectroscopy enabling nanometer-depth profiling at interfaces. *Journal of Physical Chemistry C*, *126*(26), 10818–10832. doi: 10.1021/acs.jpcc.2c01324

Berg, M. J., & Videen, G. (2011). Digital holographic imaging of aerosol particles in flight. *Journal of Quantitative Spectroscopy & Radiative Transfer*, *112*(11), 1776–1783. doi: 10.1016/j.jqsrt.2011.01.013

Berkemeier, T., Huisman, A. J., Ammann, M., Shiraiwa, M., Koop, T., & Pöschl, U. (2013). Kinetic regimes and limiting cases of gas uptake and heterogeneous reactions in atmospheric aerosols and clouds: A general classification scheme. *Atmospheric Chemistry and Physics*, *13*(14), 6663–6686. doi: 10.5194/acp-13-6663-2013

Berkemeier, T., Steimer, S. S., Krieger, U. K., Peter, T., Pöschl, U., Ammann, M., & Shiraiwa, M. (2016). Ozone uptake on glassy, semi-solid and liquid organic matter and the role of reactive oxygen intermediates in atmospheric aerosol chemistry. *Physical Chemistry Chemical Physics*, *18*(18), 12662–12674. doi: 10.5194/gmd-16-2037-2023

Berkemeier, T., Takeuchi, M., Eris, G., & Ng, N. L. (2020). Kinetic modeling of formation and evaporation of secondary organic aerosol from NO3 oxidation of pure and mixed monoterpenes. *Atmospheric Chemistry and Physics*, *20*(24), 15513–15535. doi: 10.1021/acsearthspacechem.1c00232

Berkemeier, T., Mishra, A., Mattei, C., Huisman, A. J., Krieger, U. K., & Pöschl, U. (2021). Ozonolysis of oleic acid aerosol revisited: Multiphase chemical kinetics and reaction mechanisms. *ACS Earth and Space Chemistry*, *5*(12), 3313–3323. doi: 10.1039/C6CP00634E

Berkemeier, T., Krüger, M., Feinberg, A., Müller, M., Pöschl, U., & Krieger, U. K. (2023). Accelerating models for multiphase chemical kinetics through machine learning with polynomial chaos expansion and neural networks. *Geoscientific Model Development*, *16*(7), 2037–2054. doi: 10.5194/acp-20-15513-2020

Bogan, M. J., Starodub, D., Hampton, C. Y., & Sierra, R. G. (2010). Single-particle coherent diffractive imaging with a soft x-ray free electron laser: Towards soot aerosol morphology. *Journal of Physics B: Atomic, Molecular and Optical Physics.*, *43*(19), 194013. doi: 10.1088/0953-4075/43/19/194013

Boucher, O., et al. (2013). *Clouds and aerosols climate change 2013: The physical science basis. Contribution of working group I to the fifth assessment report of the intergovernmental panel on climate change*. Cambridge, United Kingdom and New York, NY, USA: Cambridge University Press.

Briggs, D., & Grant, J. T. (2003). *Surface analysis by auger and X-ray photoelectron spectroscopy*. Chichester, UK and Manchester, UK: IMPublications.

Brown, M. A., et al. (2013). A new endstation at the Swiss light source for ultraviolet photoelectron spectroscopy, X-ray photoelectron spectroscopy, and X-ray absorption spectroscopy measurements of liquid solutions. *The Review of Scientific Instruments*, *84*(7). doi: 10.1063/1.4812786

Bührer, M., Xu, H., Hendriksen, A. A., Büchi, F. N., Eller, J., Stampanoni, M., & Marone, F. (2021). Deep learning based classification of dynamic processes in time-resolved X-ray tomographic microscopy. *Scientific Reports*, *11*(1), 24174. doi: 10.1038/s41598-021-03546-8

Chen, S., Artiglia, L., Orlando, F., Edebeli, J., Kong, X., Yang, H., et al. (2021). Impact of Tetrabutylammonium on the oxidation of bromide by ozone. *ACS Earth and Space Chemistry*, *5*(11), 3008–3021. doi: 10.1021/acsearthspacechem.1c00233

Cherukara, M. J., Zhou, T., Nashed, Y., Enfedaque, P., Hexemer, A., Harder, R. J., & Holt, M. V. (2020). AI-enabled high-resolution scanning coherent diffraction imaging. *Applied Physics Letters*, *117*(4), 044103. doi: 10.1063/5.0013065

Cheshire, T. P., Brennaman, M. K., Giokas, P. G., Zigler, D. F., Moran, A. M., Papanikolas, J. M., et al. (2020). Ultrafast relaxations in ruthenium Polypyridyl Chromophores determined by stochastic kinetics simulations. *The Journal of Physical Chemistry. B*, *124*(28), 5971–5985. doi: 10.1021/acs.jpcb.0c03110

Chi, J. W., et al. (2015). Sea salt aerosols as a reactive surface for inorganic and organic acidic gases in the Arctic troposphere. *Atmospheric Chemistry and Physics*, *15*(19), 11341–11353. doi: 10.5194/acp-15-11341-2015

Chollet, F. O. (2015). Keras. *GitHub*. Retrieved from https://github.com/fchollet/keras

Cieśla, P., Kocot, P., Mytych, P., & Stasicka, Z. (2004). Homogeneous photocatalysis by transition metal complexes in the environment. *Journal of Molecular Catalysis A: Chemical*, *224*(1), 17–33. doi: 10.1016/j.molcata.2004.08.043

Creamean, J. M., et al. (2013). Dust and biological aerosols from the Sahara and Asia influence precipitation in the Western U.S. *Science*, *339*(6127), 1572–1578.

Cremer, J. W., Thaler, K. M., Haisch, C., & Signorell, R. (2016). Photoacoustics of single laser-trapped nanodroplets for the direct observation of nanofocusing in aerosol photokinetics. *Nature Communications*, *7*(1), 10941.

Cressey, G., Henderson, C. M. B., & van der Laan, G. (1993). Use of L-edge X-ray absorption spectroscopy to characterize multiple valence states of 3d transition metals; a new probe for mineralogical and geochemical research. *Physics and Chemistry of Minerals*, *20*(2), 111–119. doi: 10.1007/BF00207204

Curtis, L., Rea, W., Smith-Willis, P., Fenyves, E., & Pan, Y. (2006). Adverse health effects of outdoor air pollutants. *Environment International*, *32*(6), 815–830.

Cziczo, D. J., Froyd, K. D., Hoose, C., Jensen, E. J., Diao, M., Zondlo, M. A., et al. (2013). Clarifying the dominant sources and mechanisms of cirrus cloud formation. *Science*, *340*(6138), 1320–1324.

Daellenbach, K. R., et al. (2020). Sources of particulate-matter air pollution and its oxidative potential in Europe. *Nature*, *587*(7834), 414–419.

D'Amato, G., Cecchi, L., D'Amato, M., & Liccardi, G. (2010). Urban air pollution and climate change as environmental risk factors of respiratory allergy: An update. *Journal of Investigational Allergology and Clinical*, *20*(2), 95–102.

Davies, J. F., & Wilson, K. R. (2015). Nanoscale interfacial gradients formed by the reactive uptake of OH radicals onto viscous aerosol surfaces. *Chemical Science*, *6*(12), 7020–7027. doi: 10.1039/C5SC02326B

Dennis-Smither, B. J., Marshall, F. H., Miles, R. E. H., Preston, T. C., & Reid, J. P. (2014). Volatility and oxidative aging of aqueous maleic acid aerosol droplets and the dependence on relative humidity. *The Journal of Physical Chemistry. A*, *118*(30), 5680–5691.

Dou, J., Luo, B., Peter, T., Alpert, P. A., Corral Arroyo, P., Ammann, M., & Krieger, U. K. (2019). Carbon dioxide diffusivity in single, levitated organic aerosol particles. *Journal of Physical Chemistry Letters*, *10*(15), 4484–4489.

Dou, J., Luo, B., Peter, T., Alpert, P. A., Arroyo, P. C., Ammann, M., & Krieger, U. K. (2021). Photochemical degradation of iron(III)-citrate/citric acid aerosol quantified with the combination of three complementary experimental techniques and a kinetic process model. *Atmospheric Chemistry and Physics*, *21*, 315–338.

Egmont-Petersen, M., de Ridder, D., & Handels, H. (2002). Image processing with neural networks—A review. *Pattern Recognition*, *35*(10), 2279–2301.

Evoy, E., Kamal, S., Patey, G. N., Martin, S. T., & Bertram, A. K. (2020). Unified description of diffusion coefficients from small to large molecules in organic–water mixtures. *The Journal of Physical Chemistry. A*, *124*(11), 2301–2308.

Fan, H., Tinsley, M. R., & Goulay, F. (2015). Effect of relative humidity on the OH-initiated heterogeneous oxidation of monosaccharide nanoparticles. *The Journal of Physical Chemistry. A*, *119*(45), 11182–11190.

Fan, L., Zhang, F., Fan, H., & Zhang, C. (2019). Brief review of image denoising techniques. *Visual Computing for Industry, Biomedicine, and Art*, *2*(1), 7.

Fierce, L., Bond, T. C., Bauer, S. E., Mena, F., & Riemer, N. (2016). Black carbon absorption at the global scale is affected by particle-scale diversity in composition. *Nature Communications*, *7*(1), 12361.

Fierce, L., Riemer, N., & Bond, T. C. (2017). Toward reduced representation of mixing state for simulating aerosol effects on climate. *Bulletin of the American Meteorological Society*, *98*(5), 971–980.

Fitzgerald, C., et al. (2016). Fluorescence lifetime imaging of optically levitated aerosol: A technique to quantitatively map the viscosity of suspended aerosol particles. *Physical Chemistry Chemical Physics*, *18*(31), 21710–21719.

Förster, J.-D., et al. (2020). MIMiX: A multipurpose *in situ* microreactor system for X-ray microspectroscopy to mimic atmospheric aerosol processing. *Atmospheric Measurement Techniques*, *13*(7), 3717–3729.

Fuzzi, S., et al. (2015). Particulate matter, air quality and climate: Lessons learned and future needs. *Atmospheric Chemistry and Physics*, *15*(14), 8217–8299.

Garvie, L. A. J., & Buseck, P. R. (1998). Ratios of ferrous to ferric iron from nanometre-sized areas in minerals. *Nature*, *396*(6712), 667–670.

Gómez Castaño, J. A., Boussekey, L., Verwaerde, J. P., Moreau, M., & Tobón, Y. A. (2019). Enhancing double-beam laser tweezers Raman spectroscopy (LTRS) for the photochemical study of individual airborne microdroplets. *Molecules*, *24*(18), 3325.

Heine, N., Houle, F. A., & Wilson, K. R. (2017). Connecting the elementary reaction pathways of Criegee intermediates to the chemical Erosion of Squalene interfaces during Ozonolysis. *Environmental Science & Technology*, *51*(23), 13740–13748.

Henke, B. L., Gullikson, E. M., & Davis, J. C. (1993). X-ray interactions: Photoabsorption, scattering, transmission, and reflection at E = 50-30,000 eV, Z = 1-92. *Atomic Data and Nuclear Data Tables*, *54*(2), 181–342.

Hinsberg, W. D. & F. A. Houle (2022). Kinetiscope. Retrieved from http://hinsberg.net/kinetiscope

Hitchcock, A. P., Morin, C., Zhang, X., Araki, T., Dynes, J., Stöver, H., et al. (2005). Soft X-ray spectromicroscopy of biological and synthetic polymer systems. *Journal of Electron Spectroscopy*, *144-147*, 259–269.

Hosny, N. A., et al. (2016). Direct imaging of changes in aerosol particle viscosity upon hydration and chemical aging. *Chemical Science*, *7*(2), 1357–1367.

Houle, F. A. (2019). Reaction–transport coupling in a nanostructured porous electrode. *Journal of Physical Chemistry C*, *123*(23), 14459–14467.

Houle, F. A., Hinsberg, W. D., & Wilson, K. R. (2015). Oxidation of a model alkane aerosol by OH radical: The emergent nature of reactive uptake. *Physical Chemistry Chemical Physics*, *17*(6), 4412–4423.

Houle, F. A., Wiegel, A. A., & Wilson, K. R. (2018a). Changes in reactivity as chemistry becomes confined to an Interface. The case of free radical oxidation of C30H62 alkane by OH. *Journal of Physical Chemistry Letters*, *9*(5), 1053–1057.

Houle, F. A., Wiegel, A. A., & Wilson, K. R. (2018b). Predicting aerosol reactivity across scales: From the laboratory to the atmosphere. *Environmental Science & Technology*, *52*(23), 13774–13781.

Houle, F. A., Miles, R. E. H., Pollak, C. J., & Reid, J. P. (2021). A purely kinetic description of the evaporation of water droplets. *The Journal of Chemical Physics*, *154*(5), 054501.

Huthwelker, T., Zelenay, V., Birrer, M., Krepelova, A., Raabe, J., Tzvetkov, G., et al. (2010). An *in situ* cell to study phase transitions in individual aerosol particles on a substrate using scanning transmission x-ray microspectroscopy. *The Review of Scientific Instruments*, *81*(11), 113706–113709.

Ingram, S., Rovelli, G., Song, Y.-C., Topping, D., Dutcher, C. S., Liu, S., et al. (2021). Accurate prediction of organic aerosol evaporation using kinetic multilayer modeling and the stokes–Einstein equation. *The Journal of Physical Chemistry. A*, *125*(16), 3444–3456.

Jacobs, M. I., Xu, B., Kostko, O., Heine, N., Ahmed, M., & Wilson, K. R. (2016). Probing the heterogeneous Ozonolysis of Squalene nanoparticles by photoemission. *The Journal of Physical Chemistry. A*, *120*(43), 8645–8656.

Jacobs, M. I., Xu, B., Kostko, O., Wiegel, A. A., Houle, F. A., Ahmed, M., & Wilson, K. R. (2019). Using nanoparticle X-ray spectroscopy to probe the formation of reactive chemical gradients in diffusion-limited aerosols. *The Journal of Physical Chemistry. A*, *123*(28), 6034–6044.

Katira, S., Garrahan, J. P., & Mandadapu, K. K. (2019). Theory for glassy behavior of Supercooled liquid mixtures. *Physical Review Letters*, *123*(10), 100602.

Kelly, S. T., Nigge, P., Prakash, S., Laskin, A., Wang, B., Tyliszczak, T., et al. (2013). An environmental sample chamber for reliable scanning transmission x-ray microscopy measurements under water vapor. *The Review of Scientific Instruments*, *84*(7), 073708.

Kilcoyne, A. L. D., et al. (2003). Interferometer-controlled scanning transmission X-ray microscopes at the advanced light source. *Journal of Synchrotron Radiation*, *10*, 125–136.

Kingma, D. P. & Ba, J. L. (2015). Adam: a method for stochastic optimization. Proceedings of the Third International Conference for Learning Representations. Retrieved from https://dblp.org/db/conf/iclr/iclr2015.html

Kirpes, R. M., Bondy, A. L., Bonanno, D., Moffet, R. C., Wang, B., Laskin, A., et al. (2018). Secondary sulfate is internally mixed with sea spray aerosol and organic aerosol in the winter Arctic. *Atmospheric Chemistry and Physics*, *18*(6), 3937–3949.

Kirpes, R. M., Bonanno, D., May, N. W., Fraund, M., Barget, A. J., Moffet, R. C., et al. (2019). Wintertime Arctic Sea spray aerosol composition controlled by sea ice lead microbiology. *ACS Central Science*, *5*(11), 1760–1767.

Kong, X., Castarède, D., Thomson, E. S., Boucly, A., Artiglia, L., Ammann, M., et al. (2021). A surface-promoted redox reaction occurs spontaneously on solvating inorganic aerosol surfaces. *Science*, *374*(6568), 747–752.

Koop, T., Bookhold, J., Shiraiwa, M., & Pöschl, U. (2011). Glass transition and phase state of organic compounds: Dependency on molecular properties and implications for secondary organic aerosols in the atmosphere. *Physical Chemistry Chemical Physics*, *13*(43), 19238–19255.

Kostko, O., Xu, B., Jacobs, M. I., & Ahmed, M. (2017). Soft X-ray spectroscopy of nanoparticles by velocity map imaging. *The Journal of Chemical Physics*, *147*(1), 013931.

Krueger, B. J., Grassian, V. H., Laskin, A., & Cowin, J. P. (2003a). The transformation of solid atmospheric particles into liquid droplets through heterogeneous chemistry: Laboratory insights into the processing of calcium containing mineral dust aerosol in the troposphere. *Geophysical Research Letters*, *30*(3), Art. no.-1148. doi: 10.1029/2002GL016563

Krueger, B. J., Grassian, V. H., Iedema, M. J., Cowin, J. P., & Laskin, A. (2003b). Probing heterogeneous chemistry of individual atmospheric particles using scanning electron microscopy and energy- dispersive X-ray analysis. *Analytical Chemistry*, *75*(19), 5170–5179.

Krueger, B. J., Grassian, V. H., Cowin, J. P., & Laskin, A. (2004). Heterogeneous chemistry of individual mineral dust particles from different dust source regions: The importance of particle mineralogy. *Atmospheric Environment*, *38*(36), 6253–6261.

Lakey, P. S. J., Wisthaler, A., Berkemeier, T., Mikoviny, T., Pöschl, U., & Shiraiwa, M. (2017). Chemical kinetics of multiphase reactions between ozone and human skin lipids: Implications for indoor air quality and health effects. *Indoor Air*, *27*(4), 816–828.

Lam, H. K., Shum, S. M., Davies, J. F., Song, M., Zuend, A., & Chan, M. N. (2019). Effects of inorganic salts on the heterogeneous OH oxidation of organic compounds: Insights from methylglutaric acid–ammonium sulfate. *Atmospheric Chemistry and Physics*, *19*(14), 9581–9593.

Lam, H. K., Xu, R., Choczynski, J., Davies, J. F., Ham, D., Song, M., et al. (2021). Effects of liquid–liquid phase separation and relative humidity on the heterogeneous OH oxidation of inorganic–organic aerosols: Insights from methylglutaric acid and ammonium sulfate particles. *Atmospheric Chemistry and Physics*, *21*(3), 2053–2066.

Laskin, A., Wietsma, T. W., Krueger, B. J., & Grassian, V. H. (2005). Heterogeneous chemistry of individual mineral dust particles with nitric acid: A combined CCSEM/EDX, ESEM, and ICP-MS study. *Journal of Geophysical Research*, *110*(D10208). doi: 10.1029/2004JD005206

Laskin, A., Moffet, R. C., Gilles, M. K., Fast, J. D., Zaveri, R. A., Wang, B., et al. (2012). Tropospheric chemistry of internally mixed sea salt and organic particles: Surprising reactivity of NaCl with weak organic acids. *Journal of Geophysical Research-Atmospheres*, *117*, D15302. doi: 10.1029/2012jd017743

Laskin, A., Laskin, J., & Nizkorodov, S. A. (2015). Chemistry of atmospheric Brown carbon. *Chemical Reviews*.

Laskin, A., Moffet, R. C., & Gilles, M. K. (2019). Chemical imaging of atmospheric particles. *Accounts of Chemical Research*, *52*(12), 3419–3431.

Lee, A. K. Y., & Chan, C. K. (2007). Heterogeneous reactions of linoleic acid and linolenic acid particles with ozone: Reaction pathways and changes in particle mass, hygroscopicity, and morphology. *The Journal of Physical Chemistry. A*, *111*(28), 6285–6295.

Lee, L., & Wilson, K. (2016). The reactive–diffusive length of OH and ozone in model organic aerosols. *The Journal of Physical Chemistry. A*, *120*(34), 6800–6812.

Leemann, S. C., Liu, S., Hexemer, A., Marcus, M. A., Melton, C. N., Nishimura, H., & Sun, C. (2019). Demonstration of machine learning-based model-independent stabilization of source properties in synchrotron light sources. *Physical Review Letters*, *123*(19), 194801.

Lehmpuhl, D. W., Ramirez-Aguilar, K. A., Michel, A. E., Rowlen, K. L., & Birks, J. W. (1999). Physical and chemical characterization of atmospheric aerosols by atomic force microscopy. *Analytical Chemistry*, *71*(2), 379–383.

Lelieveld, S., Wilson, J., Dovrou, E., Mishra, A., Lakey, P. S. J., Shiraiwa, M., et al. (2021). Hydroxyl radical production by air pollutants in epithelial lining fluid governed by Interconversion and scavenging of reactive oxygen species. *Environmental Science & Technology*, *55*(20), 14069–14079.

Li, C., He, Q., Fang, Z., Brown, S. S., Laskin, A., Cohen, S. R., & Rudich, Y. (2020). Laboratory insights into the Diel cycle of optical and chemical transformations of biomass burning Brown carbon aerosols. *Environmental Science & Technology*, *54*(19), 11827–11837.

Li, C., He, Q., Schade, J., Passig, J., Zimmermann, R., Meidan, D., et al. (2019). Dynamic changes in optical and chemical properties of tar ball aerosols by atmospheric photochemical aging. *Atmospheric Chemistry and Physics*, *19*(1), 139–163.

Li, J., & Knopf, D. A. (2021). Representation of multiphase OH oxidation of amorphous organic aerosol for tropospheric conditions. *Environmental Science & Technology*, *55*(11), 7266–7275.

Li, K., Guo, Y., Nizkorodov, S. A., Rudich, Y., Angelaki, M., Wang, X., et al. (2023). Spontaneous dark formation of OH radicals at the interface of aqueous atmospheric droplets. *Proceedings of the National Academy of Sciences of the United States of America*, *120*(15), e2220228120.

Li, Z., Smith, K. A., & Cappa, C. D. (2018). Influence of relative humidity on the heterogeneous oxidation of secondary organic aerosol. *Atmospheric Chemistry and Physics*, *18*(19), 14585–14608.

Liang, Z., Chu, Y., Gen, M., & Chan, C. K. (2022). Single-particle Raman spectroscopy for studying physical and chemical processes of atmospheric particles. *Atmospheric Chemistry and Physics*, *22*(5), 3017–3044.

Lin, B., Emami, N., Santos, D. A., Luo, Y., Banerjee, S., & Xu, B.-X. (2022). A deep learned nanowire segmentation model using synthetic data augmentation. *npj Computational Materials*, *8*(1), 88.

Liu, Y., Cain, J. P., Wang, H., & Laskin, A. (2007). Kinetic study of heterogeneous reaction of deliquesced NaCl particles with gaseous HNO3 using particle-on-substrate stagnation flow reactor approach. *The Journal of Physical Chemistry. A*, *111*(40), 10026–10043.

Liu, Y., Gibson, E. R., Cain, J. P., Wang, H., Grassian, V. H., & Laskin, A. (2008). Kinetics of heterogeneous reaction of CaCO3 particles with gaseous HNO3 over a wide range of humidity. *The Journal of Physical Chemistry. A*, *112*(7), 1561–1571.

Liu, Y., Liu, C., Ma, J., Ma, Q., & He, H. (2010). Structural and hygroscopic changes of soot during heterogeneous reaction with O3. *Physical Chemistry Chemical Physics*, *12*(36), 10896–10903.

Lohmann, U., Friebel, F., Kanji, Z. A., Mahrt, F., Mensah, A. A., & Neubauer, D. (2020). Future warming exacerbated by aged-soot effect on cloud formation. *Nature Geoscience*, *13*(10), 674–680.

Ma, Q., Ma, J., Liu, C., Lai, C., & He, H. (2013). Laboratory study on the hygroscopic behavior of external and internal C2–C4 Dicarboxylic acid–NaCl mixtures. *Environmental Science & Technology*, *47*(18), 10381–10388.

Majestic, B. J., Schauer, J. J., Shafer, M. M., Turner, J. R., Fine, P. M., Singh, M., & Sioutas, C. (2006). Development of a wet-chemical method for the speciation of Iron in atmospheric aerosols. *Environmental Science & Technology*, *40*(7), 2346–2351.

Milsom, A., Squires, A. M., Boswell, J. A., Terrill, N. J., Ward, A. D., & Pfrang, C. (2021a). An organic crystalline state in ageing atmospheric aerosol proxies: Spatially resolved structural changes in levitated fatty acid particles. *Atmospheric Chemistry and Physics*, *21*(19), 15003–15021.

Milsom, A., Squires, A. M., Woden, B., Terrill, N. J., Ward, A. D., & Pfrang, C. (2021b). The persistence of a proxy for cooking emissions in megacities: A kinetic study of the ozonolysis of self-assembled films by simultaneous small and wide angle X-ray scattering (SAXS/WAXS) and Raman microscopy. *Faraday Discussions*, *226*(0), 364–381.

Milsom, A., Squires, A. M., Ward, A. D., & Pfrang, C. (2022). The impact of molecular self-organisation on the atmospheric fate of a cooking aerosol proxy. *Atmospheric Chemistry and Physics*, *22*(7), 4895–4907.

Moffet, R. C., & Prather, K. A. (2009). In-situ measurements of the mixing state and optical properties of soot with implications for radiative forcing estimates. *Proceedings of the National Academy of Sciences of the United States of America*, *106*(29), 11872–11877.

Moffet, R. C., Tivanski, A. V., & Gilles, M. K. (2010). Scanning transmission X-ray microscopy: Applications in atmospheric aerosol research. In R. Signorell & J. Reid (Eds.), *Fundamentals and applications of aerosol spectroscopy* (pp. 419–462). Boca Raton, Florida: CRC Press.

Moffet, R. C., Furutani, H., Rödel, T. C., Henn, T. R., Sprau, P. O., Laskin, A., et al. (2012). Iron speciation and mixing in single aerosol particles from the Asian continental outflow. *Journal of Geophysical Research-Atmospheres*, *117*(D7), D07204. doi: 10.1029/2011JD016746

Monks, P. S., et al. (2009). Atmospheric composition change – Global and regional air quality. *Atmospheric Environment*, *43*(33), 5268–5350.

Mu, Q., Shiraiwa, M., Octaviani, M., Ma, N., Ding, A., Su, H., et al. (2018). Temperature effect on phase state and reactivity controls atmospheric multiphase chemistry and transport of PAHs. *Science Advances*, *4*(3), eaap7314.

Myhre, G., Myhre, C. E. L., Samset, B. H., & Storelvmo, T. (2013a). Aerosols and their relation to global climate and climate sensitivity. *National Knowledge Network*, *4*(5), 7.

Myhre, G., et al. (2013b). Anthropogenic and natural radiative forcing in *climate change 2013: The physical science basis*. In T. F. Stocker, D. Qin, G.-K. Plattner, M. Tignor, S. K. Allen, J. Boschung, et al. (Eds.), *Contribution of working group I to the fifth assessment report of the intergovernmental panel on climate change* (pp. 659–740). Cambridge, United Kingdom and New York, NY, USA: Cambridge University Press.

Nissenson, P., Knox, C. J. H., Finlayson-Pitts, B. J., Phillips, L. F., & Dabdub, D. (2006). Enhanced photolysis in aerosols: Evidence for important surface effects. *Physical Chemistry Chemical Physics*, *8*(40), 4700–4710.

O'Brien, R. E., Wang, B., Kelly, S. T., Lundt, N., You, Y., Bertram, A. K., et al. (2015). Liquid–liquid phase separation in aerosol particles: Imaging at the nanometer scale. *Environmental Science & Technology*, *49*(8), 4995–5002.

Ohigashi, T., Nagasaka, M., Horigome, T., Kosugi, N., Rosendahl, S. M., & Hitchcock, A. P. (2016). Development of in-situ sample cells for scanning transmission x-ray microscopy. *AIP Conf. Proc.*, *1741*(1), 050002.

Olson, N. E., Lei, Z., Craig, R. L., Zhang, Y., Chen, Y., Lambe, A. T., et al. (2019). Reactive uptake of isoprene Epoxydiols increases the viscosity of the Core of phase-separated aerosol particles. *ACS Earth and Space Chemistry*, *3*(8), 1402–1414.

Orlando, F., Artiglia, L., Yang, H., Kong, X., Roy, K., Waldner, A., et al. (2019). Disordered adsorbed water layers on TiO2 nanoparticles under subsaturated humidity conditions at 235 K. *Journal of Physical Chemistry Letters*, *10*(23), 7433–7438.

Paulson, S. E., Gallimore, P. J., Kuang, X. M., Chen, J. R., Kalberer, M., & Gonzalez, D. H. (2019). A light-driven burst of hydroxyl radicals dominates oxidation chemistry in newly activated cloud droplets. *Science Advances*, *5*(5), eaav7689.

Pfrang, C., Shiraiwa, M., & Pöschl, U. (2010). Coupling aerosol surface and bulk chemistry with a kinetic double layer model (K2-SUB): Oxidation of oleic acid by ozone. *Atmospheric Chemistry and Physics*, *10*(10), 4537–4557.

Pöschl, U. (2005). Atmospheric aerosols: Composition, transformation, climate and health effects. *Angewandte Chemie*, *44*(46), 7520–7540.

Pöschl, U., Rudich, Y., & Ammann, M. (2007). Kinetic model framework for aerosol and cloud surface chemistry and gas-particle interactions—Part 1: General equations, parameters, and terminology. *Atmospheric Chemistry and Physics*, *7*(23), 5989–6023.

Prabu, V., Obst, M., Hosseinkhannazer, H., Reynolds, M., Rosendahl, S., Wang, J., & Hitchcock, A. P. (2018). Instrumentation for *in situ* flow electrochemical scanning transmission X-ray microscopy (STXM). *The Review of Scientific Instruments*, *89*(6), 063702.

Prophet, A. M., Polley, K., Van Berkel, G. J., Limmer, D. T., & Wilson, K. R. (2024). Iodide oxidation by ozone at the surface of aqueous microdroplets. *Chemical Science*, *15*(2), 736 756.

Quinn, P. K., Collins, D. B., Grassian, V. H., Prather, K. A., & Bates, T. S. (2015). Chemistry and related properties of freshly Emitted Sea spray aerosol. *Chemical Reviews*, *115*(10), 4383–4399.

Raabe, J., et al. (2008). PolLux: A new facility for soft X-ray spectromicroscopy at the Swiss light source. *The Review of Scientific Instruments*, *79*(11), 113704.

Ravishankara, A. R. (1997). Heterogeneous and multiphase chemistry in the troposphere. *Science*, *276*, 1058–1065.

Reynolds, R., Ahmed, M., & Wilson, K. R. (2023). Constraining the reaction rate of Criegee intermediates with carboxylic acids during the multiphase Ozonolysis of aerosolized alkenes. *ACS Earth and Space Chemistry*, *7*(4), 901–911.

Richard, A., et al. (2010). Quantitative sampling and analysis of trace elements in atmospheric aerosols: Impactor characterization and synchrotron-XRF mass calibration. *Atmospheric Measurement Techniques*, *3*(5), 1473–1485.

Ronneberger, O., Fischer, P., & Brox, T. (2015). *U-net: Convolutional networks for biomedical image segmentation*. Springer International Publishing.

Rösner, B., et al. (2020). Soft x-ray microscopy with 7 nm resolution. *Optica*, *7*(11), 1602–1608.

Roy, K., Raabe, J., Schifferle, P., Finizio, S., Kleibert, A., van Bokhoven, J. A., & Artiglia, L. (2019). Design and performance of a new setup for spatially resolved transmission X-ray photoelectron spectroscopy at the Swiss light source. *Journal of Synchrotron Radiation*, *26*(3), 785–792.

Schneider, S. R., Lakey, P. S. J., Shiraiwa, M., & Abbatt, J. P. D. (2020). Reactive uptake of ozone to simulated seawater: Evidence for iodide depletion. *The Journal of Physical Chemistry. A*, *124*(47), 9844–9853.

Shiraiwa, M., Pfrang, C., & Pöschl, U. (2010). Kinetic multilayer model of aerosol surface and bulk chemistry (KM-SUB): The influence of interfacial transport and bulk diffusion on the oxidation of oleic acid by ozone. *Atmospheric Chemistry and Physics*, *10*(8), 3673–3691.

Shiraiwa, M., Ammann, M., Koop, T., & Pöschl, U. (2011). Gas uptake and chemical aging of semisolid organic aerosol particles. *Proceedings of the National Academy of Sciences of the United States of America*, *108*(27), 11003–11008.

Shiraiwa, M., Yee, L. D., Schilling, K. A., Loza, C. L., Craven, J. S., Zuend, A., et al. (2013). Size distribution dynamics reveal particle-phase chemistry in organic aerosol formation. *Proceedings of the National Academy of Sciences of the United States of America*, *110*(29), 11746–11750.

Signorell, R., & Winter, B. (2022). Photoionization of the aqueous phase: Clusters, droplets and liquid jets. *Physical Chemistry Chemical Physics*, *24*(22), 13438–13460.

Signorell, R., Goldmann, M., Yoder, B. L., Bodi, A., Chasovskikh, E., Lang, L., & Luckhaus, D. (2016). Nanofocusing, shadowing, and electron mean free path in the photoemission from aerosol droplets. *Chemical Physics Letters*, *658*, 1–6.

Sobanska, S., Falgayrac, G., Rimetz-Planchon, J., Perdrix, E., Brémard, C., & Barbillat, J. (2014). Resolving the internal structure of individual atmospheric aerosol particle by the combination of atomic force microscopy, ESEM–EDX, Raman and ToF–SIMS imaging. *Microchemical Journal*, *114*, 89–98.

Solomon, D., Lehmann, J., Kinyangi, J., Liang, B., Heymann, K., Dathe, L., et al. (2009). Carbon (1s) NEXAFS spectroscopy of biogeochemically relevant reference organic compounds. *Soil Science Society of America Journal*, *73*(6), 1817–1830.

Steimer, S. S., Lampimäki, M., Coz, E., Grzinic, G., & Ammann, M. (2014). The influence of physical state on shikimic acid ozonolysis: A case for *in situ* microspectroscopy. *Atmospheric Chemistry and Physics*, *14*(19), 10761–10772.

Steimer, S. S., Berkemeier, T., Gilgen, A., Krieger, U. K., Peter, T., Shiraiwa, M., & Ammann, M. (2015). Shikimic acid ozonolysis kinetics of the transition from liquid aqueous solution to highly viscous glass. *Physical Chemistry Chemical Physics*, *17*(46), 31101–31109.

Su, H., Cheng, Y., & Pöschl, U. (2020). New multiphase chemical processes influencing atmospheric aerosols, air quality, and climate in the Anthropocene. *Accounts of Chemical Research*, *53*(10), 2034–2043.

Sullivan, R. C., Guazzotti, S. A., Sodeman, D. A., Tang, Y., Carmichael, G. R., & Prather, K. A. (2007). Mineral dust is a sink for chlorine in the marine boundary layer. *Atmospheric Environment*, *41*(34), 7166–7179.

Symes, R., Sayer, R. M., & Reid, J. P. (2004). Cavity enhanced droplet spectroscopy: Principles, perspectives and prospects. *Physical Chemistry Chemical Physics*, *6*(3), 474–487.

van Aken, P. A., Liebscher, B., & Styrsa, V. J. (1998). Quantitative determination of iron oxidation states in minerals using Fe L2,3-edge electron energy-loss near-edge structure spectroscopy. *Physics and Chemistry of Minerals*, *25*(5), 323–327. doi: 10.1007/s002690050122

von Domaros, M., Lakey, P. S. J., Shiraiwa, M., & Tobias, D. J. (2020). Multiscale modeling of human skin oil-induced indoor air chemistry: Combining kinetic models and molecular dynamics. *The Journal of Physical Chemistry. B*, *124*(18), 3836–3843.

Wang, J., Zhou, J., Fang, H., Sham, T.-K., Karunakaran, C., Lu, Y., et al. (2011). Effect of humidity on individual SnO2 coated carbon nanotubes studied by *in situ* STXM. *Journal of Electron Spectroscopy*, *184*(3), 296–300.

Wang, B., O'Brien, R. E., Kelly, S. T., Shilling, J. E., Moffet, R. C., Gilles, M. K., & Laskin, A. (2015). Reactivity of liquid and semisolid secondary organic carbon with chloride and nitrate in atmospheric aerosols. *The Journal of Physical Chemistry. A*, *119*(19), 4498–4508.

Wang, Z., Wang, T., Fu, H., Zhang, L., Tang, M., George, C., et al. (2019). Enhanced heterogeneous uptake of sulfur dioxide on mineral particles through modification of iron speciation during simulated cloud processing. *Atmospheric Chemistry and Physics*, *19*(19), 12569–12585.

Wang, W., Ezell, M. J., Lakey, P. S. J., Aregahegn, K. Z., Shiraiwa, M., & Finlayson-Pitts, B. J. (2020). Unexpected formation of oxygen-free products and nitrous acid from the ozonolysis of the neonicotinoid nitenpyram. *Proceedings of the National Academy of Sciences of the United States of America*, *117*(21), 11321–11327.

Watts, B., & Ade, H. (2012). NEXAFS imaging of synthetic organic materials. *Materials Today*, *15*(4), 148–157.

Wei, G.-X., Zhang, S.-X., Li, Z., Wang, F.-L., Xue, Y.-L., Ji, T., & Liu, H.-Q. (2021). Multi-modality measurement and comprehensive analysis of hepatocellular carcinoma using synchrotron-based microscopy and spectroscopy. *Nuclear Science and Techniques*, *32*(9), 102.

Wiegel, A. A., Wilson, K. R., Hinsberg, W. D., & Houle, F. A. (2015). Stochastic methods for aerosol chemistry: A compact molecular description of functionalization and fragmentation in the heterogeneous oxidation of squalane aerosol by OH radicals. *Physical Chemistry Chemical Physics*, *17*(6), 4398–4411.

Wiegel, A. A., Liu, M. J., Hinsberg, W. D., Wilson, K. R., & Houle, F. A. (2017). Diffusive confinement of free radical intermediates in the OH radical oxidation of semisolid aerosols. *Physical Chemistry Chemical Physics*, *19*(9), 6814–6830.

Willis, M. D., & Wilson, K. R. (2022). Coupled interfacial and bulk kinetics govern the timescales of multiphase ozonolysis reactions. *The Journal of Physical Chemistry. A*, *126*(30), 4991–5010.

Won, Y., Lakey, P. S. J., Morrison, G., Shiraiwa, M., & Rim, D. (2020). Spatial distributions of ozonolysis products from human surfaces in ventilated rooms. *Indoor Air*, *30*(6), 1229–1240.

Woodward, X., Kostinski, A., China, S., Mazzoleni, C., & Cantrell, W. (2015). Characterization of dust Particles' 3D shape and roughness with nanometer resolution. *Aerosol Science and Technology*, *49*(4), 229–238.

Xiu, J., Liu, Y., Wang, B., Xue, Y., Chen, M., Ji, T., & Liu, H. (2020). Quantitative toxicological study of dose-dependent arsenic-induced cells via synchrotron-based STXM and FTIR measurement. *Analyst*, *145*(13), 4560–4568.

Xu, R., Ge, Y., Kwong, K. C., Poon, H. Y., Wilson, K. R., Yu, J. Z., & Chan, M. N. (2020). Inorganic sulfur species formed upon heterogeneous OH oxidation of Organosulfates: A case study of methyl sulfate. *ACS Earth and Space Chemistry*, *4*(11), 2041–2049.

Yan, J., et al. (2023). Kinetics and products of heterogeneous hydroxyl radical oxidation of isoprene Epoxydiol-derived secondary organic aerosol. *ACS Earth and Space Chemistry*, *7*(10), 1916–1928.

Yao, Y., Dawson, M. L., Dabdub, D., & Riemer, N. (2021). Evaluating the impacts of cloud processing on Resuspended aerosol particles after cloud evaporation using a particle-resolved model. *Journal of Geophysical Research-Atmospheres*, *126*(24), e2021JD034992. doi: 10.1029/2021JD034992

Yao, Y., Curtis, J. H., Ching, J., Zheng, Z., & Riemer, N. (2022). Quantifying the effects of mixing state on aerosol optical properties. *Atmospheric Chemistry and Physics*, *22*(14), 9265–9282.

Yu, T., Zhao, D., Song, X., & Zhu, T. (2018). NO2-initiated multiphase oxidation of SO2 by O2 on CaCO3 particles. *Atmospheric Chemistry and Physics*, *18*(9), 6679–6689.

Zelenay, V., Monge, M. E., D'Anna, B., George, C., Styler, S. A., Huthwelker, T., & Ammann, M. (2011a). Increased steady state uptake of ozone on soot due to UV/Vis radiation. *Journal of Geophysical Research*, *116*(D11), D11301. doi: 10.1029/2010JD015500

Zelenay, V., et al. (2011b). Aging induced changes on NEXAFS fingerprints in individual combustion particles. *Atmospheric Chemistry and Physics*, *11*(22), 11777–11791.

Zhang, Q., et al. (2007). Ubiquity and dominance of oxygenated species in organic aerosols in anthropogenically-influenced northern hemisphere midlatitudes. *Geophysical Research Letters*, *34*(13). doi: 10.1029/2007GL029979

Zhang, R., Khalizov, A. F., Pagels, J., Zhang, D., Xue, H., & McMurry, P. H. (2008). Variability in morphology, hygroscopicity, and optical properties of soot aerosols during atmospheric processing. *Proceedings of the National Academy of Sciences of the United States of America*, *105*(30), 10291–10296.

Zhang, R., et al. (2013). Chemical characterization and source apportionment of $PM_{2.5}$ in Beijing: Seasonal perspective. *Atmospheric Chemistry and Physics*, *13*(14), 7053–7074.

Zhang, Y., Favez, O., Canonaco, F., Liu, D., Močnik, G., Amodeo, T., et al. (2018). Evidence of major secondary organic aerosol contribution to lensing effect black carbon absorption enhancement, *npj Clim. Atmospheric Sciences*, *1*(1), 47.

Zhao, D., Song, X., Zhu, T., Zhang, Z., Liu, Y., & Shang, J. (2018). Multiphase oxidation of SO2 by NO2 on CaCO3 particles. *Atmospheric Chemistry and Physics*, *18*(4), 2481–2493.

Zhou, S., Shiraiwa, M., McWhinney, R. D., Pöschl, U., & Abbatt, J. P. D. (2013). Kinetic limitations in gas-particle reactions arising from slow diffusion in secondary organic aerosol. *Faraday Discussions*, *165*(0), 391–406.

Zhou, S., Hwang, B. C. H., Lakey, P. S. J., Zuend, A., Abbatt, J. P. D., & Shiraiwa, M. (2019). Multiphase reactivity of polycyclic aromatic hydrocarbons is driven by phase separation and diffusion limitations. *Proceedings of the National Academy of Sciences of the United States of America*, *116*(24), 11658–11663.

Zhou, L., Li, M., Cheng, C., Zhou, Z., Nian, H., Tang, R., & Chan, C. K. (2022). Real-time chemical characterization of single ambient particles at a port city in Chinese domestic emission control area — Impacts of ship emissions on urban air quality. *Science of the Total Environment*, *819*, 153117.

Ziemann, P. J., & Atkinson, R. (2012). Kinetics, products, and mechanisms of secondary organic aerosol formation. *Chemical Society Reviews*, *41*(19), 6582–6605.

11

Microanalysis Techniques to Study Atmospheric Ice Nucleation and Ice Crystal Growth

Daniel A. Knopf

ABSTRACT

Atmospheric ice crystal formation impacts the hydrological cycle and climate. Ice-nucleating particles (INPs), typically consisting of sub- and supermicrometer-sized particles, initiate heterogeneous ice nucleation processes leading to ice crystal formation. Ice nucleation commences on the nanoscale at the substrate surface and depends on ambient temperature and humidity. This chapter provides an overview of microanalysis techniques that examine the heterogeneous ice nucleation process and explore the complex morphologies of ice crystals. The first section introduces the different ice nucleation pathways, provides a brief outline of the underlying nucleation theory, and describes the different ice crystal habits. The second section provides an overview of the microanalysis techniques and experiments to study INPs or ice-nucleating substrates, including multimodal instrument approaches and in situ ice nucleation studies. Examples of microanalysis techniques include Raman, atomic force, electron, and X-ray microscopy, and the techniques' unique capabilities to examine the physical and chemical properties of the ice-nucleating substrate are discussed. The third section presents electron microscopy studies of ice crystals during growth and sublimation, displaying the morphological complexities of ice crystals. Finally, the fourth section discusses experimental requirements, including sample sizes, radiation effects, and the role of standard INPs.

11.1. INTRODUCTION

Understanding the formation of ice crystals from atmospheric aerosol particles is one of the grand challenges in the atmospheric sciences. How and which aerosol particles serve as ice-nucleating particles (INPs) is relevant for our understanding of the microphysical cloud structure that in turn impacts the hydrological cycle and climate (Boucher et al., 2013). In the atmosphere, two cloud regimes exist that contain ice particles: cirrus clouds that consist of only ice crystals, typically present at heights above 6 km and temperatures below the homogeneous freezing limit of water at ~235 K; and, at higher temperatures, mixed-phase clouds in which supercooled water droplets and ice crystals coexist (Koop et al., 2000; Murray et al., 2012; Storelvmo, 2017). Most precipitation is initiated by the formation of ice crystals (Lau & Wu, 2003; Lohmann & Feichter, 2005; Mülmenstädt et al., 2015). Furthermore, the presence of the ice phase can modulate the water vapor budget, which is of importance because it is the strongest greenhouse gas (Held & Soden, 2000; Lohmann et al., 1995). Cloud ice impacts cloud lifetime and albedo, thereby modulating the cloud's interaction with incoming shortwave and outgoing longwave radiation and, in so doing, affecting the radiative balance and, thus, climate (Baker, 1997; Baker and Peter, 2008; Boucher et al., 2013; Lohmann &

School of Marine and Atmospheric Sciences, Stony Brook University, Stony Brook, NY, USA

Microanalysis of Atmospheric Particles: Techniques and Applications, Geophysical Monograph 288, First Edition.
Edited by Joseph M. Conny and Peter R. Buseck.

DOI: 10.1002/9781119554318.ch11

Feichter, 2005; Rosenfeld et al., 2014; Storelvmo, 2017; Storelvmo et al., 2011).

The reasons that make prediction of atmospheric ice crystal formation difficult are manifold. Several different ice formation pathways exist, and each shows different dependencies on temperature (*T*), relative humidity (*RH*), and particle morphology (Hoose & Möhler, 2012; Kanji et al., 2017; Knopf et al., 2018; Knopf & Alpert, 2023; Marcolli, 2014; Pruppacher & Klett, 1997; Vali et al., 2015). Ice nucleation is considered a rare event where only a small fraction of the ambient aerosol particles, sometimes only 1 in 100,000, act as INPs (DeMott et al., 2010; Kanji et al., 2017). Finally, the INPs can be multiphase and multicomponent in nature and thus physicochemically complex, and can include inorganic, organic, biogenic, and biological matter (Cziczo et al., 2017; Kanji et al., 2017; Knopf et al., 2018; Murray et al., 2012).

Homogeneous ice nucleation proceeds from water and aqueous solution droplets and is mostly of importance for cirrus cloud formation. In this chapter, the focus is on heterogeneous ice nucleation that is facilitated by a substrate, in this case, the INP, and thus is a surface-specific process proceeding on the nanoscale. Any substrate that does not dissolve due to water uptake over time scales necessary to induce ice nucleation can serve as an INP. Under typical atmospheric thermodynamic conditions—that is, *T* and supersaturation with respect to the ice phase (S_{ice}), as will be made clear—the critical ice nucleus that initiates ice formation is typically the size of a few nanometers. As such, the study of heterogeneous ice nucleation from substrates is a truly formidable and challenging topic for the application of various microanalysis techniques.

11.1.1. Ice Nucleation Pathways

In the atmosphere, different ice formation pathways exist. This chapter focuses on primary ice formation, where ice nucleation is initiated by INPs. Secondary ice processes (SIPs) are recognized to contribute to, and in instances possibly dominate, the ambient ice crystal number concentrations (Field et al., 2017; Korolev & Leisner, 2020; Zipori et al., 2018). However, SIPs typically do not involve a substrate that facilitates the nucleation process but include, for example, shattering during droplet freezing, rime splintering, and fragmentation upon collision (Field et al., 2017; Korolev & Leisner, 2020).

Ice formation in the atmosphere proceeds by either homogeneous ice nucleation or heterogeneous ice nucleation (Knopf & Alpert, 2023). Figure 11.1 provides a schematic overview of the prevailing primary ice nucleation mechanisms (Knopf et al., 2018). Homogeneous ice nucleation involves liquid droplets composed of either

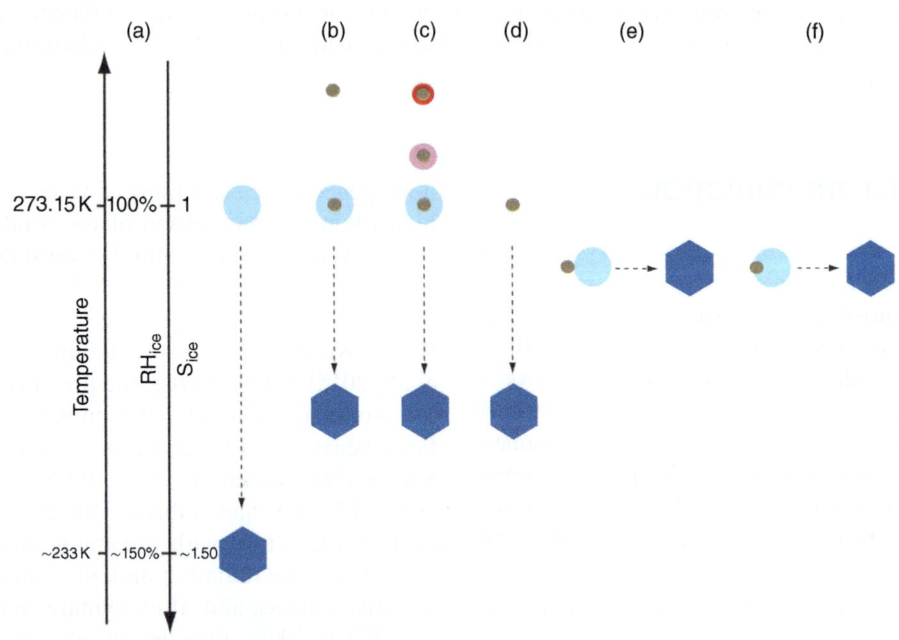

Figure 11.1 Representation of the various possible atmospheric ice nucleation pathways given for a range of temperature, relative humidity with respect to ice (RH_{ice}), and supersaturation with respect to ice (S_{ice}). Large light blue circles and dark blue hexagons display liquid water and ice, respectively. Small brown circles represent the ice-nucleating particle (INP). In (c), INP is coated by an inorganic or organic coating (red) and its aqueous solution (light red). Ice formation pathways A to F represent homogeneous ice nucleation (a), immersion freezing (b), deliquescence and water uptake followed by immersion freezing (c), deposition ice nucleation (d), contact ice nucleation (e), and inside-out freezing (f). Knopf et al. (2018)/American Chemical Society.

pure water or aqueous solutions in the absence of a substrate facilitating the nucleation process (Figure 11.1a). Hence, homogeneous ice nucleation requires lower temperatures and higher S_{ice} compared to heterogeneous ice nucleation. Micrometer-sized pure water droplets freeze homogeneously at about 235 K, which corresponds to $S_{ice} \approx 1.5$. Aqueous solution droplets will experience a freezing point depression and thus will freeze homogeneously at lower T and higher S_{ice} (Koop et al., 2000). Only the presence of INPs results in ice nucleation between the homogeneous freezing limit and the melting point of ice (273.15 K). Heterogeneous ice nucleation can also proceed below ~235 K (Knopf & Alpert, 2023), typically at S_{ice} values that are below the required S_{ice} for homogeneous ice nucleation of about 1.5 to 1.7 (Koop et al., 2000).

Ice nucleation pathway (B) displays immersion freezing, where the INP is immersed in supercooled water or aqueous solution droplets. Immersion freezing is recognized as the dominant primary ice formation pathway in mixed-phase cloud regimes (Ansmann et al., 2008; Ansmann et al., 2009; de Boer et al., 2011; Westbrook & Illingworth, 2013) where supercooled droplets and ice crystals can coexist. Pathway (C) represents water uptake and potential deliquescence of hygroscopic matter associated with the INP prior to immersion freezing. This can include amorphous organic compounds that continuously deliquesce upon increasing RH (Berkemeier et al., 2014; Knopf et al., 2018; Koop et al., 2011; Wang et al., 2012a) instead of showing a sharp solid-to-liquid transition as is typical for salts (Martin, 2000). If the condensation of liquid coincides with freezing within the supercooled temperature regime, the process is also termed *condensation freezing*. However, on a microscopic level, it likely proceeds as immersion freezing (Coluzza et al., 2017; Wex et al., 2014). Deposition ice nucleation is present in pathway (D), where ice forms on the INP from the supersaturated gas phase. Recently, this concept has been questioned, with the suggestion that in the presence of nanopores, ice formation is initiated by pore condensation freezing (David et al., 2019; Marcolli, 2014, 2020). The underlying idea is based on water in nanopores having a large negative meniscus, thereby greatly reducing the water saturation pressure. This, in turn, yields sufficiently high S_{ice} to allow for homogeneous ice nucleation to proceed in those pores. Deposition ice nucleation can contribute to ice crystal formation at lower temperatures and water-subsaturated conditions representing cirrus cloud regimes, typical of the upper troposphere (Cziczo et al., 2013; DeMott, 2002; Heymsfield et al., 2017). Pathway (E) displays contact ice nucleation where an INP collides with a supercooled droplet, thereby initiating freezing of the droplet (Ladino Moreno et al., 2013). Finally, pathway (F) represents inside-out freezing where the INP in the supercooled temperature regime is in contact with the gas phase and aqueous phase as the droplet is evaporating (Durant & Shaw, 2005; Shaw et al., 2005).

11.1.2. Ice Nucleation Kinetics

For a substrate to promote ice nucleation heterogeneously, classically, the following requirements are assumed (Hegg and Baker, 2009; Knopf et al., 2018; Pruppacher & Klett, 1997; Vali et al., 2015): (a) the substrate must be insoluble, (b) it must be larger than the critical ice nucleus, (c) its surface must have chemical bonds that can interact with and arrange water molecules, (d) it must have a crystallographic match representing the geometrical arrangement of bonds, and (e) there must be active sites of varying quality representing localized phenomena on the substrate surface. Following classical nucleation theory (CNT), for the formation of ice from supercooled water, a Gibbs free energy barrier has to be overcome to allow for the growth of a critical ice nucleus beyond the critical radius, r_{crit}. r_{crit} and the Gibbs free energy change, ΔG, can be derived as (Alpert et al., 2011; Pruppacher & Klett, 1997; Zobrist et al., 2007)

$$r_{crit} = \frac{2\sigma_{sl}(T)\cdot v_s(T)}{kT\ln S(T)} \text{ and } \Delta G(T) = \frac{16\pi\sigma_{sl}^3(T)\cdot v_s^2(T)}{3\left(kT\ln S_{ice}(T)\right)^2}, \tag{11.1}$$

where σ_{sl} is the interfacial tension between ice (solid) and water (liquid), v_s is the volume of a water molecule in ice (solid), and k is the Boltzmann constant.

Figure 11.2 outlines the behavior of the Gibbs free energy change with radius as a function of S_{ice}. For ice, subsaturated conditions (or above the ice melting temperature of 273.15 K), the activation barrier ΔG becomes infinite. As S_{ice} increases, ΔG and r_{crit} decrease, as indicated by the arrows in Figure 11.2. Clearly, changes in S_{ice} determine the likelihood of establishing and maintaining the critical ice nucleus so that nucleation and freezing commence.

Figure 11.3 displays the change of r_{crit} as a function of temperature for pure water in relation to typical particle surface areas. For typical conditions of atmospheric heterogeneous ice nucleation, the critical ice nucleus size has a radius of about 1 nm (Figure 11.3a). This implies that surface features that enable ice nucleation are in the order of 1 to 20 nm^2 (Figure 11.3b). A spherical particle of 1 µm in size has a surface area of about 3 × 10^6 nm^2, allowing for the presence of a large number of surface features that could initiate ice nucleation (Figure 11.3c). Considering these large numbers of sites, it can be assumed that it is very likely that sufficient surface features representing the ice-nucleating surface material should always be present.

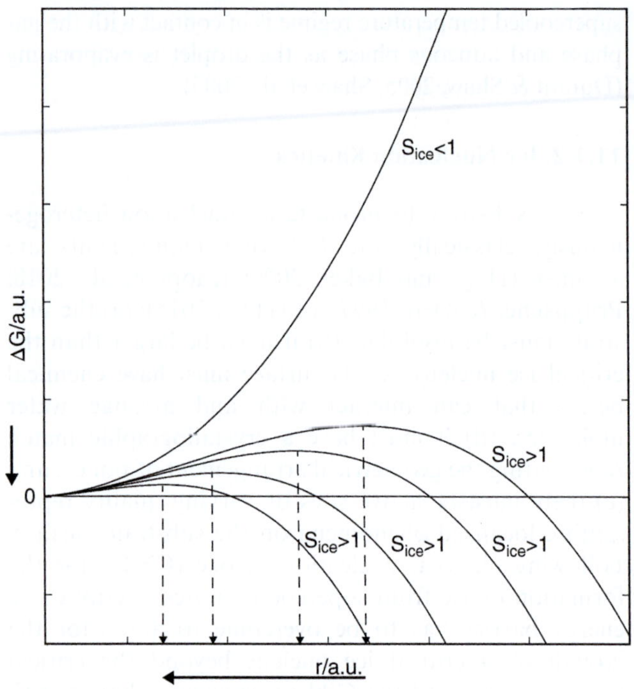

Figure 11.2 The change in Gibbs free energy necessary for the formation of a critical ice nucleus as a function of ice nucleus radius and supersaturation with respect to ice, S_{ice}. Solid black arrows indicate the decreasing Gibbs free energy barrier and decreasing radius with increasing S_{ice}. Dashed vertical arrows display how the critical radius decreases as S_{ice} increases.

Looking at these features on microscopic scales, the application of microanalysis techniques is uniquely suited to advance our understanding of the ice nucleation process.

The heterogeneous ice nucleation rate coefficient, J_{het}, for immersion freezing is defined as the number of nucleation events per surface area and time, given in units $cm^{-2}\ s^{-1}$. It can be derived using $\Delta G(T)$ and a diffusion activation energy for a water molecule to cross the water, ice nucleus interface, $\Delta F_{diff}(T)$ (Alpert et al., 2011; Knopf & Alpert, 2023; Pruppacher & Klett, 1997; Turnbull & Fisher, 1949; Zobrist et al., 2007):

$$J_{het}(T) = \frac{kT}{h} e^{\left[-\frac{\Delta F_{diff}(T)}{kT}\right]} n e^{\left[-\frac{\Delta G(T) f_{het}(m,)}{kT}\right]}. \quad (11.2)$$

The prefactor serves as an impinging frequency given by $\frac{kT}{h}$, where h is Planck's constant. n represents the number of water molecules at the INP-water interface and can be assumed to be around $10^{15}\ cm^{-2}$. $\Delta F_{diff}(T)$ depends on the diffusivity of water that can be expressed by the Vogel–Fulcher–Tammann equation (Smith & Kay, 1999). f_{het} (m, x) is the so-called *geometric factor*, where m is the compatibility factor and x represents the ratio of the radius of the substrate to the radius of the spherical critical ice nucleus (Knopf et al., 2022). x can be neglected when particles are larger than 0.1 μm. This holds for most INPs, and thus we can write $\lim_{x\to\infty} f_{het}(m,x) = \frac{m^3 - 3m + 2}{4}$. Instead of reporting m, typically the contact angle, θ, is given. θ is derived from $\cos\theta = m$ (Fletcher, 1958). A smaller θ value indicates a more efficient INP. The extreme case of an inefficient INP is represented by $\theta = 180°$, which indicates the case of homogeneous ice nucleation. In other words, for f_{het} $(m, x) = 1$, Equation (11.2) represents the case of homogeneous ice nucleation from a supercooled liquid droplet.

For the derivation of the deposition ice nucleation rate coefficients (Alpert et al., 2011; Pruppacher & Klett, 1997;

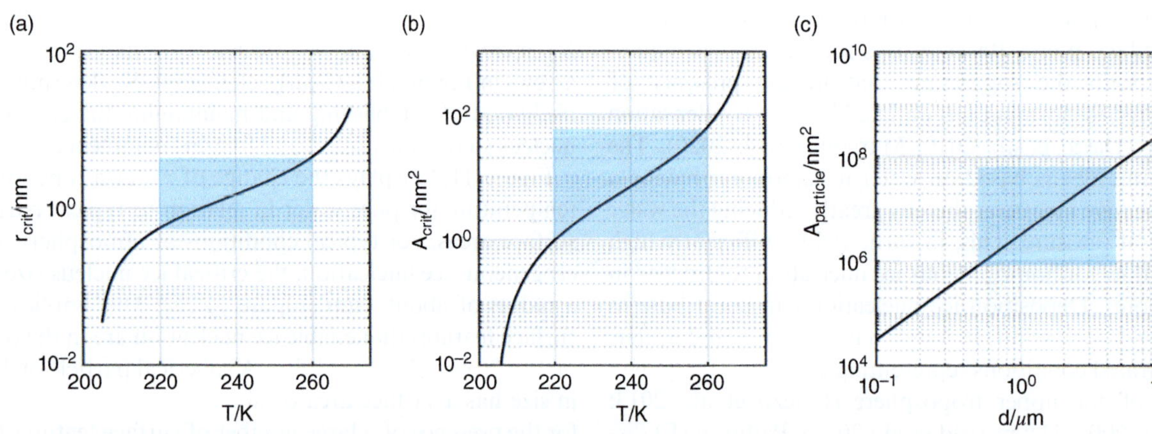

Figure 11.3 Estimates of (a) the critical ice nucleus radius (r_{crit}) and (b) its surface area (A_{crit}) as a function of temperature using Equation (11.1). For comparison, the surface area of spherical particles ($A_{particle}$) is given in (c). Shading in (a) and (b) indicates typical atmospheric conditions for heterogeneous ice nucleation. Shading in (c) indicates typical ice-nucleating particle sizes and surface areas.

Wang & Knopf, 2011), the diffusion activation energy is omitted, and n represents the number of water adsorption sites on the INP surface and can be assumed to be around 10^{15} cm^{-2}. σ_{sl} is exchanged for the interfacial tension between ice (solid) and supersaturated gas phase (vapor), σ_{sv}.

These descriptive approaches for capturing the immersion freezing and deposition ice nucleation kinetics follow CNT (Pruppacher & Klett, 1997). As such, the nucleation process is considered to be stochastic and, thus, is implicitly time dependent, resulting in $J_{het}(T)$ (Knopf et al., 2020; Knopf & Alpert, 2023). This in turn leads to the question of how many ice nucleation events have to be observed to obtain statistical significance. Based on a binomial distribution for freezing probabilities, upper and lower fiducial limits for observed nucleation events at a given confidence level can be derived (Koop et al., 1997). For example, observation of a single ice nucleation event leads to a maximum of 9.233 and minimum of 0.001 nucleation events at a confidence level of 0.999. Alpert & Knopf (2016) suggest observing at least 100–200 ice nucleation events to keep corresponding uncertainties in freezing kinetics within ±1 order of magnitude. The issue of statistical uncertainty can directly affect how to best observe particle or substrate samples using microanalysis techniques. The minimum required number of observed ice nucleation events can define the chosen sample size and field of view. If the focus is on the investigation of physicochemical surface properties that induce ice nucleation instead of nucleation kinetics, the total number of observed ice nucleation events may not be as important, although one should examine several independent samples to ensure reproducibility.

A common approach for characterizing and comparing ice nucleation efficacies of INPs is the ice nucleation active site (INAS) description (Connolly et al., 2009; Vali, 1971). This approach is based on the deterministic or singular hypothesis method. It ignores the stochastic nature of nucleation and, thus, is time independent. In this case, nucleation efficacies are expressed as INAS densities, $n_s(T)$, in units cm^{-2}, providing the number of ice nucleation events per surface area at a given temperature. This description implies that each ice nucleation event is associated with a characteristic INAS.

The concept of what establishes an INAS is a current matter of research. Characterization of an INAS requires greater resolution than is required to determine the size of the critical ice nucleus (Figure 11.3) and has not yet been achieved in situ. This is the case if the arrangement of a critical ice nucleus proceeds only at one site. However, one can envision that several ice-like clusters emerge across the given surface area that could subsequently combine to form a larger critical ice nucleus, thus surpassing r_{crit}. Clearly, the minute scale of an active site poses a challenge even for the most advanced microanalysis techniques. Application of alternative computational approaches, like molecular dynamics simulations, to model ice nucleation on substrate surfaces has yet not yielded a defined set of requirements that make up an INAS (Knopf et al., 2018; Knopf et al., 2020; Knopf & Alpert, 2023). As outlined later, however, the application of microanalysis techniques has made progress in identifying the physicochemical surface features that are involved in initiating ice nucleation on substrate surfaces.

11.1.3. Ice Crystal Habits, Morphology, Growth, and Sublimation

Subsequent to ice nucleation on the nanoscale, ice crystals grow in a humidified environment to larger sizes, reaching up to snowflake dimensions that are observable with a typical optical microscope (OM) (Bentley & Humphreys, 1931; Libbrecht, 2005; Nakaya, 1954). However, before ice crystals sediment via precipitation, the formation of micrometer-sized ice crystals and their altering morphologies under super- and subsaturated conditions impact cloud albedo and thus the climatic effect of clouds. It is well known that the shape of cirrus ice particles can have a profound impact on Earth's radiation budget (Hartmann, 2016; Jensen et al., 2009; Lawson et al., 2019; McFarquhar et al., 2002; Mishchenko et al., 1996). For example, the differences in ice crystal shapes can lead to differences in the up- and downwelling infrared irradiance of cirrus clouds by tens of watts per m^2 (Wendisch et al., 2007).

Ice crystal shapes depend on the environmental conditions T and RH (Bacon & Swanson, 2000; Bailey & Hallett, 2002, 2004; Bailey & Hallett, 2009; Pruppacher & Klett, 1997). Figure 11.4 displays the various ice crystal habits obtained from laboratory studies and field campaigns that develop for changing conditions in parameters T and S_{ice} (Bailey & Hallett, 2009). As evident in Figure 11.4, as temperature decreases, the regime of single crystals moves into the polycrystalline habits where at greater temperatures, plate-like, and at lower temperatures, columnar ice crystals dominate. Increases in S_{ice} can change crystal habits ranging from compact, thick, and thin plates to bullet rosettes and hollow-ended bullets. Figure 11.4 exemplifies the complexity of ice crystal growth and its ramifications for cloud and climate modeling.

Ice crystal and snowflake morphologies have been traditionally studied by OM, but in recent years, environmental scanning electron microscopy (ESEM) has been applied to growing and sublimating ice crystals, demonstrating, by use of greater resolution, a complex picture of linear striations, ridges, islands, steps, peaks, pits, and crevasses on the facets of the ice (Magee et al., 2014).

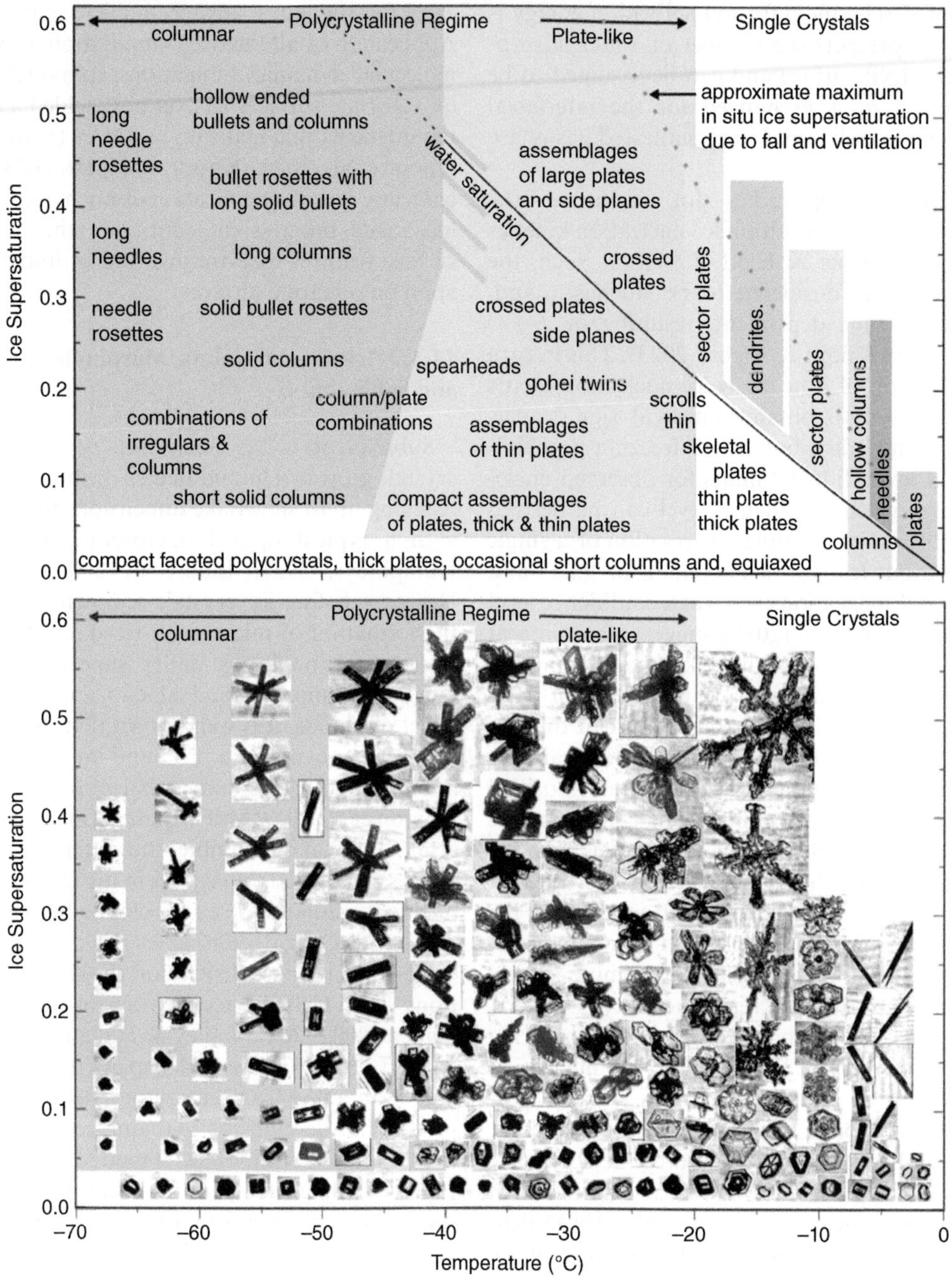

Figure 11.4 Ice crystal habit diagram derived from laboratory and field campaign observations as a function of temperature and ice supersaturation. Bailey & Hallett (2009)/with permission of the American Meteorological Society (AMS).

Magee et al. (2021) also examined cirrus ice particles using ESEM, corroborating previous ice habit studies, although the higher-resolution examination of ice crystals points to an even greater particle complexity, as shown in Figure 11.4.

Microscopic understanding of ice surfaces, including premelted layers of ice, affects many environmental processes, including growth rates and shapes of snowflakes, the sintering of snowfields, transport of impurities along grain boundaries, and thunderstorm lightning (Dash et al., 2006). Grain boundaries represent the interface between individual ice crystals. These grain boundaries change when ice crystals grow, melt, or sublimate (Cascajo-Castresana et al., 2021; Pedersen et al., 2011; Pfalzgraff et al., 2010). These processes, in turn, influence ice crystal shape, its interaction with adsorbing and reactive gas species (e.g., heterogeneous reactions), and distribution of species expelled during ice

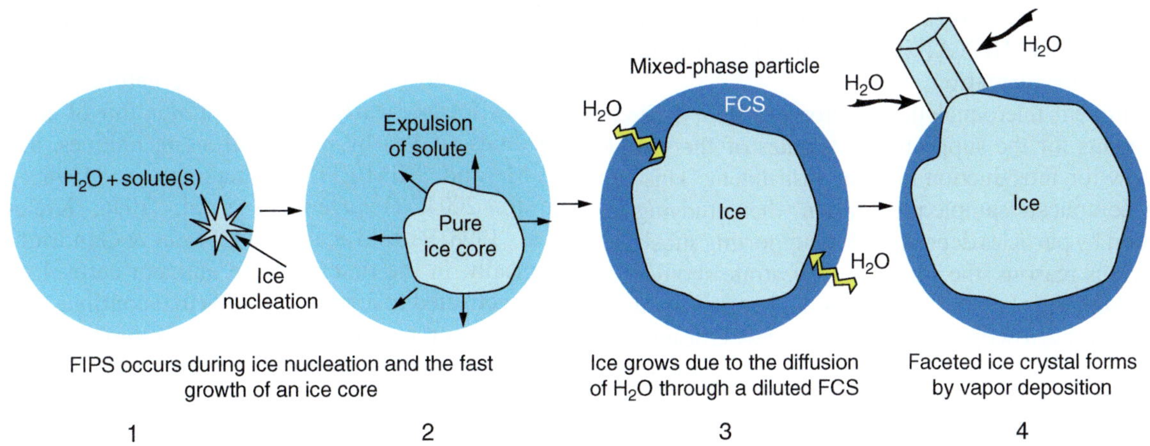

Figure 11.5 Sequences that can lead to the formation of a freeze-concentrated solution (FCS) surrounding an ice crystal (1–3). Water vapor may be taken up by the FCS, allowing the ice crystal to grow further from the diluted FCS (3). As the ice crystal disrupts the FCS, it may serve as a deposition INP (4). Bogdan (2018)/with permission of the American Chemical Society.

growth (Bogdan, 2018; Bogdan & Molina, 2017; Zobrist et al., 2008).

When ice forms from an aqueous solution, the ice crystal expels solutes, thereby generating a concentrated solution that is subsaturated with respect to ice. Figure 11.5 displays this process schematically (Bogdan, 2018). The remaining solution, also termed *freeze-concentrated solution* (FCS), may coat the ice crystal. Under atmospheric ice-forming conditions, the FCS will take up water, leading to further ice crystal growth, which then potentially acts as a nucleation site of ice formation by vapor deposition. These processes, which are especially relevant at higher altitudes and colder temperatures where freezing from aqueous solution droplets is more likely, have only received limited attention. Furthermore, these initial studies are restricted to OM (Bogdan, 2018; Bogdan et al., 2016; Bogdan & Molina, 2017).

11.2. APPLICATION OF MICROANALYSIS TECHNIQUES TO STUDY ICE FORMATION

During the last two to three decades, the field of atmospheric ice nucleation has seen the application of various microanalysis techniques to deduce the occurrence of ice nucleation and to examine the INP or ice-nucleating substrate. The study of homogeneous ice nucleation typically employs optical microscopy for the detection of the freezing of micrometer-sized droplets when temperatures are close to the homogeneous freezing limit (Knopf & Lopez, 2009; Knopf & Rigg, 2011; Riechers et al., 2013; Shardt et al., 2022; Stan et al., 2009; Tarn et al., 2021). However, when aiming for deeper supercooling, reaching temperatures beyond the homogeneous freezing limit, nanodroplets generated by a supersonic nozzle allowing for rapid cooling rates are examined for freezing temperatures and rates (Amaya & Wyslouzil, 2018; Knopf & Alpert, 2023). In these experiments, for example, Fourier transform infrared spectroscopy and small and wide angle X-ray scattering (SAXS and WAXS, respectively) are applied to probe the droplet temperature, volume, crystallinity, and structure of ice (Amaya & Wyslouzil, 2018). WAXS provided the ability to evaluate the ice structure, indicating that the ice crystals formed at 218 K and consisted of a mixture of hexagonal and cubic ice lattice structures (Andrew et al., 2017).

The majority of the heterogeneous ice nucleation studies that make use of microanalysis techniques can be divided into two approaches. One uses a multi-instrumental, also termed multimodal, analytical approach (Laskin et al., 2016; Laskin et al., 2019) where ice formation is assessed with one type of instrument, not necessarily a microanalysis technique, and the corresponding INP or ice-nucleating substrate is subsequently interrogated by microanalysis methods discussed in this book. The other approach directly applies microanalysis techniques for in situ detection of ice nucleation. In this section, we will introduce both approaches with examples of procedures and setups.

The microanalysis techniques used in studies of ice nucleation and ice crystal growth include Raman, scanning electron microscopy with energy dispersive X-ray spectroscopy (SEM/EDX or SEM/EDS), transmission electron microscopy (TEM), atomic force microscopy (AFM), and scanning transmission X-ray microscopy with near-edge X-ray absorption fine structure spectroscopy (STXM/NEXAFS). Each of these techniques has its own requirements and limitations, as discussed in the previous chapters. Relevant for studying the ice phase are sample substrates, resolution, field of view, radiation effects on the sample, and achieving uniform T and RH fields. So far, only Raman, SEM/EDX, and STXM/NEXAFS have been applied for in situ detection of ice formation.

The different microanalytical approaches have different requirements for the supporting substrates of the sample that allow for introduction into the instrument. This can further complicate sample preparation when studying ice nucleation by particles deposited on a supporting substrate. For obvious reasons, the supporting substrate should not interfere with the observed ice nucleation. In other words, it must not be hygroscopic and must be a less efficient ice-nucleating agent compared to the investigated particles or materials deposited on the supporting substrate. For the case of OM and Raman application, hydrophobically coated glass or quartz slides have been shown to minimize heterogeneous ice nucleation, even allowing the study of homogeneous freezing of aqueous droplets (Bertram et al., 2000; Knopf et al., 2002; Knopf et al., 2003; Koop et al., 1998). Silicon wafer substrates coated with a silicon nitride film are suitable for ice nucleation experiments and the application of SEM/EDX (Knopf et al., 2014; Wang et al., 2016). The few ice nucleation studies utilizing STXM/NEXAFS demonstrate that silicon nitride-coated Si frames are also suitable for studying ice nucleation in a multimodal or in situ manner (Alpert et al., 2022a; Alpert et al., 2022b; Knopf et al., 2014). Nevertheless, due to the very thin membrane films associated with the silicon nitride-coated Si frames, handling and ice nucleation experiments can present a challenge for the experimentalist.

11.2.1. Multimodal Microanalysis of Ice Nucleation

To gain an understanding of the nature of the INPs, studies focused early on the examination of ice crystal residuals (ICRs) by microanalysis techniques, including TEM and SEM/EDX (Chen et al., 1998; Cziczo et al., 2004; Heintzenberg et al., 1996; Kreidenweis et al., 1998; Petzold et al., 1998; Twohy & Gandrud, 1998). Typically, in the first step, the ambient aerosol particles were activated to form ice crystals that continue to grow in a supersaturated environment: for example, in a continuous flow diffusion chamber (Rogers, 1988). In the second step, the ice crystals are size-selected by an impactor for deposition on supporting substrates (Chen et al., 1998; Kreidenweis et al., 1998) or by a counter-flow virtual impactor (CVI) (Cziczo et al., 2017; Twohy & Gandrud, 1998). Application of these approaches—for example, Chen et al. (1998)—found that the INPs relevant for cirrus formation conditions in the upper troposphere can consist of crustal, sulfate, carbonaceous, metallic, and other unidentified materials. Clearly, these early studies using TEM and SEM advanced our understanding of the chemical complexity of atmospheric INPs.

During the Indirect and Semi-Direct Aerosol Campaign (ISDAC) near Barrow, Alaska, in the spring of 2008, Hiranuma et al. (2013) collected ICRs using CVI (Figure 11.6). The ICRs were examined by STXM/NEXAFS at the carbon K-edge to enable the speciation of the organic matter in resolution of tens of nanometers.

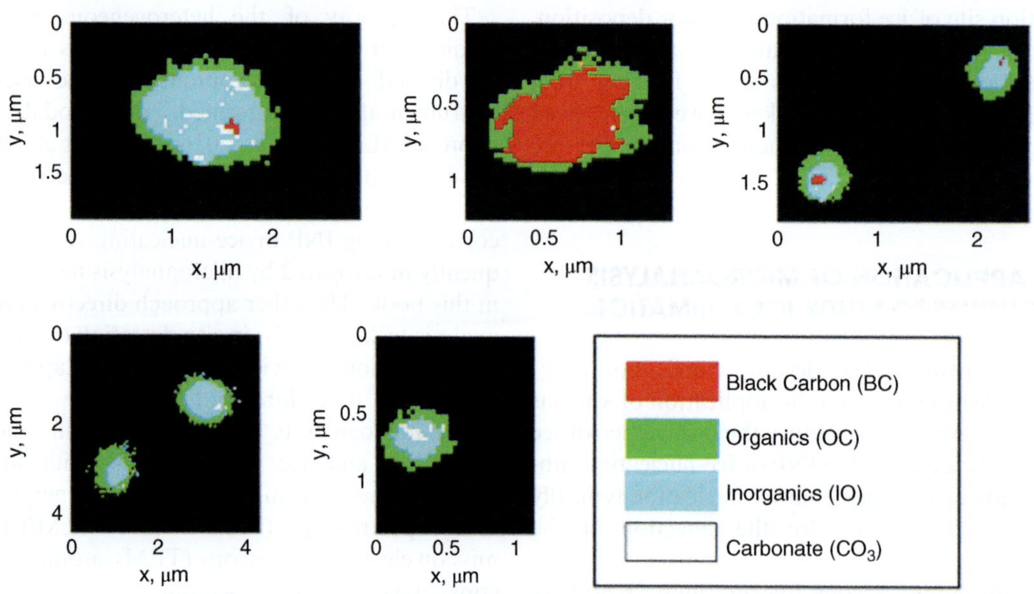

Figure 11.6 Ice crystal residuals (ICRs) collected during the Indirect and Semi-Direct Aerosol Campaign (ISDAC) by aircraft applying a counterflow virtual impactor (CVI). Panels show scanning transmission X-ray microscopy with near-edge X-ray absorption fine structure spectroscopy (STXM/NEXAFS) false-color maps of the ICRs where the color-related composition is given in the legend. The analysis indicates that the ICRs contain cores composed of inorganic, black carbon, and carbonate that are coated with organic matter. Hiranuma et al. (2013)/with permission from John Wiley & Sons.

Application of STXM demonstrated that ICRs possessed a thin organic coating composed of carboxylic functional groups (Figure 11.6). Organic matter associated with INPs has been recognized to play an important role when determining sources of atmospheric INPs (Knopf et al., 2018).

Another common approach is to collect laboratory-generated particles or substrates and field-collected particles on substrates that allow for ice nucleation experiments and microanalysis using SEM/EDX and STXM/NEXAFS (Alpert et al., 2022b; Charnawskas et al., 2017; China et al., 2017; Holden et al., 2019; Knopf et al., 2014; Knopf et al., 2021; Knopf et al., 2022; Lata et al., 2021; Wang et al., 2012a; Wang et al., 2012b). The ice nucleation experiments typically use a chamber where the particle sample is located at the bottom of a temperature-controlled cold stage (Dymarska et al., 2006; Wang & Knopf, 2011). This stage defines the particle or supporting substrate temperature. The chamber is sealed against room air, which allows for the introduction of a controlled humidified gas flow. Having control of particle *T* and *RH* permits the simulation of typical atmospheric conditions, such as mimicking of atmospheric advection that results in cloud formation. As S_{ice} increases, water uptake, immersion freezing, and deposition ice nucleation can be observed microscopically. When adjusting the chamber *RH* to achieve subsaturated conditions, water evaporates or ice sublimates. An example of this procedure using OM for the ice nucleation experiment is outlined in Figure 11.7. Application of OM typically allows for a greater field of view compared to SEM or STXM, thereby minimizing experimental artifacts such as ice formation outside of the field of view or at the fringes of the particle sample. Using this technique, radiation effects are negligible. Upon visual detection of ice formation, the particle sample area can be examined with greater magnification (Figure 11.7a). Observation of the sublimation of the ice crystals allows for pinpointing the INPs that were the source of the ice crystal formation (Figure 11.7b). Subsequently, using digital imaging analysis, the INPs

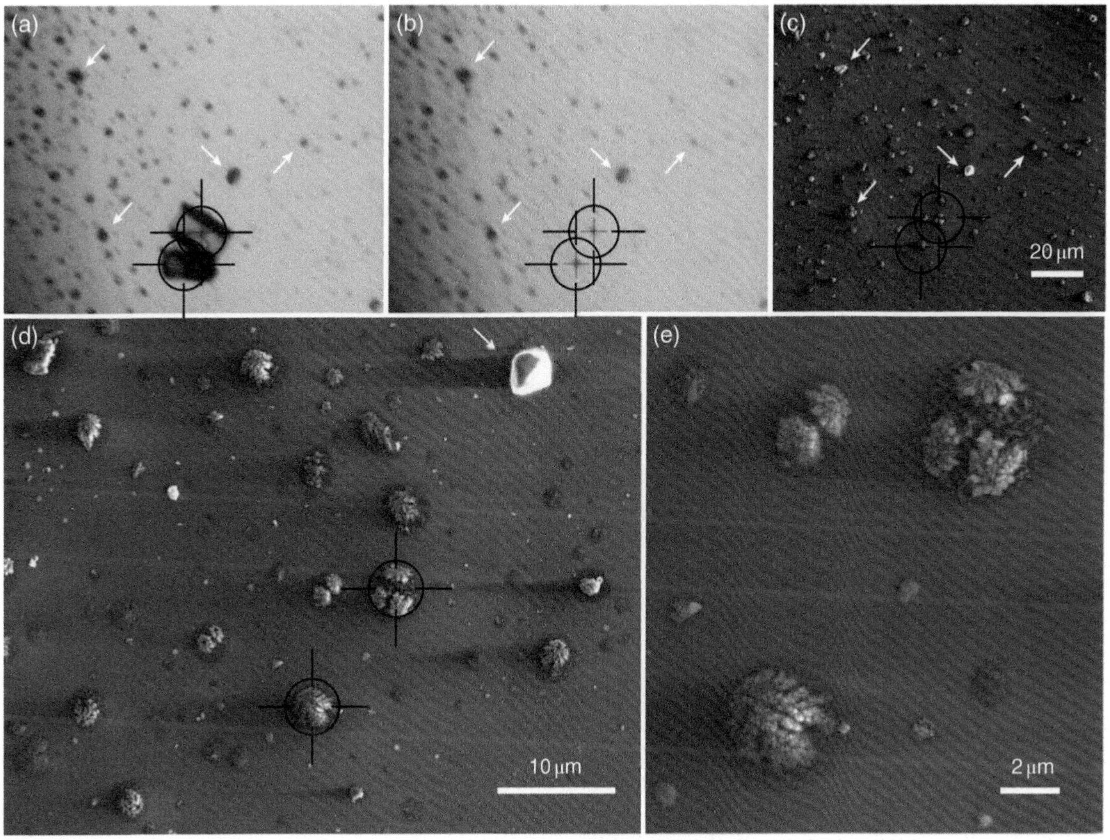

Figure 11.7 Multimodal identification of ice-nucleating particles (INPs) using optical microscopy (OM) and scanning electron microscopy (SEM). (a) The crosshairs display the position of the center of ice crystals as determined by OM. (b) Subsequent to ice crystal sublimation, the crosshairs indicate the location of INPs. (c) The same field of view as in (b) imaged by SEM. (d) and (e) Application of higher magnification using SEM to image in-detail morphology of identified INPs. The white arrows point to the same particles in (a–d). Knopf et al. (2014)/with permission of the American Geophysical Union.

can be relocated and imagined with greater resolution by the microanalysis instrumentation, here, an SEM (Figure 11.7c–e). In general, this multimodal approach allows for the morphological and chemical characterization of the particle or substrate area by SEM that initiated ice nucleation. The identified INPs, shown in Figure 11.7, are not different from other adjacent particles located on the same substrate. They demonstrate similar size, morphology, and elemental compositions as determined by EDX (not shown) as the non-INPs.

In the same ice nucleation experiment, Knopf et al. (2014) used STXM/NEXAFS to analyze INPs and non-INPs deposited on silicon nitride windows, which consist of a 100 nm thin film supported by a Si frame. The experimental challenge during the experiments was to keep the supporting substrate temperature uniformly consistent between the thin film and Si frame to avoid ice nucleation artifacts. Figure 11.8 shows a multimodal microanalysis study of ambient particles acting as INPs using OM, SEM, and STXM (Knopf et al., 2014). Figure 11.8a, b shows an image of aerosol particles as dark light-absorbing spots dispersed over the silicon nitride window as the bright square and the supporting Si frame as a darker area as visualized by OM. Panel (a) shows the emergence of an ice crystal during supersaturated conditions. Panel (b) represents the sample in (a) after sublimation of the ice crystal. Panel (c) shows the area in (a) where ice formed in greater resolution using SEM. The sample was subsequently transferred to STXM/NEXAFS to chemically speciate the organic components using the carbon K-edge (panels d–g). False color images of the INP (d) and non-INPs (e, f) reveal the presence of inorganic species,

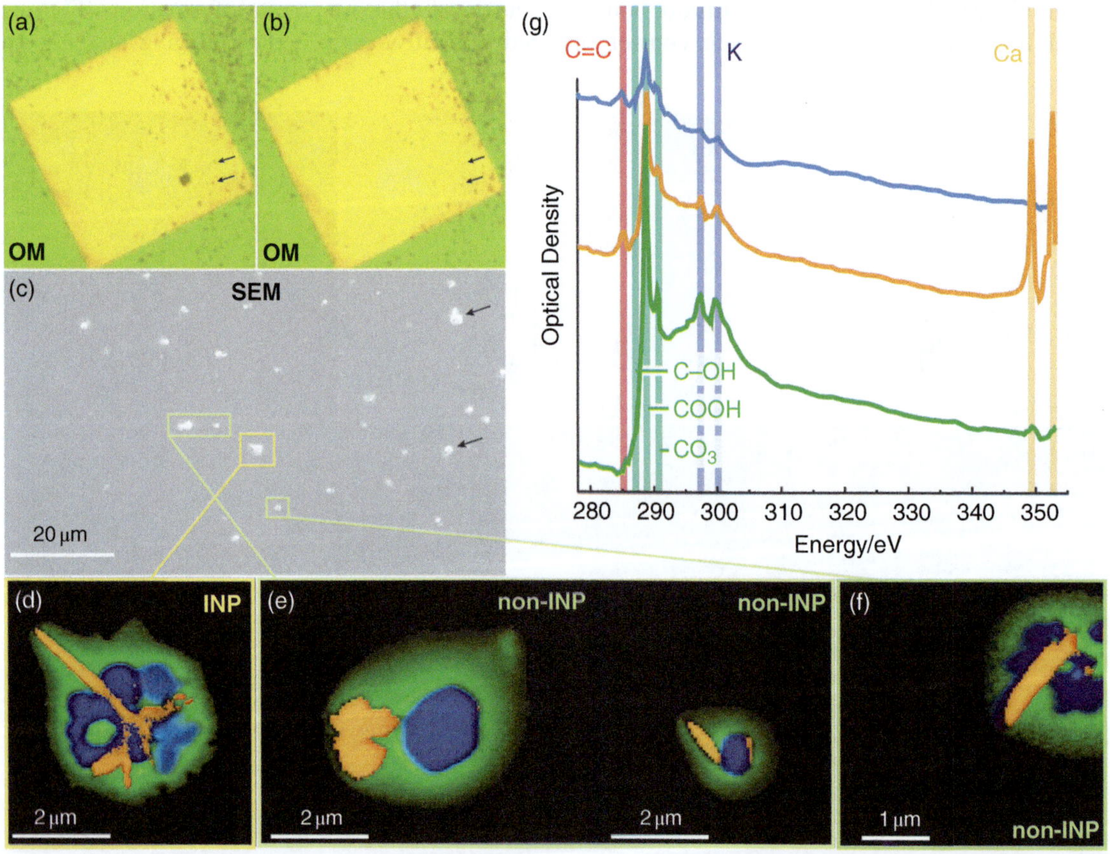

Figure 11.8 Ice-nucleating particle (INP) identification and scanning transmission X-ray microscopy with near-edge X-ray absorption fine structure spectroscopy (STXM/NEXAFS) chemical imaging of particles. (a, b) Ice formation and ice sublimation events observed in the OM experiment. (c) Scanning electron microscopy (SEM) image of the same sample area at higher magnification. The black arrows point to the same particles in (a–c). (d) Identified INP. (e, f) Non-INPs. The component images in d to f emphasize the contrast between non-carbonaceous inorganic species (blue), organic carbonaceous (green), and calcium (orange). (g) NEXAFS spectra of the dominant particle regions (inorganic, calcium, and organic) with the major functionalities highlighted. Knopf et al. (2014)/with permission of the American Geophysical Union.

calcium, and organic carbonaceous matter. The INP is surrounded by the organic carbonaceous matter. The absorption spectrum shown in panel (g) indicates that in areas dominated by inorganic species or calcium, organic functionalities are present, here COOH groups. Hence, in panels d–f, areas that are dominated by inorganic species (due to absorption through the entire particle) are also coated by organic matter. Panels d–f also demonstrate that within a resolution of about 30 nm, the INP and non-INPs cannot be physicochemically distinguished, thereby corroborating SEM/EDX observations.

The application of multimodal microanalytical techniques to study ice nucleation significantly advances our understanding of atmospheric ice crystal formation. The observations clearly demonstrate that organic matter is associated with INPs (Knopf et al., 2018). However, which physicochemical properties of the organic matter, including the presence of certain organic functionalities, surface morphology, or phase state, serve as the ice-nucleating agent or an INAS has yet to be resolved. The presence of organic matter corroborates previous CVI-based collection and chemical characterization of ICRs. Furthermore, microanalytical studies that examine the particle population on a statistically significant level and associated INPs find that the INPs cannot be physicochemically distinguished from the dominant particle types (i.e., non-INPs) in the particle population present on the substrate (Alpert et al., 2022b; China et al., 2017; Knopf et al., 2014; Knopf et al., 2022; Knopf et al., 2023b; Lata et al., 2021; Wang et al., 2012b). This emphasizes the previous point, which concerns the microscopic nature of the surface feature or INAS, which ultimately facilitates ice nucleation. Although the INP can be chemically imaged, its surface area is still orders of magnitude larger than the likely point of ice nucleation. Furthermore, the various multimodal approaches cannot resolve where on the particle ice nucleation commenced because ice formation is not monitored in situ. Despite these caveats, multimodal physicochemical characterization and chemical imaging of aerosol particles and INPs greatly advance our understanding of atmospheric ice formation. By learning the physicochemical properties of INPs, the role of chemical and physical aging of particles during atmospheric transport for ice nucleation, and the sources and emissions of aerosol particles that potentially act as INPs can be assessed, thereby improving ice crystal formation in cloud and climate models (Kanji et al., 2017; Knopf et al., 2018).

In recent studies, Friddle and Thürmer combined an OM with an ice nucleation cell that allows for control of substrate temperature and humidity. In addition, an AFM for visualization was integrated into this setup to sequentially examine the role of nano-surface steps promoting ice formation on feldspar (Friddle & Thürmer, 2019, 2020). Figure 11.9 shows a schematic of the OM-ice nucleation-AFM setup. A mixing chamber provides the desired humidity exposure of the sample. The sample is cooled by a cold nitrogen gas flow at the bottom of the cell, not shown here. An OM is used to visually detect water condensation and ice formation. An AFM is integrated into the top of the mixing cell. Figure 11.10 provides an example of an ice formation experiment where OM and AFM are collocated. This experiment indicates that ice formed on a small protrusion located on the feldspar surface. As the authors point out, the INAS cannot be directly determined due to limited resolution (Friddle & Thürmer, 2020). For example, the scale of the identified protrusion (Figure 11.10e) is larger than most atmospheric particles. During this experiment, ice was observed to rapidly

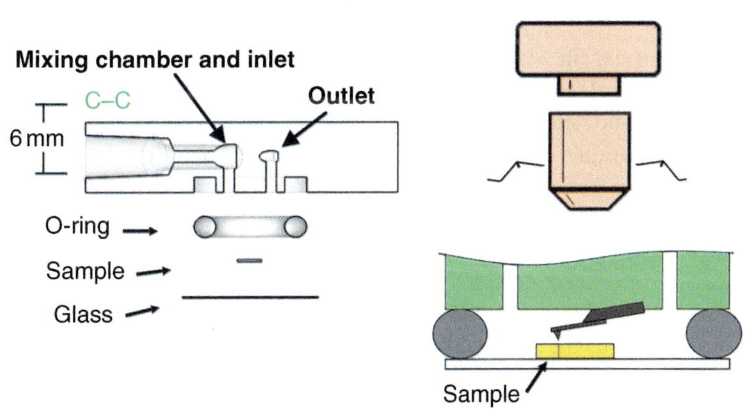

Figure 11.9 Collocated optical microscopy with atomic force microscopy (AFM) coupled to a micro-mixing chamber for humidity control and a cold stage at the bottom to control the temperature of the sample. Adapted from Friddle & Thürmer (2020).

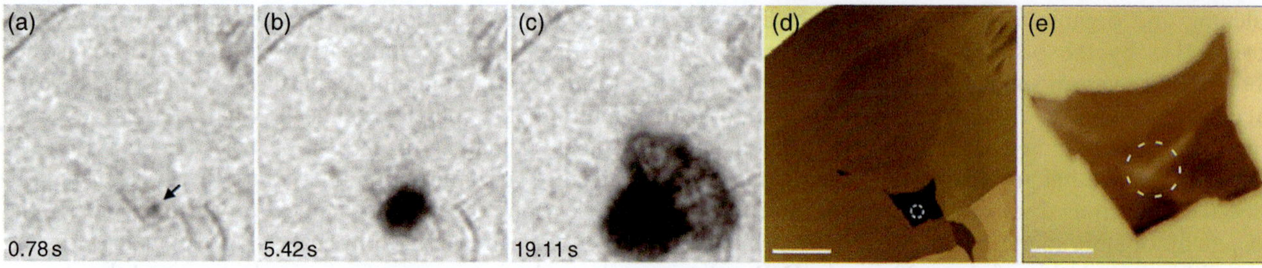

Figure 11.10 Ice formation on the feldspar surface at −29.5 °C was observed with optical microscopy (OM) (a–c) and collocated atomic force microscopy (AFM) topography (d, e). (d) AFM image of the same location in (a–c) with dashed circle around the area of nucleation. The scale bar indicates 20 μm. (e) Expanded view of the identified area of nucleation in (d) depicting a small protrusion within a larger pit. The scale bar represents 5 μm. Friddle & Thürmer (2020)/Copernicus Publications/CC BY 4.0.

spread stepwise along the feldspar surface. At −30 °C, the number of ice-bearing steps was related to step height and humidity (Friddle & Thürmer, 2019). Although humidity levels applied in these experiments reflect greatly supersaturated conditions with respect to water, S_{water}, the step height decreases with increasing S_{water}. For example, at −30 °C, for a 50% chance of a step bearing ice, a step must be taller than 1 μm at $S_{water} = 1.32$ but can be as small as 14 nm at $S_{water} = 2.32$ (Friddle & Thürmer, 2019). Typically, maximum S_{water} achieved in the atmosphere is about 1.04. How a step size of about 1 μm translates to sub- and supermicrometer-sized particles remains an open question. However, it cannot be ignored that ice was observed preferentially at steps, and so this multimodal analytical study shows that step-like features may play a role in ice formation via initial water condensation on feldspar surfaces. More research is needed to establish how such findings translate to particles that are smaller than those identified features and if ice nucleation commences for lower S_{water}.

11.2.2. In Situ Microanalysis of Ice Nucleation

Examination of ice nucleation by in situ methods involves experimental setups that couple temperature and humidity control of the particle and substrate sample with the desired microanalysis technique. Raman, SEM, and STXM have been employed to examine ice nucleation in situ. Similar to the multimodal approach discussed in the previous section, the supporting substrate on which particles are deposited onto needs to satisfy the instrument's requirements and show lower water uptake and ice nucleation efficacy compared to the sample that is being examined.

Raman Spectroscopy to Study Ice Nucleation

Raman spectroscopy is often used along with OM, making it an excellent choice to monitor phase transitions of particles, including ice nucleation and chemical composition of aerosol particles and INPs. Raman spectroscopy to study particle phase transition is currently limited to particle diameters of about 1 μm due to typically applied excitation wavelengths. Although surface-enhanced and tip-enhanced Raman spectroscopy (Bailo & Deckert, 2008; Campion & Kambhampati, 1998; Cialla et al., 2012; Zrimsek et al., 2017) can reach resolutions in the tens of nanometers, these techniques have not yet been applied to study ice nucleation in situ.

A custom-built cold stage applying confocal Raman spectroscopy was designed to detect the crystallization of aqueous HNO_3 solution droplets forming nitric acid dihydrate and nitric acid trihydrate representing a type of polar stratospheric clouds (Knopf et al., 2002). The same Raman-OM setup was subsequently applied to examine changes in composition and ice formation in aqueous $(NH_4)_2SO_4$ and H_2SO_4 droplets (Knopf et al., 2003). When applying Raman spectroscopy, care has to be taken that the laser light does not significantly warm the particle of interest. This could result in composition changes and erroneous freezing temperatures. For example, Knopf et al. (2003) monitored potential changes in the recorded vibration bands when exposing the droplets to different excitation time periods and using different laser powers. Additionally, they measured the melting point of the aqueous droplets to infer potential heating due to laser radiation. For aqueous systems, these issues may be minor, but when investigating absorbing species such as soot or certain mineral dust species, the sample or particle can significantly heat up and alter the thermodynamic conditions for nucleation (i.e., S_{ice}).

The study of atmospheric ice nucleation has involved the coupling of ice nucleation cells or environmental cells to Raman-OM, where particles or substrate can be temperature-controlled and exposed to desired levels of humidity and, in some cases, reactive gases such as ozone (Baustian et al., 2010; Baustian et al., 2012; Baustian et al., 2013; Knopf & Koop, 2006; Mael et al., 2019; Roy et al., 2021; Schill & Tolbert, 2013; Wang & Knopf, 2011). Figure 11.11a provides an example of a setup that allows exposure of deposited particles to different humidity levels and reactive gases while controlling the particle temperature (Mael et al., 2019). A humidified flow is typically generated by mixing a dry flow of an inert gas such as N_2 with a saturated N_2 gas flow. A hygrometer at the exit of the environmental cell's gas flow is applied to measure the dew point and as such the water vapor amount present in the cell (Dymarska et al., 2006; Mael et al., 2019; Wang & Knopf, 2011). As indicated in Figure 11.11b, OM allows visual inspection of the droplet or ice crystal in addition to in situ compositional examination by Raman spectroscopy.

Recording of Raman maps allows identification of the INP and its surface features that initiated heterogeneous ice nucleation (Schill & Tolbert, 2013, 2014). For example, Schill & Tolbert (2013) demonstrated that ice formed from the supersaturated gas phase on a spherical and glassy organic droplet. Another example is given in Figure 11.12, where synthetic sea-salt particles were exposed to conditions $S_{ice} > 1$ (Schill & Tolbert, 2014). Water uptake and freezing can be clearly identified in the visual recording and the associated Raman spectra (Figure 11.12a). This further corroborates that ice nucleation commenced in the immersion freezing mode. The corresponding line scan of the ice crystal (Figure 11.12b) corroborates that the immersion-freezing INP resides in the center of the ice crystal. Only the center of the line scan shows hydrate and sulfate peaks representative of the INP.

ESEM in situ Ice Nucleation Experiments

ESEM allows particles and uncoated biological and other materials to be examined in a high-pressure (up to 60 hPa) chamber with an electron beam. The gas introduced into the chamber can be water vapor, thereby opening the door for examination of environmental species and simulation of cloud-forming conditions. Reviews on the application of ESEM in material sciences and for studying the ice phase are provided by Zhang et al. (2020) and Pach & Verdaguer (2022), and a recent study applied ESEM to examine ice formation by proteins (Melillo et al., 2022).

Zimmermann et al. (2007) introduced the first application of ESEM to study heterogeneous ice nucleation. They examined silver iodide and two clay minerals for their propensity to induce condensation freezing and deposition ice nucleation. The pressure in the chamber was equal to the water's partial pressure. The lowest particle temperature examined was 250 K due to the limited regulation of the water partial pressure by the ESEM. Cooling of the substrates was achieved by means of a Peltier element. Calibration of T and RH was performed by determining the deliquescence RH of various salts at different temperatures. This pioneering study reproduced previous ice nucleation literature data. The authors noted that in some cases, the activation of particles to form ice was not reproducible—that is, in each experimental run using the same particle sample, different particles induced ice nucleation. Also, the activated fraction of particles could not be precisely determined because only a small part of the supporting substrate was imaged at the magnifications applied.

These ESEM ice nucleation studies (Zimmermann et al., 2007; Zimmermann et al., 2008) initiated further development of ESEM as a tool to examine the phase transition and ice formation by atmospheric particles. Wang et al. (2016) developed an ice nucleation (IN)-ESEM platform, shown in Figure 11.13, that allows simulation of cloud formation for temperatures as low as 190 K. With this setup, cirrus cloud conditions and polar stratospheric cloud formation can be investigated. Cooling of the sample is achieved using a cryo-cooling stage operated with liquid nitrogen (Figure 11.13a). The entire cryo-cooling stage within the ESEM chamber is insulated, and supply lines are evacuated to ensure that no water vapor condenses on the cryo-cooling stage and within the ESEM chamber except for the sample area (Figure 11.13b). The coldest spot is only the 0.2 mm^2 sample area. Avoiding water condensation outside the field of view ensures that no water vapor pressure gradients and thus erroneous RH fields could establish. The amount of water partial pressure in the ESEM chamber is independently controlled by a gas flow system that allows the mixing of a dry and humidified gas flow of N_2. The dew point of this gas mixture is continuously monitored by a chilled mirror hygrometer. Temperature and humidity were calibrated by measuring the melting points of organic compounds and the sublimation and growth of ice, respectively. Figure 11.13c shows exemplary deposition ice nucleation by a kaolinite clay mineral particle. Ice forms preferentially on the edges of the stacked kaolinite platelets and not on their basal planes. The authors suggest that the OH-terminated edge surfaces of kaolinite platelets are potential ice nucleation sites (Wang et al., 2016). Thus, the chemical differences between the edge and the basal plane surfaces might have contributed to their different propensities to nucleate ice.

In Figure 11.14, the ability of IN-ESEM to determine the thermodynamic conditions for ice nucleation initiated by kaolinite clay mineral via immersion freezing and deposition ice nucleation is demonstrated (Wang et al., 2016). The measurements are in agreement with previous OM-based measurements (Knopf et al., 2010; Wang & Knopf, 2011). Because the entire sample area

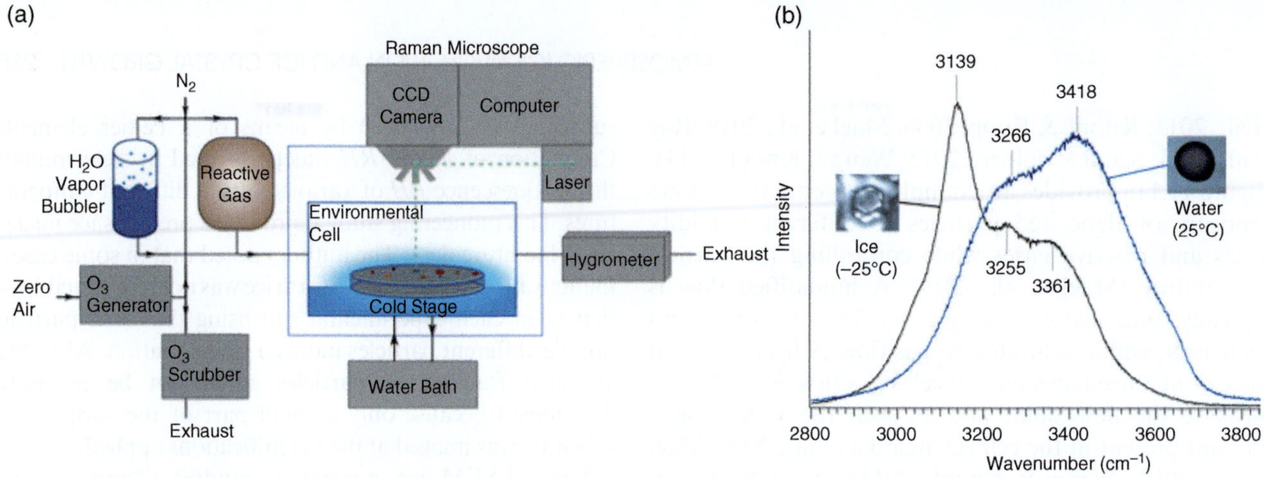

Figure 11.11 Raman microscope application to study ice nucleation. (a) An environmental cell allowing for control of the temperature of deposited aerosol particles (cold stage) and gas humidity is coupled to the Raman microscope. Humidity within the environmental cell is adjusted by mixing a dry and saturated flow of nitrogen. In addition, reactive gases can be introduced into the cell. (b) Raman spectra and corresponding optical images of ice and water at −25 and 25 °C, respectively. Mael et al. (2019)/with permission of the American Chemical Society.

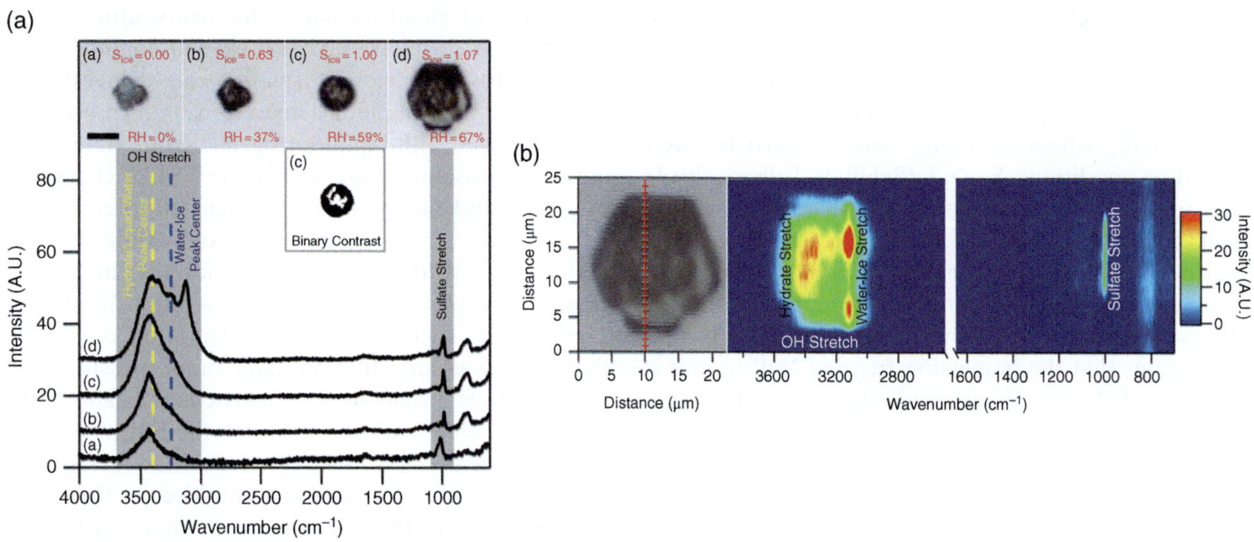

Figure 11.12 Immersion freezing from a synthetic sea-salt (SSS) particle as humidity is increased. (a) Images of an SSS particle ranging from dry conditions until ice nucleation occurred. The size bar represents 10 μm. The corresponding Raman spectra are shown below the optical images. (b) Detailed Raman spectra line maps of the frozen particle shown in (a), recorded along the red line. The SSS particle serves as immersion freezing INP, being at the center of the frozen particle as indicated by the hydrate and sulfate stretches being only present in the center region. Schill & Tolbert (2014)/with permission of the American Chemical Society.

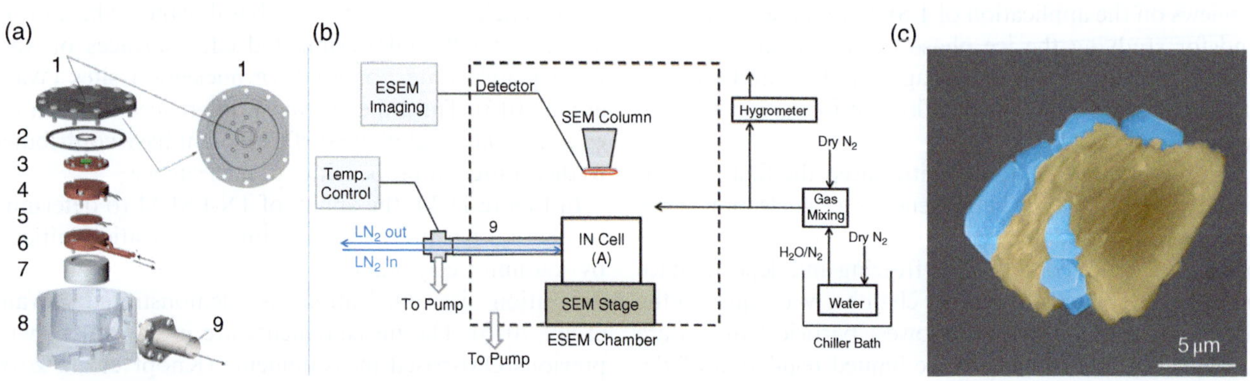

Figure 11.13 Ice nucleation cell implementation in an environmental scanning electron microscope (IN-ESEM). (a) Detailed schematic of the ice nucleation cell. (b) Overall experimental layout of the IN-ESEM system, including the IN cell within the ESEM chamber and a water vapor source (shown to the right) to control the chamber humidity. (c) Observed deposition ice nucleation by kaolinite clay mineral where ice is depicted in blue and the mineral in brown colors. Adapted from Wang et al. (2016).

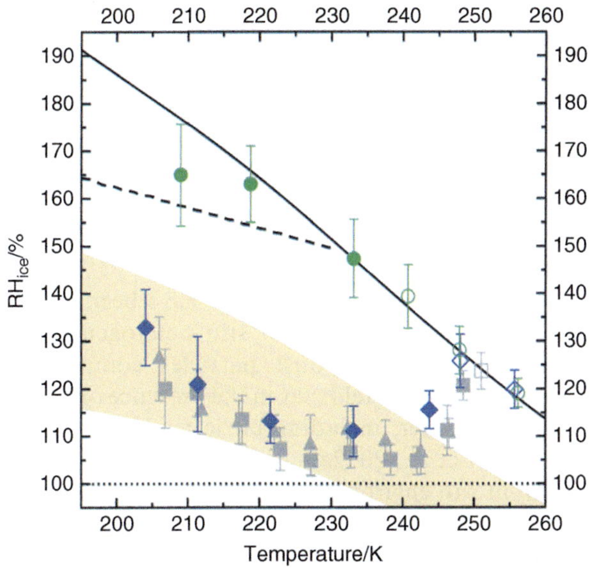

Figure 11.14 Deposition ice nucleation and immersion freezing by kaolinite clay minerals using the ice nucleation cell coupled to environmental scanning electron microscope setup (IN-ESEM) (solid and open diamonds, respectively) are given as a function of particle temperature and relative humidity with respect to ice (RH_{ice}). Observed ice formation and water condensation on blank substrates are given as solid and open circles, respectively. Gray symbols display previous literature values derived using optical microscopy (Knopf et al., 2010; Wang & Knopf, 2011). Conditions for pore condensation freezing by pores 7.5–15 nm in size are depicted by the shaded brown area (Marcolli, 2014). Solid, dotted, and dashed lines indicate the water saturation (RH = 100%), ice saturation, and homogeneous ice nucleation limit (Koop et al., 2000), respectively. Wang et al. (2016)/with permission of the Royal Society of Chemistry.

can be monitored, this IN-ESEM setup allows for INP identification and tracking when repeating ice nucleation experiments. Hence, it allows for a multimodal analytical approach. In addition to the IN-ESEM experiments, Wang et al. (2016) chemically imaged the identified INPs using STXM/NEXAFS and applied helium ion microscopy to infer the role of pores in deposition ice nucleation (shaded area in Figure 11.14). Although this study clearly identified kaolinite platelets as the location of ice nucleation, thereby advancing our understanding of which surface features enable ice nucleation, in situ observation of an INAS was not possible.

In an ice-nucleating substrate-based study, Kiselev et al. (2017) equipped an ESEM with a custom-built cold-stage and water vapor supply system to study ice nucleation by potassium-rich feldspars. The cold stage is a double-stage Peltier element coupled to the cold-water chiller, allowing temperatures as low as 213 K to be reached. The cold sample area within the ESEM chamber is 5 × 5 mm^2 in size, and the field of view in these experiments is a maximum of 700 μm in diameter. Kiselev et al. (2017) did not determine the ice nucleation onset S_{ice} from the pressure or dew point measurements but rather from measurements of the ice crystal growth rates. As shown in Figure 11.15a, b, ice nucleates in the vicinity of defect sites such as steps, cracks, and inside cavities on the (001) feldspar crystal face. In Figure 11.15c, an ice crystal forms on the (100) face of feldspar. This study shows that ice forms preferentially on high-energy (100) surface planes of feldspar. The location of ice crystals in Figure 11.15a, b can be attributed to (100) faces being exposed in cracks and cavities. Accompanying molecular-scale computer simulations corroborate that the stable

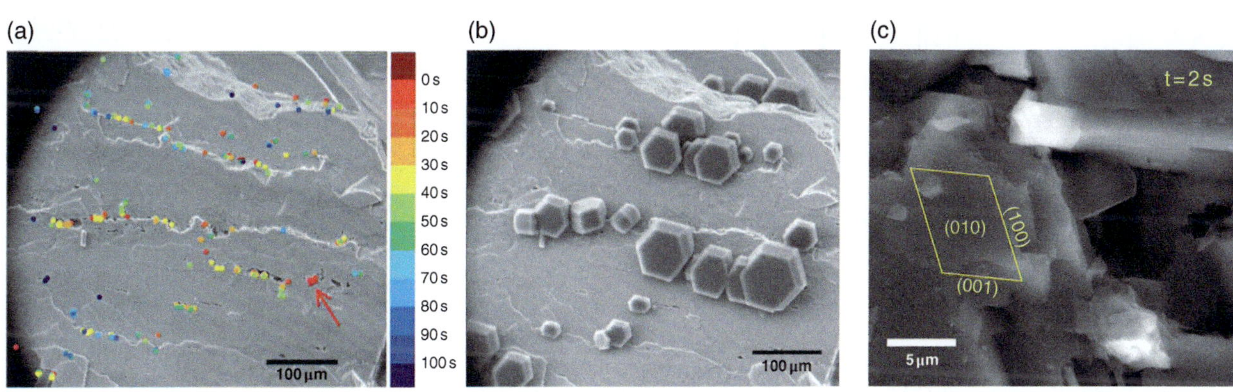

Figure 11.15 Application of an environmental scanning electron microscope (ESEM) equipped with a cold stage to study the heterogeneous ice nucleation on (001) face of feldspar weathered in carbonated water. (a) Identification of ice nucleation sites from several nucleation-evaporation cycles, plotted over the image of the bare feldspar face. The color code gives the time of nucleation (in seconds) with respect to the first detected ice crystal. The red arrow shows the location of the initial ice formation event that repeatedly occurs in all cycles. (b) Snapshot of the ice crystals nucleated at 233 K in the first nucleation-evaporation cycle. (c) Nucleation of an ice crystal (white arrow) on the inner wall of a cavity in the (010) face at 251 K. Kiselev et al. (2017)/with permission of the American Association for the Advancement of Science.

face of ice on the feldspar (100) surface is the primary prismatic plane (Kiselev et al., 2017). The (100) face of feldspar is hydroxylated, and thus, similar to the case of ice nucleation on hydroxylated side planes of kaolinite, surface OH groups seem to affect the ice nucleation propensity. This study clearly demonstrates the importance of the (100) face of feldspar for understanding its ability to initiate heterogeneous ice nucleation. However, similar to other SEM-based ice nucleation studies, the resolution is still too low to unambiguously pinpoint the exact location where the critical ice nucleus interacts with the mineral surface. Some of the identified features, such as steps in cleaved surfaces, are larger than typical atmospheric particles. Hence, how many patches of the (100) face will be present on those much smaller feldspar-containing particles? Would nanometer-size resolution discover yet other features that explain the strong ice nucleation propensity of feldspar? How do the OH groups interact with the adsorbing water and impact water structuring? Of course, these questions can be posed for any other application of microanalysis techniques to ice nucleation. It demonstrates that the application of SEM significantly advances our understanding of the fundamental interactions between water and surfaces, and as we achieve a closer look at ice nucleation, more challenging and exciting research questions arise.

STXM in situ Ice Nucleation Experiments

Soft X-rays by means of STXM have been successfully employed to observe in situ heterogeneous ice nucleation by mineral dust particles composed of either ferrihydrite or feldspar in the presence or absence of organic matter, including citric acid and xanthan gum (Alpert et al., 2022a). Alpert et al. (2022a) developed an in situ environmental ice cell or ice nucleation X-ray cell (INXCell) that can be mounted into an STXM instrument. The integration of INXCell within the STXM chamber is shown in Figure 11.16. Key

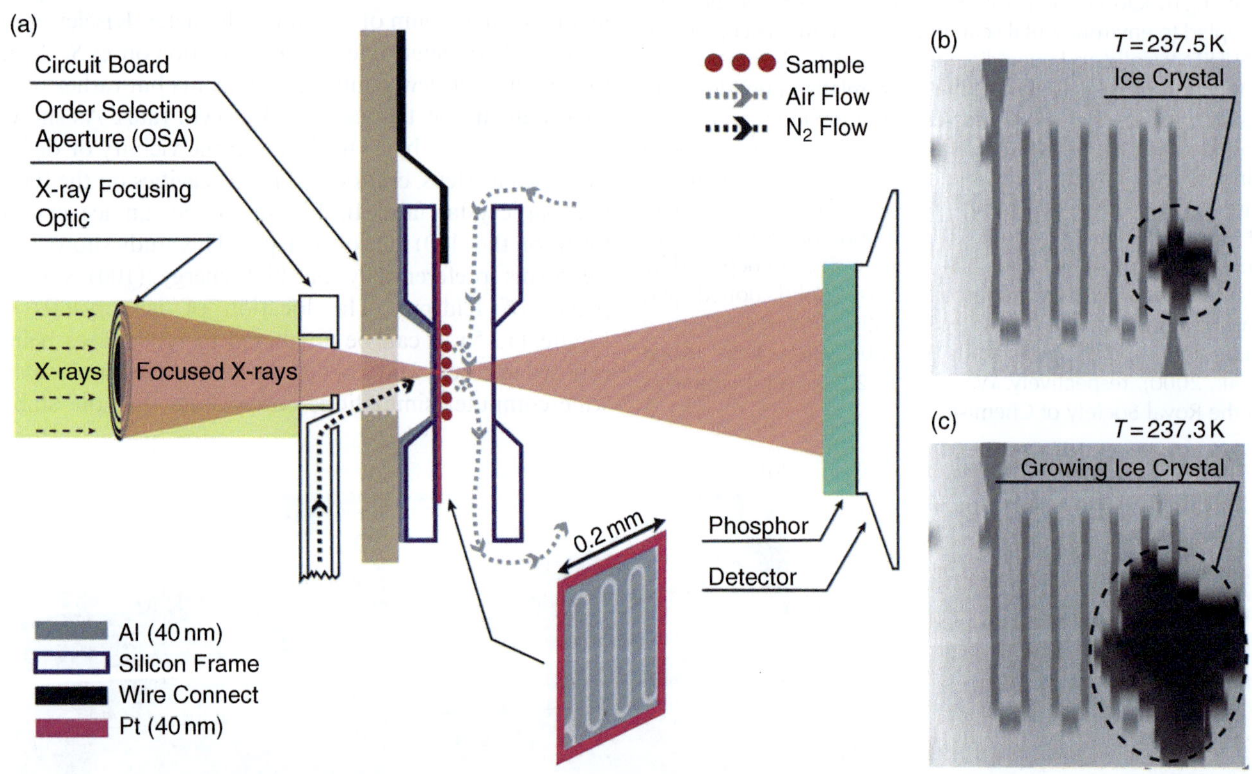

Figure 11.16 (a) Schematic configuration of the ice nucleation X-ray cell (INXCell) within the scanning transmission X-ray microscopy (STXM) chamber. The order selecting aperture (OSA) is temperature-controlled and cools a gas jet that impinges on the back of the sample holder. Particles inside the INXCell (circles) are exposed to humidified air and cooled by the impinging gas jet. The INXCell is mounted on a circuit board that is connected to the lithographically fabricated temperature sensor on the sample substrate with a thickness of 40 nm. An optical microscope image of this sensor is displayed. Transmitted X-rays were detected using a phosphor screen coupled with a photomultiplier tube (PMT). A 40 nm aluminum layer was condensed onto the reverse side of the sample membrane to increase the lateral thermal conductivity. (b, c) Coarse-mode resolution X-ray imaging (the sample area is about 200 µm across) to track ice crystal growth. Vertical bars indicate temperature sensor within the membrane. Alpert et al. (2022a)/Royal Society of Chemistry/CC BY 3.0.

features of this design are the patterned temperature sensor on the surface of the silicon nitride membrane and a temperature-controlled ordering selecting aperture (OSA), which cools the particle sample area with a jet of cooled N_2. The temperature sensor is located at exactly the same location as the deposited particles. To ensure that only the particle sample area is the coldest spot with uniform temperature in this setup, the INXCell and the particle sample are cooled separately. By guiding a cooled N_2 jet toward the particle sample area that is about 1–2 K colder than the temperature-controlled INXCell, the particles on the membrane inside, and not the silicon frame, are guaranteed to experience the coldest temperature. The INXCell has two membranes that seal the cell against the STXM vacuum chamber that is maintained at ~10^{-3} hPa, whereas inside the humidified environment of the cell, the pressure is maintained at 150 hPa. Calibration of S_{ice} was performed similarly to methods applied in OM studies (Alpert et al., 2011; Wang & Knopf, 2011) where the growth and sublimation of ice crystals are monitored to determine S_{ice} = 1. In this instance, the dew point in the INXCell equals the particle temperature.

Figure 11.16b, c displays the emergence of an ice crystal that has formed from droplets containing ferrihydrite particles serving as INPs (Alpert et al., 2022a). At constant water partial pressure, cooling the sample by 0.2 K yields rapid ice crystal growth. For initial monitoring of ice formation, a coarser resolution might be used.

The ability to visualize the ice crystal and to conduct NEXAFS on the ICRs and non-ice-nucleating particles is depicted in Figure 11.17 (Alpert et al., 2022a). Ferrihydrite particles coated with citric acid served as INPs. The shrinking ice crystal was observed using a coarse resolution and carbon pre-edge energy of 280 eV to be sensitive to the presence of inorganic matter (Figure 11.17a,b). Subsequently, a higher spatially resolved X-ray image is acquired at 288.6 eV, typical of carboxyl functional groups (Figure 11.17d). Figure 11.17e, f show NEXAFS spectra at the carbon K-edge and iron L2,3-edges, respectively, of the ICR particle and a particle that was not involved in ice formation—that is, outside the marked area of the sublimating ice crystal (Figure 11.17b–d). Both NEXAFS spectra show nearly identical spectral features, indicating the presence of ferrihydrite (Figure 11.17e)

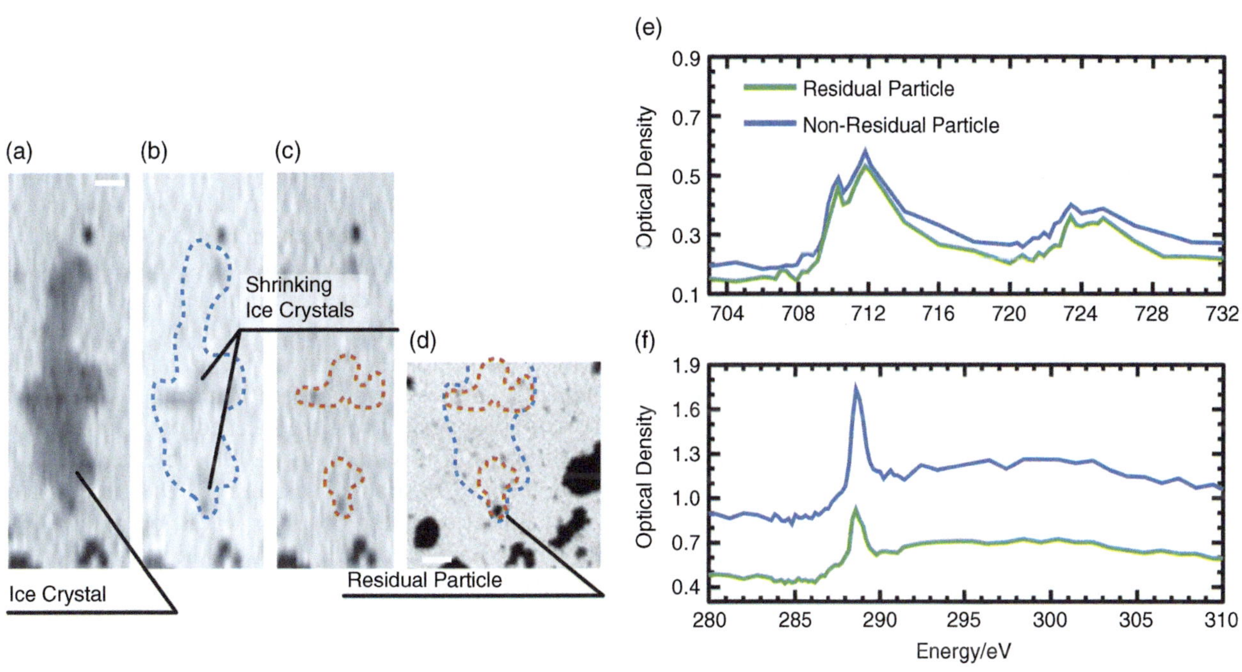

Figure 11.17 Observation of ice nucleation and spectroscopic identification of ferrihydrite particles coated with citric acid using scanning transmission X-ray microscopy with near-edge X-ray absorption fine structure spectroscopy (STXM/NEXAFS). (a–c) A sequence of X-ray images at 280 eV displaying a sublimating ice crystal. The blue and orange outlines represent the ice crystal boundaries. (d) An X-ray image at 288.6 eV showing the ice crystal boundaries, residual particles after sublimation, and organic-rich particles across the sample. The scale bar shown in (d) is 2 μm and valid for all images. NEXAFS spectra for an ice crystal residual particle and a non-ice-nucleating particle recorded at the iron L2,3-edges shown in (e) and carbon K-edge shown in (f). Alpert et al. (2022a)/Royal Society of Chemistry/CC BY 3.0.

and citric acid with the carboxyl group at 288.6 eV (Figure 11.17f). Given the resolution of ~35 nm in these experiments, no significant difference between the INP and non-INP is evident, a similar finding as in previous microanalytical studies of INPs (Alpert et al., 2022b; China et al., 2017; Knopf et al., 2014; Knopf et al., 2022; Lata et al., 2021). Alpert et al. (2022a) suggest that ice nucleation proceeded randomly with larger particles having a greater chance to serve as INPs than smaller particles.

These initial results using STXM to examine ice nucleation are an important step forward in advancing the fundamental understanding of ice nucleation. The resolution of STXM is typically in the range of 20–50 nm, meaning that it may just reach the upper limit of the surface area that determines the site of ice nucleation (Figure 11.3). STXM will likely allow for higher-resolution studies compared to SEM due to a lesser beam radiation effect on samples and water vapor. As in the case of OM- and SEM-based ice nucleation studies, a field of view to observe the entire cold sample is necessary. Consequently, instrument resolution needs to be lower than what could actually be achieved, making sample configuration and characterization critical for high-resolution chemical imaging of ice nucleation.

These examples demonstrate how microanalysis techniques have advanced our understanding of heterogeneous ice nucleation. Although the necessary resolution for monitoring the critical ice nucleus has not yet been achieved, the advancement of these microanalysis techniques has allowed us to identify physicochemical surface features and particle properties that contribute to the ice nucleation ability of atmospheric particles. Ice nucleation by kaolinite and feldspar minerals depends on cracks, cavities, specific crystal lattice faces, and hydroxylated surfaces. Organic matter, rich in carboxyl functional groups, present in laboratory-generated and ambient atmospheric particles, is involved in ice formation. INPs identified in ambient particle samples do not show unique physicochemical particle features compared to the majority of the particles in the sample. The same observation holds for laboratory-generated samples of inorganic/organic particles that serve as INPs. Although the study of INPs by microanalysis techniques is still in its infancy, the current results indicate that the search for the INP in a particle population is not one of finding the needle in the haystack but rather determining whether it is a larger or smaller piece of hay in the haystack. Clearly it is too early for a conclusion on this matter, thereby opening up exciting possibilities for new science in developing and applying microanalysis techniques to understanding atmospheric ice nucleation.

11.3. APPLICATION OF MICROANALYSIS TECHNIQUES TO STUDY ICE CRYSTAL GROWTH AND SUBLIMATION

ESEM has allowed the detailed study of ice crystal shapes under growth and sublimation conditions and when ice crystals contact each other (Cascajo-Castresana et al., 2021; Chen et al., 2017; Magee et al., 2014; Magee et al., 2021; Nair et al., 2018; Pach & Verdaguer, 2022; Pedersen et al., 2011; Pfalzgraff et al., 2010).

Variable-pressure SEM obtained high-resolution images of growing and sublimating hexagonal ice surfaces presented by Pfalzgraff et al. (2010) show "growth strands" in the center of the prismatic face, which get smoother toward its edges. During sublimation, however, "ablation strands" form on the prismatic face but close to its edge where there were deeper grooves. This study points out that the formation of ablation strands during ice crystal sublimation could result in greater reflectivity of cirrus clouds.

Using ESEM, Nair et al. (2018) examined sublimating ice crystals in detail. Under subsaturated conditions, the general sublimation behavior involved prismatic etch pits arising during the early stages, followed by ridge formation and the hollowing of basal planes. Figure 11.18 presents a suggested mechanism of the preferential crystallographic sublimation of an ice crystal that involves three key steps: initial random desorption, kink formation along the prismatic planes, and, finally, the subsequent ridge formation due to the coalescence of these kinks (Nair et al., 2018). Furthermore, the authors observed that the ridge length (λ_0 in Figure 11.18) showed a power-law relationship with the pressure deviation from saturation. For less subsaturated conditions, the ridge spacing was larger than for greater subsaturated conditions.

Pedersen et al. (2011) observed that when ice crystals grow in contact, they form an ice grain boundary that induces a surface transition on the facets in contact and does not propagate across the facet edges. A similar study by Cascajo-Castresana et al. (2021) observed the same phenomenon, as shown in Figure 11.19. These dislocations, appearing like grooves or waves, occur far (10 to 100 μm) from the contact point of the two growing ice crystals. Pedersen et al. (2011) interprets these observations as an "avalanche of dislocations" propagating away from the original dislocation. This viewpoint is supported by the theory of dislocation-mediated melting, which has suggested these continuous melting transitions (Dash, 1999). These deformations of the crystal facet likely affect the crystal's optical properties and the nature and size of the surface, thereby influencing heterogeneous (gas-to-surface) processes.

Ice crystal shapes under typical atmospheric growth and sublimation conditions were studied by Magee

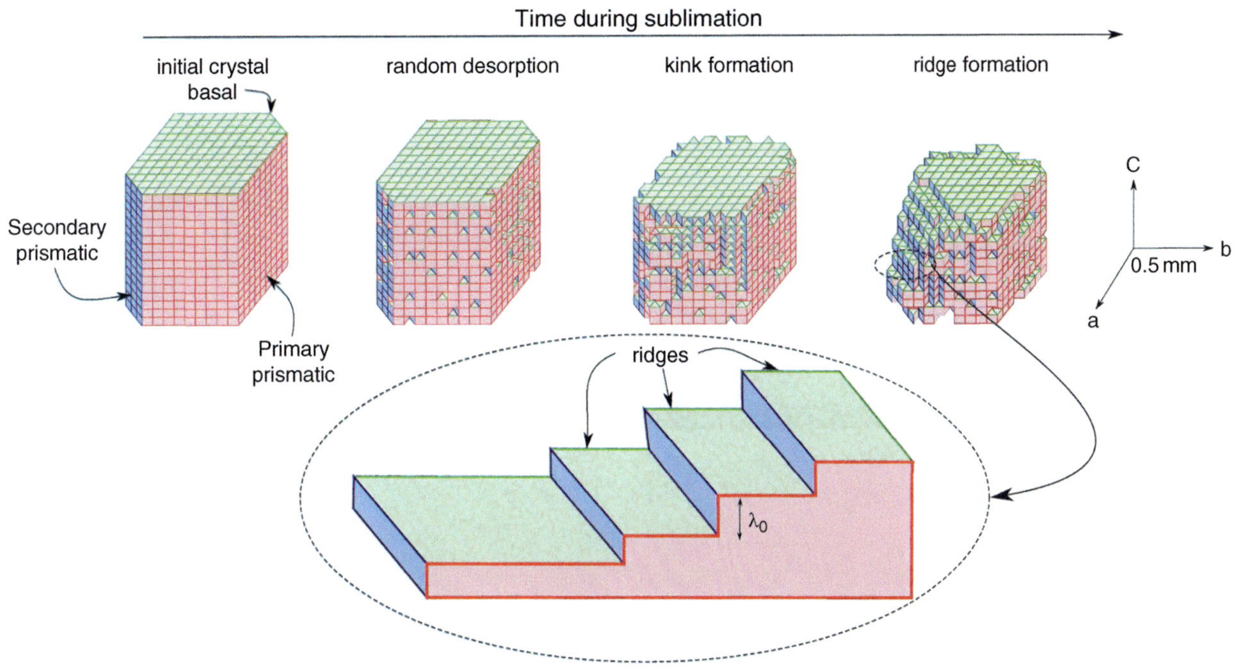

Figure 11.18 Conceptual mechanism to explain the preferential crystallographic sublimation on an ice crystal. This involves initial random removal of material, followed by sublimation of the basal planes, resulting in the hollowing of etch pits parallel to the c axis. Kinks form because the sublimation along the secondary prismatic planes is thermodynamically less favorable. The kinks adjoining the basal and prismatic planes form as adatom mobility is impeded by the prismatic plane. Ultimately, these kinks coalesce to form ridges perpendicular to the basal plane with ridge length λ_0. The basal plane remains mesoscopically smooth. Nair et al. (2018)/American Physical Society/CC BY 4.0.

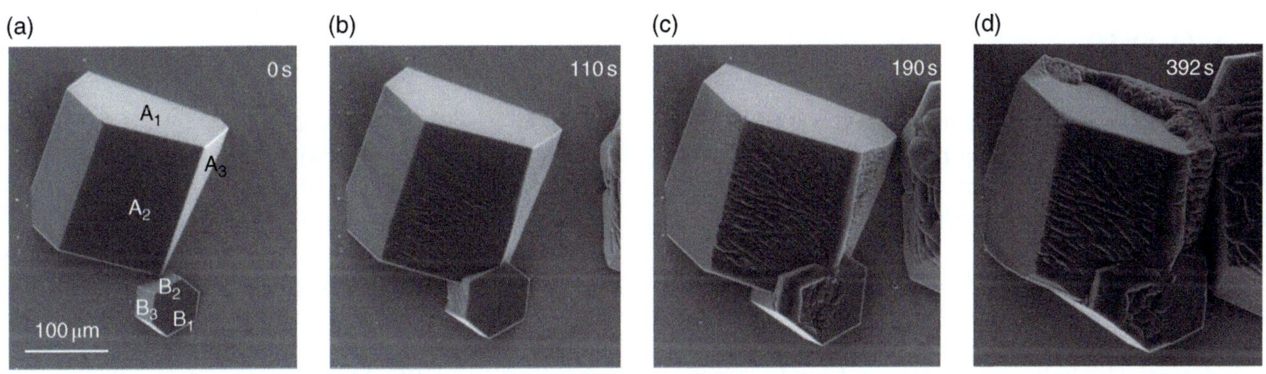

Figure 11.19 Ice crystal morphology changes upon merging of isolated crystals (crystals a and b) during growth. Upon the ice crystals making contact, the crystal planes show polycrystalline topography with a large density of grooves. Cascajo-Castresana et al. (2021)/Copernicus Publications/CC BY 4.0.

et al. (2014) due to their relevance for ice crystal optical properties. ESEM allowed them to study in detail the mesoscopic roughness of ice crystals. Figure 11.20 shows the surface roughness of ice crystals under growing and sublimating conditions (Magee et al., 2014). Figure 11.20a displays distinct surface roughness on the basal facets and prism facets of the growing crystal. Microfaceting on basal and prism facets was also observed, as shown in Figure 11.20b, although it was only observed during steady growth at temperatures below −35 °C. Ice crystal facets display very different surface roughness upon sublimation, as is evident in Figure 11.20c, d. In this case, concave and scalloped depressions develop away from the original

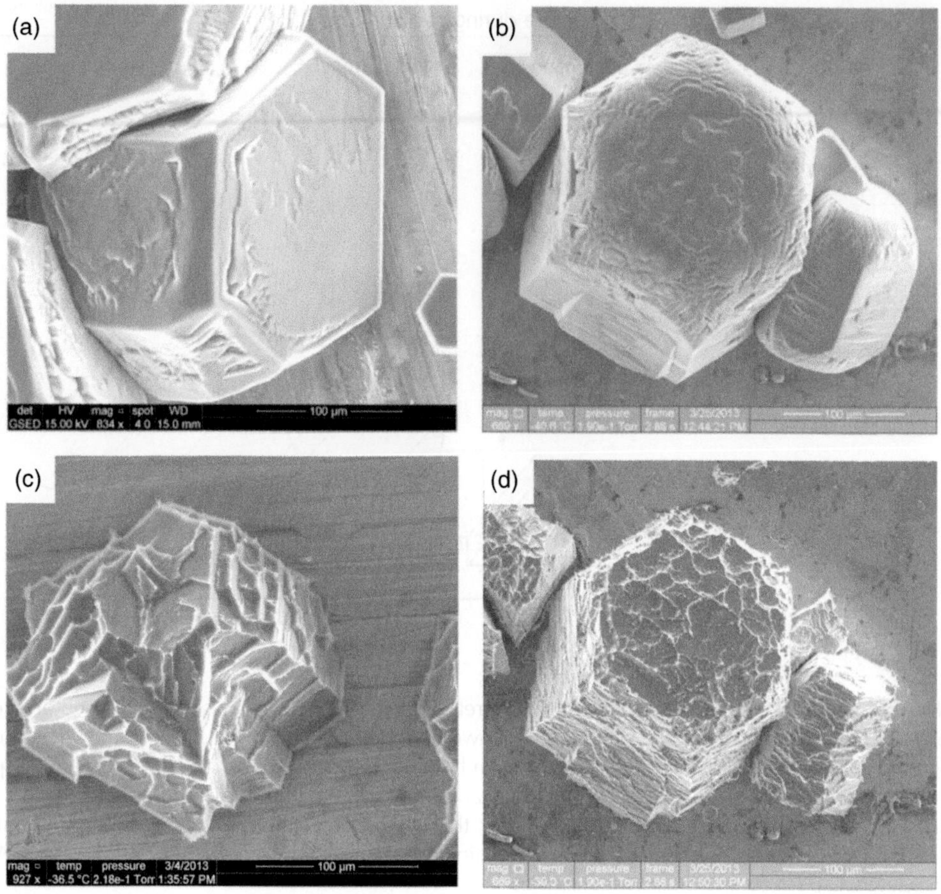

Figure 11.20 Environmental scanning electron microscopy (ESEM) images of mesoscopic surface topography of growing ice crystals (a, b) and sublimating ice crystals (c, d). Magee et al. (2014)/Copernicus Publications/

surface. Scalloping is often initiated at the site of roughness produced during previous growth. The sublimation process also seemed to occur outward from multiple centers, often at the site of a former ridge or ledge.

In a follow-up study, Magee et al. (2021) captured ice crystals in a cirrus cloud for ESEM/EDS analysis during the Ice Cryo-Encapsulation by Balloon (ICE-Ball) experiment. The balloon-borne crystal catcher payload consists of cryo-vessels with sweep tubes that open to capture and store ice crystals in small, hermetically sealed cryo-encapsulation cells. Imaging of thousands of cirrus particles revealed diverse and unanticipated particle morphologies, ice crystal habit heterogeneity, and multiple scales of mesoscopic roughening.

Aerosol particles embedded in ice crystals were also observed (Magee et al., 2021), as displayed in SEM images of ice crystals captured in cirrus clouds (Figure 11.21). An ice crystal rosette is depicted in Figure 11.21a. Contrary to the ice crystal habit diagram shown in Figure 11.4, the collected ice crystals show much greater complexity with surface pits and mesoscopic surface roughening. Figure 11.21b shows various ice crystals to which aerosol particles, visible as small white dots, are attached. Finally, an iron-rich aerosol particle (Figure 11.21c) adhering in the shallow hollow of a trigonal single ice crystal was identified by means of EDS analysis.

Application of ESEM for ice crystal growth and sublimation studies clearly has generated detailed insights into the evolution of ice crystals and demonstrated that ice crystal habits on the microscale can be tremendously complex. Classical ice crystal habit diagrams, such as shown in Figure 11.4, obtained typically with OM, do not capture the intricate nature of ice crystal facet surfaces that can include mesoscopic roughening, surface pits, strands, and grooves. All these features can influence ice crystal's optical properties and physical interaction and chemical reactivity with gas-phase species.

Figure 11.21 Captured cirrus ice crystals for cryo-scanning electron microscopy (cryo-SEM) and energy-dispersive X-ray spectroscopy (EDS). (a) A rosette with mixed-aspect crystals and several geometric surface pits, and high mesoscopic roughening. (b) Ice crystals with embedded aerosol particles (white spots on crystal faces). (c) Shallow hollowed trigonal ice particle with iron-rich embedded aerosol. The EDS spectrum of the iron-rich aerosol is given on the right-hand side of (c). Magee et al. (2021)/Copernicus Publications/CC BY 4.0.

11.4. EXPERIMENTAL REQUIREMENTS AND CHALLENGES

Microanalysis techniques to study ice nucleation and ice crystal growth have only gained traction in the last 15 years. Thus, this research area is expected to continue to evolve. The recent successful quantitative observation of ice nucleation in an STXM instrument is a clear indication of the advancement of microanalyses in studying atmospheric ice nucleation and ice crystal formation. In this section, the sample requirements, radiation effects, and the role of standard INPs are briefly summarized.

11.4.1. Sample Preparation

When studying ice nucleation, sample-supporting substrates suitable for the various microanalysis techniques must not interfere with the ice nucleation process. In other words, the supporting substrate should be more hydrophobic and less ice nucleation active compared to the INPs or substrates being examined. This is specifically true when studying homogeneous ice nucleation, which requires substrates that remain inert with regard to ice formation for temperatures as low as ~235 K (Knopf, 2003; Knopf & Lopez, 2009; Knopf & Rigg, 2011). When studying colder homogeneous freezing temperatures, a nonsubstrate approach is most suitable (Amaya & Wyslouzil, 2018; Laksmono et al., 2015). Water droplets-oil emulsions have been employed to study ice nucleation, allowing the observation of a large number of micrometer-sized droplets separated by oil (Knopf et al., 2020; Reicher et al., 2018; Riechers et al., 2013; Shardt et al., 2022; Stan et al., 2009; Tarn et al., 2021). However, currently, this approach excludes the application of microanalysis techniques described here.

For Raman studies, hydrophobically coated glass or quartz slides are typically employed (Knopf, 2003; Knopf et al., 2002; Knopf et al., 2003). Silicon nitride-coated silicon wafer chips are suitable for studies of heterogeneous ice nucleation by deposited particles (Knopf et al., 2022; Lata et al., 2021; Wang et al., 2016). For STXM applications, silicon nitride-coated Si frames can be used for ice nucleation studies, although due to membrane thickness of only ~100 nm, sample handling is difficult, and ensuring temperature homogeneity across the membrane is challenging (Alpert et al., 2022a; Knopf et al., 2014).

11.4.2. Sample Size for Ice Nucleation Studies

The sample size affects the experimental procedure and results in various ways. The sensitivity of ice nucleation is determined by how many particles can be observed at the same time. For example, in a sample with 100,000 deposited particles, observing one ice nucleation event places the detection limit at 10^{-5}. This is a very low detection limit representing an inherent strength of the particle-on-substrate approach when studying ice nucleation. If freezing kinetics are to be determined, the greater the number of observed ice nucleation events, the smaller the uncertainty due to the stochastic nature of the nucleation process (Alpert & Knopf, 2016; Knopf et al., 2020; Knopf & Alpert, 2023; Koop, 2004; Koop et al., 1997; Pruppacher & Klett, 1997). This requirement necessitates the repetition of experiments to increase the number of observed ice nucleation events.

As discussed in the case of multimodal and in situ ice nucleation studies, typically the entire sample area is observed to determine which particle or at which location ice crystals formed initially. This allows for subsequent in-detail examination and relocation of the particle or surface feature that initiated the ice nucleation. However, if the sample area is large, necessitating a large field of view, instrument resolution is usually low, thereby eliminating the main capability of the instrument. For this reason, a compromise in the design of the sample stage and experimental procedure must be made to ensure that the entire temperature-controlled sample area can be visualized. The very first experiment should establish that ice indeed forms only on the sample stage on which the supporting substrate is mounted. As such, the design of the cold stage must be the coldest area in the experimental setup. If ice forms outside the sample area, the humidity field above the sample can be distorted. The same holds true when a specific sample area is investigated for ice formation at greater magnification but an ice crystal forms outside the field of view. In this case, the humidity above the visualized sample section may not be the one set for the observed conditions. In other words, as soon ice forms and the ice crystal grows, it depletes the water's partial pressure, resulting in pressure differences across the sample. At 1 atm and 253 K and a diffusion coefficient of water vapor of about 0.18 cm²/s, the steady-state concentration profile around a 10 μm particle establishes after 3.4×10^{-7} s (Seinfeld & Pandis, 1998). This concentration profile will then govern the flux of water molecules toward the sink, here, the low-pressure or water vapor-depleted region around the ice crystal. Hence, the adjustment of the concentration profile and subsequent transport of water molecules can be assumed to occur more rapidly than the detection of the resulting effects. Those include the evaporation of droplets or the necessity of a significant increase in humidity to initiate the next ice nucleation event. This effect is especially crucial for densely populated particle samples. For this reason, if the thermodynamic conditions under which ice nucleation proceeds are crucial measurement parameters, typically only the first ice nucleation event is considered in the analysis (Alpert et al., 2022a; Dymarska et al., 2006; Knopf et al., 2014; Wang et al., 2016; Wang & Knopf, 2011).

11.4.3. Beam-Induced Radiation Effects

The energy deposited onto the substrate or particle by the excitation source can impact the temperature of the particle and, in the worst-case scenario, result in compositional alterations. Significant warming of the sample can alter the sample's propensity to initiate ice nucleation. For example, at 253 K, a sample warming by 1 K can alter S_{ice} by 10%. Hence, careful corroboration of sample warming upon radiation exposure has to be conducted for different beam energies before examining ice nucleation processes. This will also depend on the absorbing nature of the sample being examined.

Laser light as an excitation source for Raman scattering can deposit energy and thus heat onto a particle or substrate sample. This depends on the absorbing nature of the examined material. The radiation effect causing sample warming can be assessed by measuring the melting points of organic liquids and ice for different laser power and exposure times (Knopf et al., 2003). Changes in spectral features, such as the surface area of a vibration band, for different laser exposures at constant temperature can also indicate potential heating and compositional changes of the sample.

When using ESEM to examine surface features serving as sites for ice nucleation and to determine nucleation kinetics, the electron beam can affect these measurements by its effect on water. The electron beam used in SEM application may induce the dissociation of water molecules when studying water-substrate interactions (Bjornehohn et al., 2016; Hodgson & Haq, 2009; Pach & Verdaguer, 2022; Tamtogl et al., 2021). Higher beam energies needed to achieve greater resolution will consequently yield more water dissociation, likely impacting the humidity field and the secondary electron emission and detection. Such radiation effects are the likely main reasons why resolution when studying ice nucleation can be limited. This phenomenon raises the question of whether an electron beam approach can be used to image the location of the critical ice nucleus at nano- and subnano-scales to resolve the physicochemical nature of an INAS.

AFM might provide an alternative method to study atmospheric ice nucleation and ice crystal growth in the future (Thürmer & Nie, 2013). This method does not suffer from beam radiation effects. Recent developments demonstrate the capability of AFM to monitor the growth of 2D ice layers (Ma et al., 2020) and multilayer ice (Peng et al., 2022). These studies apply non-contact AFM and qPlus-AFM that use functionalized tips that allow direct imaging of the chemical structure of molecules with single-bond resolution (Peng et al., 2022). Furthermore, Chasnitsky et al. (2023) developed a new AFM setup that allows imaging of ice crystal surfaces in supercooled water containing antifreeze agents at the sub-10 nm scale. Successful application of AFM to ice nucleation and growth will depend on cantilever tip design, operating mode, and low noise sensors (Peng et al., 2022).

11.4.4. Standard Ice-Nucleating Particles

INPs with defined ice nucleation properties and exactly determined surface area would greatly benefit ice nucleation research by allowing assessment of our predictive understanding of ice nucleation and evaluation of ice nucleation instrument performance. Two ice nucleation measurement intercomparison studies that included up to 17 different instruments have been conducted in recent years (DeMott et al., 2018; Hiranuma et al., 2015). Illite NX and potassium feldspar mineral dust, two natural soil dusts, and bacterial matter served as test INPs to assess the performance of the various instrumentation. In the case of illite INPs, differences in INAS density of about three orders of magnitude among the instruments were observed (Hiranuma et al., 2015). One of the many outcomes of the intercomparison study described by DeMott et al. (2018) was the requirement to know the accurate size of INPs being probed by the instruments. Clearly, even when employing commercially available particles derived from natural resources (like minerals) on a nanoscale resolution, the particles will likely not exactly resemble each other, leading to further uncertainties in the evaluation of ice nucleation instrumentation and our understanding of ice nucleation processes.

Artificially generated standard surfaces and particles with well-controlled nanoscale physicochemical properties will allow the systematic study of ice nucleation. Microanalysis techniques will play a crucial role in characterizing the chemical nature of such standard surfaces and in determining the surface area accurately. Ice-nucleating substrate surfaces where specific chemical or morphological features are located within nanoscale resolution will provide the means to evaluate the fundamental processes of ice nucleation (Knopf et al., 2020) and consequently improved prediction of ice crystal formation in cloud models (Knopf et al., 2023a). This approach has currently only been attempted in molecular dynamics simulations (Cabriolu & Li, 2015; Cox et al., 2015; Croteau et al., 2010; Lupi et al., 2014; Qiu et al., 2017; Sosso et al., 2016; Zielke et al., 2016). Design of artificial substrates will allow the region where ice nucleation commences to be confined, thereby decreasing the necessary field of view and increasing the observational resolution. This should facilitate the optimization of the microanalytical detection method.

Assuming the existence of a standard INP that has an accurately defined ice nucleation efficiency and exactly known surface area, what would immersion freezing results look like when employing these INPs? Applying a Monte Carlo method of 10^5 freezing simulations, Knopf et al. (2020) discuss the expected results of isothermal and constant cooling rate immersion freezing experiments when using INPs with exactly the same surface area. Figure 11.22 shows these sensitivity results for isothermal (where the temperature is held constant) and constant cooling rate (where the sample of droplets are cooled at a constant rate) experiments using droplets that contain one INP with the same surface area. As time passes or the droplet sample is cooled, more and more of the initial liquid droplets freeze, thereby changing the unfrozen (UNF) and frozen fraction (FF) of droplets in the sample. Because UNF and FF are exponential functions of J_{het}, a log-linear dependency of UNF with time is expected for isothermal freezing experiments, and a sigmoidal curve for FF is expected for constant cooling rate experiments (Figure 11.22a, b) (Knopf et al., 2020; Pruppacher & Klett, 1997). It is evident that a smaller

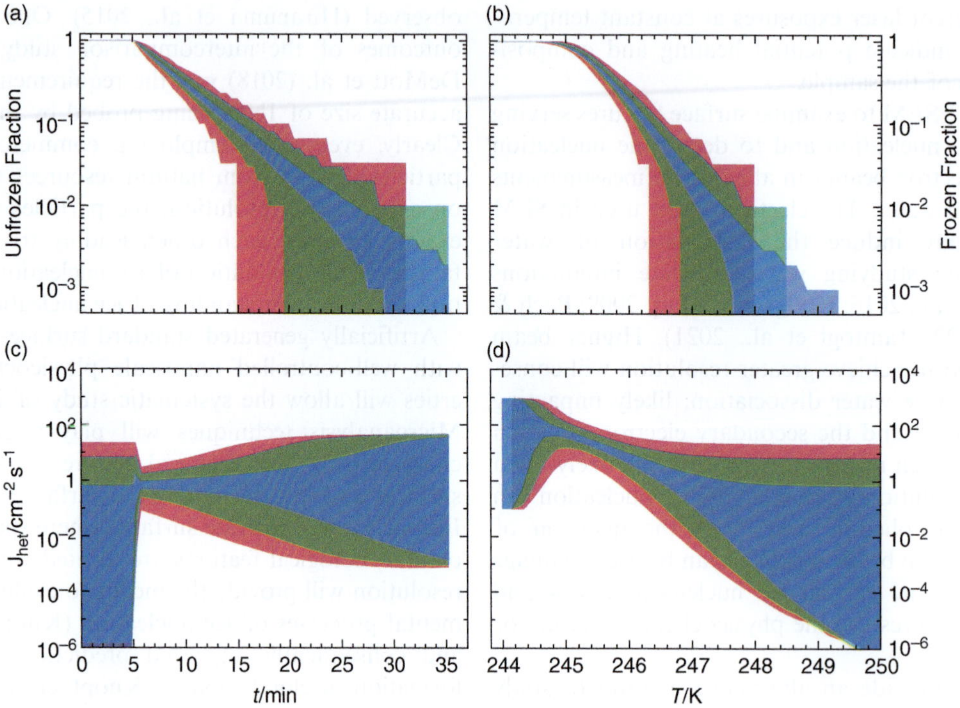

Figure 11.22 Application of standard INPs that have exactly the same surface area and ice nucleation efficiency in immersion freezing simulations using 30, 100, and 1,000 observed freezing events of droplets depicted as red, green, and blue colors, respectively. Isothermal and constant cooling rate freezing experiments are depicted in (a, c) and (b, d), respectively. Sensitivity of uncertainties in unfrozen fraction, frozen fraction, and heterogeneous ice nucleation rate coefficients, J_{het}, are displayed. The shading in unfrozen and frozen fraction curves in (a) and (b), respectively, represent 5th and 95th percentiles. The shadings in (c) and (d) represent the upper and lower fiducial limits of the derived J_{het} values. Adapted from Knopf et al. (2020).

number of observed ice nucleation events results in a larger uncertainty of UNF and FF. Hence, the stochastic nature of ice nucleation would still impose significant uncertainties—that is, data scatter—in the experimental results. Thus, even with a standard INP at hand, a sufficiently large number of freezing events have to be observed to minimize uncertainties. Figure 11.22c, d show the corresponding J_{het} values for isothermal and constant cooling rate experiments, respectively. The derived J_{het} values represent constant J_{het} values for isothermal freezing experiments and a log-linear dependency of J_{het} on temperature for the constant cooling rate experiments. The greater the number of observed freezing events, the smaller the uncertainty in J_{het}. Because statistically, most of the droplets freeze around the median of the UNF or FF, the corresponding uncertainties are lowest at that point. The few freezing events that occur at the shortest and longest times or warmest and coldest temperatures yield the greatest uncertainties. Hence, even when applying INPs with perfectly controlled and characterized surface area, some scatter in the freezing data can be expected if immersion freezing follows CNT. Hence, standard INPs will allow for careful evaluation of the application of CNT or INAS based parameterizations and determine their scalability to atmospheric processes.

ACKNOWLEDGMENTS

This study was supported by the Atmospheric System Research Program of the U.S. Department of Energy (DOE), Office of Science, Office of Biological and Environmental Research (OBER), Climate and Environmental Sciences Division (CESD). D.A.K. acknowledges support by the U.S. DOE grant DE-SC0021034. D.A.K. is indebted to Josephine Aller and Peter Alpert for proofreading this manuscript.

AVAILABILITY STATEMENT

All data and methods discussed in this book chapter have been published previously, as given in the references. All data and methods needed to draw the conclusions presented in this chapter are shown in this chapter.

REFERENCES

Alpert, P. A., & Knopf, D. A. (2016). Analysis of isothermal and cooling-rate-dependent immersion freezing by a unifying stochastic ice nucleation model. *Atmospheric Chemistry and Physics*, *16*(4), 2083–2107. doi: 10.5194/acp-16-2083-2016

Alpert, P. A., Aller, J. Y., & Knopf, D. A. (2011). Initiation of the ice phase by marine biogenic surfaces in supersaturated gas and supercooled aqueous phases. *Physical Chemistry Chemical Physics*, *13*(44), 19882–19894. doi: 10.1039/C1CP21844A

Alpert, P. A., Boucly, A., Yang, S., Yang, H., Kilchhofer, K., Luo, Z., et al. (2022a). Ice nucleation imaged with X-ray spectro-microscopy. *Environmental Science: Atmospheres*. doi: 10.1039/d1ea00077b

Alpert, P. A., Kilthau, W. P., O'Brien, R. E., Moffet, R. C., Gilles, M. K., Wang, B., et al. (2022b). Ice-nucleating agents in sea spray aerosol identified and quantified with a holistic multi-modal freezing model. *Science Advances*, *8*(44), eabq6842. doi: 10.1126/sciadv.abq6842

Amaya, A. J., & Wyslouzil, B. E. (2018). Ice nucleation rates near similar to 225 K. *The Journal of Chemical Physics*, *148*(8), 9. doi: 10.1063/1.5019362

Amaya, A. J., Pathak, H., Modak, V. P., Laksmono, H., Duane Loh, N., Sellberg, J. A., et al. (2017). How cubic can ice be?. *The journal of Physical Chemistry Letter*, *8*(14), 3216–3222. doi: 10.1021/acs.jpclett.7b01142

Ansmann, A., Tesche, M., Althausen, D., Müller, D., Freudenthaler, V., Heese B., et al. (2008). Influence of Saharan dust on cloud glaciation in southern Morocco during SAMUM, *Journal of Geophysical Research*, 112, D04210. doi: 10.1029/2007JD008785

Ansmann, A., Tesche, M., Seifert, P., Althausen, D., Engelmann, R., Fruntke, J., et al. (2009). Evolution of the ice phase in tropical altocumulus: SAMUM lidar observations over Cape Verde, *Journal of Geophysical Research*, 113. doi: 10.1029/2008JD011659

Bacon, N. J., & Swanson, B. D. (2000). Laboratory measurements of light scattering by single levitated ice crystals. *Journal of the Atmospheric Sciences*, *57*(13), 2094–2104.

Bailey, M., & Hallett, J. (2002). Nucleation effects on the habit of vapour grown ice crystals from −18 to −42 degrees C. *Quarterly Journal of the Royal Meteorological Society*, *128*(583), 1461–1483. doi: 10.1002/qj.200212858304

Bailey, M., & Hallett, J. (2004). Growth rates and habits of ice crystals between −20 degrees and −70 degrees C. *Journal of the Atmospheric Sciences*, *61*(5), 514–544.

Bailey, M. P., & Hallett, J. (2009). A comprehensive habit diagram for atmospheric ice crystals: Confirmation from the laboratory, AIRS II, and other Field studies. *Journal of the Atmospheric Sciences*, *66*(9), 2888–2899. doi: 10.1175/2009jas2883.1

Bailo, E., & Deckert, V. (2008). Tip-enhanced Raman scattering. *Chemical Society Reviews*, *37*(5), 921–930. doi: 10.1039/b705967c

Baker, M. B. (1997). Cloud microphysics and climate. *Science*, *276*(5315), 1072–1078.

Baker, A., & Peter, T. (2008). Small-scale cloud processes and climate. *Nature Geoscience*, *451*(17), 299–300.

Baustian, K. J., Wise, M. E., & Tolbert, M. A. (2010). Depositional ice nucleation on solid ammonium sulfate and glutaric acid particles. *Atmospheric Chemistry and Physics*, *10*, 2307–2317.

Baustian, K. J., Cziczo, D. J., Wise, M. E., Pratt, K. A., Kulkarni, G., Hallar, A. G., & Tolbert, M. A. (2012). Importance of aerosol composition, mixing state, and morphology for heterogeneous ice nucleation: A combined field and laboratory approach. *Journal of Geophysical Research*, *117*, D06217. doi: 10.1029/2011jd016784

Baustian, K. J., Wise, M. E., Jensen, E. J., Schill, G. P., Freedman, M. A., & Tolbert, M. A. (2013). State transformations and ice nucleation in amorphous (semi-)solid organic aerosol. *Atmospheric Chemistry and Physics*, *13*(11), 5615–5628. doi: 10.5194/acp-13-5615-2013

Bentley, W. A., & Humphreys, W. J. (1931). *Snow crystals*, 226 pp. New York, London: McGraw-Hill.

Berkemeier, T., Shiraiwa, M., Pöschl, U., & Koop, T. (2014). Competition between water uptake and ice nucleation by glassy organic aerosol particles. *Atmospheric Chemistry and Physics*, *14*(22), 12513–12531. doi: 10.5194/acp-14-12513-2014

Bertram, A. K., Koop, T., Molina, L. T., & Molina, M. J. (2000). Ice formation in (NH4)(2)SO4-H2O particles. *The Journal of Physical Chemistry. A*, *104*(3), 584–588.

Bjornehohn, E., et al. (2016). Water at interfaces. *Chemical Reviews*, *116*(13), 7698–7726. doi: 10.1021/acs.chemrev.6b00045

Bogdan, A. (2018). Ice clouds: Atmospheric ice nucleation concept versus the physical chemistry of freezing atmospheric drops. *The Journal of Physical Chemistry. A*, *122*(39), 7777–7781. doi: 10.1021/acs.jpca.8b07926

Bogdan, A., & Molina, M. J. (2017). Physical chemistry of the freezing process of atmospheric aqueous drops. *The Journal of Physical Chemistry. A*, *121*(16), 3109–3116. doi: 10.1021/acs.jpca.7b02571

Bogdan, A., Molina, M. J., & Tenhu, H. (2016). Freezing and glass transitions upon cooling and warming and ice/freeze-concentration-solution morphology of emulsified aqueous citric acid. *European Journal of Pharmaceutics and Biopharmaceutics*, *109*, 49–60. doi: 10.1016/j.ejpb.2016.09.012

Boucher, O., et al. (2013). Clouds and aerosols. In T. F. Stocker, D. Qin, G.-K. Plattner, M. Tignor, S. K. Allen, J. Boschung, et al. (Eds.), *Climate change 2013: The physical science basis. Contribution of working group I to the fifth assessment report of the intergovernmental panel on climate change*. Cambridge University Press, Cambridge, United Kingdom and New York, NY, USA.

Cabriolu, R., & Li, T. S. (2015). Ice nucleation on carbon surface supports the classical theory for heterogeneous nucleation. *Physical Review E*, *91*(5), 052402. doi: 10.1103/PhysRevE.91.052402

Campion, A., & Kambhampati, P. (1998). Surface-enhanced Raman scattering. *Chemical Society Reviews*, *27*(4), 241–250. doi: 10.1039/a827241z

Cascajo-Castresana, M., Morin, S., & Bittner, A. M. (2021). The ice-vapour interface during growth and sublimation. *Atmospheric Chemistry and Physics*, *21*(24), 18629–18640. doi: 10.5194/acp-21-18629-2021

Charnawskas, J. C., et al. (2017). Condensed-phase biogenic-anthropogenic interactions with implications for cold cloud formation. *Faraday Discussions*, *200*, 164–195. doi: 10.1039/C7FD00010C

Chasnitsky, M., Cohen, S. R., Rudich, Y., & Braslavsky, I. (2023). Atomic force microscopy imaging of ice crystal surfaces formed in aqueous solutions containing ice-binding proteins. *Journal of Crystal Growth*, *601*, 126961. doi: 10.1016/j.jcrysgro.2022.126961

Chen, Y. L., Kreidenweis, S. M., McInnes, L. M., Rogers, D. C., & DeMott, P. J. (1998). Single particle analyses of ice nucleating aerosols in the upper troposphere and lower stratosphere. *Geophysical Research Letters*, *25*(9), 1391–1394.

Chen, X., Shu, J. P., & Chen, Q. (2017). Abnormal gas-liquid-solid phase transition behaviour of water observed with in situ environmental SEM. *Scientific Reports*, *7*, 10. doi: 10.1038/srep46680

China, S., et al. (2017). Ice cloud formation potential by free tropospheric particles from long-range transport over the northern Atlantic Ocean. *Journal of Geophysical Research*, *122*(5), 3065–3079. doi: 10.1002/2016jd025817

Cialla, D., Marz, A., Bohme, R., Theil, F., Weber, K., Schmitt, M., & Popp, J. (2012). Surface-enhanced Raman spectroscopy (SERS): Progress and trends. *Analytical and Bioanalytical Chemistry*, *403*(1), 27–54. doi: 10.1007/s00216-011-5631-x

Coluzza, I., et al. (2017). Perspectives on the future of ice nucleation research: Research needs and unanswered questions identified from two international workshops. *Atmosphere*, *8*(8). doi: 10.3390/atmos8080138

Connolly, P. J., Möhler, O., Field, P. R., Saathoff, H., Burgess, R., Choularton, T., & Gallagher, M. (2009). Studies of heterogeneous freezing by three different desert dust samples. *Atmospheric Chemistry and Physics*, *9*(8), 2805–2824. doi: 10.5194/acp-9-2805-2009

Cox, S. J., Kathmann, S. M., Slater, B., & Michaelides, A. (2015). Molecular simulations of heterogeneous ice nucleation. I. Controlling ice nucleation through surface hydrophilicity. *The Journal of Chemical Physics*, *142*(18), 184704. doi: 10.1063/1.4919714

Croteau, T., Bertram, A. K., & Patey, G. N. (2010). Water adsorption on kaolinite surfaces containing trenches. *The Journal of Physical Chemistry. A*, *114*(5), 2171–2178. doi: 10.1021/jp910045u

Cziczo, D. J., Murphy, D. M., Hudson, P. K., & Thomson, D. S. (2004). Single particle measurements of the chemical composition of cirrus ice residue during CRYSTAL-FACE. *Journal of Geophysical Research*, *109*(D4), D04201. doi: 10.1029/2003JD004032

Cziczo, D. J., Froyd, K. D., Hoose, C., Jensen, E. J., Diao, M. H., Zondlo, M. A., et al. (2013). Clarifying the dominant sources and mechanisms of cirrus cloud formation. *Science*, *340*(6138), 1320–1324. doi: 10.1126/science.1234145

Cziczo, D. J., Ladino, L. A., Boose, Y., Kanji, Z. A., Kupiszewski, P., Lance, S., et al. (2017). Measurements of ice nucleating particles and ice residuals. In *Ice formation and evolution in clouds and precipitation: Measurement and modeling challenges* (pp. 8.1–8.13). American Meteorological Society. doi: 10.1175/AMSMONOGRAPHS-D-16-0008.1

Dash, J. G. (1999). History of the search for continuous melting. *Reviews of Modern Physics*, *71*(5), 1737–1743. doi: 10.1103/RevModPhys.71.1737

Dash, J. G., Rempel, A. W., & Wettlaufer, J. S. (2006). The physics of premelted ice and its geophysical consequences. *Reviews of Modern Physics*, *78*(3), 695–741. doi: 10.1103/RevModPhys.78.695

David, R. O., Marcolli, C., Fahrni, J., Qiu, Y., Perez Sirkin, Y. A., Molinero, V., et al. (2019). Pore condensation and freezing is responsible for ice formation below water saturation for porous particles. *PNAS*, *116*, 8184–8189. doi: 10.1073/pnas.1813647116

de Boer, G., Morrison, H., Shupe, M. D., & Hildner, R. (2011). Evidence of liquid dependent ice nucleation in high-latitude stratiform clouds from surface remote sensors. *Geophysical Research Letters*, *38*, L01803. doi: 10.1029/2010gl046016

DeMott, P. J. (2002). Laboratory studies of cirrus cloud processes. In D. K. Lynch, K. Sassen, D. O. Starr, & G. Stephens (Eds.), *Cirrus* (pp. 102–136). New York: Oxford University Press.

DeMott, P. J., Prenni, A. J., Liu, X., Kreidenweis, S. M., Petters, M. D., Twohy, C. H., et al. (2010). Predicting global atmospheric ice nuclei distributions and their impacts on climate. *Proceedings of the National Academy of Sciences of the United States of America*, *107*(25), 11217–11222. doi: 10.1073/pnas.0910818107

DeMott, P. J., et al. (2018). The fifth international workshop on ice nucleation phase 2 (FIN-02): Laboratory intercomparison of ice nucleation measurements. *Atmospheric Measurement Techniques*, *11*. doi: 10.5194/amt-11-6231-2018

Durant, A. J., & Shaw, R. A. (2005). Evaporation freezing by contact nucleation inside-out. *Geophysical Research Letters*, *32*(20), L20814. doi: 10.1029/2005gl024175

Dymarska, M., Murray, B. J., Sun, L. M., Eastwood, M. L., Knopf, D. A., & Bertram, A. K. (2006). Deposition ice nucleation on soot at temperatures relevant for the lower troposphere. *Journal of Geophysical Research*, *111*(D4), D04204. doi: 10.1029/2005jd006627

Field, P. R., et al. (2017). Secondary ice production: Current state of the science and recommendations for the future. In *Ice formation and evolution in clouds and precipitation: measurement and modeling challenges* (pp. 7.1–7.20). American Meteorological Society. doi: 10.1175/AMSMONOGRAPHS-D-16-0014.1

Fletcher, N. H. (1958). Size effect in heterogeneous nucleation. *The Journal of Chemical Physics*, *29*(3), 572–576.

Friddle, R. W., & Thürmer, K. (2019). How nanoscale surface steps promote ice growth on feldspar: Microscopy observation of morphology-enhanced condensation and freezing. *Nanoscale*, *11*(44), 21147–21154. doi: 10.1039/c9nr08729j

Friddle, R. W., & Thürmer, K. (2020). Mapping ice formation to mineral-surface topography using a micro mixing chamber with video and atomic-force microscopy. *Atmospheric Measurement Techniques*, *13*(5), 2209–2218. doi: 10.5194/amt-13-2209-2020

Hartmann, D. L. (2016). Tropical anvil clouds and climate sensitivity. *Proceedings of the National Academy of Sciences of the United States of America*, *113*(32), 8897–8899. doi: 10.1073/pnas.1610455113

Hegg, D. A., & Baker, M. B. (2009). Nucleation in the atmosphere. *Reports on Progress in Physics*, *72*(5), 056801. doi: 10.1088/0034-4885/72/5/056801

Heintzenberg, J., Okada, K., & Strom, J. (1996). On the composition of non-volatile material in upper tropospheric aerosols and cirrus crystals. *Atmospheric Research*, *41*(1), 81–88. doi: 10.1016/0169-8095(95)00042-9

Held, I. M., & Soden, B. J. (2000). Water vapor feedback and global warming. *Annual Review of Environment and Resources*, *25*, 441–475.

Heymsfield, A. J., et al. (2017). Cirrus clouds. *Meteorological Monographs*, *58*, 2.1–2.26. doi: 10.1175/amsmonographs-d-16-0010.1

Hiranuma, N., Brooks, S. D., Moffet, R. C., Glen, A., Laskin, A., Gilles, M. K., et al. (2013). Chemical characterization of individual particles and residuals of cloud droplets and ice crystals collected on board research aircraft in the ISDAC 2008 study. *Journal of Geophysical Research*, *118*(12), 6564–6579. doi: 10.1002/jgrd.50484

Hiranuma, N., et al. (2015). A comprehensive laboratory study on the immersion freezing behavior of illite NX particles: A comparison of 17 ice nucleation measurement techniques. *Atmospheric Chemistry and Physics*, *15*(5), 2489–2518. doi: 10.5194/acp-15-2489-2015

Hodgson, A., & Haq, S. (2009). Water adsorption and the wetting of metal surfaces. *Surface Science Reports*, *64*(9), 381–451. doi: 10.1016/j.surfrep.2009.07.001

Holden, M. A., Whale, T. F., Tarn, M. D., O'Sullivan, D., Walshaw, R. D., Murray, B. J., et al. (2019). High-speed imaging of ice nucleation in water proves the existence of active sites. *Science Advances*, *5*(2), 10. doi: 10.1126/sciadv.aav4316

Hoose, C., & Möhler, O. (2012). Heterogeneous ice nucleation on atmospheric aerosols: A review of results from laboratory experiments. *Atmospheric Chemistry and Physics*, *12*(20), 9817–9854. doi: 10.5194/acp-12-9817-2012

Jensen, E. J., et al. (2009). On the importance of small ice crystals in tropical anvil cirrus. *Atmospheric Chemistry and Physics*, *9*(15), 5519–5537.

Kanji, Z. A., Ladino, L. A., Wex, H., Boose, Y., Burkert-Kohn, M., Cziczo, D. J., & Krämer, M. (2017). Overview of ice nucleating particles. In *Ice formation and evolution in clouds and precipitation: Measurement and modeling challenges* (pp. 1.1–1.33). American Meteorological Society. doi: 10.1175/AMSMONOGRAPHS-D-16-0006.1

Kiselev, A., Bachmann, F., Pedevilla, P., Cox, S. J., Michaelides, A., Gerthsen, D., & Leisner, T. (2017). Active sites in heterogeneous ice nucleation—The example of K-rich feldspars. *Science*, *355*(6323), 367–371. doi: 10.1126/science.aai8034

Knopf, D. A. (2003). *Thermodynamic properties and nucleation processes of upper tropospheric and lower stratospheric aerosol particles* (Ph.D. thesis). Swiss Federal Institute of Technology, Zurich.

Knopf, D. A., & Alpert, P. A. (2023). Atmospheric ice nucleation. *Nature Reviews Physics*, 5, 203–217. doi: 10.1038/s42254-023-00570-7

Knopf, D. A., & Koop, T. (2006). Heterogeneous nucleation of ice on surrogates of mineral dust. *Journal of Geophysical Research*, *111*(D12), D12201. doi: 10.1029/2005jd006894

Knopf, D. A., & Lopez, M. D. (2009). Homogeneous ice freezing temperatures and ice nucleation rates of aqueous ammonium sulfate and aqueous levoglucosan particles for relevant atmospheric conditions. *Physical Chemistry Chemical Physics*, *11*(36), 8056–8068. doi: Doi 10.1039/B903750k

Knopf, D. A., & Rigg, Y. J. (2011). Homogeneous ice nucleation from aqueous inorganic/organic particles representative of biomass burning: Water activity, freezing temperatures, nucleation rates. *The Journal of Physical Chemistry. A*, *115*(5), 762–773. doi: 10.1021/jp109171g

Knopf, D. A., Koop, T., Luo, B. P., Weers, U. G., & Peter, T. (2002). Homogeneous nucleation of NAD and NAT in liquid stratospheric aerosols: Insufficient to explain denitrification. *Atmospheric Chemistry and Physics*, *2*, 207–214.

Knopf, D. A., Luo, B. P., Krieger, U. K., & Koop, T. (2003). Thermodynamic dissociation constant of the bisulfate ion from Raman and ion interaction modeling studies of aqueous sulfuric acid at low temperatures. *The Journal of Physical Chemistry. A*, *107*(21), 4322–4332.

Knopf, D. A., Wang, B., Laskin, A., Moffet, R. C., & Gilles, M. K. (2010). Heterogeneous nucleation of ice on anthropogenic organic particles collected in Mexico City. *Geophysical Research Letters*, *37*, L11803. doi: 10.1029/2010GL043362

Knopf, D. A., Alpert, P. A., Wang, B., O'Brien, R. E., Kelly, S. T., Laskin, A., et al. (2014). Microspectroscopic imaging and characterization of individually identified ice nucleating particles from a case field study. *Journal of Geophysical Research*, *119*, 10365–10381. doi: 10.1002/2014JD021866

Knopf, D. A., Alpert, P. A., & Wang, B. (2018). The role of organic aerosol in atmospheric ice nucleation: A review. *ACS Earth and Space Chemistry*, *2*(3), 168–202. doi: 10.1021/acsearthspacechem.7b00120

Knopf, D. A., Alpert, P. A., Zipori, A., Reicher, N., & Rudich, Y. (2020). Stochastic nucleation processes and substrate abundance explain time-dependent freezing in supercooled droplets. *npj Climate and Atmospheric Science*, *3*(2), 1–9. doi: 10.1038/s41612-020-0106-4

Knopf, D. A., et al. (2021). Aerosol–ice formation closure: A southern Great Plains Field campaign. *Bulletin of the American Meteorological Society*, *102*(10), E1952–E1971. doi: 10.1175/BAMS-D-20-0151.1

Knopf, D. A., et al. (2022). Micro-spectroscopic and freezing characterization of ice-nucleating particles collected in the marine boundary layer in the eastern North Atlantic. *Atmospheric Chemistry and Physics*, *22*(8), 5377–5398. doi: 10.5194/acp-22-5377-2022

Knopf, D. A., Silber, I., Riemer, N., Fridlind, A. M., & Ackerman, A. S. (2023a). A 1D model for nucleation of ice from aerosol particles: An application to a mixed-phase Arctic stratus cloud layer. *Journal of Advances in Modeling Earth Systems*, *15*(10), 26. doi: 10.1029/2023ms003663

Knopf, D. A., et al. (2023b). Physicochemical characterization of free troposphere and marine boundarylayer ice-nucleating particles collected by aircraft in the eastern North Atlantic. *Atmospheric Chemistry and Physics*, *23*(15), 8659–8681. doi: 10.5194/acp-23-8659-2023

Koop, T. (2004). Homogeneous ice nucleation in water and aqueous solutions. *Zeitschrift für Physikalische Chemie*, *218*(11), 1231–1258.

Koop, T., Luo, B. P., Biermann, U. M., Crutzen, P. J., & Peter, T. (1997). Freezing of HNO3/H2SO4/H2O solutions at stratospheric temperatures: Nucleation statistics and experiments. *The Journal of Physical Chemistry. A*, *101*(6), 1117–1133.

Koop, T., Ng, H. P., Molina, L. T., & Molina, M. J. (1998). A new optical technique to study aerosol phase transitions: The nucleation of ice from H2SO4 aerosols. *The Journal of Physical Chemistry. A*, *102*(45), 8924–8931.

Koop, T., Luo, B. P., Tsias, A., & Peter, T. (2000). Water activity as the determinant for homogeneous ice nucleation in aqueous solutions. *Nature*, *406*(6796), 611–614. doi: 10.1038/35020537

Koop, T., Bookhold, J., Shiraiwa, M., & Pöschl, U. (2011). Glass transition and phase state of organic compounds: Dependency on molecular properties and implications for secondary organic aerosols in the atmosphere. *Physical Chemistry Chemical Physics*, *13*(43), 19238–19255. doi: 10.1039/c1cp22617g

Korolev, A., & Leisner, T. (2020). Review of experimental studies of secondary ice production. *Atmospheric Chemistry and Physics*, *20*(20), 11767–11797. doi: 10.5194/acp-20-11767-2020

Kreidenweis, S. M., Chen, Y., Rogers, D. C., & DeMott, P. J. (1998). Isolating and identifying atmospheric ice-nucleating aerosols: A new technique. *Atmospheric Research*, *46*(3–4), 263–278.

Ladino Moreno, L. A., Stetzer, O., & Lohmann, U. (2013). Contact freezing: A review of experimental studies. *Atmospheric Chemistry and Physics*, *13*(19), 9745–9769. doi: 10.5194/acp-13-9745-2013

Laksmono, H., et al. (2015). Anomalous behavior of the homogeneous ice nucleation rate in "no-Man's land". *Journal of Physical Chemistry Letters*, *6*(14), 2826–2832. doi: 10.1021/acs.jpclett.5b01164

Laskin, A., Gilles, M. K., Knopf, D. A., Wang, B., & China, S. (2016). Progress in the analysis of complex atmospheric particles. *Annual Review of Analytical Chemistry*, *9*(1), 117–143. doi: 10.1146/annurev-anchem-071015-041521

Laskin, A., Moffet, R. C., & Gilles, M. K. (2019). Chemical imaging of atmospheric particles. *Accounts of Chemical Research*, *52*(12), 3419–3431. doi: 10.1021/acs.accounts.9b00396

Lata, N. N., Zhang, B., Schum, S., Mazzoleni, L., Brimberry, R., Marcus, M. A., et al. (2021). Aerosol composition, mixing state, and phase state of free tropospheric particles and their role in ice cloud formation. *ACS Earth and Space Chemistry*, *5*(12), 3499–3510. doi: 10.1021/acsearthspacechem.1c00315

Lau, K. M., & Wu, H. T. (2003). Warm rain processes over tropical oceans and climate implications. *Geophysical Research Letters*, *30*(24), 2290. doi: 10.1029/2003gl018567

Lawson, R. P., et al. (2019). A review of ice particle shapes in cirrus formed in situ and in anvils. *Journal of Geophysical Research*, *124*(17–18), 10049–10090. doi: 10.1029/2018jd030122

Libbrecht, K. G. (2005). The physics of snow crystals. *Reports on Progress in Physics*, *68*(4), 855–895. doi: 10.1088/0034-4885/68/4/r03

Lohmann, U., & Feichter, J. (2005). Global indirect aerosol effects: A review. *Atmospheric Chemistry and Physics*, *5*, 715–737.

Lohmann, U., Roeckner, E., Collins, W. D., Heymsfield, A. J., McFarquhar, G. M., & Barnett, T. P. (1995). The role of water vapor and convection during the central equatorial Pacific experiment from observations and model simulations. *Journal of Geophysical Research*, *100*(D12), 26229–26245.

Lupi, L., Hudait, A., & Molinero, V. (2014). Heterogeneous nucleation of ice on carbon surfaces. *Journal of the American Chemical Society*, *136*(8), 3156–3164. doi: 10.1021/ja411507a

Ma, R. Z., et al. (2020). Atomic imaging of the edge structure and growth of a two-dimensional hexagonal ice. *Nature*, *577*(7788) 60-+. doi: 10.1038/s41586-019-1853-4

Mael, L. E., Busse, H., & Grassian, V. H. (2019). Measurements of immersion freezing and heterogeneous chemistry of atmospherically relevant single particles with Micro-Raman spectroscopy. *Analytical Chemistry*, *91*(17), 11138–11145. doi: 10.1021/acs.analchem.9b01819

Magee, N. B., Miller, A., Amaral, M., & Cumiskey, A. (2014). Mesoscopic surface roughness of ice crystals pervasive across a wide range of ice crystal conditions. *Atmospheric Chemistry and Physics*, *14*(22), 12357–12371. doi: 10.5194/acp-14-12357-2014

Magee, N., et al. (2021). Captured cirrus ice particles in high definition. *Atmospheric Chemistry and Physics*, *21*(9), 7171–7185. doi: 10.5194/acp-21-7171-2021

Marcolli, C. (2014). Deposition nucleation viewed as homogeneous or immersion freezing in pores and cavities. *Atmospheric Chemistry and Physics*, *14*(4), 2071–2104. doi: 10.5194/acp-14-2071-2014

Marcolli, C. (2020). Technical note: Fundamental aspects of ice nucleation via pore condensation and freezing including Laplace pressure and growth into macroscopic ice. *Atmospheric Chemistry and Physics*, *20*(5), 3209–3230. doi: 10.5194/acp-20-3209-2020

Martin, S. (2000). Phase transitions of aqueous atmospheric particles. *Chemical Reviews*, *100*, 3403–3453.

McFarquhar, G. M., Yang, P., Macke, A., & Baran, A. J. (2002). A new parameterization of single scattering solar radiative properties for tropical anvils using observed ice crystal size and shape distributions. *Journal of the Atmospheric Sciences*, *59*(16), 2458–2478. doi: 10.1175/1520-0469(2002)059<2458:Anposs>2.0.Co;2

Melillo, J. H., Nikulina, E., Iriarte-Alonso, M. A., Cerveny, S., & Bittner, A. M. (2022). Electron microscopy and calorimetry of proteins in supercooled water. *Scientific Reports*, *12*(1), 11. doi: 10.1038/s41598-022-20430-1

Mishchenko, M. I., Rossow, W. B., Macke, A., & Lacis, A. A. (1996). Sensitivity of cirrus cloud albedo, bidirectional reflectance and optical thickness retrieval accuracy to ice particle shape. *Journal of Geophysical Research-Atmospheres*, *101*(D12), 16973–16985. doi: 10.1029/96jd01155

Mülmenstädt, J., Sourdeval, O., Delanoe, J., & Quaas, J. (2015). Frequency of occurrence of rain from liquid-, mixed-, and ice-phase clouds derived from A-train satellite retrievals. *Geophysical Research Letters*, *42*(15), 6502–6509. doi: 10.1002/2015gl064604

Murray, B. J., O'Sullivan, D., Atkinson, J. D., & Webb, M. E. (2012). Ice nucleation by particles immersed in supercooled cloud droplets. *Chemical Society Reviews*, *41*(19), 6519–6554. doi: 10.1039/C2CS35200A

Nair, M., Husmann, A., Cameron, R. E., & Best, S. M. (2018). In situ ESEM imaging of the vapor-pressure-dependent sublimation-induced morphology of ice. *Physical Review Materials*, *2*(4), 7. doi: 10.1103/PhysRevMaterials.2.040401

Nakaya, U. (1954). *Snow crystals, natural and artificial*, 510 pp. Cambridge: Harvard University Press.

Pach, E., & Verdaguer, A. (2022). Studying ice with environmental scanning Electron microscopy. *Molecules*, *27*(1), 17. doi: 10.3390/molecules27010258

Pedersen, C., Mihranyan, A., & Stromme, M. (2011). Surface transition on ice induced by the formation of a grain boundary. *PLoS One*, *6*(9), 4. doi: 10.1371/journal.pone.0024373

Peng, J. B., Guo, J., Ma, R. Z., & Jiang, Y. (2022). Water-solid interfaces probed by high-resolution atomic force microscopy. *Surface Science Reports*, *77*(1). doi: 10.1016/j.surfrep.2021.100549

Petzold, A., Strom, J., Ohlsson, S., & Schroder, F. P. (1998). Elemental composition and morphology of ice-crystal residual particles in cirrus clouds and contrails. *Atmospheric Research*, *49*(1), 21–34. doi: 10.1016/s0169-8095(97)00083-5

Pfalzgraff, W. C., Hulscher, R. M., & Neshyba, S. P. (2010). Scanning electron microscopy and molecular dynamics of surfaces of growing and ablating hexagonal ice crystals. *Atmospheric Chemistry and Physics*, *10*(6), 2927–2935. doi: 10.5194/acp-10-2927-2010

Pruppacher, H. R., & Klett, J. D. (1997). *Microphysics of clouds and precipitation*. Dordrecht: Kluwer Academic Publishers.

Qiu, Y. Q., Odendahl, N., Hudait, A., Mason, R., Bertram, A. K., Paesani, F., et al. (2017). Ice nucleation efficiency of Hydroxylated organic surfaces is controlled by their structural fluctuations and mismatch to ice. *Journal of the American Chemical Society*, *139*(8), 3052–3064. doi: 10.1021/jacs.61312210

Reicher, N., Segev, L., & Rudich, Y. (2018). The WeIzmann Supercooled droplets observation on a microarray (WISDOM) and application for ambient dust. *Atmospheric Measurement Techniques*, *11*, 233–248. doi: 10.5194/amt-11-233-2018

Riechers, B., Wittbracht, F., Hutten, A., & Koop, T. (2013). The homogeneous ice nucleation rate of water droplets produced in a microfluidic device and the role of temperature uncertainty. *Physical Chemistry Chemical Physics*, *15*(16), 5873–5887. doi: 10.1039/c3cp42437e

Rogers, D. C. (1988). Development of a continuous flow thermal gradient diffusion chamber for ice nucleation studies. *Atmospheric Research*, *22*(2), 149–181. doi: 10.1016/0169-8095(88)90005-1

Rosenfeld, D., Sherwood, S., Wood, R., & Donner, L. (2014). Climate effects of aerosol-cloud interactions. *Science*, *343*(6169), 379–380. doi: 10.1126/science.1247490

Roy, P., Mael, L. E., Hill, T. C. J., Mehndiratta, L., Peiker, G., House, M. L., et al. (2021). Ice nucleating activity and residual particle morphology of bulk seawater and sea surface microlayer. *ACS Earth and Space Chemistry*, *5*(8), 1916–1928. doi: 10.1021/acsearthspacechem.1c00175

Schill, G. P., & Tolbert, M. A. (2013). Heterogeneous ice nucleation on phase-separated organic-sulfate particles: Effect of liquid vs. glassy coatings. *Atmospheric Chemistry and Physics*, *13*(9), 4681–4695. doi: 10.5194/acp-13-4681-2013

Schill, G. P., & Tolbert, M. A. (2014). Heterogeneous ice nucleation on simulated sea-spray aerosol using Raman microscopy. *Journal of Physical Chemistry C*, *118*(50), 29234–29241. doi: 10.1021/jp505379j

Seinfeld, J. H., & Pandis, S. N. (1998). *Atmospheric chemistry and Pysics. From air pollution to climate change*, 1326 pp. New York: John Wiley.

Shardt, N., Isenrich, F. N., Waser, B., Marcolli, C., Kanji, Z. A., deMello, A. J., & Lohmann, U. (2022). Homogeneous freezing of water droplets for different volumes and cooling rates. *Physical Chemistry Chemical Physics*, *24*(46), 28213–28221. doi: 10.1039/d2cp03896j

Shaw, R. A., Durant, A. J., & Mi, Y. (2005). Heterogeneous surface crystallization observed in undercooled water. *The Journal of Physical Chemistry. B*, *109*(20), 9865–9868.

Smith, R. S., & Kay, B. D. (1999). The existence of supercooled liquid water at 150 K. *Nature*, *398*(6730), 788–791.

Sosso, G. C., Tribello, G. A., Zen, A., Pedevilla, P., & Michaelides, A. (2016). Ice formation on kaolinite: Insights from molecular dynamics simulations. *The Journal of Chemical Physics*, *145*(21), 10. doi: 10.1063/1.4968796

Stan, C. A., Schneider, G. F., Shevkoplyas, S. S., Hashimoto, M., Ibanescu, M., Wiley, B. J., & Whitesides, G. M. (2009). A microfluidic apparatus for the study of ice nucleation in supercooled water drops. *Lab on a Chip*, *9*(16), 2293–2305. doi: 10.1039/b906198c

Storelvmo, T. (2017). Aerosol effects on climate via mixed-phase and ice clouds. In R. Jeanloz & K. H. Freeman (Eds.), *Annual Review of Earth and Planetary Sciences* (Vol. *45*, pp. 199–222). doi: 10.1146/annurev-earth-060115-012240

Storelvmo, T., Hoose, C., & Eriksson, P. (2011). Global modeling of mixed-phase clouds: The albedo and lifetime effects of aerosols. *Journal of Geophysical Research*, *116*, D05207. doi: 10.1029/2010jd014724

Tamtogl, A., Bahn, E., Sacchi, M., Zhu, J. D., Ward, D. J., Jardine, A. P., et al. (2021). Motion of water monomers reveals a kinetic barrier to ice nucleation on graphene. *Nature Communications*, *12*(1), 8. doi: 10.1038/s41467-021-23226-5

Tarn, M. D., Sikora, S. N. F., Porter, G. C. E., Shim, J. U., & Murray, B. J. (2021). Homogeneous freezing of water using microfluidics. *Micromachines*, *12*(2), 23. doi: 10.3390/mi12020223

Thürmer, K., & Nie, S. (2013). Formation of hexagonal and cubic ice during low-temperature growth. *Proceedings of the National Academy of Sciences of the United States of America*, *110*(29), 11757–11762. doi: 10.1073/pnas.1303001110

Turnbull, D., & Fisher, J. C. (1949). Rate of nucleation in condensed systems. *The Journal of Chemical Physics*, *17*(1), 71–73. doi: 10.1063/1.1747055

Twohy, C. H., & Gandrud, B. W. (1998). Electron microscope analysis of residual particles from aircraft contrails. *Geophysical Research Letters*, *25*(9), 1359–1362. doi: 10.1029/97gl03162

Vali, G. (1971). Quantitative evaluation of experimental results on heterogeneous freezing nucleation of supercooled liquids. *Journal of the Atmospheric Sciences*, *28*(3), 402–409.

Vali, G., DeMott, P. J., Möhler, O., & Whale, T. F. (2015). Technical note: A proposal for ice nucleation terminology. *Atmospheric Chemistry and Physics*, *15*(18), 10263–10270. doi: 10.5194/acp-15-10263-2015

Wang, B., & Knopf, D. A. (2011). Heterogeneous ice nucleation on particles composed of humic-like substances impacted by O3. *Journal of Geophysical Research*, *116*, D03205. doi: 10.1029/2010jd014964

Wang, B., Lambe, A. T., Massoli, P., Onasch, T. B., Davidovits, P., Worsnop, D. R., & Knopf, D. A. (2012a). The deposition ice nucleation and immersion freezing potential of amorphous secondary organic aerosol: Pathways for ice and mixed-phase cloud formation. *Journal of Geophysical Research*, *117*, D16209. doi: 10.1029/2012jd018063

Wang, B., Laskin, A., Roedel, T., Gilles, M. K., Moffet, R. C., Tivanski, A. V., & Knopf, D. A. (2012b). Heterogeneous ice nucleation and water uptake by field-collected atmospheric particles below 273 K. *Journal of Geophysical Research*, *117*, D00V19. doi: 10.1029/2012JD017446

Wang, B., Knopf, D. A., China, S., Arey, B. W., Harder, T. H., Gilles, M. K., & Laskin, A. (2016). Direct observation of ice nucleation events on individual atmospheric particles. *Physical Chemistry Chemical Physics*, *18*(43), 29721–29731. doi: 10.1039/C6CP05253C

Wendisch, M., Yang, P., & Pilewskie, P. (2007). Effects of ice crystal habit on thermal infrared radiative properties and forcing of cirrus. *Journal of Geophysical Research-Atmospheres*, *112*(D8), D08201. doi: 10.1029/2006jd007899

Westbrook, C. D., & Illingworth, A. J. (2013). The formation of ice in a long-lived supercooled layer cloud. *Quarterly Journal of the Royal Meteorological Society*, *139*(677), 2209–2221. doi: 10.1002/qj.2096

Wex, H., DeMott, P. J., Tobo, Y., Hartmann, S., Roesch, M., Clauss, T., et al. (2014). Kaolinite particles as ice nuclei: Learning from the use of different kaolinite samples and different coatings. *Atmospheric Chemistry and Physics*, *14*(11), 5529–5546. doi: 10.5194/acp-14-5529-2014

Zhang, Z. Y., Zhou, Y. B., Zhu, X. L., Fei, L. F., Huang, H. T., & Wang, Y. (2020). Applications of ESEM on materials science: Recent updates and a look forward. *Small Methods*, *4*(2), 21. doi: 10.1002/smtd.201900588

Zielke, S. A., Bertram, A. K., & Patey, G. N. (2016). Simulations of ice nucleation by kaolinite (001) with rigid and flexible surfaces. *The Journal of Physical Chemistry. B*, *120*(8), 1726–1734. doi: 10.1021/acs.jpcb.5b09052

Zimmermann, F., Ebert, M., Worringen, A., Schutz, L., & Weinbruch, S. (2007). Environmental scanning electron microscopy (ESEM) as a new technique to determine the ice nucleation capability of individual atmospheric aerosol particles. *Atmospheric Environment*, *41*(37), 8219–8227. doi: 10.1016/j.atmosenv.2007.06.023

Zimmermann, F., Weinbruch, S., Schutz, L., Hofmann, H., Ebert, M., Kandler, K., & Worringen, A. (2008). Ice nucleation properties of the most abundant mineral dust phases. *Journal of Geophysical Research*, *113*, D23204. doi: 10.1029/2008jd010655

Zipori, A., Reicher, N., Erel, Y., Rosenfeld, D., Sandler, A., Knopf, D. A., & Rudich, Y. (2018). The role of secondary ice processes in mid-latitude continental clouds. *Journal of Geophysical Research*, *123*(22), 12762–12777. doi: 10.1029/2018JD029146

Zobrist, B., Koop, T., Luo, B. P., Marcolli, C., & Peter, T. (2007). Heterogeneous ice nucleation rate coefficient of water droplets coated by a nonadecanol monolayer. *Journal of Physical Chemistry C*, *111*(5), 2149–2155. doi: 10.1021/Jp066080w

Zobrist, B., Marcolli, C., Pedernera, D. A., & Koop, T. (2008). Do atmospheric aerosols form glasses? *Atmospheric Chemistry and Physics*, *8*(17), 5221–5244.

Zrimsek, A. B., Chiang, N. H., Mattei, M., Zaleski, S., McAnally, M. O., Chapman, C. T., et al. (2017). Single-molecule chemistry with surface- and tip-enhanced Raman spectroscopy. *Chemical Reviews*, *117*(11), 7583–7613. doi: 10.1021/acs.chemrev.6b00552

12

Studying Aerosol Hygroscopicity Using Environmental Electron Microscopy

Evelyn Freney

ABSTRACT

Aerosol particles play an essential role in climate, air quality, and human health. It is therefore vital to understand their composition, morphology, and size distribution and how these properties influence their reactivity in the atmosphere. This chapter highlights the valuable contributions of environmental scanning and transmission electron microscopy (E-SEM/TEM) to the study of aerosol particles in the atmosphere, with particular reference to aerosol–cloud interactions. This chapter begins with a brief overview detailing how E-SEM/TEM has been instrumental in examining the interactions between aerosol particles and atmospheric gases, particularly water vapor, overcoming the limitations inherent in conventional online methods. Challenges and future prospects are discussed, recognizing the potential for developing synergies between E-SEM/TEM and other analytical tools. This collaborative approach is crucial to answering key questions about aerosol–cloud interactions, leading the way for a more complete understanding of these complex atmospheric processes.

12.1. INTRODUCTION

The description of an aerosol particle and its related mixing state requires characterizing a number of different properties, including chemical composition (elemental and molecular) and physical properties (size, state, and morphology). This information is essential for understanding the physicochemical reactivity of an aerosol particle and how reactivity changes throughout its lifetime in the atmosphere.

Aerosol particles in the atmosphere have an important impact on several aspects of our daily lives. These include visibility, plant damage, human health, and climate. Both direct and indirect effects are important for climate. The direct effects of aerosol particles rely principally on their size and chemical composition, whereas aerosol indirect effects are complex and depend on microphysical processes in which aerosol particles interact with atmospheric water vapor to form liquid droplets (Lohmann & Feichter, 2005) or ice crystals (DeMott et al. 2003). These interactions play an important role in precipitation initiation and hence influence Earth's water cycle. They are equally responsible for modifying the radiative properties of clouds by affecting cloud lifetime and cloud albedo. However, the uncertainties related to these processes remain among the largest when trying to understand the impact of aerosol particles on climate (IPCC, 2023).

The behavior of aerosol particles in a humid environment is well studied, and a wide range of studies provide detailed information on the relative humidity (RH) at which crystalline particles will take on water to become a droplet in a process called *deliquescence* and the reverse process where aerosol particles become crystalline again, known as *efflorescence* (Mikhailov et al., 2009).

As highlighted in Mikailov et al. (2009), understanding aerosol-particle hygroscopicity is essential for many reasons, particularly when considering surface reactions

Laboratory of Physical Meteorology, UMR 6016 CNRS/UBP, University of Clermont Auvergne, Aubière, France

Microanalysis of Atmospheric Particles: Techniques and Applications, Geophysical Monograph 288, First Edition.
Edited by Joseph M. Conny and Peter R. Buseck.

DOI: 10.1002/9781119554318.ch12

with other atmospheric species. For these reactions, particle viscosity and phase state are important and can change as the solute concentration is altered during the uptake of water. Once activated into a cloud droplet, aqueous-phase chemistry can be important for changing particle chemical and physical properties (Feingold & Kreidenweis, 2000).

A wide range of methods are used to study water uptake by aerosol particles. These range from Fourier transform infrared spectroscopy (FTIR) via electrodynamic balances to mobility spectrometers (Biskos et al., 2006). These methods provide accurate measurements, especially for particle deliquescence RH (DRH) and efflorescence RH (ERH), and some, such as mobility spectrometers, are capable of sampling large quantities of aerosol particles in a continuous manner and often in autonomous sampling (Holmgren et al., 2014). Nevertheless, with these methods, it is difficult to assess the impact of the mixing state and morphology of particles in a humid environment.

Currently, the majority of experiments focusing on aerosol-particle behavior in a humid environment use either a hygroscopic tandem differential mobility analyzer (Holmgren et al., 2014) or a cloud condensation nuclei counter (Schmale et al., 2018). To understand how particles act as ice-nucleating sites in the atmosphere, a number of systems, both online and offline, are commonly used (Demott et al., 2018; Lacher et al., 2024). These instrument developments have evolved in recent years to allow autonomous sampling over extended periods of time, with the advantage of providing information on diurnal and seasonal variability. They are generally run in parallel with aerosol size distribution measurements and aerosol chemical composition measurements to provide information on trends in aerosol hygroscopic properties and their links to aerosol physical and chemical properties (Schmale et al., 2018; Asmi et al., 2012). These studies have led to the development of a number of parameterizations that can be used and tested in atmospheric models (Kim et al., 2021). However, these measurements can not provide information of the microscopic processes occurring at the surface of the particles. Likewise, they are unable to provide detailed information on how aerosol mixing state can impact aerosol hygroscopic properties.

As the body of research on aerosol hygroscopic properties grows, our comprehension of the primary physical and chemical mechanisms governing the formation of cloud droplets has also advanced. With the Kelvin effect playing a pivotal role, where particle size is a dominant variable when determining the formation of cloud droplets from aerosol particles. However, the influence of organic aerosol particles on cloud droplet formation (CCN) is important and is the focus of many studies (Topping & McFiggans, 2012; Petters & Petters, 2016).

In colder atmospheres, the formation of ice crystals by aerosol particles becomes important. The numbers of these ice-nucleating particles (INPs) in the atmosphere are typically low (for example, one per liter of air [1/l]), but they play an important role in the initiation of precipitation (Lohmann & Feichter, 2005). It is generally considered that larger insoluble aerosol particles, having high surface areas, are the best INPs and that hygroscopic aerosol particles commonly exhibit a lower capacity for forming ice crystals (Berkemeier et al., 2014). As a result of the low number concentrations of INPs in the atmosphere, it is difficult to provide accurate measurements of ice-nucleating particles and therefore to understand the physical and chemical interactions that are involved in the between aerosol particles and ice crystals (Lacher et al., 2024). In order to improve this understanding, microscopic-scale studies are essential.

12.2. ENVIRONMENTAL CELLS

In the previous chapters, we have seen how different forms of microscopy can provide essential measurements of aerosol particles and their structure, mixing state, and morphology. However, it was not until the environmental cell was coupled to microscopic tools that it became possible to study ambient aerosol properties under different environments of temperature and humidity.

The ability to observe dynamic processes in real time with near-atomic resolution offers a wealth of information for understanding the structural aspects of gas/solid reactions. For this reason, environmental cells coupled with electron microscopes have applications in a large number of scientific domains. These range from material science to life sciences (i.e., to study biological samples in situ) and to atmospheric sciences where the aim is to study how aerosol particles transform from solid particles to liquid droplets.

Environmental cells were initially developed to overcome the limitations of conventional electron microscopy. The first reported observations of hydrated samples in electron microscopy were published by Abrams and McBain (1944). In the late 1980s, environmental electron microscopy was developed, defined as a microscope capable of having pressures high enough at 0 °C to maintain water in its liquid phase (640 Pa, compared to the pressure in conventional electron microscopy of less than 1.3×10^{-4} Pa). The motivation behind these initial developments was to observe liquid and single-particle specimens. Detailed descriptions of the environmental scanning electron microscopy (ESEM) instrument can be found in Danilatos (1988), Sharma and Weiss (1998), and, more recently, in Goldstein et al. (2017). In environmental microscopes, there is generally a series of differential apertures employed to maintain a high vacuum

(1.3×10^{-4} Pa) in the electron gun and along the electron-beam path and a higher pressure in the sample/specimen chamber. Both environmental SEM and TEM systems can be operated at pressures of approximately 18 Torr at 18 °C (Goldstein et al., 2017; Wise et al., 2005). If higher pressures are used, water vapor tends to condense on the stage, making the RH reading less accurate. Additionally, as the electron beam passes from the high-vacuum zone to the high-pressure zone of the sample chamber, interactions occur with the gas atoms. These interactions increase the probability of elastic scattering events, leading to unfocused or lower-quality images.

12.2.1. Environmental Electron Microscopy in Atmospheric Science

Environmental Scanning Electron Microscopy

The first application of ESEM to atmospheric aerosol particles was by Ebert et al. (2002). In this study, typical atmospheric salts NaCl, $(NH_4)_2SO_4$, Na_2SO_4, and NH_4NO_3 with diameters ranging from 100 to 20 μm were exposed to RH values ranging from 15% at 5 °C up to 100% (Figure 12.1). The study by Ebert et al. illustrated how ESEM could provide accurate measurements of DRH and ERH for a series of common atmospheric salts and showed how the technique could be used to study structural changes to individual aerosol particles. They were equally able to illustrate how a thin film of water is visible on a particle's surface prior to droplet formation and that the film is stable for up to 30 minutes. The study also illustrated how the hygroscopic properties of soot particles could be studied even under supersaturated conditions. Such observations were not possible with techniques such as FTIR, electrodynamic balances, or tandem differential mobility analyzers.

An advantage of using ESEM for atmospheric studies is that it is possible to investigate interactions with water vapor on a single-particle scale and to observe structural changes to the aerosol particles. An additional advantage of using ESEM is that one can probe the chemical composition of individual particles using energy dispersive

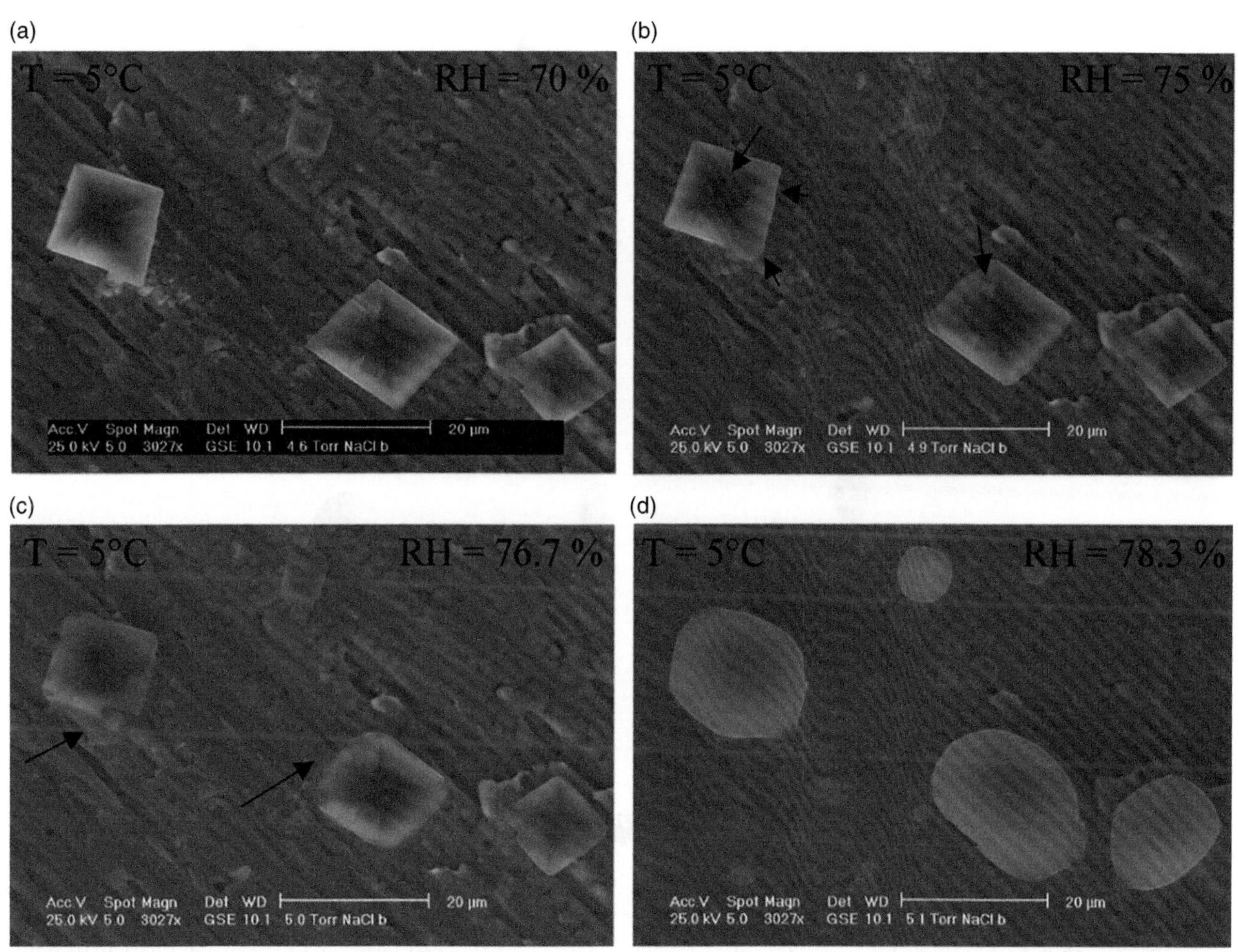

Figure 12.1 Environmental scanning electron microscopy images of NaCl particles exposed to relative humidities from 70% up to 78%. Ebert et al. (2002)/with permission of ELSEVIER.

X-ray spectroscopy (EDXS), providing the qualitative chemical composition of a number of elements. This was illustrated in a subsequent study applied to mixtures of atmospheric salts (Hoffman et al., 2004), showing that internally mixed particles can undergo continuous hygroscopic growth from humidities ranging from 5% up to 90%. Another study exposed aerosol particles to mixtures of nitric acid and water to represent different atmospheric environments (Krueger et al. 2003). They were able to demonstrate how calcium nitrate formed following the reaction of ambient calcite and sea salt with nitric acid (Laskin et al., 2005). Shortly following these initial studies using ESEM methods, ETEM was also applied to atmospheric research.

Environmental Transmission Electron Microscopy

Wise et al. (2005) first demonstrated the accuracy and reliability of ETEM to study the behaviour of atmospheric aerosol particles using NaCL and KCl particles. In this work, it was also noted that similar to observations of Ebert et al. (2002), a thin layer of water commonly formed on the surface of aerosol particles prior to the onset of deliquescence (Figure 12.2). The study by Wise et al. established that ETEM is a powerful analytical tool, capable of resolving chemical and morphological features at a nanometer scale, as well as studying aerosol-particle behavior in an atmospheric environment.

This initial study was followed by a number of studies on laboratory-generated aerosol particles (Wise

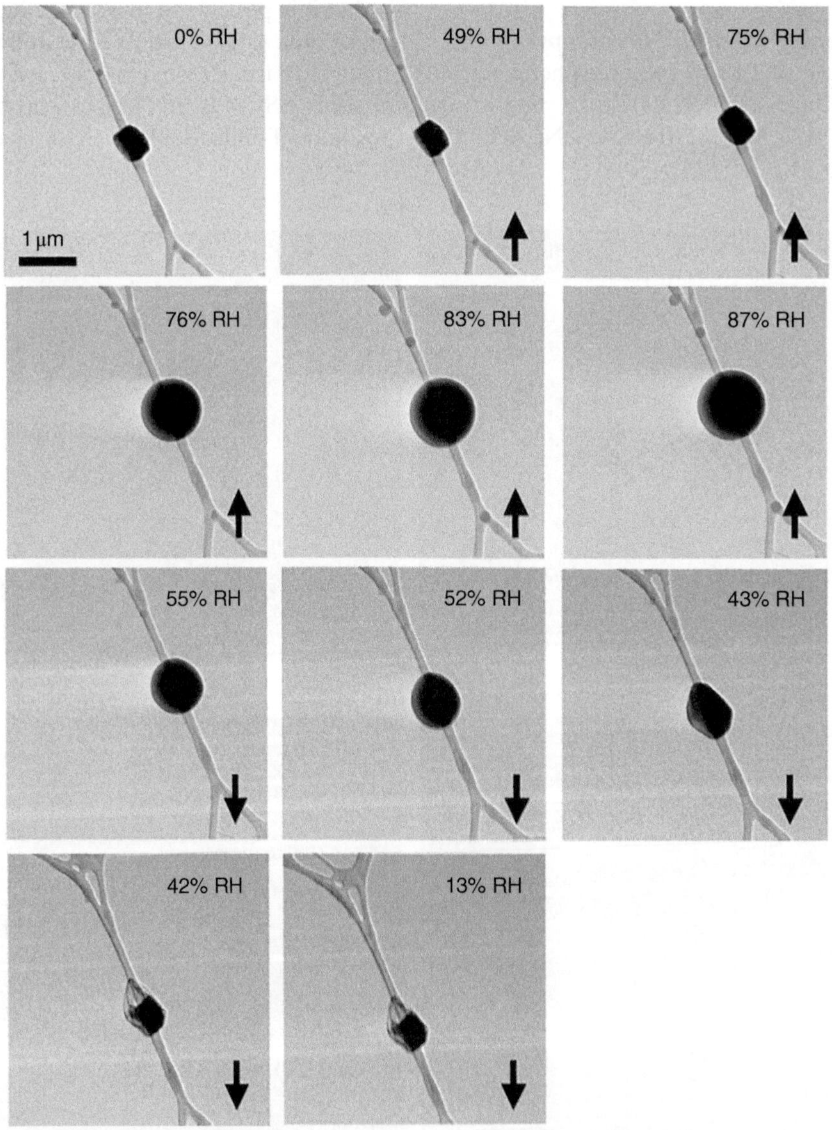

Figure 12.2 Environmental transmission electron microscopy images of NaCl particles exposed to relative humidities ranging from 0% up to 87%, showing their deliquescence and efflorescence cycles. Wise et al. (2005) / with permission of Taylor and Francis.

et al., 2008), those collected aboard aircraft, and those sampled from ground-based measurements (Wise et al., 2008). These studies demonstrated how internal mixing of aerosol particles (soluble/insoluble) can influence the climatic impact of aerosol particles. Specifically, when insoluble aerosol particles combine with soluble particles, they will grow to CCN-active diameters and hence require lower water supersaturations for activation. This aspect was explored by studying the impact of coatings on aerosol hygroscopic properties (Semeniuk et al., 2007) and by exploring how biomass-burning aerosol particles behave in environments of varying humidity (Freney et al., 2009), as well as for understanding the water uptake of ambient marine aerosol particles (Wise et al., 2009). These studies illustrated that many marine aerosol particles are composed of a well-defined salt core and a coating containing a mixture of organic compounds and other crystalline salts. This coating absorbs water at humidities much lower than that of the DRH of NaCl, causing these particles to increase in size prior to the typical DRH of NaCl.

As a result of these investigations, it became possible to observe the morphology of internally mixed particles in a humid environment. Subsequent work delved into the impact of internal mixtures on both soluble and insoluble materials, revealing that insoluble substances commonly remained suspended inside the droplet, resulting in a core-shell type structure. Using discreet dipole approximations (DDA) (Draine and Flatau (2008)), it was possible to study the changing radiative properties as a function of RH (Freney et al., 2010, Adachi et al., 2011). This work showed an increase in scattering of up to 68% (for supermicron droplets) when different solid structures are considered inside a droplet.

12.3. APPLICATION TO MORE COMPLEX AEROSOL-CLOUD INTERACTIONS: ICE NUCLEATION

Although there are outstanding questions regarding the role of aerosol-particle hygroscopicity, composition, and mixing state in the formation of cloud droplets, the major formation routes for liquid droplets (via physical properties) are well characterized. However, this is not the case for the formation of ice crystals on or in atmospheric particles. Ice nucleation occurs on a very small fraction of aerosol particles in the atmosphere but has an important impact on the precipitation cycle and therefore the climate (Kanji et al., 2017).

A range of both online and offline methods exist to quantify the ice-nucleating particle number concentration in the atmosphere and to relate these concentrations to aerosol-particle compositions and physical properties (Demott et al., 2018; Lacher et al., 2024). However, despite remaining key scientific questions that must be answered (Bartels-Rausch, 2013), the physical and chemical mechanisms governing the formation of ice crystals in the atmosphere are still unknown. Zimmerman et al. (2008) presented the first images of ice-crystal growth using an environmental scanning electron microscope on typical atmospheric particle types such as kaolinite and montmorillonite, as well as on efficient ice-nucleation compounds such as silver iodide. In that work, the samples were exposed to atmospheric water vapor at 250 K and 270 K. These measurements highlighted specific processes related to the initiation of ice-crystal growth, showing how surface steps, cracks, and cavities, referred to as surface defects, can be nucleating sites for ice. This highlighted that initial ice nucleation is dependent on aerosol-particle surface properties and particle type. This was confirmed later by Kiselev et al. (2017), who demonstrated how electron microscopy can be used to observe the onset of ice nucleation (Figure 12.3a, b), followed by rapid ice-crystal growth (Figure 12.3c, d). Using the size resolution of the electron microscopy, it was possible to observe that the initiation of ice-crystal growth occurred at steps between the different facades of the feldspar sample (Figure 12.3e, f). This study illustrates how particle size and surface area are important for the initiation of ice-crystal growth in clouds. The study also illustrated that under the same environmental conditions (RH, T), the formation of ice crystals depends on specific particle properties.

Although ice-nucleation growth has been observed using ETEM, there has been no attempt to control it and make it reproducible. Having the possibility to perform these measurements in a controlled manner would provide promising results. In the following section, we provide examples that illustrate how electron microscopy, when combined with online measurements, provides complementary measurements and detailed insights into aerosol–cloud interactions.

12.4. COMPLEMENTARY INSTRUMENTS AND TECHNIQUES FOR THE MICROSCOPIC INVESTIGATION OF AEROSOL-CLOUD INTERACTIONS

12.4.1. Electrodynamic Balance

The electrodynamic balance has proven to be useful for studying aerosol hygroscopic properties. Initial studies showed how single droplets could be trapped in a substrate-free space and exposed to temperatures ranging from +15 °C down to below −30 °C (Davis & Ray 1980). Other experiments investigated the freezing probability of a droplet when in contact with another particle and were used to determine a contact freezing probability (Vortisch et al., 2000; Hoffmann et al., 2013). This technique has

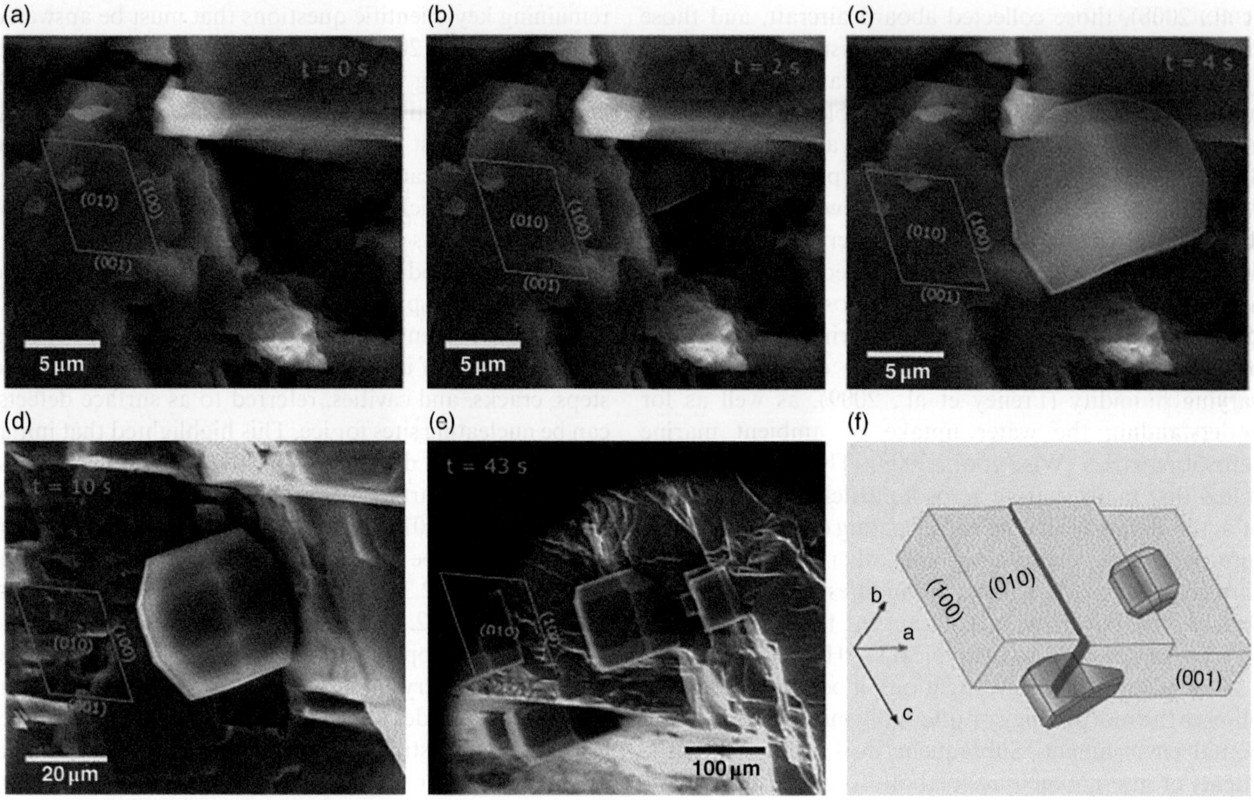

Figure 12.3 Ice nucleation on the face of a feldspar dust particle at different magnifications. The images show the growth of ice crystals on the step between different faces of the feldspar sample. Kiselev et al. (2017) with permission of the American Association for the Advancement of Science.

also been shown to be useful for investigating secondary ice formation in the atmosphere (Keinert et al., 2020).

12.4.2. Atomic Force Microscopy

Atomic force microscopy has been used to probe and understand the processes of humidification of individual aerosol particles. The method has the advantage of not requiring a vacuum so that particles can be exposed to humidities in an ambient environment. Additionally, no electron beam is used, thereby avoiding potential damage to particle surfaces and modification of their properties. The technique can also be used to probe particles over a wide range of diameters. Its utility was demonstrated by Laskina et al. (2015) in comparing the deliquescence and efflorescence relative humidities of aerosol particles with diameters of 100 nm and 6 μm, demonstrating that the size of particles is important in the initial update of water.

Atomic force microscopy uses a sharp tip about 1 nm in radius to scan over the surface of the particle through either direct contact or tapping to image the particle. However, the need to scan over the full surface area for each image acquired makes the technique slow and less optimal for studying phase transitions of aerosol particles.

12.5. CONCLUSION

Representing aerosol particles within atmospheric models is challenging as a result of their dynamic nature, with chemical and physical attributes undergoing modifications during transport. These evolving properties constitute key sources of uncertainties in small- to medium-scale models. To reduce these uncertainties, an effective approach involves observations of real atmospheric particles in controlled environments.

Techniques that provide information on particle surface properties, detailed morphology, and mixing state are essential for understanding atmospheric processes involving aerosols. Single-particle measurements are capable of providing accurate information on which particle populations are likely to be cloud active. Using electron microscopy, it has been demonstrated that ice nucleation on atmospheric particles is strongly dependent on particle morphology, size, and surface defects/impurities.

Using electron microscopy together with online instruments (having high time resolution measurements) can provide essential information on aerosol composition, mixing state, and morphology. Adachi et al. (2022)

demonstrated the usefulness of combining electron microscopy analysis of aerosol particles collected on TEM filters with measurements of ice-nucleation properties using a cold-stage device coupled to an optical microscope. Likewise, Mahrt et al. (2023) combined electron microscopy measurements with online ice-nucleation measurements in a continuous flow chamber to demonstrate the importance of wood charcoal as the source of potential INPs.

Environmental electron microscopy provides valuable insights into the behavior of aerosol particles in different atmospheric environments. Continued exploration of the capabilities of these instruments will continue to advance our understanding of aerosol particles.

AVAILABILITY STATEMENT

All data discussed in this chapter can be sourced using the cited references. No unpublished data are used.

REFERENCES

Abrams, I., & McBain, J. W. (1944). A closed cell for electron microscopy. *Science*, *100*(2595), 273–274. doi: 10.1126/science.100.2595.273

Adachi, K., Dibb, J. E., Scheuer, E., Katich, J. M., Schwarz, J. P., Perring, A. E., et al. (2022). Fine ash-bearing particles as a major aerosol component in biomass burning smoke. *Journal of Geophysical Research: Atmospheres*, *127*(2). doi: 10.1029/2021jd035657

Adachi, K., Freney, E. J., & Buseck, P. R. (2011). Shapes of internally mixed hygroscopic aerosol particles after deliquescence, and their effect on light scattering. *Geophysical Research Letters*, *38*. doi: 10.1029/2011gl047540

Asmi, E., Freney, E., Hervo, M., Picard, D., Rose, C., Colomb, A., & Sellegri, K. (2012). Aerosol cloud activation in summer and winter at puy-de-Dôme high altitude site in France. *Atmospheric Chemistry and Physics*, *12*(23), 11589–11607. doi: 10.5194/acp-12-11589-2012

Bartels-Rausch, T. (2013). Ten things we need to know about ice and snow. *Nature*, *494*, 27–29. doi: 10.1038/494027a

Berkemeier, T., Shiraiwa, M., Pöschl, U., & Koop, T. (2014). Competition between water uptake and ice nucleation by glassy organic aerosol particles. *Atmospheric Chemistry and Physics*, *14*(22), 12513–12531. doi: 10.5194/acp-14-12513-2014

Biskos, G., Malinowski, A., Russell, L. M., Buseck, P. R., & Martin, S. T. (2006). Nanosize effect on the deliquescence and the efflorescence of sodium chloride particles. *Aerosol Science and Technology*, *40*(2), 97–106. doi: 10.1080/02786820500484396

Danilatos, G. D. (1988). Foundations of environmental scanning electron microscopy. In *Advances in electronics and electron physics* (pp. 109–250). doi: 10.1016/s0065-2539(08)60902-6

Davis, E. J., & Ray, A. (1980). Single aerosol particle size and mass measurements using an electrodynamic balance. *Journal of Colloid and Interface Science*, *75*(2), 566–576. doi: 10.1016/0021-9797(80)90478-6

DeMott, P. J., Sassen, K., Poellot, M. R., Baumgardner, D., Rogers, D. C., Brooks, S. D., et al. (2003). African dust aerosols as atmospheric ice nuclei. *Geophysical Research Letters*, *30*(14). doi: 10.1029/2003gl017410

DeMott, P. J., Möhler, O., Cziczo, D. J., Hiranuma, N., Petters, M. D., Petters, S. S., et al. (2018). The fifth international workshop on ice nucleation phase 2 (FIN-02): Laboratory intercomparison of ice nucleation measurements. *Atmospheric Measurement Techniques*, *11*(11), 6231–6257. doi: 10.5194/amt-11-6231-2018

DeMott, P. J., Prenni, A. J., McMeeking, G. R., Sullivan, R. C., Petters, M. D., Tobo, Y., et al. (2015). Integrating laboratory and field data to quantify the immersion freezing ice nucleation activity of mineral dust particles. *Atmospheric Chemistry and Physics*, *15*(393–409), 2015. doi: 10.5194/acp-15-393-2015

Draine, B. T., & Flatau, P. J. (2008). Discrete-dipole approximation for periodic targets : Theory and tests. *Journal of the Optical Society of America*, *25*(11), 2693. doi: 10.1364/josaa.25.002693

Ebert, M., Inerle-Hof, M., & Weinbruch, S. (2002). Environmental scanning electron microscopy as a new technique to determine the hygroscopic behaviour of individual aerosol particles. *Atmospheric Environment*, *36*(39–40), 5909–5916. doi: 10.1016/s1352-2310(02)00774-4

Feingold, G., & Kreidenweis, S. (2000). Does cloud processing of aerosol enhance droplet concentrations? *Journal of Geophysical Research: Atmospheres*, *105*, 24351–24361. doi: 10.1029/2000jd900369

Freney, E., Martin, S. T., & Buseck, P. R. (2009). Deliquescence and efflorescence of potassium salts relevant to biomass-burning aerosol particles. *Aerosol Science and Technology*, *43*(8), 799–807. doi: 10.1080/02786820902946620

Freney, E., Adachi, K., & Buseck, P. R. (2010). Internally mixed atmospheric aerosol particles: Hygroscopic growth and light scattering. *Journal of Geophysical Research*, *115*(D19). doi: 10.1029/2009jd013558

Goldstein, J. I., Newbury, D. E., Michael, J. R., Ritchie, N. W. M., Scott, J. H. J., & Joy, D. C. (2017). Variable pressure scanning Electron microscopy (VPSEM). In *Scanning electron microscopy and X-ray microanalysis* (pp. 173–185). doi: 10.1007/978-1-4939-6676-9_12

Hoffman, R. C., Laskin, A., & Finlayson-Pitts, B. J. (2004). Sodium nitrate particles: Physical and chemical properties during hydration and dehydration, and implications for aged sea salt aerosols. *Journal of Aerosol Science*, *35*(7), 869–887. doi: 10.1016/j.jaerosci.2004.02.003

Hoffmann, N., Kiselev, A., Rzesanke, D., Duft, D., & Leisner, T. (2013). Experimental quantification of contact freezing in an electrodynamic balance. *Atmospheric Measurement Techniques*, *6*(9), 2373–2382. doi: 10.5194/amt-6-2373-2013

Holmgren, H. D., Sellegri, K., Hervo, M., Rose, C., Freney, E., Villani, P., & Laj, P. (2014). Hygroscopic properties and mixing state of aerosol measured at the high-altitude site Puy de Dôme (1465 m a.s.l.), France. *Atmospheric Chemistry and Physics*, *14*(18), 9537–9554. doi: 10.5194/acp-14-9537-2014

IPCC (2023). In C. W. Team, H. Lee, & J. Romero (Eds.), *Climate change 2023: Synthesis report*. Contribution of Working Groups I, II and III to the Sixth Assessment Report of the Intergovernmental Panel on Climate Change (pp. 35–115). Geneva, Switzerland: IPCC. doi: 10.59327/IPCC/AR6-9789291691647

Kanji, Z. A., Ladino, L. A., Wex, H., Boose, Y., Burkert-Kohn, M., Cziczo, D. J., & Krämer, M. (2017). Overview of ice nucleating particles. *Meteorological Monographs*, *58*, 1.1–1.33. doi: 10.1175/amsmonographs-d-16-0006.1

Keinert, A., Spannagel, D., Leisner, T., & Kiselev, A. (2020). Secondary ice production upon freezing of freely falling drizzle droplets. *Journal of the Atmospheric Sciences*, *77*(8), 2959–2967. doi: 10.1175/jas-d-20-0081.1

Kim, A.-H., Yum, S. S., Chang, D. Y., & Park, M. (2021). Optimization of the sulfate aerosol hygroscopicity parameter in WRF-Chem. *Geoscientific Model Development*, *14*, 259–273. doi: 10.5194/gmd-14-259-2021

Kiselev, A., Bachmann, F., Pedevilla, P., Cox, S. J., Michaelides, A., Gerthsen, D., & Leisner, T. (2017). Active sites in heterogeneous IcE nucleation—The example of K-rich feldspars. *Science*, *355*(6323), 367–371. doi: 10.1126/science.aai8034

Krueger, B. J., Grassian, V. H., Iedema, M. J., Cowin, J. P., & Laskin, A. (2003). Probing heterogeneous chemistry of individual atmospheric particles using scanning electron microscopy and energy-dispersive x-ray analysis. *Analytical Chemistry*, *75*(19), 5170–5179. doi: 10.1021/ac034455t

Lacher, L., Adams, M. P., Barry, K., Bertozzi, B., Bingemer, H., Boffo, C., et al. (2024). The Puy de Dôme ICe nucleation Intercomparison campaign (PICNIC): Comparison between online and offline methods in ambient air. *Atmospheric Chemistry and Physics*, *24*, 2651–2678. doi: 10.5194/acp-24-2651-2024

Laskin, A., Iedema, M. J., Ichkovich, A., Graber, E. R., Taraniuk, I., & Rudich, Y. (2005). Direct observation of completely processed calcium carbonate dust particles. *Faraday Discussions*, *130*, 453. doi: 10.1039/b417366j

Laskina, O., Morris, H. S., Grandquist, J. R., Qin, Z., Stone, E. A., Tivanski, A. V., & Grassian, V. H. (2015). Size matters in the water uptake and hygroscopic growth of atmospherically relevant multicomponent aerosol particles. *Journal of Physical Chemistry A*, *119*(19), 4489–4497. doi: 10.1021/jp510268p

Lohmann, U., & Feichter, J. (2005). Global indirect aerosol effects: A review. *Atmospheric Chemistry and Physics*, *5*(3), 715–737. doi: 10.5194/acp-5-715-2005

Mahrt, F., Rösch, C., Gao, K., Dreimol, C., Zawadowicz, M. A., & Kanji, Z. A. (2023). Physicochemical properties of charcoal aerosols derived from biomass pyrolysis affect their ice-nucleating abilities at cirrus and mixed-phase cloud conditions. *Atmospheric Chemistry and Physics*, *23*(2), 1285–1308. doi: 10.5194/acp-23-1285-2023

Mikhailov, E., Vlasenko, S. S., Martin, S. T., Koop, T., & Pöschl, U. (2009). Amorphous and crystalline aerosol particles interacting with water vapor: Conceptual framework and experimental evidence for restructuring, phase transitions and kinetic limitations. *Atmospheric Chemistry and Physics*, *9*(24), 9491–9522. doi: 10.5194/acp-9-9491-2009

Petters, S. S., & Petters, M. D. (2016). Surfactant effect on cloud condensation nuclei for two-component internally mixed aerosols. *Journal of Geophysical Research: Atmospheres*, *121*(4), 1878–1895. doi: 10.1002/2015jd024090

Schmale, J., Henning, S., Decesari, S., Henzing, B., Keskinen, H., Sellegri, K., et al. (2018). Long-term cloud condensation nuclei number concentration, particle number size distribution and chemical composition measurements at regionally representative observatories. *Atmospheric Chemistry and Physics*, *18*(4), 2853–2881. doi: 10.5194/acp-18-2853-2018

Semeniuk, T. A., Wise, M. E., Martin, S. T., Russell, L. M., & Buseck, P. R. (2007). Water uptake characteristics of individual atmospheric particles having coatings. *Atmospheric Environment*, *41*(29), 6225–6235. doi: 10.1016/j.atmosenv.2007.04.001

Sharma, R., & Weiss, K. (1998). Development of a TEM to study in situ structural and chemical changes at an atomic level during gas-solid interactions at elevated temperatures. *Microscopy Research and Technique*, *42*(4), 270–280. doi: 10.1002/(sici)1097-0029(19980915)42:4

Topping, D. O., & McFiggans, G. (2012). Tight coupling of particle size, number and composition in atmospheric cloud droplet activation. *Atmospheric Chemistry and Physics*, *12*, 3253–3260. doi: 10.5194/acp-12-3253-2012

Vortisch, H., Krämer, B., Weidinger, I. M., Wöste, L., Leisner, T., Schwell, M., et al. (2000). Homogeneous freezing nucleation rates and crystallization dynamics of single levitated sulfuric acid solution droplets. *Physical Chemistry Chemical Physics*, *2*(7), 1407–1413. doi: 10.1039/a908225e

Wise, M. E., Biskos, G., Martin, S. T., Russell, L. M., & Buseck, P. R. (2005). Phase transitions of single salt particles studied using a transmission Electron microscope with an environmental cell. *Aerosol Science and Technology*, *39*(9), 849–856. doi: 10.1080/02786820500295263

Wise, M. E., Martin, S. T., Russell, L. M., & Buseck, P. R. (2008). Water uptake by NACl particles prior to deliquescence and the phase rule. *Aerosol Science and Technology*, *42*(4), 281–294. doi: 10.1080/02786820802047115

Wise, M. E., Freney, E., Tyree, C. A., Allen, J. O., Martin, S. T., Russell, L. M., & Buseck, P. R. (2009). Hygroscopic behavior and liquid-layer composition of aerosol particles generated from natural and artificial seawater. *Journal of Geophysical Research*, *114*(D3). doi: 10.1029/2008jd010449

Zimmermann, F., Weinbruch, S., Schütz, L., Hofmann, H., Ebert, M., Kandler, K., & Worringen, A. (2008). Ice nucleation properties of the most abundant mineral dust phases. *Journal of Geophysical Research*, *113*, D23204.

AUTHOR INDEX

Microanalysis of Atmospheric Particles: Techniques and Applications, Geophysical Monograph 288, First Edition.
Edited by Joseph M. Conny and Peter R. Buseck.

DOI: 10.1002/9781119554318.authorindex

SUBJECT INDEX

Note: Page numbers with figures in *italic* and tables in **bold**.

Microanalysis of Atmospheric Particles: Techniques and Applications, Geophysical Monograph 288, First Edition.
Edited by Joseph M. Conny and Peter R. Buseck.

DOI: 10.1002/9781119554318.subjectindex